PREFACE

The second in a series of International Symposia on Nitrogen Fixation was held in Salamanca, Spain on 13-17 September 1976 under the sponsorship and auspices of the Consejo Superior de Investigaciones Cientificas, the University of Salamanca and the Charles F. Kettering Foundation. This volume contains the formal presentations made at that time.

Continuing in the same mode as that initiated at the First Symposium in Pullman, Washington, USA, in 1974, a multidisciplinary format was used in an attempt to cross the boundaries of the various sciences involved. We hope that intellectual exchange and interaction both within and between disciplines occurred and that this forum for discussion and communication will help speed progress in this area of research so vital to mankind.

Salamanca, Spain
Sept. 19, 1976

W.E. NEWTON
J.R. POSTGATE
C. RODRIGUEZ-BARRUECO

CONTENTS

INTRODUCTION

The importance of the biological fixation of nitrogen

I am going to try to explain in non-technical terms why symposia such as this one are arranged from time to time, what they hope to achieve, and what practical implications the findings may have.

The need for symposia of this type arises in the first place because of the supreme importance of the element nitrogen to all the forms of life that occur on Earth. Two of the most essential constituents of the cells of plants or animals, namely nucleic acids (which carry the hereditary characters) and proteins (which among other functions provide the bases for enzymes) contain nitrogen, as does also the characteristic pigment of plants cells - chlorophyll. Thus a bacterium, a fungus, a green plant or an animal of any kind cannot grow and function unless it can acquire nitrogen in an acceptable form.

There is of course a vast amount of elemental nitrogen associated with the Earth - in the gaseous form in the atmosphere, in the dissolved condition in the oceans, in both forms in the soil, and in an occluded state in certain rocks, so that it might be thought that we were better placed with regard to nitrogen than to some other essential elements such as phosphorus. However a severe limitation arises because typical plants and also animals can of themselves make no use of elemental nitrogen: thus gaseous nitrogen is always present within the airspaces in plant tissues, but is not utilised, while again our lungs absorb only the oxygen from inhaled air, and reject the nitrogen. It is only in the combined form that most organisms can utilise nitrogen. Green plants grow satisfactorily if their roots are supplied with nitrates or ammonium salts, and under natural conditions they can usually obtain these from the soil, or from the water if they are aquatics. Starting with these simple compounds, the enzymes in their cells are able to build up proteins and the other complex nitrogenous substances required for the plant's continued growth. Animals do not have such synthetic powers and they acquire their nitrogen chiefly by consuming, in their food, protein which has originally been synthesised by a plant. Here it may be observed that despite notable efforts, biochemists have not yet succeeded in effecting the total synthesis of protein in the laboratory.

So we see that the supply of nitrogen for living organisms in general goes back to the nitrogenous salts taken up by plants, and the origin of these salts must now be considered. If we think of a tract of land which carries an established natural flora and fauna, in the short term nitrogenous salts are returned more or less constantly to the soil as a result of the decay of dead plants or animals, since through the action of soil micro-organisms the organic nitrogenous substances in such dead materials are broken down with the eventual production of ammonium salts and nitrates. Thus the nitrogen is constantly cycling, in a combined form, between the soil and the living organisms.

But losses usually occur during this cycling. Thus a loss of soil nitrogen from a particular area can occur through drainage, while again various soil micro-organisms decompose nitrate for their own purposes, with the evolution of gaseous nitrogen. Under agricultural conditions, where a part of the plant growth is removed as a crop, the depletion of soil nitrogen is much more rapid. So if fertility is to be maintained it is necessary for there to be some compensating process, either natural or man-devised. Here it may be noted that unlike the situation with other essential elements for plant growth, such as phosphorus, sulphur or magnesium, combined forms of nitrogen are rarely present in underlying rocks and minerals, so that no replenishment of soil nitrogen occurs from that source.

This is where we come to nitrogen fixation. Certain primitive organisms - namely particular bacteria and particular blue-green algae - have the enzymic equipment to bring elemental nitrogen from the atmosphere into a combined form, namely ammonia. The term *fixation* is applied to this process. The bacterial or algal cells use the ammonia to build up their proteins and other nitrogenous constituents. The chemical energy required for this fixation comes from sunlight - directly in the case of blue-green algae and indirectly in the case of most of the nitrogen-fixing bacteria since they are saprophytes. Such nitrogen-fixing organisms are present in the free-living state in soils and in natural waters. When their cells die and decay, the newly-fixed nitrogen enters into general circulation.

But now in a considerable number of instances these nitrogen-fixing micro-organisms, particularly the bacteria, grow in a more or less close association with a higher plant, usually with its roots. The bacterium derives carbohydrates formed in photosynthesis from the plant and uses these in its growth and nitrogen fixation, while the higher plant takes over for its own use at least some, and in many cases most, of the products of fixation. These associations are generally called symbioses, but sometimes I feel that it is more helpful to think of them as higher plants with captive or slave micro-organisms which provide the plant with a private source of combined nitrogen. Moreover in some way the higher plant seems to be able to induce the micro-organism to fix far more nitrogen than it requires for its own growth. In recent years the number of known examples of these associations has increased considerably, and I shall concentrate on those which are most widespread.

The longest-known are those in which a nitrogen-fixing organism, usually a bacterium, is attracted to the roots of the plant and is induced to invade them. The invaded tissues swell up to form what are called root nodules, easily visible to the naked eye. The cells of these nodules become filled with bacteria, though the latter do not spread to any other part of the plant. The products of the fixation effected by the bacteria pass quickly into the plant in the form of amino acids, and because of this a nodulated plant is able to grow healthily in a soil or other rooting medium which is devoid of combined nitrogen such as nitrate. When a nodulated plant eventually dies, the nitrogen which has been fixed in the nodules becomes incorporated into the soil and makes it more fertile.

The best-known nodule-forming plants are the legumes, such as clovers, peas, beans, grams and acaias, and - by a fortunate coincidence for the human race - in addition to the fixation of nitrogen associated with them, many legumes have the further valuable feature that their seeds provide acceptable, protein-rich food for humans, and their foliage fodder for animals. As a result legumes have long been used in agriculture as food providers and soil improvers, though the extent of the soil improvement depends on how much of the crop growth is harvested.

Some other plants also bear nitrogen-fixing root nodules, inhabited by a different bacterium from that present in legume nodules. Alders, casuarinas and myricas - like *Myrica faya* of the Canary Islands - are examples. The number of these plants - about 160 species according to present information - is small compared with the legumes, but many of them occur abundantly in their appropriate climatic areas. Like the legumes they are soil improvers, but those so far known do not produce edible products for humans, though some, such as *Purshia*, are important browse plants for animals. In addition the fact that they are all shrubs or trees makes them unsuited to conventional agriculture, but there are definite openings for their exploitation in forestry, and forest scientists in several countries are investigating how this can best be achieved.

In recent years evidence has come forward to show that other higher plants set up a less specialised association between their roots and nitrogen-fixing bacteria. No nodules form, and the bacteria proliferate instead on the surface of the root or within the air-spaces of the outer tissues of the root. These are called rhizosphere associations and they seem to operate in the same general way as nodulated plants - the bacteria use carbohydrates which leak out from the root surface or from the cells of the outer tissues, while a proportion of the nitrogen fixed is appropriated by the higher plant. Associations of this kind have been detected in a number of tropical grasses which are important as fodder for animals, also in rice, in grasses and weeds of temperate regions, in plants (including mangroves) which are rooted in mud at the edge of water, and in the water hyacinth. In some instances there is evidence that the plants gain very significant amounts of fixed nitrogen, though in others information on this aspect is awaited. So it seems possible that in their natural places of growth - though not in semi-sterile conditions in a greenhouse - a considerable number of plants have, through these rhizosphere associations, direct access to newly-fixed nitrogen, and are correspondingly less dependent on soil nitrates. Nonetheless the widely held belief that the nitrogen supply is commonly the limiting factor in the growth of natural and agricultural vegetation must not be forgotten.

Some other associations, this time involving blue-green algae, can be mentioned briefly. Thus practically all cycads - gymnospermous trees which in earlier eras were abundant over many parts of the world - bear root nodules inhabited by such an alga; fixation proceeds in the nodules, conveying benefit to the cycad and eventually to the soil. The water fern *Azolla* has leaf cavities occupied by a nitrogen-fixing blue-green alga, and there is good evidence that in rice-fields, where the fern often grows abundantly, the fixation contributes substantially to fetility. In some lichens the algal symbiont is a blue-green which fixes nitrogen.

Now so far I have indicated that the free-living nitrogen-fixing organisms and those living in association with a higher plant serve to replenish the supplies of combined nitrogen in the biosphere - to compensate for losses arising in ways already mentioned. This topping-up has been going on for a very long time, and although it is generally believed that Earth's atmosphere at an earlier stage contained substantial amounts of ammonia, that is so long ago that it can be said with some confidence now that most of the nitrogen incorporated in the vast quantities of proteins and other compounds in the biosphere was originally fixed by a bacterium or a blue-green alga. So under the natural system these organisms fulfil an essential role in facilitating the growth of all other forms of life, and this is why biological fixation of nitrogen has been ranked as equalling photosynthesis in importance. It should be noted again that biological fixation is driven by solar energy, and further that it proceeds at normal temperatures. In passing it may also be noted that it is uncertain whether the original acquisition of the nitrogen-fixing property by these organisms was related to a scarcity of combined nitrogen on Earth at some stage in its evolution, or whether the enzyme nitrogenase - which can catalyse other reactions besides nitrogen fixation - originally discharged some different function.

However it is not only through biological fixation that soil nitrogen can be increased. The success - in the form of greatly improved crops - which attended the use of phosphate and potash fertilisers in the 19th century stimulated a desire to have something in a bag that would provide combined nitrogen and replace the rather inconvenient organic materials previously used for the purpose, and perhaps make rigid crop rotations unnecessary. In due course industrial chemists found means of combining atmospheric nitrogen into such forms as ammonia, ammonium salts and urea, which are acceptable to crop plants. So now there was industrial fixation of nitrogen as well as the biological kind. The output of this industry has risen by leaps and bounds over the years, and according to FAO statistics total world production in 1974 was 41 million metric tons of nitrogen, which is expected to be doubled by 1978. Most countries now have nitrogen-fixing factories, and more are being built all the time. Table 1 shows the use that is made of fertiliser nitrogen in different parts of the world.

Now this symposium is not directly concerned with industrial fixation, though it is possible that the study of the chemistry of the biological process - which is on our programme - will lead eventually to improvements in the industrial process. But the rapid increase in the use of "bag" nitrogen is on all our minds, and though I can really only speak for myself, I am fairly certain that most of us have mixed feelings on the matter. It is appreciated that industrial fixation is a positive step towards better crops and the relief of hunger, but one wonders whether the world can really afford to do it that way as a permanent policy. Such fixation involves a considerable usage of energy, recent publications indicating that for

Table 1

Fertiliser usage in kg N per hectare of arable land in 1973/4*

Means for continents

Africa	5.3
North & Central America	35.6
South America	9.7
Asia	19.5
Europe	78.3

Means for selected countries

Egypt	133	India	11
Kenya	12	Japan	155
South Africa	18	U.K.	122
Nigeria	0.2	Spain	34
U.S.A.	43	Netherlands	476
Brazil	13	U.S.S.R.	27

*The data shown are taken from the Annual Fertiliser Review, 1974, published by the Food and Agriculture Organisation of the United Nations, Rome, 1975.

the fixation of unit weight of nitrogen the energy yielded by nearly twice that weight of oil is required. Thus, including the cost of transport, the fertiliser is rather expensive, perhaps too much so for developing countries. It is realised that not all farmers want to grow legumes, and that the growing of good legume crops is not just a matter of sowing seeds and leaving the rest to Nature; first it must be ascertained whether the bacteria appropriate to a particular legume are present in the soil, and if they are not then there must be a resort to seed inoculation, which has its problems. We also appreciate that biological fixation is a relatively slow process, so that the 250 kg of nitrogen per hectare which can be fixed by a clover stand in one year is probably only a small part of one day's production by a nitrogen factory. However in the long term it will be more sensible to make maximum use of the biological process in order to avoid the depletion of fossil fuels and the damage to the environment that may result from the liberal application of fertiliser nitrogen. In the short term the exploitation of biological fixation is again necessary because of the cost of fertiliser nitrogen and the only limited supplies that are available in some regions.

This symposium, then, is primarily concerned with biological fixation, and many aspects will be discussed. A better understanding of the chemical mechanism is likely to result from the deliberations, while the discussion of physiological and genetical aspects of the legume association may point the way towards increasing the amount of fixation by legume crops. The latest work on rhizosphere associations will be presented, and may indicate possibilities for exploitation in connection with crops other than legumes. Recent studies of non-legume nodules and blue-green algae will be described, and methods considered for turning the fixation associated with them to greater practical advantage. Genetic experiments in which the property of fixation can be transferred to bacteria which normally do not have it will be considered.

The meeting will also bring benefits of another kind, since I am sure that the participants will return to their laboratories and research institutes re-invigorated by their discussions and by their sojourn in this very pleasant and historic city.

Professor G. Bond

DEDICATION

Plenary Session

The participants in the symposium were welcomed by governmental, civic and academic representatives, and by the chairman and co-chairman of the organising committee.

The chairman of the organising committee referred to the death earlier in the year of Professor O.N. Allen, and was sure that all participants would wish him to express their esteem for the distinguished and life-long work of Professor Allen on nitrogen fixation, especially in legumes. One who had for so long been a leading participant in international meetings would be sadly missed from the present one.

The plenary lecture was then delivered by Professor G. Bond.

CHEMISTRY OF DINITROGEN

G.J. LEIGH

Agricultural Research Council, Unit of Nitrogen Fixation, University of Sussex, Brighton BN1 9QJ, U.K.

INTRODUCTION

The chemistry of dinitrogen is growing, and it would no longer be true to describe this molecule as unreactive. However, chemists are still collecting facts rather than interpreting them and do not yet understand the chemistry of N_2 to any great degree. A series of reactants and products have been identified which may eventually divulge details of the thermodynamics of the processes which involve them. We know exceedingly little about the mechanisms of the processes which give rise to the products. This discussion will concentrate on how dinitrogen can be brought into reaction and the kinds of products formed, together with speculation concerning the kinds of reaction which may occur. In the context of the Symposium, I shall concentrate on reduction reactions.

REACTION OF N_2 WITH H_2

On the basis of the sums of approximate bond energies, H is negative (i.e., reaction proceeds spontaneously). Taking the H-H bond dissociation energy as 430 $kJmole^{-1}$, the N≡N as 940 $kJmole^{-1}$, the N≡N as 410 $kJmole^{-1}$, the N-N as 159 $kJmole^{-1}$ and the N-H as 389 $kJmole^{-1}$, the following result is obtained.

$$3H_{2(g)} + N_{2(g)} \longrightarrow 2NH_{3(g)}$$

$$\Delta H = -104\ kJmole^{-1}$$

3x430 940 2334

A more detailed examination of this process, assuming the widely held belief of our biologically-inclined colleagues that two-electron steps are important in the reduction of dinitrogen to ammonia (Stiefel, 1973; Newton *et al.*, 1976), shows the following.

$$N_{2(g)} + H_{2(g)} \longrightarrow N_2H_{2(g)}$$

$\Delta H = +182$ kJmole^{-1}

940 430

$$N_2H_{2(g)} + H_{2(g)} \longrightarrow N_2H_{4(g)}$$

$\Delta H = -97$ kJmole^{-1}

1188 430 1715

$$N_2H_{4(g)} + H_{2(g)} \longrightarrow 2NH_{3(g)}$$

$\Delta H = -189$ kJmole^{-1}

1715 430 2334

In these formal gas phase reactions, the formations of both N_2H_2 and N_2H_4 are endothermic. This activation energy barrier to the reaction of dinitrogen is additional to any kinetic barriers. The energy of this reduction system as a function of the stage of the reaction is shown in Fig. 1. It is unlikely that consideration in free energy terms rather than enthalpy terms would change the diagram appreciably.

It is the aim of the inorganic chemist to overcome these activation energy problems. In theory, this might be accomplished in the following ways: (A) carry out the reaction in (aqueous) solution to take advantage of solvation effects; (B) modify the reaction route; and (C) attempt a combination of these two.

The N_2+$3H_2$ reaction in aqueous solution

Following Shilov (1974), the enthalpies of formation (in kJmole^{-1}) of the various components of the system in the gas phase, in neutral water and in acid solution are as follows.

	N_2	N_2H_2	N_2H_4	$2NH_3$
Gas		205	96	-92
Water		142	33	-159
Acid			-8	-267

These data suggest that free N_2H_2 is an unlikely intermediate in any reaction scheme, but that a four-electron jump to hydrazine is exothermic in acid solution. It is worth noting here that a metal-dinitrogen complex as the starting point, rather than dinitrogen itself, appears to make free N_2H_2 even less likely. It also threatens to upset the positive balance obtained with hydrazine. This situation occurs because the

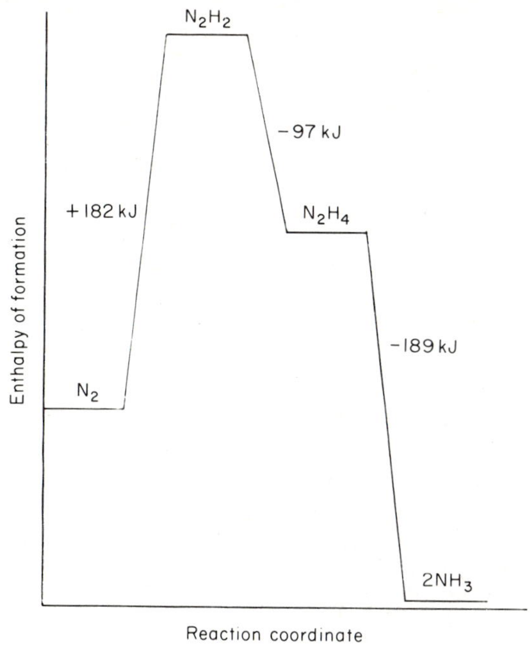

Fig. 1. *Energy profile for the reaction* $N_2 + 3H_2 \rightarrow 2NH_3$, *considered in steps* $N_2 + H_2 \rightarrow N_2H_2 + H_2 \rightarrow N_2H_4 + H_2 \rightarrow 2NH_3$.

complex would be even lower on the energy scale than free N_2, thus increasing the enthalpy of activation.

Again adapting Shilov (1974), these data may be expressed in an alternative way. Let us consider the permissible range of redox potentials which we can employ to reduce dinitrogen (Fig. 2).

The dotted lines represent the limits at which water is reduced. The vertical axis presents the redox potential E°(red) for the indicated reactions. For these reactions to take place in water, they must be such that water is less easily reduced. The horizontal axis represents an applied reducing potential E°. Hence, any dinitrogen-reducing reaction which operates in aqueous solution must have a redox potential below the horizontal dotted line and it must be reduced by a reducing agent with a potential to the right of the vertical dotted line. Further, because E° (red) must be greater than E°, the inclined line represents a third boundary below which the reducer-reductant couple must fall. In practice, these constraints show that only couples within the shaded area are capable of reducing dinitrogen in

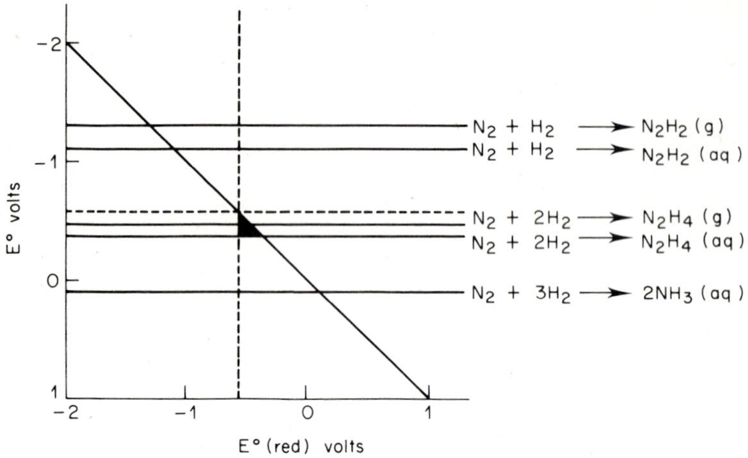

Fig. 2. *Redox relationships in dinitrogen reduction.*

aqueous solution, which is undoubtedly why the number of reductants is limited. When we add that these are purely thermodynamic considerations with kinetic factors yet to enter into our discussions, the utility of this method is clearly rather limited. Suffice it to say, however, that hydrazine is invariably produced in those aqueous reducing systems which approach genuine catalytic behaviour.

Modification of the reaction route

An alternative way of modifying the reaction route is by complexing the reactants at each stage to a metal. Unfortunately data are not available on a complete series of complexes which may be denoted as:

$$M\text{-}N_2\text{-}M$$
$$M\text{-}N_2H_2\text{-}M$$
$$M\text{-}N_2H_4\text{-}M$$
$$M\text{-}NH_3 \quad NH_3\text{-}M$$

where M represents a metal ion together with its other ligands. The nearest we can approach is probably a series of complexes due to Sellman (1972): $[C_5H_5(CO)_2MnN_2]$, $[C_5H_5(CO)_2MnN_2H_2Mn(CO)_2C_5H_5]$, $[C_5H_5(CO)_2MnN_2H_4]$, $[C_5H_5(CO)_2MnNH_3]$. There are analogous chromium (Sellman *et al.*, 1976), molybdenum (Sellman *et al.*, 1975) and tungsten (Sellman *et al.*, 1973) complexes which parallel the required series even more closely, e.g. $[\{C_5H_5$

$(CO)_5Cr\}_2N_2$, $[\{(CO)_5Cr\}_2(N_2H_2)]$, $[\{(CO)_5Cr\}_2(N_2H_4)]$ and $[(CO)_5Cr(NH_3)]$, but the required measurements have not been made. It should also be emphasised that in none of these cases has an N_2-complex been converted to a complex of any reduced species, rather the reduced species may be oxidised to N_2.

Binder and Sellman (1973) have measured the N(1S) binding energies of the manganese compounds listed above. Remarkably, these binding energies are almost invariant as are the manganese $(2P_{3/2})$ energies. A simple-minded interpretation is that the charges on nitrogen and manganese do not change throughout the series. There is certainly no evidence of progressive reduction, and no sense of running down-hill from N_2H_2 to $2NH_3$. Indeed, the results are even more difficult to interpret because various kinetic factors may interfere. Thus, the reaction of $RuCl_3.3H_2O$ with hydrazine to give $[Ru(NH_3)_5(N_2)]^{2+}$ occurs in part because the ruthenium provides a mechanistic pathway for the breakdown of the thermodynamically unstable hydrazine to dinitrogen and ammonia. Similarly, $[C_5H_5(CO)_2Mn(N_2H_4)]$ yielded $[C_5H_5(CO)_2Mn(N_2)]$ and $[C_5H_5(CO)_2Mn(NH_3)]$ when attempts were made to convert it to $[\{C_5H_5(CO)_2Mn\}_2(N_2H_4)]$ (Sellman *et al.*, 1976). These reactions cannot be rationalised simply in terms of hydrazine exhibiting its well recognised properties as a reducing agent. The observation of ammonia in these systems is thus no criterion of easy *reduction* from N_2H_4 or N_2H_2 to NH_3.

Reduction of dinitrogen mediated by metal ions in aqueous solution

It is thus clear that complexing of N_2 and its successive reduction products, N_2H_2 and N_2H_4, does not represent an easy reduction sequence. It is possible, however, that the combined effects of solvation and complexing could lead to the required sequence. In fact, there is remarkably little definitive evidence for the mechanisms of reduction of dinitrogen in aqueous solution catalysed by transition metal ions. The available information is reviewed below.

Schrauzer (1976) has published a tremendous amount of work concerning a system supposedly analogous to nitrogenase, which converts dinitrogen to ammonia. It is based upon a complex, $Na_2[Mo_2O_4(cysteinate)_2]5H_2O$, or a mixture of sodium molybdate plus cysteine in aqueous solution, buffered to pH 7-10, together with a reducing agent (sodium borohydride or sodium dithionite). The copious experimental data are interpreted (Schrauzer and Doemeny, 1971; Schrauzer *et al.*, 1974) as demonstrating the presence in solution of a reduced Mo^{IV} species (*1*), which is oxidised during fixation to species (*2*) (Fig. 3).

However, the N_2-trapping species (*1*) is present in very low concentration ($\sim$1% of total molybdenum) (Schrauzer, 1976)

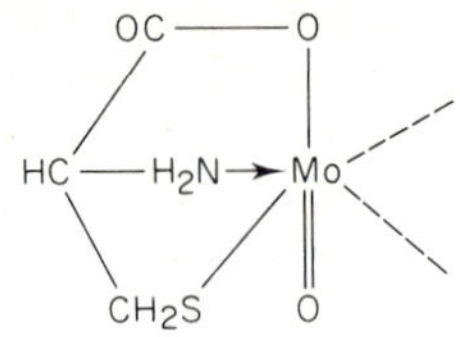

(1)

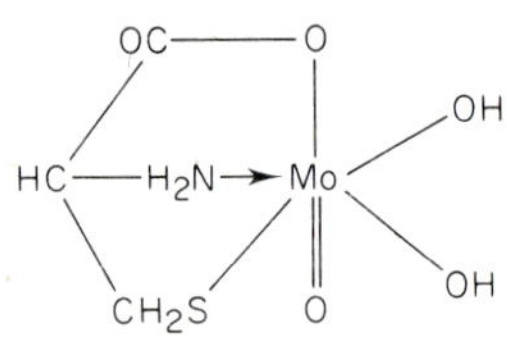

(2)

Fig. 3. *Proposed intermediates in Schrauzer systems.*

and the rate of production of ammonia is very small <1 mmol N_2 reduced per hour per mole of Mo (Schrauzer and Doemeny, 1971; Schrauzer *et al.*, 1974). Some complex isotope-labelling experiments suggest that N_2 is reduced initially to N_2H_2, which is believed to be generated by binding N_2 side-on to molybdenum. The N_2 is subsequently hydrolysed to N_2H_2. However, reduction of N_2H_2 does not occur. Rather, it is supposed to disproportionate in part to N_2H_4 plus N_2 and, to a larger extent, to N_2 plus H_2. The reduction of the N_2H_4 to NH_3 is then reasonably facile. This analysis rests to a great extent on the presumption that N_2 and C_2H_2 are reduced analogously and at the same site, which is, at the very least, questionable in the natural system (Smith *et al.*, 1976).

More recently, Schrauzer has produced another molybdenum system which fixes N_2. Thus, $[Mo(CN)_4O_2]^{4-}$ may be reduced by sodium borohydride, sodium dithionite or ferredoxin model compounds in aqueous solution to give an intermediate which reduces both cyanide and N_2 (Schrauzer *et al.*, 1975; Schrauzer *et al.*, 1976). Yields of ammonia are considerably higher than in the molybdothiol systems and hydrazine is also observed. The diimide mechanism is again favoured. However, the interpretation of these results is not straightforward. For example, the turnover rate reaches about 10 moles of N per mole "active molybdenum catalyst" (incidentally, chemists in general now seem to be able to talk of inactive catalysts and still keep straight faces) per minute, which is about 20% of the turnover rate of nitrogenase. However, neither nitrogenase nor the "active molybdenum catalyst" are well-defined species, so this can only be a semi-quantitative statement.

Schrauzer has proposed a detailed scheme based on his data. It cannot be challenged in detail because it is rather hypothetical. However, two comments are not out of place.

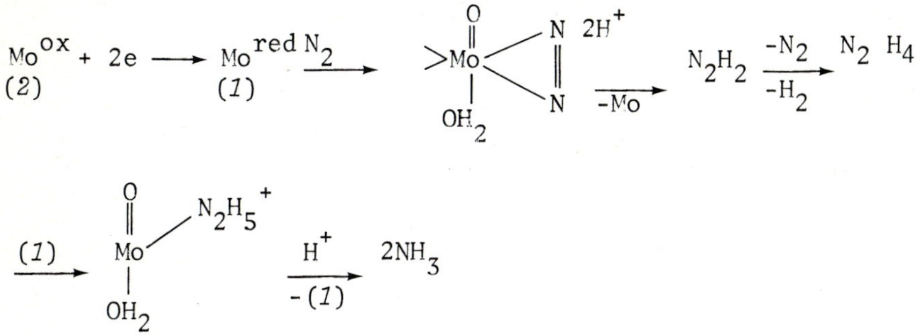

Firstly, side-on bonded N_2 is not known in any stable complex apart from $[\{C_6H_5(NaOEt_2)_2\{(C_6H_5)_2Ni\}N_2NaLi_6\text{-}(OEt)_4 OEt_2\}_2]$ (Jonas *et al.*, 1976), which is air-sensitive and prepared under non-aqueous conditions, and has the structure shown schematically below (Fig. 4). The N-N bond length is 1.34Å,

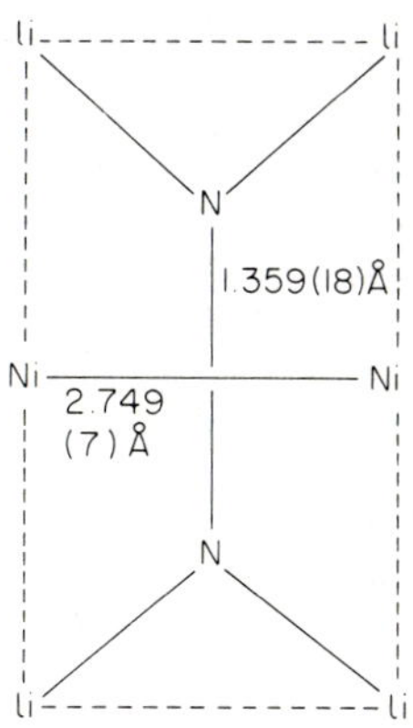

Fig. 4. *Schematic representation of part of the molecule of* $[\{C_6H_5(Na\cdot OEt_2)_2\{(C_6H_5)_2Ni\}_2N_2NaLi_6(OEt)_4\cdot OEt_2\}_2]$ *showing principal interactions between nickel, lithium and nitrogen atoms.*

which is consistent with an N-N double bond (corresponding, should one wish to view it so, to coordinated N_2^{2-}). The N_2 in the nickel complex can apparently be protonated to N_2H_4 in some unknown manner. This much of the chemistry is consistent with Schrauzer's scheme. Against it is the peculiar binding of N_2 across a Ni-Ni bond, which contrasts with the picture of one molybdenum atom being responsible for binding N_2.

Secondly, although N_2H_2 has been clearly implicated by Schrauzer, there is no evidence or suggestion that the N_2H_2 is bound between *two* metal atoms. Indeed, the molybdenum is believed to influence the decomposition of the N_2H_2, but the interaction between molybdenum and the N_2H_2 is never explained. This binding contrasts with the model considered at the outset and differs from the models cited in the discussion below.

The other considerable body of work on N_2 reduction in aqueous solution comes from Shilov and his collaborators (Shilov, 1976). They have used a variety of systems, some heterogeneous and some homogeneous. For example, a molybdenum (V) or molybdenum (VI) compound (e.g. $MoCl_5$ or sodium molybdate) in strongly alkaline medium is reduced by titanium (III) chloride or vanadium (II) salts to give a heterogeneous mass which converts N_2 to N_2H_4 (Denisov *et al.*, 1970a). Maximum yield (about 100 moles N_2H_4 per atom molybdenum in a few minutes) is obtained when 0.5 moles of magnesium ions/mole titanium are present. In contrast to the Schrauzer systems, which operate at normal temperatures and pressures, best yields are obtained at about 100°C and 100-150 atmospheres N_2 pressure.

The mechanism of such reductions are not at all clear. Shilov believes that a molybdenum (III)-dinitrogen complex is formed and indeed, a molybdenum (III)-carbonyl can be isolated from such systems with carbon monoxide (Shilov *et al.*, 1975). Both Mo(III) and N_2 together inhibit the evolution of H_2 from water by titanium (III) (Denisov *et al.*, 1975) and it is postulated that the reduction source is a Ti(III)-Mg(II)-hydroxide, in which Mo(III) ions effectively prevent interaction of the reducing centres with water (hence, inhibit dihydrogen evolution) and in which dinitrogen is bound as a bridge between two molybdenums.

If a vanadium (II)/magnesium (II) system is made strongly alkaline, then it rapidly converts dinitrogen to hydrazine even in the absence of molybdenum (Denisov *et al.*, 1970b). The conversion is quantitative in a matter of minutes at 10-20 atmospheres dinitrogen pressure. In this system, four V^{2+} ions are

$$4V(OH)_2 + N_2 + 4H_2O \xrightarrow{Mg(OH)_2} 4V(OH)_3 + N_2H_4$$

involved in some transition state with a bridging dinitrogen again postulated (Denisov *et al.*, 1973; Denisov *et al.*, 1974).

The addition of pyrocatechol makes this system homogeneous, but the pH becomes much more critical (maximum yield at about pH 10) and magnesium is not necessary. The product, in this case, is ammonia and not hydrazine and there is no sign that hydrazine is an intermediate. The mechanism is again assumed to involve a binuclear vanadium system (Nikonova *et al.*, 1972; Nikonova *et al.*, 1974; Shilov, 1976).

We thus have another system which certainly works, is genuinely catalytic, and is obscure in its functioning. Note, however, that diimide is *not* postulated as an intermediate (c.f. Schrauzer system). Indeed, Shilov indicates that the initial product in the molybdenum-containing systems is a hydrazine derivative which he writes

$$Mo{=}N{-}\overset{H}{N}{-}Mo^{+},$$

which reacts with water to form hydrazine. Further (c.f. Schrauzer), Mo(III) is implicated not Mo(IV). Although the postulated mechanism differs from both our theoretical pattern and from Schrauzer's scheme, it contains the interesting idea that the metal ion which traps N_2 is responsible for reducing it. We shall return to this concept later.

Where does this leave the postulated pattern upon which we based our discussion? Compounds with dinitrogen bridging between metal atoms are certainly well known (Chatt *et al.*, 1970) with many containing two different metal atoms. Only one, the classical $[(NH_3)_5Ru(N_2)Ru(NH_3)_5]^{4+}$, is soluble in aqueous solution (Harrison *et al.*, 1968). Few, apart from the nickel complex alluded to above, are convertible to hydrazine or ammonia. Recently, Bercaw's group (Manriquez *et al.*, 1976) has prepared such a complex from the reduction of dichlorobis(pentamethylcyclopentadienyl)-zirconium under dinitrogen. This system is air-sensitive and non-aqueous, unlike most of those discussed so far. The X-ray structure analysis of the reduction product shows that it contains three dinitrogen molecules (Fig. 5).

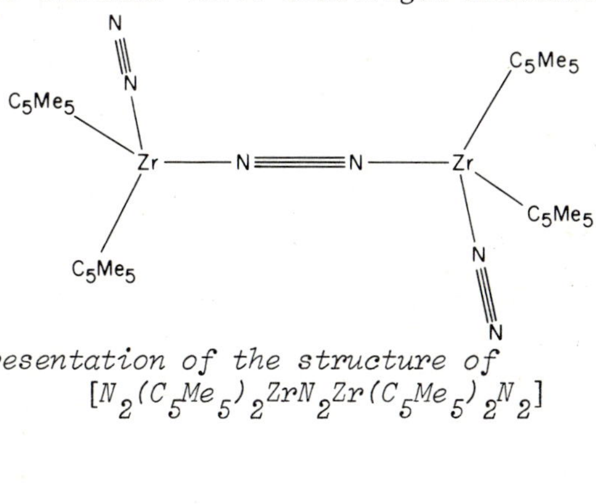

Fig. 5. *Representation of the structure of* $[N_2(C_5Me_5)_2ZrN_2Zr(C_5Me_5)_2N_2]$

The N-N bond lengths are 1.116(8), 1.114(7) (terminal) and 1.182(5) (bridging) Å. Thus, the bridging dinitrogen is partially reduced, but not as much as in the nickel complex. Protonation of this complex yields one mole of hydrazine and two moles of dinitrogen and it is reasonable to assume that it is the more reduced, bridging dinitrogen which protonates. Further, exchange reactions with $^{15}N_2$ show that only two dinitrogens are replaced, presumably the terminal dinitrogens, which at first sight, provides an unequivocal conversion of bridging dinitrogen to hydrazine. However, protonation of the bridge-labelled species shows that one-half mole of hydrazine comes from the bridging dinitrogen and the other one-half mole originates in a terminal dinitrogen. Moreover, protonation of the formally analogous $[\{(CO)(C_5Me_5)_2\text{-}Zr\}_2N_2]$ gives no ammonia whatsoever.

A symmetrical intermediate, $[(C_5Me_5)_2Zr(N_2H)_2]$ which is not binuclear is invoked to explain these results. It does not indicate why the carbonyl is unproductive. We can, however, draw important conclusions from these results. These are: (1) solvation by water is not necessary for converting dinitrogen to hydrazine and ammonia; (2) the fact that a binuclear species is formed from N_2 is no guarantee that such a binuclear species will persist throughout a reaction sequence; and (3) the reactivity of the binuclear species may well be reflected by the length of the N-N bond (which should suggest the degree of reduction of the N-N triple bond brought about by complexation), but it is equally possible that the bond length may be no more a reliable indicator of reactivity than the N-N bond length in terminal dinitrogen complexes.

NITRIDING SYSTEMS

Nevertheless, the evidence for binuclear intermediates is more compelling than our discussion may have implied so far. Binuclear intermediates have been implicated in particular reactions which are related to the zirconium system (Manriquez *et al.*, 1976) and which actually predate it. Most of the current chemical work in nitrogen fixation was stimulated by the work of Volpin and Shur (1964), which first became widely known about 1965. They showed that in a reaction mixture of a transition metal compound (generally a halide) and a strong reducing agent (generally a Grignard reagent) in a non-aqueous solvent, such as ether, an active intermediate is formed, which could trap N_2 and split it, so that, upon addition of aqueous acid, ammonia is generated (Volpin, 1972).

It has been known for many years that dinitrogen reacts with some metals and it was relatively straightforward to infer that the Volpin reactions involved reduction of the transition metal derivative to finely divided metal, which reacted with N_2 to form nitrides. The efficiency of the various transition elements in these nitrogen-fixing systems falls along the First Transition Series from titanium to iron, which is very poor, and cobalt and nickel, which do not form ammonia at all Volpin systems (Volpin *et al.*, 1965a; Volpin *et al.*, 1965b; Volpin *et al.*, 1966; Volpin and Shur, 1970). This reactivity parallels the heat of formation of the nitrides of the various elements (Leigh, 1971). This observation was, at first sight, unexpected because the formation of nitrides involves reduction of dinitrogen, and as titanium is much less rich in valence electrons than iron, it should be a less effective reducing agent. However, titanium (IV), which has no valence electrons, is the preferred oxidation state of titanium under aqueous aerobic conditions and it is the ionisation energy of the valence electrons (which increases from titanium to iron) and not the number of electrons which seems to control reactivity.

It is not clear how nitriding occurs. In the case of lithium and calcium, it is believed that the primary interaction involves electron transfer from metal to dinitrogen, forming the metastable ion N_2^- (Roberts and Tompkins, 1959). The chemisorption of N_2 on metal surfaces has been extensively studied, primarily as a consequence of the importance of the Haber process, with iron receiving particular emphasis. Most Haber catalysts are mixtures of iron plus promoters and all the evidence would suggest that the rate-determining step is the chemisorption of dinitrogen. However, dinitrogen species have not been identified on such catalysts, so that dissociation of dinitrogen to nitrogen atoms occurs, presumably initiated by electron transfer to dinitrogen from the catalyst. For a recent review of this area, see Emmett (1975).

Crystalline metals behave differently. On crystalline iron, molecular chemisorption takes place on (111) and the molecules are stable (Brill *et al.*, 1967) and do not split easily to give adatoms. The function of the promoters in Haber catalysts must be concerned with the splitting of adsorbed dinitrogen molecules, but how they operate is not clear. In whatever manner they operate, it seems that the mechanism holds few lessons for those seeking to explain the function of nitrogenase or the reactions we have discussed above. The sites of adsorption of molecules on iron (111) appears to be the cavities between the hexagonally-packed iron atoms. There is no dinitrogen complex with an analogous structure, although carbonyl complexes with some similar characteristics are known (Fischer and Palm, 1958) (Fig. 6). Again, it does not seem possible to extrapolate from here to the systems under discussion.

Information concerning the mechanism of the Volpin reactions must therefore be gleaned from the systems themselves,

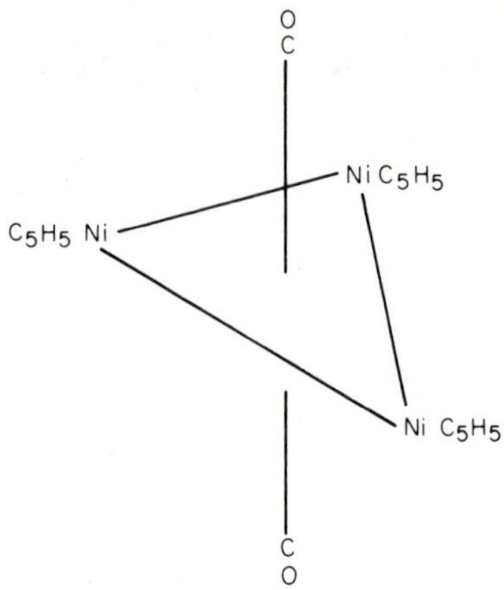

Fig. 6. *Representation of the structure* $[(C_5H_5)_3Ni_3(CO)_2]$.

but the search for intermediates in the Volpin systems has not been very rewarding. Binuclear species have been suggested on the basis of kinetic and some analytical data in the iron systems, e.g. $[FeCl_3(PPh_3)_2]$ plus Pr^iMgBr (Borodko *et al.*, 1971). A complex stable at -50° was isolated and formulated as:

$$[(PPh_3)_2\underset{\displaystyle Pr^i}{\overset{\displaystyle H}{Fe}}-N{=}N-\underset{\displaystyle Pr^i}{Fe}(PPh_3)_2]$$

This species reacts with hydrogen chloride at -60° to yield hydrazine. An intermediate with bridging diimide has been postulated. Similarly, an intermediate binuclear species has been postulated in the system $FeCl_3$/PhLi which can likewise lead to hydrazine, although ammonia is believed to be formed in this case by a competitive nitriding reaction (Broitman *et al.*, 1972; Broitman *et al.*, 1971).

A favourite subject for investigation is the system based upon bis(cyclopentadienyl)titanium. Volpin showed that the system $[(C_5H_5)_2TiCl_2]$/Grignard reagent is particularly efficient at fixing nitrogen. Undoubtedly, this is a nitriding reaction, but some intermediates have been isolated. The reaction of $[(C_5H_5)_2TiCl_2]$ with an alkyl-Grignard reagent leads to reduction and metal-carbon bond cleavage and intermediates are difficult to isolate (Nechiporenko *et al.*, 1965). Aryl-Grignard reagents are much better behaved. For example PhMgBr produces $[(C_5H_5)_2TiPh]$ which reacts with dinitrogen in, say, toluene solution at -40° to yield $[(C_5H_5)_2PhTiN_2TiPh(C_5H_5)_2]$. This complex is relatively stable, but the dinitrogen briding the titanium atoms cannot be protonated (Teuben, 1973). There is evidence that very unstable alkyl analogues of the phenyl complexes are formed in the appropriate reaction mixtures (Shilov *et al.*, 1971). However, they react with further Grignard reagent to form $[(C_5H_5)_2TiN_2Ti(C_5H_5)_2]$ (Borodko *et al.*, 1972; Borodko *et al.*, 1973), which is paramagnetic and has $\nu(N_2)$ at 1255cm^{-1}, which is exceedingly low and suggestive of less than a double bond. This complex, which is unstable, reacts with methanolic hydrogen chloride at -60° to give hydrazine and dinitrogen (1:1) and with ethereal hydrogen chloride at -60° to form ammonia and dinitrogen (1:2). These results suggest an initial protonation to diimide which then decomposes to form N_2 plus N_2H_4 or N_2 plus NH_3 depending upon the conditions (Borodko *et al.*, 1972).

There are two other complexes which appear to be related. Van Tamelen *et al.* (1969) have described the very slow reaction of titanocene $[(C_5H_5)_2Ti]$ with dinitrogen in benzene solution over several weeks to give a green solution, which supposedly contains a species $[\{(C_5H_5)_2TiN_2\}_2]$, $\nu(N_2)$ at 1960cm^{-1}, which can be reduced to give nitride. Unfortunately, the careful work of Brintzinger and Bercaw (1970) showed that titanocene is not as simple a compound as once thought, so that van Tamelen's dinitrogen-containing intermediate must be something other than that postulated. Bercaw *et al.* (1972) did succeed in isolating the unstable "true" titanocene. This complex reacts with dinitrogen at -80° to form a blue adduct $[(C_5H_5)_2TiN_2Ti(C_5H_5)_2]$, which appears different to that isolated by Shilov's group (for example, it exhibits no $\nu(N_2)$). Reduction with lithium naphthalene produces a nitride-type species that hydrolyses to give ammonia.

The relevance of these observations to the Volpin system is not clear. Maskill and Pratt (1968) studied the rate of fixation in the system $[(C_5H_5)_2TiCl_2]$/EtMgBr and concluded that species containing TiN_2Ti groups are significant intermediates. Shilov's group (Borodko *et al.*, 1973) envisage the subsequent reduction of their binuclear species by an excess of Grignard reagent as producing the species $[(C_5H_5)_2Ti{=}N{-}\underset{\displaystyle MgCl}{\underset{|}{N}}{-}Ti(C_5H_5)_2]$ and then $[(C_5H_5)_2TiN(MgCl)_2]$. These species can explain the formation of ammonia formed upon their hydrolyses, but are not altogether consistent with the earlier view that the $[(C_5H_5)_2TiCl_2]$/EtMgBr reaction proceeds with cleavage of the cyclopentadienyl groups from the titanium (Nechiporenko *et al.*, 1965). It is, however, consistent with further observations of Ungurenasu and Streba (1972), who postulate species such as $[(C_5H_5)_2Ti{-}\underset{\displaystyle Li}{\underset{|}{N}}{-}\underset{\displaystyle Li}{\underset{|}{N}}{-}Ti(C_5H_5)_2]$

and 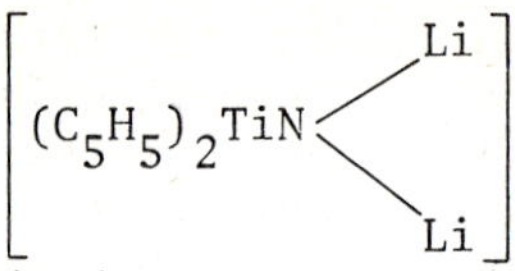 during fixation of nitrogen by $[(C_5H_5)_2TiCl_2]$/,i/Hg and also with further observations by Brintzinger's group (Bercaw *et al.*, 1972). The latter workers have been able to isolate a yellow solid from the reaction of $[(C_5H_5)_2TiCl_2]$ and EtMgCl in ether at room temperature with the stoichiometric formula $\{(C_5H_5)_2TiEt\}_2 \cdot 6MgCl_2 \cdot 7Et_2O$. This product reacts directly with 130 atm N_2 in diethyl ether at room temperature to give a black solid, which is probably a nitride and is formulated as $[\{(C_5H_5)_2Ti\}_3(N)_2]$. Treatment with hydrogen chloride yields two ammonia molecules for every three titanium atoms and regenerates $[(C_5H_5)_2TiCl_2]$.

It seems unequivocal, therefore, that, in at least some non-aqueous systems, dinitrogen forms binuclear complexes with one metal atom bound at either end. In appropriate circumstances, these species can be reduced further to nitrides which hydrolyse to ammonia. Certain complexes hydrolyse to hydrazine without need for further reduction. In no case is the sequence of reactions or the reaction mechanism clear. In aqueous systems, there is circumstantial evidence for binuclear intermediates again, but none for stepwise protonation in two-proton steps as we originally postulated. Evidence is now beginning to accumulate that dinitrogen can also react when it is bound at one end only and it does so in ways quite distinct from that implied by two-proton steps. Further, the structures of the intermediates are much better defined than most of those described above.

MONONUCLEAR N_2 COMPLEXES

The coordination of dinitrogen in the most commonly observed fashion, i.e. end-on and at one end only, results in its polarisation (Leigh *et al.*, 1970). This polarisation should encourage reactivity that is not possible in the non-polarised free dinitrogen or even in symmetrically coordinated bridging dinitrogen. At its simplest, the reactivity engendered is Lewis basicity and consequently the formation of adducts with Lewis acids is possible. The classical example is $[ReCl(N_2)(PMe_2Ph)_4]$, in which the dinitrogen is believed to be polarised in the sense $Re-\overset{\delta+}{N}-\overset{\delta-}{N}$. Although there are many adducts of the dinitrogen complex (Chatt *et al.*, 1970), there is no evidence that adduct formation promotes any subsequent reactivity of the dinitrogen towards protonation.

The critical observation in this area is that $[W(N_2)_2(dppe)_2]$ ($dppe=Ph_2PCH_2CH_2PPh_2$) reacts with acetyl chloride in benzene solution to give a product formulated as [WCl(N_2COCH_3)(dppe)$_2$](Chatt *et al.*, 1972). This compound contains the grouping W-N=N-C(=O)-CH_3 and has a structure which is apparently analogous to that of [MoCl(N_2COPh)(dppe)$_2$] which has been determined by X-ray structure analysis (Fig. 7). This molybdenum

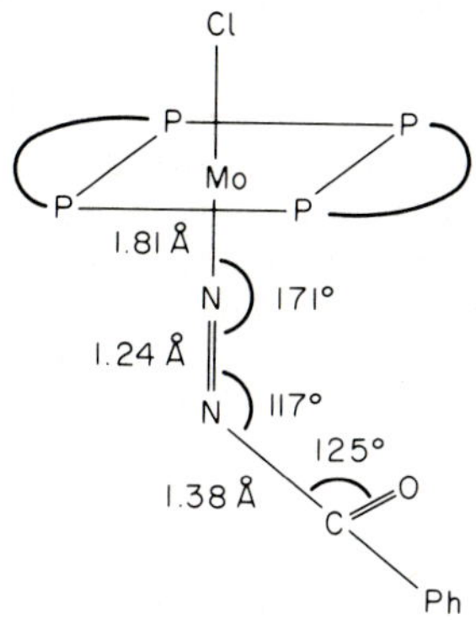

Fig. 7. *Representation of the structure of* [*MoCl*(N_2*COPh*)(*dppe*)$_2$].

complex, obtained by reaction of [Mo(PhCN)(N_2)-(dppe)$_2$] with PhCOCl (Tatsumi *et al.*, 1976), contains a linear Mo-N-N system with an N-N separation of 1.24 Å, which is nearer a double than a triple bond length. From this observation has sprung a whole series of reactions of monocoordinated dinitrogen with electrophilic reagents. Of these, the simplest is H+. Thus, [W(N_2)$_2$(dppe)$_2$] reacts with hydrogen chloride to yield [WCl_2(N_2H_2)(dppe)$_2$](Chatt *et al.*, 1974). One molecule of dinitrogen is evolved as gas and it is unlikely that the remaining dinitrogen reacts other than in a monocoordinated form. It is possible to postulate reaction paths such as 2[W(N_2)$_2$(dppe)$_2$] ⟶ 2[W(N_2)(dppe)$_2$] + 2N_2↑ ⟶ [(N_2)(dppe)$_2$W(N_2)W(dppe)$_2$] $\xrightarrow{HCl}$ [Cl_2W(N_2H_2)(dppe)$_2$] + [W(N_2)(dppe)$_2$] etc. but these appear unlikely and unnecessarily complex. Only one mole of dinitrogen is lost per tungsten and yields of the N_2H_2 adduct are greater than 60%. The structure of the N_2H_2 complex is believed to be a function of the other ligands. Simple electron counting suggests that the eighteen-electron rule is fulfilled by the following different structures for N_2H_2:

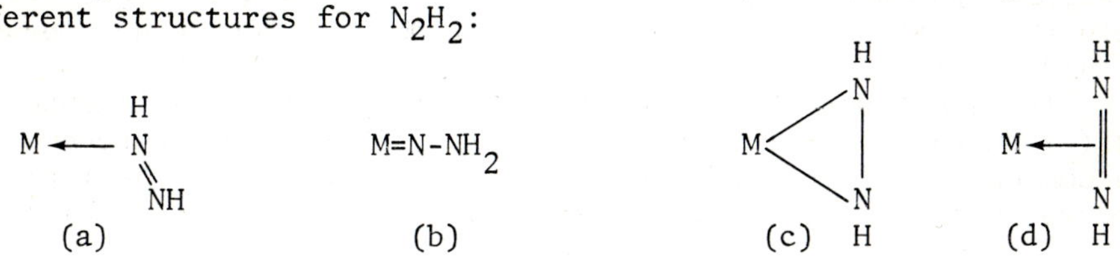

In (a), N_2H_2 is a two-electron donor; in (b), it is a four-electron donor dianion; and in (c) and (d), which in real terms are undistinguishable by physical and chemical methods, the N_2H_2 is a four- or a two-electron donor, respectively. Two series of N_2H_2 complexes have now been isolated, represented by the following two compounds:

$[ClW(N_2H_2)(dppe)_2]^+$ and $[Cl_2W(N_2H_2)(dppe)_2]$

(*3*) (*4*)

To fulfill the electron-counting requirements, N_2H_2 should be a four-electron donor in (*3*) but only a two-electron donor in (*4*). Infra-red and ^1H n.m.r. spectroscopy demonstrate quite clearly that N_2H_2 has a different structure in (*3*) compared to (*4*). The X-ray structure analysis of (*3*) is consistent with the ligand arrangement (b) (Heath *et al.*, 1974) (Fig. 8). Further, treatment

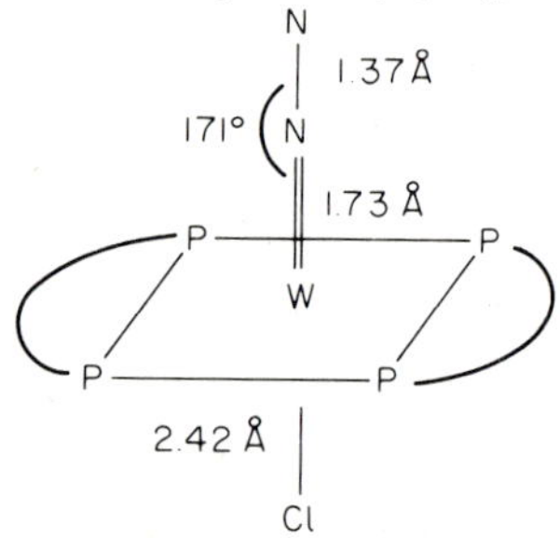

Fig. 8. *Representation of the structure of* $[WCl(NNH_2)(dppe)_2]^+$, *amido-hydrogen atoms being omitted.*

of $[WCl_2(N_2H_2)(dppe)_2]$ (*4*) with base affords $[WCl(N_2H)(dppe)_2]$ (*5*), which can be reprotonated to give (*4*) (Chatt, 1975). The complex (*5*) may well be an intermediate in the formation of (*4*) from $[W(N_2)_2(dppe)_2]$, so that a probable reaction sequence is:

$$[W(N_2)_2(dppe)_2] \longrightarrow [WCl(N_2H)(dppe)_2] \longrightarrow [WCl_2(N_2H_2)(dppe)_2]$$

(*6*) (*5*) (*4*)

Note that the sequence of metal oxidation states in the series (*6*)$\rightarrow$(*5*)$\rightarrow$(*4*) is 0$\rightarrow$II$\rightarrow$IV, i.e., the favoured two-electron sequence, but only one of each pair of electrons goes into the dinitrogen. The behaviour of the molybdenum homologue is similar.

It is not possible to protonate the hydrazido(2-) ligand in (*4*) further. Brulet and van Tamelen (1975) have succeeded in converting N_2 in $[Mo(N_2)_2(dppe)_2]$ to ammonia, but their procedure involves heating the complex at reflux in a high-boiling solvent, such as N-methylpyrrolidone, before protonation. The course of this reaction is completely obscure. However, analogous complexes, such as $[M(N_2)_2(PMe_2Ph)_4]$ (M=Mo or W), can be protonated further to give ammonia (Chatt *et al.*, 1975). Thus, $[W(N_2)_2(PMe_2Ph)_4]$ reacts with sulphuric acid in methanol solution (in certain circumstances, solvent methanol alone is acid enough to protonate the dinitrogen) to give 0.94 mole N_2, 0.02 mole hydrazine and 1.8 mole ammonia. The molybdenum homologue yields 1.6 - 1.7 moles N_2, <0.1 mole hydrazine and 0.6 - 0.7 mole ammonia.

These yields of products obviously depend on the metal-dinitrogen complex used, and the molybdenum and tungstem complexes do show significant differences in chemical behaviour. Thus, upon acid treatment of the tungsten complex, ammonia generation appears to occur immediately. The molybdenum complex reacts in a more complex fashion (Chatt, Pearman and Richards, unpublished) and on acid treatment yields directly one mole of dinitrogen but no free ammonia and a further undefined molybdenum complex. Upon treatment with base, this complex yields 2/3 mole ammonia and 2/3 mole dinitrogen, which is consistent with the decomposition of diimide according to the equation:

$$3N_2H_2 \longrightarrow 2NH_3 + 2N_2$$

It seems unlikely that free diimide is involved because diimide more usually decomposes to N_2 plus N_2H_4, so the metal also influences the decomposition path. Borodko *et al.* (1972) have already noted how solvent influences the reaction path in titanium systems.

It is clear that there is no unique reaction pathway for the protonation of dinitrogen. By choice of solvent and reagents, it is possible to isolate other complexes from the $[W(N_2)_2(PMe_2Ph)_4]$-hydrogen chloride reaction, such as $[WCl_3(N_2H_3)(PMe_2Ph)_2]$. This product can be protonated directly to ammonia plus very little hydrazine (Chatt, Pearman and Richards, unpublished). Evidently, the monodentate phosphine, which dissociates from the metal more readily than the bidentate dppe, allows new kinds of reaction to take place. Thus, a series of reaction steps based, for the first time, on firm empirical evidence, can be postulated. Schematically, these are represented in Fig. 9. No complexes containing only one nitrogen atom have been isolated from these protonation systems. However, complexes containing the imido-group are known (Chatt and Dilworth, 1975) and these protonate readily to ammonia.

It is worth summarising the position here. It is apparent that dinitrogen can be activated by coordination to a transition metal. The coordination need not be at both ends, mono -

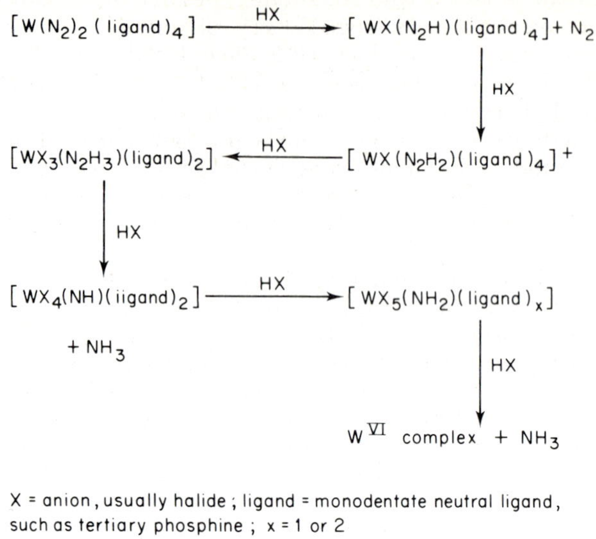

Fig. 9. *Suggested scheme for stepwise protonation of* [$W(N_2)_2$ *(ligand)*$_4$] *to give ammonia.*

coordinated dinitrogen being at least as reactive as bridging dinitrogen. There is no hard evidence for side-on bonded dinitrogen in the usual systems. Solvation is not a necessary contributor to producing reactivity. There is more than one reduction path, but diimide or hydrazido(2-) seem established as intermediates. There is no empirical evidence for two-electron jumps; single-electron steps are much more likely. The pattern seems to suggest coordination of N_2 to a metal followed by successive protonation, electroneutrality being maintained by an equivalent flow of electrons into the dinitrogen from the metal which becomes oxidised in the process (Fig. 10).

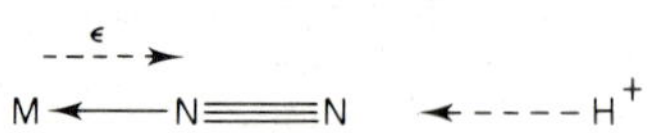

Fig. 10. *Representation of suggested mechanism for protonation of coordinated dinitrogen.*

I leave others to argue the implications for the mechanism of enzymic fixation. I would only say that I have found Occam's razor a very useful tool, especially for removing excessive verbiage, and would, therefore, base my models on the simplest requirements consistent with the empirical observations - that is, end-on coordination of dinitrogen at a single metal site with single-electron reduction steps.

This would be a good place to end this review, except that a new and growing field of reactivity of dinitrogen, which is of a different kind, although still apparently based upon the electrophilic properties of attacking reagents, would be omitted. The complexes $[M(N_2)_2(dppe)_2]$ react with acyl halides to form complexes $[MCl(N_2COR)(dppe)_2]$. They also react with alkyl halides, RX (X = Cl, Br or I), the ease of reaction apparently falling in the sequence I > Br > Cl. The products of these reactions are alkyldiazenido-complexes, $[MXN_2R(dppe)_2]$, and they can be protonated further to give $[MXN_2HR(dppe)_2]$, which are alkylhydrazido(2-) derivatives, homologues of the hydrazido(2-) complexes described above. Only one simple dialkyl complex has been isolated, $[WBr(N_2Me_2)(dppe)_2]Br$ (Day *et al.*, 1975 Diamantis *et al.*, 1975; George and Iske, 1976; Chatt, 1975). These reactions occur in benzene solution, but whereas $[Mo(N_2)_2(dppe)_2]$ reacts in the dark, $[W(N_2)_2(dppe)_2]$ requires irradiation by at least diffuse daylight. This observation is one of several differences which molybdenum and tungsten chemistry are beginning to show. Others have already been alluded to when discussing the protonation reactions.

Amines have not yet been produced in quantity from these reactions yet. In fact, $[WBr(N_2Me)(dppe)_2]$ and $[WBr(N_2Me_2)(dppe)_2]Br$, when treated with N-methylpyrrolidone (as described by van Tamelen) and then sulphuric acid, produce only small amounts of methylamine or dimethylamine, about 20% of the expected yields. Evidently, N-C bond cleavage (producing ammonia eventually) is quite facile (Chatt, Leelamani and Leigh, unpublished).

Methylene dibromide in benzene also induces different reactions with molybdenum- and tungsten-bis(dinitrogen) complexes. The former produces $[MoBr(N_2CH_2Br)(dppe)_2]$ and the latter a diazomethane complex $[WBr(CH_2N_2)(dppe)_2]Br$ (Ben-Shoshan *et al.*, 1976). Reaction of either dinitrogen complex with methyl bromide in tetrahydrofuran produces a novel complex of diazobutanol, $[MBr(N_2CHCH_2CH_2CH_2OH)(dppe)_2]Br$ (Bevan *et al.*, 1976). The methyl bromide is essential for this reaction, the methyl group being evolved as methane.

We are only part way to explaining these reactions and the differences between molybdenum and tungsten chemistry. The X-ray structure analyses summarised in the Table show that the residue of the dinitrogen in all these products is essentially in the form of an NN doubly bonded unit, whatever the formal bonded arrangement, so to that extent the products are similar (Bevan *et al.*, 1976). X-ray photoelectron spectroscopy of $[M(N_2)_2(dppe)_2]$ (Leigh, unpublished) suggests that, to a first approximation, the dinitrogens are in similar electronic conditions. However, X.P.S. is rather a crude probe, so that other

	N(1s)eV	P(2p)eV
$[Mo(N_2)_2(dppe)_2]$	398.6 (broad)	131.0
$[W(N_2)_2(dppe)_2]$	399.1 (broad)	131.2

approaches may be more fruitful. The redox potentials for the couples $[M(N_2)_2(dppe)_2]/[M(N_2)_2(dppe)_2]^+$ measured against the saturated calomel electrode are -0.362 volts (M = Mo) and -0.351 volts (M = W)(Elson and Leigh, unpublished). These values are

Table

	W-N/A	N-N/A	M-N-N°
$[WCl(N\text{-}NH_2)(dppe)_2]^+$	1.73(1)	1.37(2)	171(1)
$[WBr\{N\text{-}NH(Me)\}(dppe)_2]^+$	1.768(14)	1.32(2)	174(1)
$[WBr\{N\text{-}N{=}CH(CH_2)_3OH\}\text{-}(dppe)_2]^+$	1.778(14)	1.306(22)	172(1)

so similar as to be considered identical, which means that, if any electron excitation process from the metal atoms is involved, the excitation starts from the highest occupied molecular orbital which is at the same energy level regardless of the metal. In fact, $[M(N_2)_2(dppe)_2]^+$ itself is unlikely to be involved in any alkylation reactions because it decomposes faster than the alkylation process by a factor of about 10^3.

Because the tungsten systems require irradiation, differences in the u.v. spectra of Mo and the W complexes might be expected. They are, however, very similar, with the lowest energy charge transfer band being at about 380nm in both cases (Leigh and Pickett, unpublished). Consequently, the energies of the lowest unoccupied molecular orbitals must be very similar too. This treatment, of course, begs the question of how these orbitals are constituted. They may indeed have nothing to do with dinitrogen and be constituted from metal and phosphorus orbitals. Nevertheless, there is no evidence to suggest that excitation of electrons from the metal atom is a significant step in these reactions of N_2.

We do have preliminary evidence that the alkylation reactions of dinitrogen involve free radicals (Devan *et al.*, unpublished). All the reaction products described above can be rationalised on the basis of the significant step being a metal-assisted homolysis of an alkyl-halogen bond, which requires radiation in tungsten systems. The reactions might be described as follows:

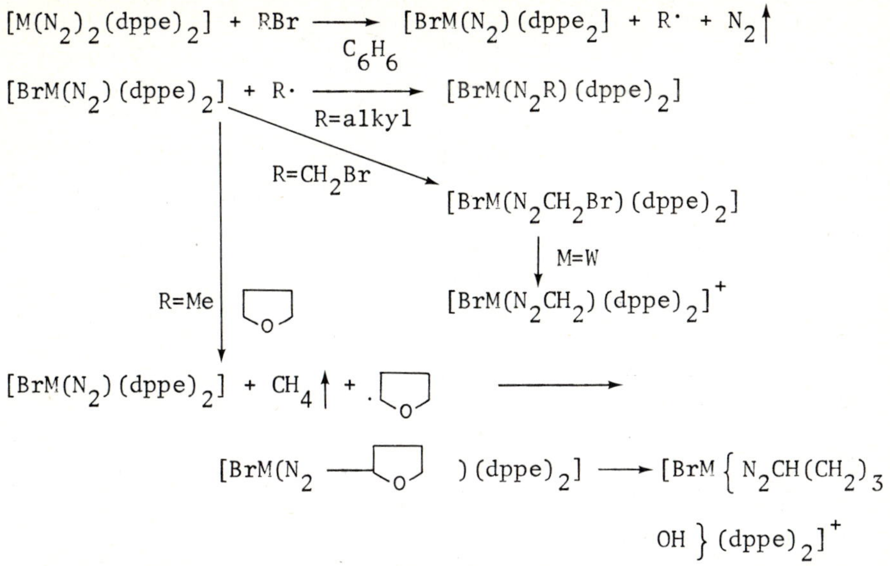

All this work points to a new kind of reactivity of coordinated dinitrogen with electrophiles and with free radicals.

The chemistry of coordinated dinitrogen is broader and more complex than ever expected. No firm patterns have emerged and we are, as yet, unable to make simple predictions. It would be unwise to extrapolate from this mass of data to a conclusion about how nitrogenase must function. The bacteria may well know something more than we do. The comfort is that they have had much longer to think about it.

ACKNOWLEDGEMENTS

I am indebted to my colleagues in the Agricultural Research Council, Unit of Nitrogen Fixation, and the School of Molecular Sciences, whose insights, ideas and criticisms have helped me to develop the themes presented here. In particular, I should like to mention Professor J. Chatt and Dr.'s J. R. Dilworth, C.J. Pickett and R.L. Richards. They deserve a considerable amount of any credit ensuing. The infelicities of expression and the errors are mine.

REFERENCES

Ben-Shoshan, R., Chatt, J., Hussain, W. and Leigh, G.J. (1976). *J.Organometal.Chem.*, 112: C9-C10.

Bercaw, J.E., Marvich, R.H., Bell, L.G. and Brintzinger, H.H. (1972). *J.Am.Chem.Soc.*, 94: 1219-1238.

Bevan, P.C., Chatt, J., Head, R.A., Hitchcock, P.B. and Leigh, G.J. (1976). *J.Chem.Soc.Chem.Comm.*, 509-510.
Binder, H. and Sellmann, D. (1973). *Angew.Chem.internat.Edit.*, 12: 1017-1019.
Borodko, Yu.G., Broitman, M.O., Kachapina, L.M., Shilov, A.E. and Ukhin, L.Yu. (1971). *Chem.Comm.*, 1185-1186.
Borodko, Yu.G., Ivleva, I.N., Kachapina, L.M., Salienko, S.I., Shilova, A.K. and Shilov, A.E. (1972). *J.Chem.Soc.Chem. Comm.*, 1178.
Borodko, Yu.G., Ivleva, I.N., Kachapina, L.M., Kvashina, E.F., Shilova, A.K. and Shilov, A.E. (1973). *J.Chem.Soc.Chem. Comm.*, 169-170.
Brill, R., Richter, E.R. and Ruch, E. (1967). *Angew.Chem.internat.Edit.*, 6: 882-883.
Brintzinger, H.H. and Bercaw, J.E. (1970). *J.Am.Chem.Soc.*, 92: 6182-6185.
Broitman, M.O., Denisov, N.T., Shuvalova, N.I. and Shilov, A.E. (1971).
Broitman, M.O., Vorontsova, T.A. and Shilov, A.E. (1972). *Kin. i Kat.*, 13: 61-64.
Brulet, C.R. and van Tamelen, E.E. (1975). *J.Am.Chem.Soc.*, 97: 911-912.
Chatt, J., Dilworth, J.R., Leigh, G.J. and Richards, R.L. (1970). *Chem.Comm.*, 955-956.
Chatt, J., Heath, G.A. and Leigh, G.J. (1972). *J.Chem.Soc.Chem. Comm.*, 444-445.
Chatt, J., Heath, G.A. and Richards, R.L. (1974). *J.Chem.Soc., Dalton Trans.*, 2074-2082.
Chatt, J. (1975). *J.Organometal.Chem.*, 100: 17-28.
Chatt, J. and Dilworth, J.R. (1975). *J.Chem.Soc.Chem.Comm.*, 983-984.
Chatt, J., Pearman, A.J. and Richards, R.L. (1975). *Nature*, 253; 39-40.
Day, V.W., George, T.A. and Iske, S.D. (1975). *J.Am.Chem.Soc.*, 97: 4127-4128.
Denisov, N.T., Shuvalov, V.F., Shuvalova, N.I., Shilova, A.K. and Shilov, A.E. (1970a). *Kin.i Kat.*, 11: 813-814; *Dokl. Akad.Nauk S.S.S.R.*, 195: 879-891.
Denisov, N.T., Efimov, O.N., Shuvalova, N.I., Shilova, A.K. and Shilov, A.E. (1970b). *Russ.J.Phys.Chem.*, 44: 1693-1694.
Denisov, N.T., Shuvalova, N.I. and Shilov, A.E. (1973). *Kin.i Kat.*, 14: 1325-1326.
Denisov, N.T., Shuvalova, N.I., Ivleva, I.N. and Shilov, A.E. (1974). *Zhur.Fiz.Khim.*, 48: 2238-2242.
Denisov, N.T., Shilov, A.E., Shuvalova, N.I. and Panova, T.P. (1975). *Kin.i Kat.Lett.*, 2: 237-241.
Diamantis, A.A., Chatt, J., Leigh, G.J. and Heath, G.A. (1975). *J.Organometal.Chem.*, 84: C11-C12.

Emmett, P.H. (1975). *In:* The Physical Basis for Heterogeneous Catalysis (E. Drauglis and R.I. Jaffee, eds), pp. 3-34. Plenum Press, London and New York.

Fischer, E.O. and Palm, C. (1958). *Chem.Ber.*, 91: 1725-1731.

George, T.A. and Iske, S.D. (1976). *In:* Proc.1st Internat.Symp. on Nitrogen Fixation (W.E. Newton and C.J. Nyman, eds), Vol. I, pp. 27-32. Washington State University Press, Pullman, Wash.

Heath, G.A., Mason, R. and Thomas, K.M. (1974). *J.Am.Chem.Soc.*, 96: 259-260.

Harrison, D.E., Weissberger, E. and Taube, H. (1968). *Science*, 159: 320-322.

Jonas, K., Brauer, D.J., Kruger, C., Roberts, P.J. and Tsay, Y.-H. (1976). *J.Am.Chem.Soc.*, 98: 74-81.

Leigh, G.J., Murrell, J.N., Bremser, W. and Proctor, W.G. (1970). *Chem.Comm.*, 1661.

Leigh, G.J. (1971). *In:* The Chemistry and Biochemistry of Nitrogen Fixation (J.R. Postgate, ed.), pp. 19-56. Plenum Press, London and New York.

Manriquez, J.M., Sanner, R.D., Marsh, R.E. and Bercaw, J.E. (1976). *J.Am.Chem.Soc.*, 98: 3042-3044.

Maskill, R. and Pratt, J.M. (1968). *J.Chem.Soc.(A).*, 1914-18.

Nechiporenko, G.N., Tabrina, G.M., Shilova, A.K. and Shilov, A.E. (1965). *Dokl.Akad.Nauk S.S.S.R.*, 164: 1062-1065.

Newton, W.E., Corbin, J.L. and McDonald, J.W. (1976). *In:* Proc. 1st Internat.Sympos. on Nitrogen Fixation (W.E. Newton and C.J. Nyman, eds), Vol. I, pp. 53-74. Washington State University Press, Pullman, Wash.

Nikonova, L.A., Ovcharenko, A.G., Efimov, O.N., Avilov, V.A. and Shilov, A.E. (1972). *Kin.i Kat.*, 13: 1602-1603.

Nikonova, L.A., Pershikova, N.I., Bodeiko, M.V., Olinik, L.G., Sokolov, D.N. and Shilov, A.E. (1974). *Dokl.Akad.Nauk S.S.S.R.*, 216: 140-141.

Roberts, M.W. and Tompkins, F.C. (1959). *Proc.Roy.Soc.(London)*, 215A: 369-377.

Schrauzer, G.N. and Doemeny, P.A. (1971). *J.Am.Chem.Soc.*, 93: 1608-1618.

Schrauzer, G.N., Keefer, G.W., Tano, K. and Doemeny, P.A. (1974). *J.Am.Chem.Soc.*, 96: 641-652.

Schrauzer, G.N., Robinson, P.R., Moorehead, E.L. and Vickery, T.M. (1975). *J.Am.Chem.Soc.*, 97: 7069-7076.

Schrauzer, G.N. (1976). *In:* Proc.1st Int. Sympos. on Nitrogen Fixation (W.E. Newton and C.J. Nyman, eds), Vol. I, pp. 79-116. Washington State University Press, Pullman, Wash.

Schrauzer, G.N., Robinson, P.R., Moorehead, E.L. and Vickery, T.M. (1976). *J.Am.Chem.Soc.*, 98: 2815-2820.

Sellmann, D. (1972). *J.Organometal.Chem.*, 44: C46-C48.

Sellmann, D., Brandl, A. and Endell, R. (1973). *Angew.Chem.Internat.Edit.*, 12: 1019.
Sellmann, D., Brandl, A. and Endell, R. (1975). *J.Organometal. Chem.*, 97: 229-243.
Sellmann, D., Brandl, A. and Endell, R. (1976). *J.Organometal. Chem.*, 111: 303-329.
Shilov, A.E., Shilova, A.K., Kvashina, E.F. and Vorontsova, T.A. (1971). *Chem.Comm.*, 1590-1591.
Shilov, A.E. (1974). *Russ.Chem.Rev.*, 43: 378-398.
Shilov, A.E., Shilova, A.K. and Vorontsova, T.A. (1975). *Kin. i Kat.Lett.*, 2: 143-148.
Shilov, A.E. (1976). *In:* Proc.1st Int. Sympos. on Nitrogen Fixation (W.E. Newton and C.J. Nyman, eds), Vol. I, pp. 42-52. Washington State University Press, Pullman, Wash.
Smith, B.E., Thorneley, R.N.F., Eady, R.R. and Mortenson, L.E. (1976). *Biochem.J.*, 157: 439-447.
Stiefel, E.I. (1973). *Proc.Nat.Acad.Sci.U.S.A.*, 70: 988-992.
Tatsumi, T., Hidai, M. and Uchida, Y. (1976). *Inorg.Chem.*, 14: 2530-2534.
Teuben, J.H. (1973). *J.Organometal.Chem.*, 57: 159-167.
Ungurenasu, C. and Streba, E. (1972). *J.Inorg.Nuclear Chem.*, 34: 3753-3758.
van Tamelen, E.E., Fechter, R.B., Schneller, S.W., Boche, G., Greeley, R.H. and Akermark, B. (1969). *J.Am.Chem.Soc.*, 91: 1551-1552.
Volpin, M.E. and Shur, V.B. (1964). *Dokl.Akad.Nauk S.S.S.R.*, 156: 1102-1104.
Volpin, M.E., Ilatovskaya, M.A., Larikov, E.I., Khidekel, M.L., Shvetsov, Yu.A. and Shur, V.B. (1965a). *Dodl.Akad.Nauk S.S.S.R.*, 164: 331-333.
Volpin, M.E., Belyi, A.A. and Shur, V.B. (1965b). *Izvest.Akad. Nauk S.S.S.R.*, 2225-2226.
Volpin, M.E., Chapovskaya, N.K. and Shur, V.B. (1966). *Izvest. Akad.Nauk S.S.S.R.*, 1083-1084.
Volpin, M.E. and Shur, V.B. (1970). *Organometallic Reactions 1*, 55-117.
Volpin, M.E. (1972). *Pure Appl.Chem.*, 30: 607-626.

REDUCTION OF MOLECULAR NITROGEN TO HYDRAZINE AT TITANIUM AND ZIRCONIUM

J.E. BERCAW

Division of Chemistry and Chemical Engineering, California Institute of Technology, Pasadena, California 91125, U.S.A.
Contribution No. 5361

INTRODUCTION

The relatively high efficiency of titanocene-based N_2-reducing systems has stimulated much research since their discovery over a decade ago (Volpin and Shur, 1966). It has been demonstrated that titanocene, $[(C_5H_5)_2Ti]_n$(n=1 or 2), is an intermediate common to many of these reaction mixtures, and that this complex absorbs N_2 in a rapid, reversible reaction to yield the blue binuclear dinitrogen complex $[(C_5H_5)_2Ti]_2N_2$ (van Tamelen, 1970; Borodko *et al.*, 1972; Bercaw *et al.*, 1972; Ungurenasu and Streba, 1972; van Tamelen *et al.*, 1972). Other investigators have found that Ti(III) complexes of the type $(C_5H_5)_2$ TiR(R=C_6H_5, $CH_2C_6H_5$, *i*-C_3H_7) also reversibly absorb N_2 to form diamagnetic, blue binuclear dinitrogen complexes $[(C_5H_5)_2TiR]_2$ N_2 (Teuben, 1973; van der Weij and Teuben, 1976; Shilov *et al.*, 1969, 1971). Both types of dinitrogen complexes liberate ammonia in good yield when treated first with sodium naphthalenide, then with HCl. Furthermore, under appropriate conditions $[(C_5H_5)_2$ $Ti]_2N_2$ may be protonated directly to yield either NH_3 and N_2 or N_2H_4 and N_2.

It has not yet been possible to isolate a crystalline dinitrogen complex of titanocene suitable for definitive structural characterization. The nature of these "active" dinitrogen complexes and their relationship to other structurally characterized, but inert, dinitrogen complexes have thus been subjects of considerable interest and speculation. The difficulties encountered when attempting to isolate stable crystalline forms of $[(C_5H_5)_2Ti]_2N_2$ and $[(C_5H_5)_2TiR]_2N_2$ may be traced, at least in part, to facile cyclopentadienyl-to-titanium hydrogen transfer inherent to these systems. We have therefore investigated the closely related pentamethylcyclopentadienyl

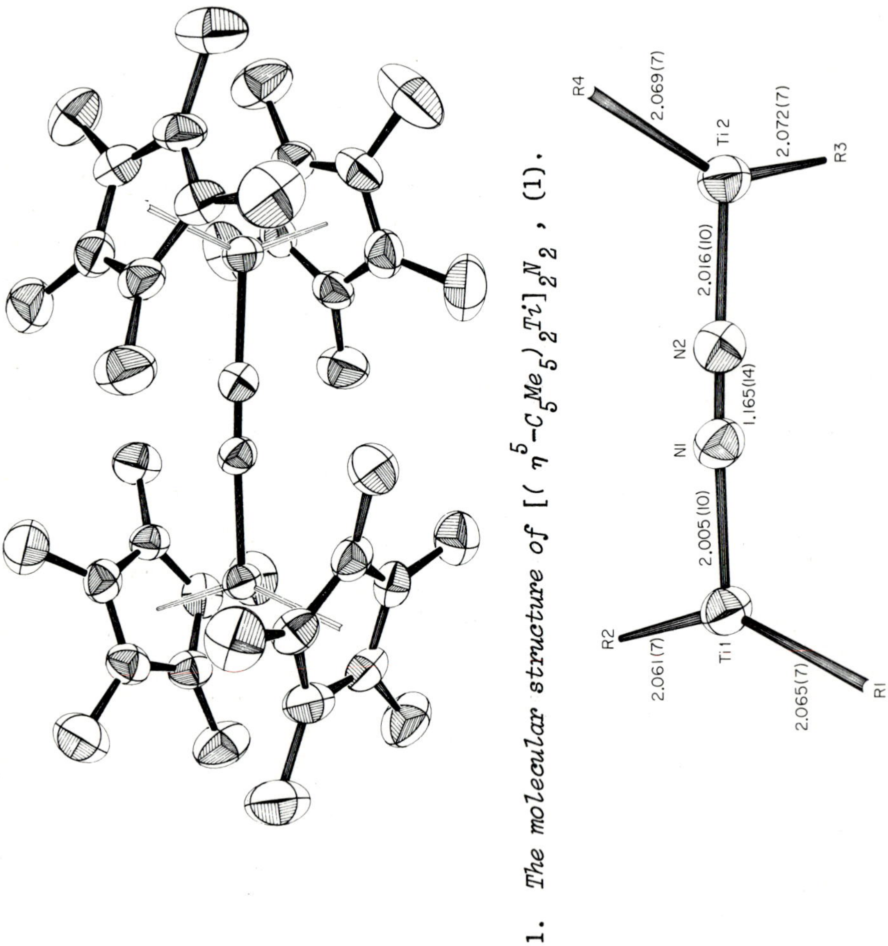

Fig. 1. *The molecular structure of* $[(\eta^5\text{-}C_5Me_5)_2Ti]_2N_2$, (1).

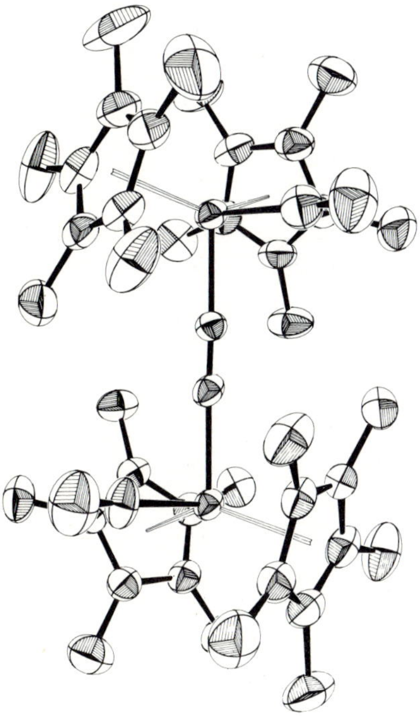

Fig. 2. *The molecular structure of* $[(\eta^5\text{-}C_5Me_5)_2ZrN_2]_2N_2$, (3).

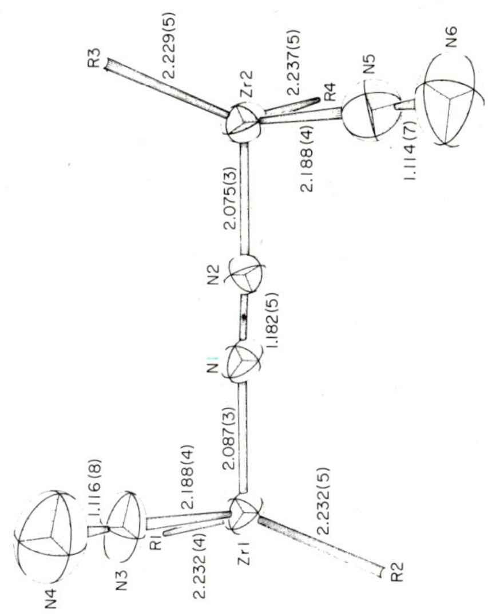

derivatives. We find compounds containing the (η^5-$C_5Me_5)_2M$ (M=Ti, Zr) moiety are often more stable and much more amenable to study than their (η^5-C_5H_5) analogs (Bercaw and Brintzinger, 1971; Bercaw *et al.*, 1974; Bercaw, 1974; Manriquez and Bercaw, 1974; McKenzie *et al.*, 1975; Manriquez *et al.*, 1976; Sanner *et al.*, 1976a, 1976b).

SYNTHESIS AND CHARACTERIZATION OF DINITROGEN COMPLEXES OF (η^5-$C_5Me_5)_2Ti$ AND (η^5-$C_5Me_5)_2Zr$

The starting materials, (η^5-$C_5Me_5)_2TiCl_2$ and (η^5-$C_5Me_5)_2ZrCl_2$, were prepared similarly according to equations (1) and (2).

$$2\ Li(C_5Me_5) + TiCl_3 \longrightarrow \longrightarrow (\eta^5\text{-}C_5Me_5)_2TiCl_2 + 2LiCl + \tfrac{1}{2}H_2 \quad (1)$$

$$2\ Li(C_5Me_5) + ZrCl_4 \longrightarrow (\eta^5\text{-}C_5Me_5)_2ZrCl_2 + 2LiCl \quad (2)$$

Because direct reduction of (η^5-$C_5Me_5)_2TiCl_2$ proceeded at a rate comparable to that for the decomposition of the reduced products, an indirect route (equations (3) - (6)) was employed to generate (η^5-$C_5Me_5)_2Ti$. Exposure of (η^5-$C_5Me_5)_2Ti$ to N_2 in pentane at 0°C yields crystalline [(η^5-$C_5Me_5)_2Ti]_2N_2$ (**1**).

$$(\eta^5\text{-}C_5Me_5)_2TiCl_2 + 2LiCH_3 \longrightarrow (\eta^5\text{-}C_5Me_5)_2Ti(CH_3)_2 + 2LiCl \quad (3)$$

$$(\eta^5\text{-}C_5Me_5)_2Ti(CH_3)_2 \longrightarrow (\eta^5\text{-}C_5Me_5)(C_5Me_4CH_2)TiCH_3 + CH_4 \quad (4)$$

$$(\eta^5\text{-}C_5Me_5)(C_5Me_4CH_2)TiCH_3 + 2H_2 \longrightarrow (\eta^5\text{-}C_5Me_5)_2TiH_2 + CH_4 \quad (5)$$

$$(\eta^5\text{-}C_5Me_5)_2TiH_2 \rightleftharpoons (\eta^5\text{-}C_5Me_5)_2Ti + H_2 \quad (6)$$

1 is a dark blue complex (λ_{max}=512,642 nm) and exhibits no absorption in the infra-red attributable to an NN stretching mode. **1** is paramagnetic (μ_{eff}=2.18(1) BM per Ti), and a single

$$2(\eta^5\text{-}C_5Me_5)_2Ti + N_2 \rightleftharpoons [(\eta^5\text{-}C_5Me_5)_2Ti]_2N_2 \quad (7)$$

resonance is observed in the ^1H nmr spectrum which is paramagnetically shifted 60.9 ppm downfield of TMS at -55°C.

The structure of **1**, as determined by single crystal x-ray diffraction methods (Sanner *et al.*, 1976b), is shown in Fig. 1. The structure consists of two staggered (η^5-$C_5Me_5)_2Ti$ units bridged by N_2 in a linear Ti-N N-Ti arrangement. N—N distances for the two molecules in the asymmetric unit (space group P1) are 1.165(14) and 1.155(14) Å, somewhat longer than those for other binuclear dinitrogen complexes (Sellman, 1974). The conventional end-on N_2 bonding mode is of special significance in view of earlier suggestions that either a side-on (Bercaw *et al.*, 1974; Bercaw, 1974) or doubly bent (azo)(Borodko *et al.*, 1972) coordination should be favoured for complexes of this type.

On cooling toluene solutions of $(\eta^5\text{-}C_5Me_5)_2Ti$ or 1 below *ca.* -20°C under 1 atmosphere N_2, the purple-blue (λ_{max}=578 nm) complex, $[(\eta^5\text{-}C_5Me_5)_2TiN_2]_2N_2$ (**2**), is formed. Although **2** is not sufficiently stable for its isolation as a stable crystalline material, it has been studied *in situ* below -50°C. ^1H and ^{15}N nmr and ir spectra for **1** are indicative of a structure identical to that for $[(\eta^5\text{-}C_5Me_5)_2ZrN_2]_2N_2$ (see below).

Reduction of $(\eta^5\text{-}C_5Me_5)_2ZrCl_2$ with sodium amalgam in toluene under N_2 leads to $[(\eta^5\text{-}C_5Me_5)_2ZrN_2]_2N_2$ (3) in 50-60% yield.

$$2(\eta^5\text{-}C_5Me_5)_2ZrCl_2 + 4Na \xrightarrow{N_2} [(\eta^5\text{-}C_5Me_5)_2ZrN_2]_2N_2 + 4NaCl \qquad (8)$$

The structure of **3**, as determined by single-crystal x-ray diffraction methods (Sanner *et al.*, 1976a), is illustrated in Fig. 2. The binuclear structure consists of two $(\eta^5\text{-}C_5Me_5)_2$ $ZrN{\equiv}N$ moieties bridged by a third dinitrogen ligand. Terminal and bridging dinitrogen ligands are bonded end-on in essentially linear $Zr\text{-}N{\equiv}N$ and $Zr\text{-}N{\equiv}N\text{-}Zr$ arrangements. The two terminal N-N bond lengths (1.116(8) and 1.114(7) Å) are within the range of those found previously for mononuclear dinitrogen complexes (1.11±0.01 Å)(Sellman, 1974), only slightly expanded from that for free N_2(1.0976 Å). The bridging dinitrogen ligand, however, exhibits a significantly longer N-N distance (1.182(5)Å), indicative of a substantial reduction in bond order.

The infra-red spectrum of **3** (Nujol mull) exhibits two strong bands at 2041 and 2006 cm^{-1} and a band of medium intensity at 1556 cm^{-1}, all of which shift upon substitution with doubly labeled $^{15}N_2$ to 1972, 1937, and 1515 cm^{-1}, respectively. This spectrum is interpreted on the basis of a coupling of the symmetric combination of terminal N-N stretches (2041 cm^{-1}) with the bridging N-N mode (1556 cm^{-1}), the latter gaining intensity from the former. The strongest band (2006 cm^{-1}) is attributed to the antisymmetric combination of terminal N-N modes. The observation of two terminal N-N stretching frequencies is significant in that it indicates substantial electronic coupling of the two $(\eta^5\text{-}C_5Me_5)_2ZrN$ N units *via* the $\mu\text{-}N_2$. The nearly orthogonal spatial relationship of the two terminal N_2 ligands eliminates the possibility of through-space dipolar coupling.

The ^1H nmr spectrum of **3** at 5°C (toluene-d_8) shows the expected two singlets attributable to the pairwise equivalent (η^5-C_5Me_5) rings. ^{15}N nmr **3** identical with that in the crystalline state. Thus, at -28°C (toluene-d_8), the 18.25 MHz ^{15}N nmr spectrum for **3**-($^{15}N_2$)$_3$ consists of two doublets attributable to the two ^{15}N nuclei of the two equivalent terminal dinitrogen ligands ($^1J_{^{15}N-^{15}N}$ = 6.2 Hz) centred at 89.8 and 160.4 ppm upfield of a third singlet resonance due to the two ^{15}N nuclei of the μ-N_2. The ^{15}N nmr spectrum for **3**-($^{15}N\equiv^{14}N$)$_3$ exhibits the same spectrum with the exception that the two upfield doublets now appear as the expected singlets.

Variable temperature ^1H and ^{15}N nmr experiments are in accord with (η^5-C_5Me_5) ring site exchange, involving step-wise dissociation-association of the two terminal dinitrogen ligands of **3**. In the temperature range -26°C to +45°C, the two singlets (1.769 δ and 1.788 δ) in the ^1H nmr spectrum steadily broaden, coalesce at +11°C and finally, give rise to a single line which narrows to ~ +45°(1.796 δ). Over the same temperature range, the two doublets due to the ^{15}N nuclei of the two equivalent terminal dinitrogen ligands for **3**-($^{15}N_2$)$_3$ steadily broaden and move slightly upfield toward the chemical shift of free $^{15}N_2$ in toluene-d_8. The signal due to the μ-N_2 remains a sharp singlet over the entire range of temperatures.

These ^1H and ^{15}N nmr data can be fitted to the process shown in Figs 3 and 4. Thus (Fig. 3), a terminal N_2 dissociates to generate the intermediate (η^5-C_5Me_5)$_2$Zr(N_2)-N_2-Zr(η^5-C_5Me_5)$_2$, which may then add N_2 to either side of the trigonal Zr centre. A series of two such sequences (Fig. 4) may effect complete (η^5-C_5Me_5) ring exchange between the two types of sites (α, eclipsed with N_2, and β), and also subjects the ^{15}N nuclei to a weighted average of each of three sites, two bonded to Zr plus that for free N_2 in toluene-d_8.

N_2 REDUCTION TO HYDRAZINE

Treatment of **2** or **3** with anhydrous HCl in toluene at -80°C yields (η^5-C_5Me_5)$_2$TiCl$_2$ or (η^5-C_5Me_5)$_2$ZrCl$_2$, respectively, N_2, N_2H_4, NH_3 and a small amount of H_2 (Table I). The stoichiometry for the reaction with **3** is in close agreement with equation (9), wherein the four reducing equivalents available

$$[(\eta^5\text{-}C_5Me_5)_2ZrN_2]_2N_2 + 4\,HCl \longrightarrow 2(\eta^5\text{-}C_5Me_5)_2ZrCl_2 + 2N_2 + N_2H_4 \quad (9)$$

in the dimer are utilized in the reduction of one of the three N_2 ligands to N_2H_4. Identical treatment of **1** with HCl again leads to hydrazine, but in much lower yield (Table I). This

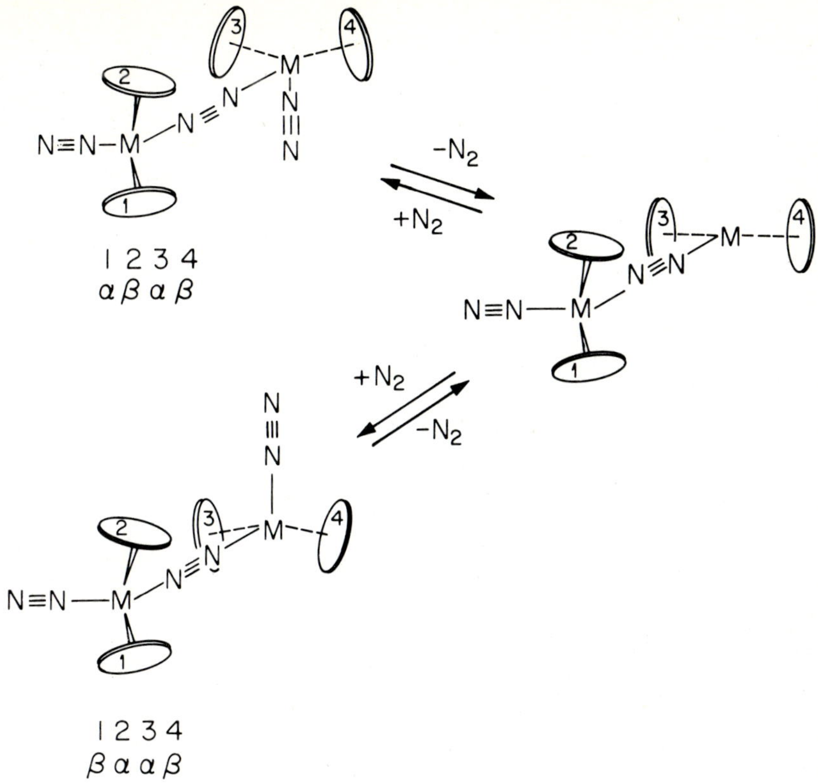

Fig. 3. *Rearrangement for* **3** *which exchanges (η^5-C_5Me_5) rings 1 and 2 between the two types of sites (α and β).*

Table I

Product distribution for the reactions of **1**, **2** and **3** with HCl in toluene at -80°.

	N_2/dimer released	N_2/dimer reduced	N_2H_4/dimer	$2NH_3$/dimer
$[(\eta^5\text{-}C_5Me_5)_2Ti]_2N_2$ (**1**)	0.79	0.21	0.15	0.07
$[(\eta^5\text{-}C_5Me_5)_2TiN_2]_2N_2$ (**2**)	2.28	0.72	0.66	0.03
$[(\eta^5\text{-}C_5Me_5)_2ZrN_2]_2N_2$ (**3**)	2.02	0.98	0.86	0.12

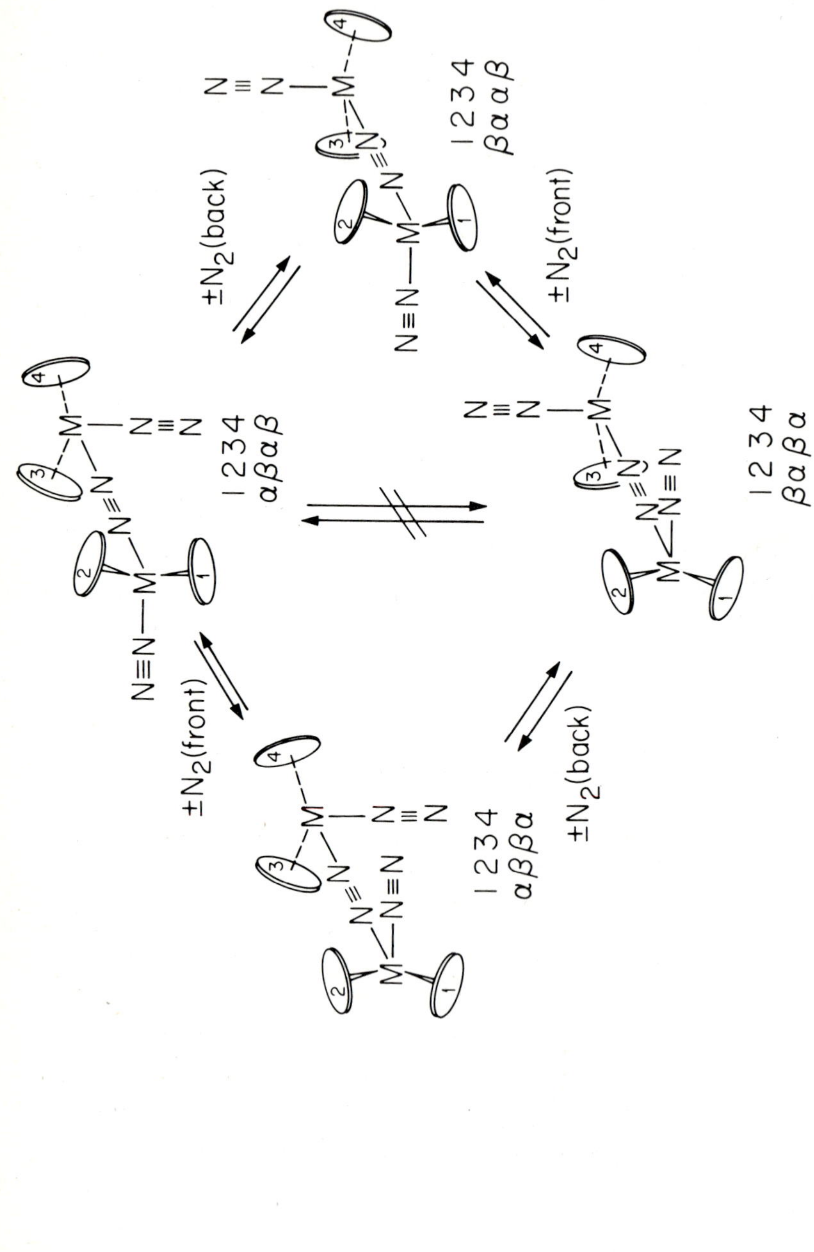

Fig. 4. *Sequence which interchanges all four (* η^5*-*C_5Me_5*) rings for 3.*

observation presents some intriguing questions concerning the mechanism of hydrazine formation in the reaction of **2** or **3** with HCl; in particular, which of the two types of N_2 (terminal or bridging) is reduced to hydrazine.

The high lability of the terminal dinitrogen ligands for **3** observed in the nmr experiments suggested that it should be possible to prepare **3** labeled specifically with ^{15}N. Table II summarizes the results of exchange of free $^{14}N_2$ with **3**-$(^{15}N_2)_3$. The first three entries of Table II demonstrate that after 15 minutes exposure to free $^{14}N_2$, the composition of the gas phase above the solution is in close agreement with that predicted for complete exchange of two of the three dinitrogen ligands of **3**. Thus under these conditions, exchange occurs between free N_2 and only the two equivalent terminal dinitrogen ligands. Incomplete exchange is observed after 5 minutes, permitting an estimate of the terminal N_2 exchange half-life of 2.6(1) minutes ($\tilde{p}N_2$ = 1 atm). On the basis of the experiment conducted at reduced pressure, the exchange rate appears proportional to free N_2 concentration ($t_{1/2}$= 5.0(3)min, $\tilde{p}N_2$ = 0.50 atm).

The results of these exchange experiments allow an accurate assessment of the extent of ^{15}N labeling in both terminal and bridge positions for **3**, and thus should allow a determination of which of the three dinitrogen ligands of **3** is reduced to hydrazine in the reaction with HCl. Following exchange of free $^{14}N_2$ with the terminal positions of **3**-$(^{15}N_2)_3$, a 20 M excess of HCl was added at -196°C. On warming to -80°C, an immediate reaction accompanied toluene melting. Evolved N_2 and hydrazine (after IO_3^- oxidation to N_2) were subsequently analyzed by mass spectrometry. The results (Table III) clearly indicate that the dinitrogen ligand which is reduced to hydrazine is *not exclusively* the -N_2 as might be anticipated. Rather, the composition of both evolved N_2 and hydrazine are in close agreement with those expected for the reaction proceeding according to equation (10).

$$[(\eta^5\text{-}C_5Me_5)_2Zr(N_2{}^t)]_2(N_2{}^b)+4HCl \longrightarrow (\eta^5\text{-}C_5Me_5)_2ZrCl_2+3/2N_2{}^t$$
$$+1/2N_2{}^b+1/2N_2{}^tH_4+1/2N_2{}^bH_4 \qquad (10)$$

Furthermore, we have found that $[(\eta^5\text{-}C_5Me_5)_2Zr(CO)]_2N_2$, (**4**), may be obtained *via* treatment of **3** with CO at -20°C. On the basis of its ir spectrum (ν(CO) 1902 (ms), 1860(s), ν(NN) 1682 (ms)) and nmr ((toluene-d_8)s, δ 1.80 (30H); s, δ 1.82 (30H)), the structure of **4** appears to be identical with **3** with CO substituted for the terminal N_2 ligands. Significantly, treatment of **4** with HCl as before yields 1 mole of N_2 with no detectable N_2H_4 or NH_3. This observation and the results of the

Table II

Results of N_2 exchange of $[(C_5Me_5)_2Zr(^{15}N_2)]_2(^{15}N_2)$ (99.36% $^{15}N{\equiv}N^{15}$) with free natural N_2 (99.32% $^{14}N{\equiv}N^{14}$) in toluene at -23° (estimated standard deviation

mmols $[(C_5Me_5)_2Zr(^{15}N_2)]_2(^{15}N_2)$	mmols $^{14}N_2$ added	$\tilde{p}_{N_2}$ (atm)	exposure time (min)	$\begin{Bmatrix} X(^{14}N_2) \\ X(^{15}N_2) \end{Bmatrix}^{calc.}$ [a]	$\begin{Bmatrix} X(^{14}N_2) \\ X(^{15}N_2) \end{Bmatrix}^{obs}$	mols N_2 exchanged per mol dimer
0.1255(2)	2.106(1)	1.0	15	{0.887, 0.106}	{0.886(1), 0.107(1)}	2.02(2)
0.1906(2)	2.164(1)	1.0	30	{0.844, 0.149}	{0.841(1), 0.152(1)}	2.04(2)
0.1091(2)	2.338(1)	1.0	60	{0.908, 0.0848}	{0.906(1), 0.0863(10)}	2.04(2)
0.1217(2)	2.154(1)	1.0	5	{0.892, 0.101}	{0.921(1), 0.0728(10)}	1.44(2)
0.1141(2)	1.130(1)	0.50	15	{0.826, 0.167}	{0.846(1), 0.147(1)}	1.77(2)

[a] Mol fractions $^{14}N{\equiv}N^{14}$ and $^{15}N{\equiv}N^{15}$ expected in gas phase for complete exchange of terminal N_2 ligands only.

Table III

Results of N_2 labelling experiments for the reaction:
$[(C_5Me_5)_2ZrN_2]N_2 + 4HCl \longrightarrow 2(C_5Me_5)_2ZrCl_2 + 2N_2 + N_2H_4$
(estimated standard deviations).

$X(^{15}N_2)$ in two positions of $[(C_5Me_5)_2ZrN_2]_2N_2$		$X(^{15}N_2)$ in N_2 evolved on HCl addition		$X(^{15}N_2)$ in hydrazine obtained on HCl addition	
terminal	bridge	calculated[a]	observed[b]	calculated[a]	observed[b]
0.107(5)	0.9936(1)	0.328(4)	0.368(1)	0.550(3)	0.528(1)
0.152(5)	0.9936(1)	0.362(4)	0.388(1)	0.573(3)	0.547(1)
0.086(5)	0.9936(1)	0.313(4)	0.329(1)	0.540(3)	0.488(1)
0.350(5)	0.9936(1)	0.510(4)	0.580(1)	0.671(3)	0.665(1)
0.270(5)	0.9936(1)	0.446(4)	0.525(1)	0.629(3)	0.634(1)
0.394(5)	0.9936(1)	0.539(4)	0.619(1)	0.691(3)	0.682(1)
0.079(5)[c]	0.9936(1)[c]	0.308(4)[c]	0.335(1)[c]	0.536(3)[c]	0.522(1)[c]

[a] Calculated for reaction proceeding 100% according to equation 2.

[b] No $^{15}N{\equiv}N^{14}$ was formed in reaction.

[c] DCl substituted for HCl.

N_2 labelling experiments are indications that the terminal N_2 ligands are playing roles beyond that of mere spectators in the reaction of **3** with HCl.

The data require a reaction sequence mediated by a symmetric species.in which one terminal N_2 and the μ-N_2 have become equivalent. While a number of mechanisms satisfying this requirement could be formulated, we favour one involving protonation of a terminal dinitrogen of **3**, loss of the other terminal N_2, and generation of the symmetric reaction intermediate (η^5-$C_5Me_5)_2Zr(N_2H)_2$ (**5**)(Scheme I). Consistent with the labelling experiments, **5** would then lead to 1 mole each of N_2 and N_2H_4. Generation of the neutral, monomeric species **5** from **3** would require a formal two-electron transfer to the N_2-bearing Zr, accompanied by release of the other Zr in the fully oxidised state, *i.e.* as (η^5-$C_5Me_5)_2ZrCl_2$. Strong electronic coupling of the two Zr(II) centres through the μ-N_2 of **3**, as indicated by its structure, ir, and visible spectra, should facilitate such a Zr-to-Zr charge transfer.

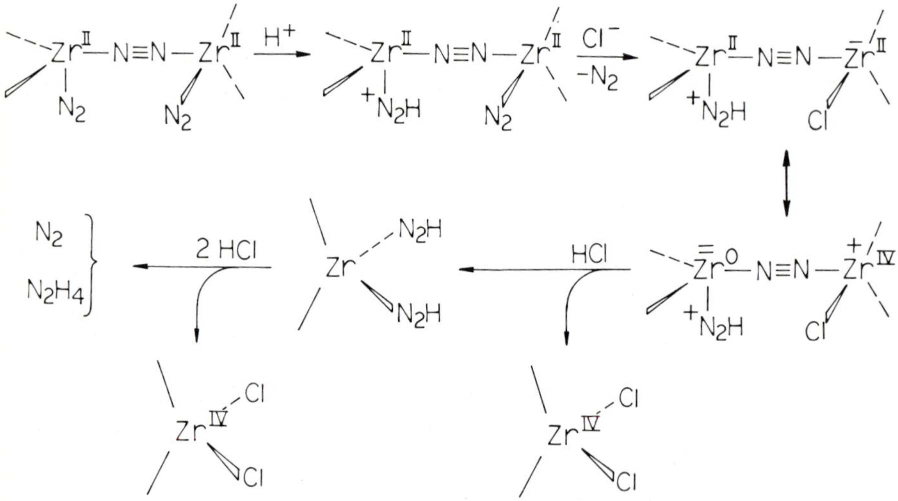

Fig. 5.

Substitution of *p*-toluenesulfonic, trifluoroacetic or acetic acid for HCl lowers the yields of reduced N_2 to 66%, 81% and 42% respectively. Solvent substantially affects the reaction with HCl as evidenced by a variation in yield from 100% in hexane, 98% in toluene, and 68% in methanol. Curiously, HCl in diethyl ether leads to only 41% reduction of one of the three N_2 ligands of **3**. However, the implications of these preliminary data are not yet clear because the hydrazine/ammonia distributions for the produc-s have not yet been measured. The formation of ammonia in the reaction of **3** with HCl in toluene is believed to arise from interference of a reaction of hydrazine with **3** during the latter stages of the reaction. This theory gains support from the treatments of **3** with excess anhydrous N_2H_4 at -80°C in toluene when N_2(3 mol) and NH_3(2 mol) are produced instantaneously.

CONCLUSIONS

Bis(pentamethylcyclopentadienyl) derivatives of Ti and Zr are useful congeners for their (C_5H_5) analogs by virtue of enhanced stability and crystallizability. The structures of $[(\eta^5\text{-}C_5Me_5)_2Ti]_2N_2$ and $[\eta^5\text{-}C_5Me_5)_2ZrN_2]_2N_2$ show "conventional" linear M-N≡N-M (Ti, Zr) and Zr-N≡N arrangements. It therefore appears that the same linear Ti-N≡N arrangement is most likely for the parent compound $[(\eta^5\text{-}C_5Me_5)_2Ti]_2N_2$, rather than the doubly-bent or side-on TiN_2Ti arrangements suggested earlier. It should also be pointed out that the complex $(\eta^5\text{-}C_5Me_5)_2TiN_2$, reported earlier to be present in both end-on and side-on N_2 bonded isomers (Bercaw *et al.*, 1974; Bercaw, 1974), probably does not, in fact, exist. Rather, this compound is the titanium analog of $[(\eta^5\text{-}C_5Me_5)_2ZrN_2]_2N_2$. The apparent 1:1 stoichiometry obtained previously arose because, under the conditions originally used to measure the N_2 content, the solution undoubtedly contains a mixture of $[(\eta^5\text{-}C_5Me_5)_2Ti]_2N_2$ and $[(\eta^5\text{-}C_5Me_5)_2TiN_2]_2N_2$. No evidence appears to remain, therefore, in support of any dinitrogen complex of titanocene (or zirconocene) containing N_2 bonded side-on to the metal.

The facile, nearly stoichiometric protonation of N_2 to N_2H_4 observed for $[(\eta^5\text{-}C_5Me_5)_2TiN_2]_2N_2$ and $[(\eta^5\text{-}C_5Me_5)_2ZrN_2]_2N_2$ stands in contrast to the behaviour of $[(\eta^5\text{-}C_5Me_5)_2Ti]_2N_2$. It may be that the small amounts of ammonia and hydrazine formed in the reaction of $[(\eta^5\text{-}C_5Me_5)_2Ti]_2N_2$ with HCl results from some $[(\eta^5\text{-}C_5Me_5)_2TiN_2]_2N_2$ generated from the former compound and evolved N_2 during the latter stages of the reaction. Thus, N_2 reduction is most efficient for those compounds formally containing *two* dinitrogen ligands per metal. The intermediate monomeric species $(\eta^5\text{-}C_5Me_5)_2Zr(N_2H)_2$, implicated by the ^{15}N-labelling studies, bears similarity to intermediates suggested by Chatt *et al.* (1975, 1975a, 1975b) for the protonation of $L_4M(N_2)_2$($L=PR_3$, M=Mo,W). These Mo and W compounds represent the only other structurally characterized dinitrogen complexes capable of liberating NH_3 and/or N_2H_4 on simple protonation. That these also contain *two* dinitrogen ligands may well be no coincidence.

The strong coupling of the two Zr(II) centres through the μ-N_2 of **3** also appears to play a deciding role in the reduction of N_2 to N_2H_4 and implicates a μ-N_2 as a possible site for electron transfer at the active site of nitrogenase. The mechanism by which (η^5-$C_5Me_5)_2Zr(N_2H)_2$ decomposes to N_2 and N_2H_4 is still not clear, particularly with regard to the possible intermediacy of free N_2H_2. We have noted that H_2 does *not* inhibit the production of hydrazine in the reaction of [(η^5-$C_5Me_5)_2ZrN_2]_2N_2$ with HCl. Exactly how this observation relates to the competitive H_2 inhibition of N_2 reduction and the N_2-catalyzed HD formation from D_2 by nitrogenase (Jackson *et al.*, 1968; Newton *et al.*, 1976) also remains to be clarified.

ACKNOWLEDGEMENTS

I wish to thank Robert D. Sanner, Donald R. McAlister and especially Juan M. Manriquez for their participation in this research. I also wish to acknowledge the support of the National Science Foundation.

REFERENCES

Bercaw, J.E. (1974). *J.Am.Chem.Soc.*, 96: 5087-5095.
Bercaw, J.E. and Brintzinger, H.H. (1971). *J.Am.Chem.Soc.*, 93: 2046-2048.
Bercaw, J.E., Marvich, R.H., Bell, L.G. and Brintzinger, H.H. (1972). *J.Am.Chem.Soc.*, 94: 1219-1238.
Bercaw, J.E., Rosenberg, E. and Roberts, J.D. (1974). *J.Am.Chem. Soc.*, 96: 612-613.
Borodko , Y.G., Ivleva, I.N., Kachapina, L.M., Salienko, S.I., Shilova, A.K. and Shilov, A.E. (1972). *J.Chem.Soc.,Chem. Comm.*, 1178.
Chatt, J. (1975). *J.Organometal.Chem.*, 100: 17-28.
Chatt, J., Pearman, A.J. and Richards, R.L. (1975a). *Nature*, 253: 39-40.
Chatt, J., Pearman, A.J. and Richards, R.L. (1975b). *J.Organometal.Chem.*, 101: C45-C47.
Jackson, E.K., Parshall, G.W. and Hardy, R.W.F. (1968). *J.Biol. Chem.*, 243: 4952-4958.
Manriquez, J.M. and Bercaw, J.E. (1974). *J.Am.Chem.Soc.*, 96: 6229-6230.
Manriquez, J.M., Sanner, R.D., Marsh, R.E. and Bercaw, J.E. (1976). *J.Am.Chem.Soc.*, 98: 3042-3044.
McKenzie, T.C., Sanner, R.D. and Bercaw, J.E. (1975). *J.Organometal.Chem.*, 102: 457-466.
Newton, W.E., Corbin, J.L. and McDonald, J.W. (1976). *In:* Proceedings of the First International Symposium on Nitrogen Fixation (W.E. Newton and C.J. Nyman, eds), Vol. 1, pp. 53-74. Washington State University Press, Pullman, Washington.

Sanner, R.D., Manriquez, J.M., Marsh, R.E. and Bercaw, J.E. (1976a). *J.Am.Chem.Soc.,* in press.
Sanner, R.D., Duggan, D.M., McKenzie, T.C., Marsh, R.E. and Bercaw, J.E. (1976b). *J.Am.Chem.Soc.,* in press.
Sellman, D. (1974). *Angew.Chem.internat.Edit.,* 14: 639-469.
Shilov, A.E., Shilova, A.K. and Kvashina, E.F. (1969). *Kinet.i Katal.,* 10: 1164 (English translation).
Shilov, A.E., Kvashina, E.F. and Vorontsova, T.A. (1971). *Chem. Comm.,* 1590-1591.
Teuben, J.H. (1973). *J.Organometal.Chem.,* 57: 159-167.
Ungurenasu, C. and Streba, E. (1972). *J.Inor.Nucl.Chem.,* 34: 3753-3758.
van der Weij, F.W. and Teuben, J.H. (1976). *J.Organometal.Chem.,* 105: 203-207.
van Tamelen, E.E. (1970). *Accounts Chem.Res.,* 3: 361- and literature cited therein.
van Tamelen, E.E., Cretney, W., Klaentschi, N. and Miller, J.S. (1972). *J.Chem.Soc., Chem.Comm.,* 481.
Volpin , M.E. and Shur, V.B. (1966). *Nature,* 209: 1236-1237 (1964). *Dokl.Akad.Nauk S.S.S.R.,* 1102-1104.

DINITROGEN FIXATION IN HOMOGENEOUS PROTIC MEDIA

L.A. NIKONOVA and A.E. SHILOV

Institute of Chemical Physics, Chernogolovka, Moscowskaya Oblast, USSR

INTRODUCTION

The results of our investigations of N_2 reduction in protic media obtained up until 1974 were presented at the First Symposium on Nitrogen Fixation (Shilov, 1976). This process was shown to be more favourable thermodynamically in protic than in aprotic media because the reaction products (NH_3, N_2H_4) and possible intermediate complexes are solvated, forming hydrogen bonds with proton-containing solvent molecules whereas N_2 itself has almost no basic properties. At the same time, strong reducing agents cannot be used in protic media since they will react with the solvent molecule producing H_2. Suitable reducing agents turned out to be compounds of Ti(III), Cr(II) and V(II). The first two reagents reduce N_2 only in the presence of Mo(III) which, like V(II), apparently forms intermediate complexes with N_2 susceptible to further reduction of the now activated dinitrogen molecule. The reaction products are hydrazine and (or) ammonia (Shilov, 1974).

The reaction proceeds in alkaline media. The first systems studied were hydroxides and reduced N_2 heterogeneously. The activity of these heterogeneous N_2-reducing systems in alkaline media is connected with two important properties: (1) sufficiently high reducing potential for the transformation of N_2 into hydrazine; and (2) availability of several electrons from ions surrounding the activated N_2 molecule to perform to a hydrazine derivative, e.g.,

$$Ti^{III}...Mo^{III}...N{\equiv}N...Mo^{III}...Ti^{III} \xrightarrow{H_2O} Ti^{IV}...Mo^{IV}\begin{matrix}H\\ N{=}Mo^{IV}...\\ {=}N\end{matrix}$$

$$Ti^{IV} + OH^-.$$

Thus, N_2 reduction proceeds without the intermediate formation of free, thermodynamically unfavoured diimide, N_2H_2.

Elucidation of the reaction mechanism and the structure of the intermediate complex is difficult with heterogeneous systems.

Therefore, it was very encouraging to find (Nikonova *et al.*, 1972) a homogeneous system containing V(II) complexes with catechol and other aromatic dioles, which turned out to be capable of reducing N_2 effectively to ammonia in aqueous and alcoholic solution. Up to the present time, these systems are unique among abiological homogeneous model systems as far as reaction rates and yields of NH_3 are concerned.

The present report deals with results obtained for N_2 reduction in these systems.

REACTION CONDITIONS AND PRODUCTS

For catechol in aqueous media, the reaction proceeds in the pH range 8.5 to 13.5 with the maximum yields being observed at pH 10 (Fig. 1). The reduction of N_2 takes place at room

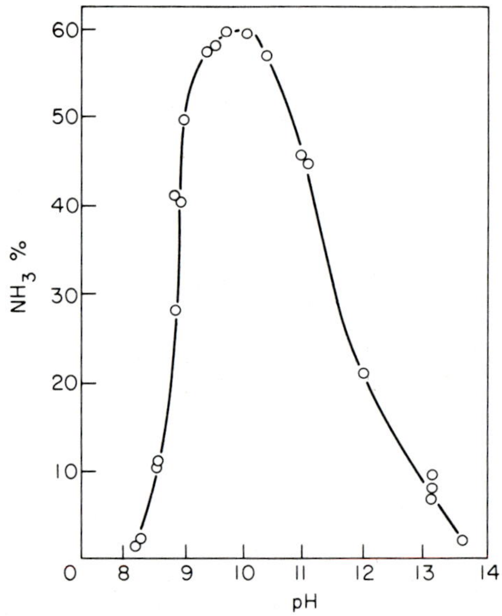

Fig. 1. *Dependence of NH_3 yield on pH in N_2 reduction by V(II)-catechol complex in water at room temperature.*

temperature or below. In alcoholic solutions at room temperature and atmospheric pressure of dinitrogen, the yield of ammonia reaches 60% with respect to the expenditure of V(II) (Fig. 2). In aqueous solutions, such yields are observed only at elevated pressures. In the absence of dinitrogen, a parallel reaction of V(II) oxidation by the solvent proceeds with the evolution of H_2. The yields of NH_3 increase with the increasing dinitrogen pressure (Fig. 2) due to the inhibition of

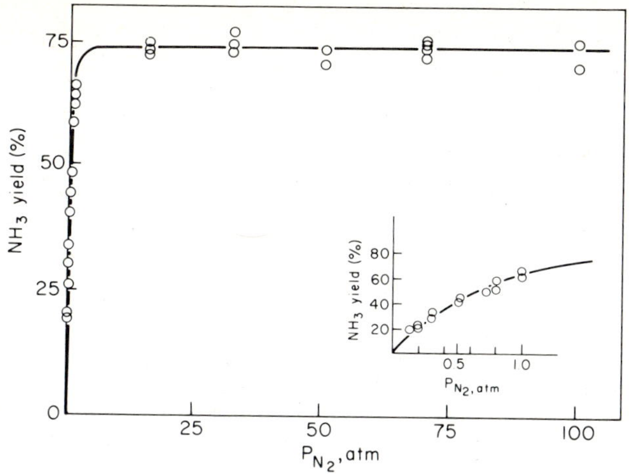

Fig. 2. *Dependence of NH_3 yield on N_2 pressure. Total solution volume 5 ml; amounts of V(II), catechol and NaOH 0.25, 2.5 and 2.75 mmoles respectively; t = 25°; solvent methanol.*

the competing reaction of dihydrogen formation. At no pressure, however, does the yield of ammonia exceed 75% of the reducing agent used, 25% of the V(II) electrons being accounted for in the parallel production of dihydrogen. This process of dihydrogen evolution is coupled with the reduction of N_2 because at dinitrogen pressures corresponding to the limiting yields of ammonia both the rate of dihydrogen formation and the rate of ammonia formation increase linearly with the rise of dinitrogen pressure. Thus, the stoichiometry of the reaction corresponds to equation (1)(Nikonova *et al.*, 1974, 1976).

$$8V^{2+} + N_2 + 8H_2O \longrightarrow 8V^{3+} + 2NH_3 + H_2 + 8OH^- \qquad (1)$$

Interestingly, this result is similar to biological dinitrogen fixation where ammonia formation and dihydrogen evolution are parallel processes and, at sufficiently high N_2 pressures, the minimum H_2 yield corresponds to 25% of the electrons of the reducing agent. Also similarly to the biological dinitrogen fixation, carbon monoxide inhibits the formation of ammonia but not the formation of dihydrogen (the rate of H_2 evolution is even increased in the presence of CO). Acetylene is quantitatively reduced to ethylene. With C_2D_2 in H_2O, cis-dideuteroethylene is observed. The range of pH for the reduction of acetylene is much broader than that for dinitrogen (from pH 5 to concentrated alkalis).

Some data comparing biological dinitrogen fixation with N_2 reduction by the V(II)-catechol complex are summarized in Table I. A number of similarities are obvious and they permit us to consider the V(II)-catechol complex as a model for the nitrogenase active centre. Therefore, the study of the mechanism of N_2 reduction in this model system might throw some light on the mechanism of enzyme functioning.

Table I

Comparison of the reactions of the N_2-ase active centre and the V(II)-catechol complex

Reactions and conditions	N_2-ase	V(II)-catechol complex
Reduction of N_2	Product: NH_3 ambient temperatures and pressures	Product: NH_3 ambient temperatures and pressures
	Reducing agent: (*in vitro*) $S_2O_4^{2}$ (reaction catalytic)	Reducing agent: the complex itself (reaction noncatalytic)
	Yield per e 75% NH_3, 25% H_2	Yield per e (saturating N_2 pressures) in water: 65% NH_3 in methanol: 75% NH_3; 25% H_2
	pH 6 - 8	pH 8.5 - 13.5
H_2 evolution	Quantitative in the absence of other substrates	Quantitative in the absence of other substrates
Inhibition by CO	CO inhibits N_2 fixation but not H_2 evolution	CO inhibits N_2 fixation but not H_2 evolution
Reduction of C_2H_2	Ethylene is the only product cis-$C_2H_2D_2$ is formed from C_2H_2 in D_2O or C_2D_2 in H_2O	Ethylene is only product at pH 13 cis-$C_2H_2D_2$ is formed from C_2D_2 in H_2O
Reduction of N_2H_4	Product: NH_3 with parallel H_2 formation	NH_3 is only product

KINETIC STUDY OF DINITROGEN REDUCTION

The kinetics of N_2 reduction was followed spectrophotometrically in methanol solution, where ammonia yields are sufficiently high at N_2 pressures lower than atmospheric. Reactions were carried out in the presence of Li methylate which was used to establish definite alkalinity of the solution. This system was homogeneous, and rates of dihydrogen evolution and dinitrogen reduction were smaller than those observed when Na-methylate was used under similar conditions (Nikonova *et al.*, 1976; Luneva *et al.*, 1976a, 1976b).

Changes in the electronic spectrum of the V(II)-catechol complex under N_2 in the course of the reaction are shown in Fig. 3. Isobestic points at 22120 cm^{-1} and 17280 cm^{-1} show that the system contains only two kinds of species in noticeable amounts, i.e., those of the initial V(II) and final V(III)-catechol complexes. Optical density at 15440 cm^{-1} was used to

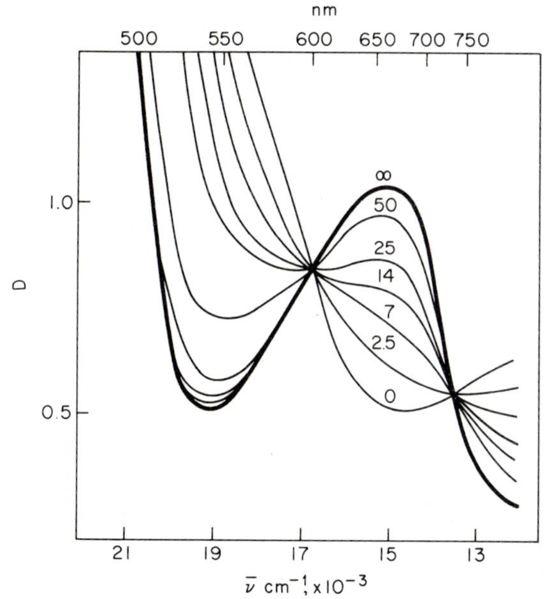

Fig. 3. *Electronic spectra of V(II)-catechol system under N_2 (P_{N_2}960 mm Hg) in methanol. Figures on the curves correspond to the reaction times in minutes.*

measure the concentration of V(III) complex. Kinetic curves of the change in V(II) concentration for two N_2 concentrations and for no N_2 are shown in Fig. 4. The linear dependence of initial

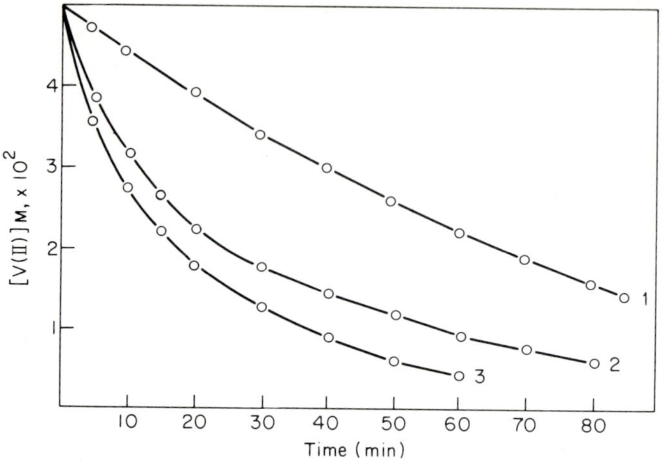

Fig. 4. *Kinetic curves of the V(II) oxidation by* N_2*: 1,* P_{N_2} *= 0 mmHg; 2,* P_{N_2} *= 740 mmHg; 3,* P_{N_2} *= 960 mmHg.*

rate of V(II) oxidation on N_2 concentration indicates that the reaction is first order with respect to N_2. In the absence of N_2, parallel H_2 evolution (reaction (2)) takes place. Kinetic

$$2V(II) + 2H_2O \longrightarrow 2V(III) + H_2 + 2OH^- \qquad (2)$$

investigations, using both the rates of H_2 evolution and the rates of V(III) formation, showed that reaction (2) is 0.5 order with respect to V(II) (Fig. 5). Knowing the rate constant of reaction (2), one can calculate its rate at every point of the kinetic curve in the presence of N_2 and thus find that part of the rate of V(II) oxidation corresponding to N_2 reduction. The plot of the rate found in such a way versus $[V(II)]^2$ is linear (Fig. 6), indicating that the reaction is second order with respect to V(II).

Thus, kinetic studies of the oxidation of V(II)-catechol complex in an atmosphere of dinitrogen showed the kinetic equation of this reaction to have the following form:

$$\frac{d[V(II)]}{dt} = k_1 [V(II)]^2 [N_2] + k_2 [V(II)]^{1/2}$$

The first term pertains to the reduction of dinitrogen to ammonia and the second to the parallel (and independent) reaction

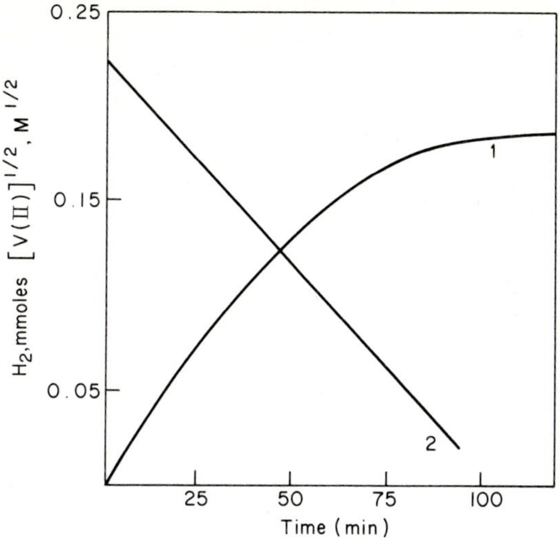

Fig. 5. *Kinetics of H_2 evolution in water solution of V(II)-catechol complex at 25°C: 1, $[H_2]_t$ vs time; 2, $[V(II)]_t^{1/2}$ vs time; $[V(II)]_t$ calculated as $[V(II)]_o-2[H_2]_t$. Total solution volume 7.5 ml. Initial concentrations of V(II), catechol and NaOH were 0.05, 0.5 and 0.6 M respectively (pH 10.6).*

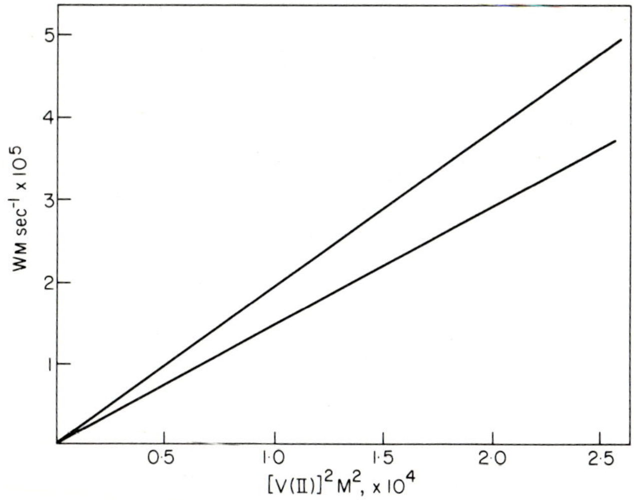

Fig. 6. *Dependence of dinitrogen reduction rate on the concentration of V(II)-catechol complex: 1, P_{N_2} = 740 mm Hg; 2, P_{N_2} = 960 mm Hg.*

of H_2 evolution from the solvent protons. Hydrazine, which could be the intermediate reaction product, was found to be easily reduced to ammonia in the V(II)-catechol solutions. When the reaction of dinitrogen reduction is abruptly stopped at its initial stages by adding excessive acid or oxidizer ($VOSO_4$) into the reaction mixture, a small quantity of hydrazine was found, which seemed to confirm its intermediate formation. The kinetic study of hydrazine reduction with the help of a similar procedure (the reaction being stopped by adding acid or V(IV) has, however, shown its rate constant to be smaller by at least two orders of magnitude than it would have been had the hydrazine, found in the system when dinitrogen is being reduced, been the only intermediate during the formation of ammonia (Luneva *et al.*, 1976b). From this result, it can be inferred that, if the main pathway of the reaction proceeds *via* hydrazine, it has some distinct properties as compared with free hydrazine in that it does not leave the coordination sphere of vanadium in the complex.

The second order with respect to vanadium concentration for the reaction of the reduction of N_2 and, at the same time, the 0.5 order for the dihydrogen formation are indicative of the polynuclear structure of the transition state during the reduction of dinitrogen. H_2 formation from the protons of the solvent requires two electrons. The comparison of kinetic orders shows that 8 electrons happen to be available for the reduction of dinitrogen in the transition state. It is natural to assume that stoichiometric equation (1) reflects the reaction mechanism. It is possible, for example, to assume that, at pH 8.5 to 13.5, a particle containg 4 vanadium atoms is present in the solution. This particle forms a complex with N_2 which undergoes an equilibrium transformation into a hydrazine derivative of the following type:

Irreversible reduction to ammonia becomes possible on collision with another tetranuclear particle. The N-N bond is, in this case, split and an ammonia derivative is formed which is then hydrolyzed to NH_3. The two "extra" electrons are, at the same time, utilized for the formation of the H_2 molecule. An alternative possibility exists in the reaction of two binuclear particles with the $V^{2+} \xrightarrow{-2e} V^{4+}$ transformation when dinitrogen is being reduced to ammonia and H_2 formed. The subsequent reaction of V^{4+} with the excess of $V^{2+} + V^{4+} \longrightarrow 2V^{3+}$ will result in the formation of the final V(III) complex observed. Kinetic data, therefore, undoubtedly indicate that, at least, a tetranuclear (and, which is more probable taking into account the energy requirements, an 8-nuclear) transition state is responsible for N_2 reduction.

The 0.5 order for the formation of dihydrogen can be explained by the equilibrium dissociation of the complex into two reactive particles (each of them, probably, including $2V^{2+}$ ions). The absence or the slow rate of the reaction in the undissociated state indicates that the active centre is inaccessible to solvent molecules due to the organic ligands surrounding the V^{2+} ions. These ligands appear to be an important factor for dinitrogen to become a more reactive substrate than for the proton-containing molecules of the solvent.

The catalytic role of CO in H_2 formation and its inhibition of N_2 reduction can be explained by the assumption of complex formation of the tetramer with CO, which can readily react with H_2O (possibly *via* dissociation into dimers).

OTHER SOLUBLE V(II) COMPLEXES REDUCING N_2

Only aromatic compounds containing two OH or SH groups in the ortho-position form V(II) complexes soluble in alkaline aqueous or methanol media. With phenol, resorcinol and hydroquinone, V^{2+} forms heterogeneous mixtures which are inactive with respect to dinitrogen reduction. The data on N_2 reduction by some soluble V(II)-complexes with some analogs of catechol are presented in Table II (Isaeva and Nikonova, 1976).

Table II

Dinitrogen Reduction in Aqueous V(II)-polyphenol-NaOH(KOH) Systems (P_{N_2} = 100 atm, t = 25°C)

Polyphenol	pH of max. yield	NH_3 yield (%)
Catechol	10.3	61.0
4-methylcatechol	11	58.0
Protocatechuic acid	6-13	1
Pyrogallol	13.5	59.5
Gallic acid	12.9	46.8
3.4-Dimercaptotoluene*	6-13	1
2-Mercaptophenol*	6-13	1

*) In methanol or aqueous methanol

V(II) complexes with pyrogallol and gallic acid are capable of reducing dinitrogen at much higher values of pH than for the catechol complexes (pH < 13). The complex with protocatechuic acid, as distinct from those of gallic acid which is similar to it), is altogether inactive with respect to N_2. On substitution of one or both OH groups in catechol by SH, the complexes produced are incapable to reduce N_2. These data, as well as the narrow range of pH within which V(II) complexes with catechol form ammonia from dinitrogen, are indicative of the requirement that only definite types of complexes should be present in the system, evidently, in the form of associates.

Conclusions, drawn from the consideration of N_2 reduction by V(II)-catechol can be compared with the reaction of the nitrogenase active centre. One can visualize the following mechanism for nitrogenase functioning on the ground of analogy with V(II)-catechol system, where hydrazine is not formed as an intermediate and where the stoichiometry is similar to that of nitrogenase reaction (i.e. 25% of the electrons are expended for H_2 formation). There are several (at least 6) Fe atoms together with 2 atoms of Mo in the active centre of nitrogenase capable of accepting electrons from a reducing agent. The number of electrons necessary to be transferred to the active centre is different for the various substrates and therefore, the reduction potential of the active enzyme is also different. The situation with dinitrogen can be represented as the functioning of a siphon (Fig. 7), which starts operating for N_2 only when

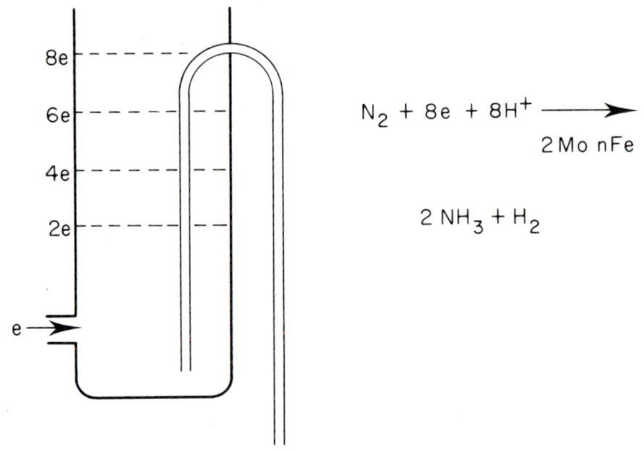

Fig. 7. *Scheme of 8-electron N_2 reduction in N_2-ase and model systems.*

the 8-electron level is reached. H_2 is formed along with NH_3 from the two "extra" electrons. For other substrates, the reaction might start from a different level, e.g., the two electron level is possibly sufficient for C_2H_2 and therefore H_2 is not formed. Further investigations of both model system and nitrogenase will show whether this hypothesis is adequate.

REFERENCES

Isaeva, S.A. and Nikonova, L.A. (1976). *Izv.Akad.Nauk S.S.S.R., Ser.Khim.*, in press.

Luneva, N.P., Nikonova, L.A. and Shilov, A.E. (1976a). *React. Kinet.Catal.Lett.*, 5: N2-

Luneva, N.P., Nikonova, L.A. and Shilov, A.E. (1976b). *Kinet. i katal.*, in press.

Nikonova, L.A., Ovcharenko, A.G., Efimov, O.N., Avilov, V.A. and Shilov, A.E. (1972). *Kinet. i katal*, 13: 1602.

Nikonova, L.A., Pershikova, N.I., Bodeyko, M.V., Oleinik, L.G., Sokolov, D.N. and Shilov, A.E. (1974). *Dokl.Akad.Nauk S.S. S.R.*, 216: 140-

Nikonova, L.A., Isaeva, S.A., Pershikova, N.I. and Shilov, A.E. (1976). *J.Mol.Cat.*, 1: 367-

Shilov, A.E. (1974). *Uspekhi Khim.*, 43: 863-

Shilov, A.E. (1976). *In:* Proceedings of the First International Symposium on Nitrogen Fixation (W.E. Newton and C.J. Nyman, eds), pp. 42-52. Washington State University Press, Pullman, Washington.

SYNTHESES, PROPERTIES AND REACTIONS OF DINITROGEN, DIIMINE, HYDRAZINE AND AMMONIA COMPLEXES OF TRANSITION METALS

D. SELLMANN

Laboratorium für Anorganische und Analytische Chemie, Gesamthochschule Paderborn, Paderborn, West Germany.

INTRODUCTION

There is still no conclusive proof that in nitrogenase the transition metals molybdenum and iron interact with molecular nitrogen on its route to ammonia. Several findings, however, point to such an interaction and additionally at a stepwise reduction of dinitrogen, in the course of which the initially formed dinitrogen complex - it may be mono- or binuclear - is reduced stepwise via a diimine and hydrazine to ammonia complexes, roughly according to the Fig. 1 (Dalton and Mortenson, 1972). Herein, some studies, which are aimed mainly at the

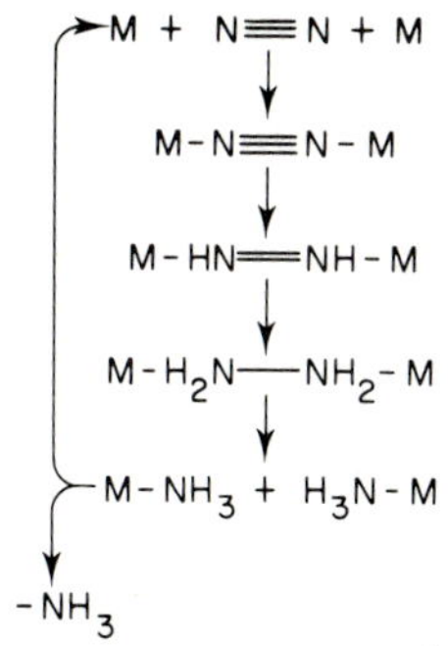

Fig. 1. *Stepwise reduction of dinitrogen to ammonia.*

following points, are reported: (1) Synthesis and characterization of series of transition metal complexes with N≡N, HN=NH, H_2N-NH_2 and NH_3 ligands; (2) Clearcut reactions of these nitrogen ligands, together with the dependence of their reactivity upon the transition metal centres; and (3) Conclusions which may be drawn from these studies with respect to N_2 fixation.

The complexes chosen for investigation are of the metal-carbonyl type, because such complexes allow reactions to be monitored and reaction products characterised easily, very often by IR spectroscopy alone. We started with hydrazine complexes because they represent one of the intermediate stages of Fig. 1 and had been only scantily investigated with respect to their redox behaviour.

SYNTHESIS OF HYDRAZINE COMPLEXES

Some years ago, uncharged transition metal carbonyl hydrazine complexes were unknown. They can be obtained, however, in yields of nearly 100% by reacting tetrahydrofuran (THF) complexes with hydrazine, e.g. by the following equations (1-3) (Sellmann, 1970, 1971a). Depending on the amount of hydrazine

$$(OC)_5M\text{-}THF + N_2H_4 \xrightarrow[THF]{20^\circ C} (OC)_5M\text{-}N_2H_4 + THF \quad (M = Cr, Mo, W) \tag{1}$$

$$C_5H_5M(CO)_2THF + N_2H_4 \xrightarrow[THF]{20^\circ C} C_5H_5M(CO)_2N_2H_4 + THF \quad (M = Mn, Re) \tag{2}$$

$$2(OC)_5M\text{-}THF + N_2H_4 \xrightarrow[THF]{20^\circ C} [(OC)_5M]_2N_2H_4 + 2THF \quad (M = Cr, 2) \tag{3}$$

added, the reactions yield mononuclear complexes, e.g., $(OC)_5CrN_2H_4$ and $C_5H_5Mn(CO)_2N_2H_4$, as well as binuclear complexes, e.g. $[(OC)_5Cr]_2N_2H_4$ containing hydrazine as a bidentate bridging ligand.

REDOX REACTIONS OF HYDRAZINE COMPLEXES

Disproportionation

In some cases, unexpectedly, attempts to synthesize binuclear complexes failed. Thus, it was not possible to obtain a complex of the formula $[C_5H_5Mn(CO)_2]_2N_2H_4$ at temperatures between +20°C and -50°C. Instead, the complexes $C_5H_5Mn(CO)_2N_2$

$$2\ C_5H_5Mn(CO)_2THF + N_2H_4 \xrightarrow{THF} [C_5H_5Mn(CO)_2]_2N_2H_4 + 2THF \tag{4}$$

and $C_5H_5Mn(CO)_2NH_3$ are formed, signifying that the hydrazine-bridged complex $[C_5H_5Mn(CO)_2N_2H_4$ exists at best as an intermediate, which disproportionates immediately after it has formed (Sellmann, unpublished). A similar reaction is found with molybdenum. In this case, however, it is possible to trap $[(OC)_5Mo]_2N_2H_4$ by lowering the reaction temperature to -30°C (Sellmann *et al.*, 1975).

These reactions show that the reactivity of hydrazine is influenced in a quite unexpected manner by the different metal centres, although the complexes themselves look very similar otherwise. If hydrazine functions as a bridging ligand between metal centres of high electron density, e.g. $[C_5H_5Mn(CO)_2]$, it is so highly labilised that it disproportionates immediately to dinitrogen and ammonia.

Reduction

With respect to reduction, coordinatively bound hydrazine again shows different behaviour, which depends on the metal centre as well as on the bridging or non-bridging function of hydrazine. Thus, the hydrazine ligand of $C_5H_5Mn(CO)_2N_2H_4$ as well as of $(OC)_5CrN_2H_4$ is easily reduced in acidic media, e.g. by H_2SO_3 leading to the corresponding ammonia complexes (Sellmann, unpublished).

$$C_5H_5Mn(CO)_2N_2H_4 \xrightarrow{\text{Reduction}} C_5H_5Mn(CO)_2NH_3 \quad (5a)$$

$$(OC)_5CrN_2H_4 \xrightarrow{\text{Reduction}} (OC)_5CrNH_3 \quad (5b)$$

In $[(OC)_5Cr]_2N_2H_4$, however, the N_2H_4 ligand withstands all attempts at reduction in acidic media, regardless of whether mild or strong reducing agents are used (Sellmann *et al.*, 1976).

Oxidation

Synthesis of N_2 complexes The hydrazine complexes proved to be excellent starting compounds for synthesizing N_2 complexes. The fast oxidation, e.g., with H_2O_2/Cu^{2+}, yields N_2 complexes of manganese, rhenium and chromium (Sellmann, 1971b; Sellmann, 1972a; Sellmann and Maisel, 1972a).

$$C_5H_5M(CO)_2N_2H_4 + 2H_2O_2 \xrightarrow{THF/Cu^{2+}} C_5H_5M(CO)_2N_2 + 4H_2O \; (M = Mn, Re) \quad (6)$$

$$C_6H_6Cr(CO)_2N_2H_4 + 2H_2O_2 \xrightarrow{THF/Cu^{2+}} C_6H_6Cr(CO)_2N_2 + 4H_2O \quad (7)$$

Provided hydrazine complexes are obtainable, this method may become valuable for testing metal complexes with respect to N_2 coordination. $C_5H_5Mn(CO)_2N_2$, which is also formed according to equation (8), illustrates this point (Sellmann, 1972b).

$$C_5H_5Mn(CO)_2THF + N_2 \underset{}{\overset{THF}{\rightleftharpoons}} C_5H_5Mn(CO)_2N_2 + THF \quad (8)$$

This equilibrium is shifted far to the left side and contains $C_5H_5Mn(CO)_2N_2$ only in minor quantities. The fast oxidation of $C_5H_5Mn(CO)_2N_2H_4$ in THF, however, yields large quantities of the N_2 complex because the rate of oxidation of the N_2H_4 complex is much faster than that of the subsequent dissociation of the N_2 complex. In other words, the oxidation according to equation (6) generates the N_2 complex in a non-equilibrium concentration allowing its easy detection, isolation and characterization. In other cases, the equilibrium is shifted to the side of the N_2 complexes, which therefore may be synthesized directly from N_2 (Sellmann and Maisel, 1972b).

$$C_6H_3(CH_3)_3Cr(CO)_2THF + N_2 \xrightarrow[20°C]{THF} C_6H_3(CH_3)_3Cr(CO)_2N_2 + THF \quad (9)$$

Synthesis of N_2H_2 complexes Chemical intuition leads one to expect the oxidation of N_2H_4 complexes to proceed at least in two steps, the first of which should yield complexes with diimine (N_2H_2) as ligand, for instance according to equation (10).

$$C_5H_5Mn(CO)_2N_2H_4 \longrightarrow C_5H_5Mn(CO)_2HN{=}NH \longrightarrow C_5H_5Mn(CO)_2N{\equiv}N \quad (10)$$

As diimine, HN=NH, is assumed to be also the first reduction product of N_2 in biological N_2 fixation, its structure and properties demand special attention. Free N_2H_2, which decomposes above -180°C and has been obtained in a pure state only quite recently (Wiberg *et al.*, 1974), may have the following structures among others.

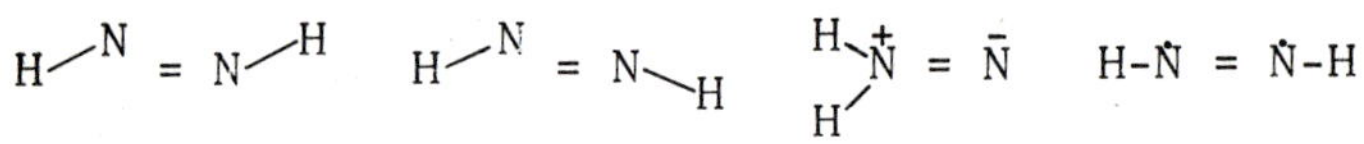

1 *2* *3* *4*

They represent trans- and cis-diimine (*1* and *2*), iso-diimine (*3*) (an analogue of formaldehyde) and triplet-diimine (*4*) (an analogue of dioxygen). In all cases, however, the two N atoms are linked by a genuine double bond.

The sometimes observed stabilization of unstable molecules by coordination to transition metal complexes made it worthwhile to attempt the stabilisation of diimine, especially with the complexes that gave N_2, N_2H_4 and NH_3 derivatives before. However, the oxidation of $C_5H_5Mn(CO)_2N_2H_4$ does not yield the expected mononuclear diimine complex $C_5H_5Mn(CO)_2HN{=}NH$, instead very low yields of the binuclear diimine complex $[C_5H_5Mn(CO)_2]_2N_2H_2$ are obtained, according to reaction (11) (Sellmann, 1973).

$$C_5H_5Mn(CO)_2N_2H_4 + 2H_2O_2 \xrightarrow[-40°C]{THF/Cu^{2+}}$$

$$\text{trans-}C_5H_5(OC)_2Mn{-}HN{=}NH{-}Mn(CO)_2C_5H_5 \quad (11)$$

$$+ \text{ byproducts, e.g. } C_5H_5Mn(CO)_2N_2, \quad C_5H_5Mn(CO)_2NH_3, \; N_2$$

Because a mononuclear hydrazine complex yields a binuclear diimine complex with the accompanying evolution of N_2, the diimine complex must be formed by several reaction steps, including reaction (2).

$$C_5H_5Mn(CO)_2HN{=}NH + C_5H_5Mn(CO)_2N_2 \longrightarrow [C_5H_5Mn(CO)_2]_2N_2H_2 + N_2 \quad (12)$$

As diimine gets stabilized when it acts as a bridging ligand, diimine-bridged complexes may also be obtained in a one-step reaction starting from hydrazine-bridged complexes (Sellmann *et al.*, 1976).

$$[(OC)_5Cr]_2N_2H_2 + 2H_2O_2 \xrightarrow[20^\circ C]{THF/Cu^{2+}} cis\text{-}(OC)_5Cr\text{-}HN{=}NH\text{-}Cr(CO)_5 + 2H_2O \quad (13)$$

In the case of chromium, oxidising agents other than H_2O_2 may be used, e.g. $Pb(OAc)_4$, p-tosylazide and even molecular oxygen.

The isolation of diimine complexes from the reaction mixtures depends crucially upon the reaction conditions, e.g. temperature, pH of the reaction medium, concentration and rapidity of the work-up. These conditions become particularly important in the synthesis of $[(OC)_5Mo]_2N_2H_2$ which represents the most labile of the diimine complexes (Sellmann *et al.*, 1975).

SPECTROSCOPIC PROPERTIES AND STRUCTURE OF DIIMINE COMPLEXES

The structure of the diimine ligand in $[C_5H_5(OC)_2Mn]_2N_2H_2$ and $[(OC)_5Cr]_2N_2H_2$ as well as in the analogous rhenium, molybdenum and tungsten complexes can be derived from their IR, ^1H-NMR, UV and mass spectra. In the case of the chromium complex, an X-ray structure determination confirms the spectroscopic results with respect to the diimine structure of N_2H_2.

Especially informative are the IR spectra. Thus, in accordance with the selection rules of group theory, the trans-diimine structure of N_2H_2 in $[C_5H_5Mn(CO)_2]_2N_2H_2$ follows from the observation of three N_2H_2 IR absorptions, whereas $[(OC)_5Cr]_2N_2H_2$, with five N_2H_2 IR bands, has a cis-diimine structure of the N_2H_2 ligand. Table I shows the frequencies and assignments of the respective vibrations. The frequencies given in Table I belong to almost pure diimine vibrations as proved by both deuteration and ^{15}N experiments and by the fact that substitution of manganese and chromium by their group homologues shifts the IR frequencies to only a negligible extent.

Table I

IR frequencies and assignments of ligating diimine

trans-$[C_5H_5Mn(CO)_2]_2N_2H_2$		cis-$[(OC)_5Cr]_2N_2H_2$	
frequency $[cm^{-1}]$	assignment	frequency $[cm^{-1}]$	assignment
3250	νas (NH)	3480	νas (NH)
1338	δas (HNNH)	3250	νs (NH)
1040	τ (HNNH)	1415	ν (N=N)
		1352	δas (HNNH)
		1105	δs (HNNH)

Unfortunately, $[C_5H_5Mn(CO)_2]_2N_2H_2$, $[(OC)_5Cr]_2N_2H_2$ and the other diimine complexes are X-ray amorphous. They crystallize from tetrahydrofuran as solvates and lose the THF solvate molecules as soon as the mother liquor is removed, resulting in a complete breakdown of the crystal lattice. For these reasons, an X-ray structure determination has been carried out as yet only for [$(CO)_5Cr$ $_2N_2H_2$. 2THF] at -30°C. It proves the diimine structure of N_2H_2 but, in contrast to the unsolvated complex, the solvated one contains not cis- but trans-diimine, as is shown in Fig. 2 (Huttner *et al.*, 1974). This complex is the first and only compound in which the existence of diimine has been confirmed both chemically and by X-ray crystallography. The N-N distance (125 pm) and HNN angle (122.3°) correspond well with the values expected for sp^2 hybridized N atoms.

The different IR spectroscopic and X-ray structure results regarding $[(OC)_5Cr]_2N_2H_2$ and [$(OC)_5Cr$ $_2N_2H_2$. 2THF] may be due to a cis-trans equilibrium of this complex in solution, from which the "more symmetric" transform crystallizes preferentially as the solvate. Removal of the solvate molecules with concomitant breakdown of the crystal lattice leads to the cis form observed in the IR spectrum.

While the IR spectra of $[C_5H_5M(CO)_2]_2N_2H_2$ (M = Mn, Re) and $[(OC)_5M]_2N_2H_2$ (M = Cr, Mo, W) differ with respect to the N_2H_2 ligand, their ^{1}H-NMR and UV spectra look very similar. The ^{1}H-NMR spectra always show a single NH resonance in the range of 16,000 - 20,000 cm^{-1} (Table II).

COMPARISON OF LIGATING N_2, N_2H_2, N_2H_4 and NH_3

The complexes described above are particularly suited to a comparison of the electronic character of the individual N ligands with respect to their binding properties and electron-donating abilities because they represent the only series with

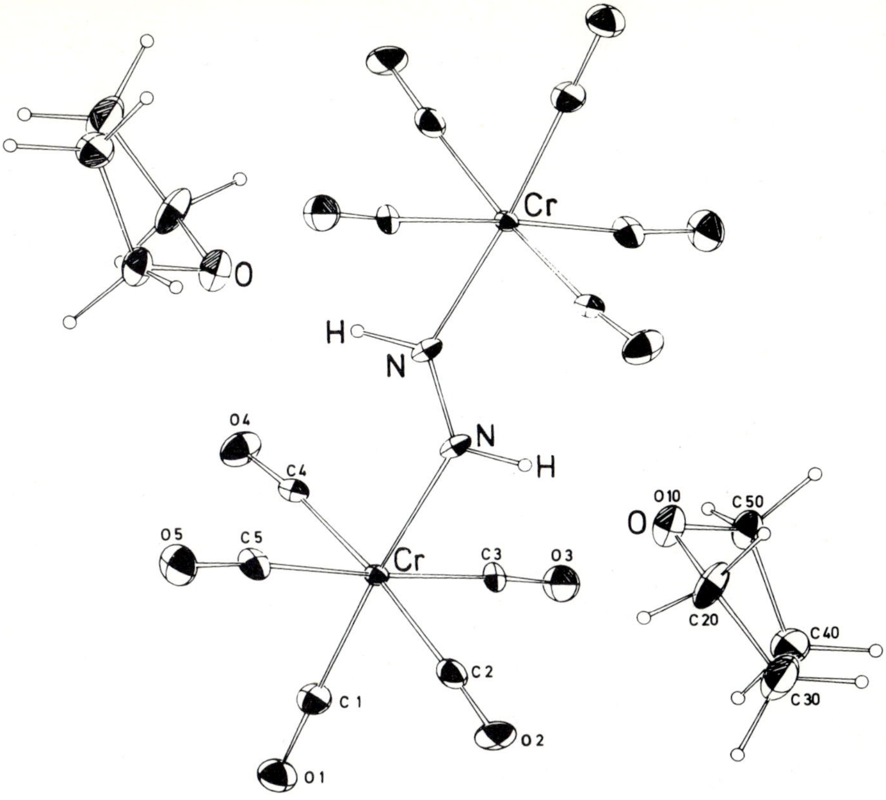

Fig. 2. *Structure of* [*(OC)$_5$Cr $_2$N$_2$H$_2$. 2 THF*]

N_2, N_2H_2, N_2H_4 and NH_3 coordinated to identical metal centres. On comparing the UV spectra of the N_2H_2, N_2H_4 and NH_3 complexes, the special character of diimine in the class of nitrogen - hydrogen compounds becomes apparent. In Fig. 3, the UV spectra of $[(OC)_5Cr]_2N_2H_2$, $[(OC)_5Cr]_2N_2H_4$, $(OC)_5CrN_2H_4$ and $(OC)_5CrNH_3$ are shown (Sellmann *et al.*, 1976). The absorption curves of the hydrazine and ammonia complexes are typical of pentacarbonyl-amine metal complexes. The diimine complex, however, shows an additional strong absorption at 18,870 cm^{-1}. It is obviously characteristic of ligating diimine and is responsible for the intense dark colour of the diimine complexes. These complexes are usually red-purple to black-blue in contrast to the corresponding yellow-to-orange hydrazine and ammonia

Table II

N_2H_2 proton chemical shifts and
UV absorptions of diimine complexes

Complex	$\delta_{N\underline{H}}$ [ppm]	λ_{max} [cm^{-1}]	ϵ_{max}
$[C_5H_5Mn(CO)_2]_2N_2H_2$	14.3	16,666	24 500
$[C_5H_5Re(CO)_2]_2N_2H_2$	15.2	16,949	>19 000
$[(OC)_5Cr]_2N_2H_2$	16.2	18,870	15 000
$[(OC)_5Mo]_2N_2H_2$	16.5	19,565	>10 900
$[(OC)_5W]_2N_2H_2$	16.4	18,500	23 000

complexes, which certainly points to the formation of multi-centre -molecular orbitals between the two metals and the two nitrogen atoms.

The different electronic character and influence of the individual nitrogen ligands is also obvious in the IR and ^{1}H-NMR spectra. Table III shows some selected data for the manganese and tungsten complexes. The CO frequencies in Table III show ammonia and hydrazine again to behave very similarly but diimine is distinctly different probably because it can accept electron density from the metal centres *via* back donation into its anti-bonding π-molecular orbitals. This effect occurs to an even greater extent in the case of N_2, of course. Even more striking are the differences in the chemical shifts of the NH protons, which cover the range from 1-6 ppm downfield in the case of hydrazine and ammonia to more than 16 ppm in the case of diimine. This difference clearly shows an increasing de-shielding of the protons on going from ammonia to diimine and is best explained by assuming acidic properties for the diimine protons.

REACTIONS OF N_2H_2 COMPLEXES

Two reactions of the diimine complexes can be most easily explained by the acidity of the diimine proto-s: (A) the characteristically rapid H-D exchange in protic media; and (B) their disproportionation. Both reactions are base-catalysed (Sellmann *et al.*, 1975).

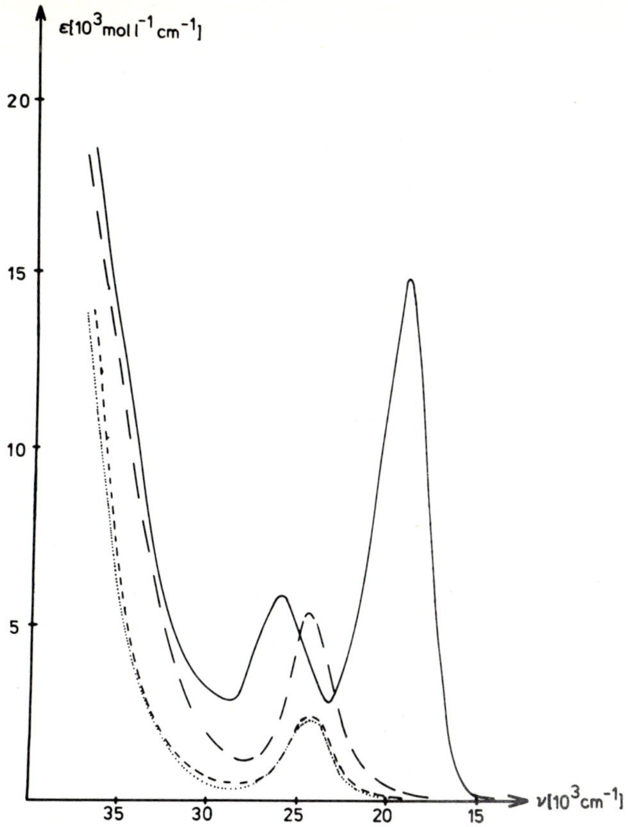

Fig. 3. *UV spectra of* $(OC)_5CrNH_3$ .... , $(OC)_5CrN_2H_4$ ---- , $[(OC)_5Cr]_2N_2H_4$ — — — — , *and* $[(OC)_5Cr]_2N_2H_2$ ——— *in THF.*

H-D Echange On dissolving the diimine complexes in deuterated solvents, e.g., CD_3OD, rapid and reversible conversion to deutero-diimine complexes occurs, e.g. equation (14).

$$[(OC)_5Cr]_2N_2H_2 + 2CD_3OD \rightleftharpoons [(OC)_5Cr]_2N_2D_2 + 2CD_3OH \qquad (14)$$

The rate of this reaction can be measured by ^1H-NMR spectrometry. By comparing the halflives of the deuteration of the corresponding diimine, hydrazine and ammonia complexes, it can be seen that the deuteration of the diimine complex occurs characteristically fast. Table IV shows the values for the chromium complexes which have been studied most thoroughly.

Table III

CO frequencies and proton chemical shifts of manganese and tungsten N_2, N_2H_2, N_2H_4 and NH_3 complexes

Complex	ν(CO) [cm^{-1}]			δ N<u>H</u>[ppm]
	A_1	B_1	$(A_1+B_1)/2$	
$C_5H_5Mn(CO)_2N_2$	1980	1923	1951	-
$[C_5H_5Mn(CO)_2]_2N_2$	1950	1904	1927	-
$[C_5H_5Mn(CO)_2]_2N_2H_2$	1915	1882	1899	14.3
$C_5H_5Mn(CO)_2N_2H_4$	1918	1842	1880	2.0
$C_5H_5Mn(CO)_2NH_3$	1918	1840	1879	1.4
	$A_1^{(2)}$	E	$A_1^{(1)}$	
$[(OC)_5W]_2N_2H_2$	2054	1948	1920	16.4
$[(OC)_5W]_2N_2H_4$	2066	1932	1897	6.6
$(OC)_5WN_2H_4$	2068	1920	1888	4.8; 3.7
$(OC)_5WNH_3$	2068	1915	1884	3.1

Table IV

Halflives of deuteration of $[(OC)_5Cr]N_xH_y$ complexes with an excess of CD_3OD

	$\tau_{1/2}$	δ N<u>H</u>[ppm]
$[(OC)_5Cr]_2N_2H_2$	45 s	16.2
$[(OC)_5Cr]_2N_2H_4$	15 min	6.0
$(OC)_5CrNH_3$	>> 60 h	1.9

Clearly, the hydrazine and ammonia protons exchange much slower than the diimine protons. Actually, for $(OC)_5CrNH_3$, no H-D exchange can be observed at all in neutral media at room temperature, only by addition of catalytic amounts of base, e.g., Na OCH_3, can H-D exchange be accomplished. The observed trend in exchange rates from the NH_3 to the N_2H_2 complex corresponds well with the increase in acidity expected from both the ^{1}H-NMR shifts and reasons of bonding and structure inferred from the analogously increasing acidity in the series $C_2H_6 \longrightarrow C_2H_4 \longrightarrow C_2H_2$.

The H-D exchange of $[(OC)_5Cr]_2N_2H_4$ is catalysed by bases. On adding acids, e.g. DCl, the exchange is stopped completely while on addition of the more basic D_2O instead of CD_3OD, the exchange is accelerated to a half-life of less than 15 secs. This result can be explained by equilibrium (15), which is shifted completely to the left in the presence of strong acids. A

$$[(OC)_5Cr]_2N_2H_2 + B \rightleftharpoons [\ (OC)_5Cr\ _2N_2H]^- + BH^+ \qquad (15)$$

(B = base or solvent)

further acceleration of the H-D exchange by addition of stronger bases than water could not be accomplished as then a second reaction, disproportionation, becomes predominant.

Disproportionation Even catalytic amounts of a strong base, e.g. $NaOCH_3$, cause the immediate disproportionation of the diimine complexes, e.g., according to equation (16). The stoi-

$$2[(OC)_5Cr]_2N_2H_2 + 2THF \xrightarrow[21°C,\ 0.05\ mole\ NaOCH_3]{THF/CH_3OH}$$

$$[(OC)_5Cr]_2N_2H_4 + 2(OC)_5CrTHF + N_2 \qquad (16)$$

chiometry of the reaction, the catalysis by bases, the mechanism of the H-D exchange and the high polarity of the solvents point to the ionic mechanism outlined in Fig. 4.

The first step of the reaction involves the deprotonation of the diimine complex by the base B to produce the monoanion Ia. Ia then reduces another molecule of I to the hydrazido complex anion IIa by hydride transfer and yields the neutral N_2 complex, $[(OC)_5Cr]_2N_2$, simultaneously. The anion IIa is a very strong base itself (*cf.* $N_2H_3^-$ in NaN_2H_3) and deprotonates another molecule I (or eventually, the corresponding acid BH^+ (= CH_3OH)) yielding the neutral hydrazine complex II and generating another monoanion Ia (or the base CH_3O^-). The existence of the intermediate Ia explains the necessity of only catalytic amounts of base to trigger the whole process. (The N_2 complex, $[(OC)_5$

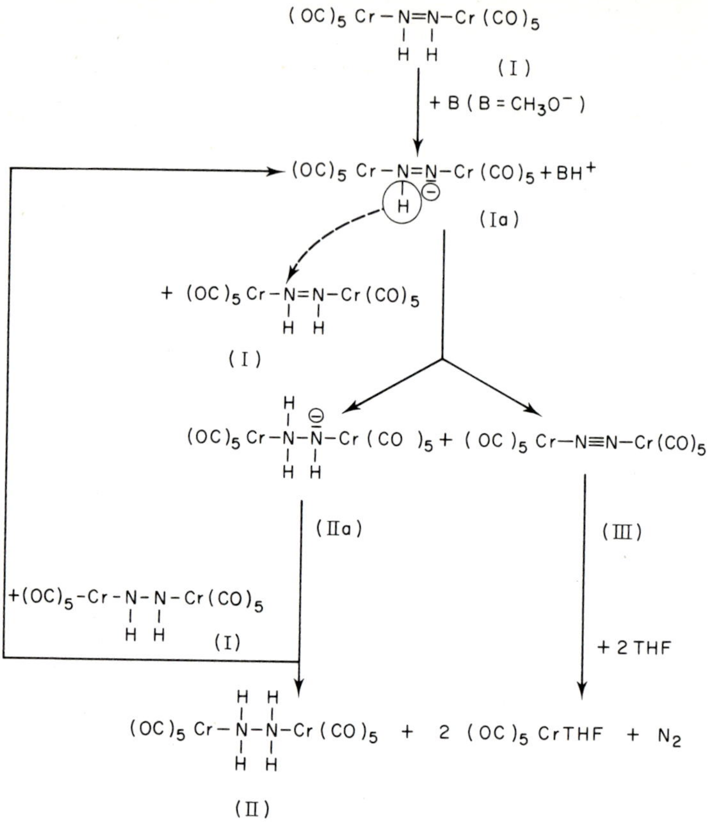

Fig. 4. *Base-catalysed disproportionation of* $[(OC)_5Cr]_2N_2H_2$

$Cr]_2N_2$, goes on reacting to give $(OC)_5CrTHF$ and N_2.)

$[C_5H_5Mn(CO)_2]_2N_2H_2$ also undergoes a base-catalysed disproportionation. In this case, the N_2-bridged complex, $[C_5H_5Mn(CO)_2]_2N_2$, can be isolated as easily-decomposing green crystals. However, the second product, the hydrazine bridged complex $[C_5H_5Mn(CO)_2]_2N_2H_4$, decomposes immediately, yielding $C_5H_5Mn(CO)_2N_2$ and $C_5H_5Mn(CO)_2NH_3$ (see section C.1.), and cannot be isolated . Weak bases too, e.g. H_2O, cause slow disproportionation of $[(OC)_5Cr]_2N_2H_2$. .This result shows that the rate of disproportionation also depends on the strength of the catalysing base. In the presence of weak bases like H_2O, the reverse reaction of $[(OC)_5Cr\ _2N_2H]^-$ with BH+ to give the neutral complex, $[(OC)_5Cr]_2N_2H_2$ (according to equation (15)), is competing with the irreversible disproportionation (equation (16))

Such a competition, however, is no longer possible in the presence of strong bases like CH_3O^- because of the much lower concentration of BH^+. Furthermore, the rate of disproportionation is strongly influenced by the metal centres. Whereas $[(OC)_5Cr]_2N_2H_2$ and $[(OC)_5W]_2N_2H_2$ disproportionate with comparable rates, $[(OC)_5Mo]_2N_2H_2$ reacts much faster and even H_2O causes its complete disproportionation within a few minutes.

Comparison of Diimine Complexes Regarding the N_2H_2 molecule itself, it should be pointed out that it represents a special case among the azo-compounds not only because of its extremely easy decomposition but, even more, because of its two NH bonds, which enable it to undergo successive cation-anion reactions as shown in the disproportionation of $[(OC)_5Cr]_2N_2H_2$. The common azo compounds with N-C bonds, e.g. azo methane, cannot react this way and decompose according to a radical mechanism.

$$R\text{-}N{=}N\text{-}R \longrightarrow R^{\cdot} + N_2 \; {}^{\cdot}R$$

The completely different properties of complexes like $[C_5H_5M(CO)_2]_2N_2H_2$ (M=Mn, Re) and $[(OC)_5M]_2N_2H_2$ (M=Cr, Mo, W) in comparison with complexes like $[MCl_2(diphos)_2N_2H_2]$ (M=Mo, W), for which also a diimine structure of the N_2H_2 ligand has been proposed, are amazing. Whereas the first type is always binuclear and deeply coloured, shows N_2H_2 proton resonance at low fields and is highly sensitive to bases, the latter ones are mononuclear and yellow to yellow-orange and react with bases to give N_2H complexes (equation (17)), from which the starting

$$[WBr_2(N_2H_2)(diphos)_2] + NEt_3 \longrightarrow [WBr(N_2H)(diphos)_2] + [NEt_3H]Br \qquad (17)$$

complex can be regained on addition of acid (Chatt *et al.*, 1975).

CONCLUSION

Regarding the various series of N_2, N_2H_2, N_2H_4 and NH_3 complexes, both strong similarities and striking differences are found. For instance, the chromium, molybdenum and tungsten-pentacarbonyl complexes generally show the same reactions and spectroscopic properties but their reactivities are highly different, with molybdenum forming by far the most reactive complexes. In fact, even $Mo(CO)_6$ shows a much higher reactivity than $Cr(CO)_6$ or $W(CO)_6$ and forms hydrazine and ammonia complexes under normal conditions, e.g., equation (18). As this reaction certainly obeys a S_{N1} mechanism, it shows, by the way,

$$Mo(CO)_6 + N_2H_4 \xrightarrow[20°C]{THF/H_2O} (OC)_5MoN_2H_4 + CO \qquad (18)$$

that Mo(0) centres may well be stable in aqueous media (Sellmann *et al.*, 1975).

A serious objection to the assumption of diimine as an intermediate stage of N_2 fixation is its high enthalpy of formation, which amounts to more than 48 kcal/mole. It has been estimated, however, that this value may be lowered to less than 15 kcal/mole by coordination to transition metals and the formation of hydrogen bridges. Furthermore, the N_2-dependent D_2-H exchange in nitrogenases, which may occur between D_2 and a hydrogen of a nitrogenase-bound diimine, is still under discussion (Dalton and Mortenson, 1972). Indeed, the structure and reactions of $[(OC)_5Cr]_2N_2H_2$ show that N_2H_2 does form hydrogen bridges and has easily exchangeable hydrogen atoms.

Because basic ammonia is the end-product of N_2 fixation and diimine complexes undergo a base-catalysed disproportionation, it could well be that nitrogenase-bound diimine undergoes an ammonia-catalysed disproportionation. In other words, N_2 fixation could require the input of electrons only for the first step of N_2 reduction from dinitrogen to diimine with the subsequent reaction steps occurring via disproportionation reactions. The easy disproportionation of N_2H_4 in the non-isolable $[C_5H_5Mn(CO)_2]_2N_2H_4$ shows that N_2H_4 also can be labilized by coordination to appropriate metal centres. The generally observed high reactivity of the molybdenum complexes suggests that nitrogenases may prefer molybdenum rather than iron at the active site.

REFERENCES

Chatt, J., Pearman, A. J. and Richards, R. L. (1975). *J.Organometal.chem.*, 101: 045-

Dalton, H. and Mortenson, L.E. (1972). *Bacteriol.Rev.*, 36: 231-

Huttner, G., Gartzke, W. and Allinger, K. (1974). *Angew.Chem. internat.Edit.*, 13: 822-

Sellmann, D. (1970). *Zeitschr.Naturf.*, 25b: 890-

Sellmann, D. (1971a). *Zeitschr.Naturf.*, 26b: 1070-

Sellmann, D. (1971b). *Angew.Chem.internat.Edit.*, 10: 919-

Sellmann, D. (1972a). *J.Organometal.Chem.*, 36: C27-

Sellmann, D. and Maisel, G. (1972a). *Zeitschr.Naturf.*, 27b: 465-

Sellmann, D. (1972b). *Angew.Chem.internat.Edit.*, 11: 534-

Sellmann, D. and Maisel, G. (1972b). *Zeitschr.Naturf.*, 27b: 718-

Sellmann, D. (1973). *J.Organometal.Chem.*, 44: C22-

Sellmann, D., Brandl, A. and Endell, R. (1975). *J.Organometal. Chem.*, 90: 309-
Sellmann, D., Brandl, A. and Endell, R. (1975). *J. Organometal. Chem.*, 97: 229-
Sellmann, D., Brandl, A. and Endell, R. (1976). *J.Organometal. Chem.*, 111: 303-
Wiberg, N., Fischer, G. and Bachhuber, H. (1974). *Chem.Ber.*, 107: 1456-

THE MECHANISMS OF NITROGEN FIXATION

E. I. STIEFEL

Charles F. Kettering Research Laboratory, 150 East South College Street, Yellow Springs, Ohio 45387, U.S.A.

The word mechanism evokes a variety of different images in practitioners of the diverse disciplines which are involved in nitrogen fixation research. For the purpose of this paper, we find it convenient to first distinguish between the stoichiometric and intimate mechanisms of reaction (Langford and Gray, 1966). By stoichiometric mechanism, we mean the delineation of the elementary steps which occur in a chemical or enzymological reaction. These steps include such aspects as the direction of electron flow, the stoichiometry and type of inhibition, allosteric effects and, in general, those gross features of the reaction which are conventionally established by considerations of stoichiometry, kinetics and thermodynamics. For example, Fig. 1 shows a stoichiometric mechanism for cytochrome P_{450} action. We note the absence of chemical formulas, names alone

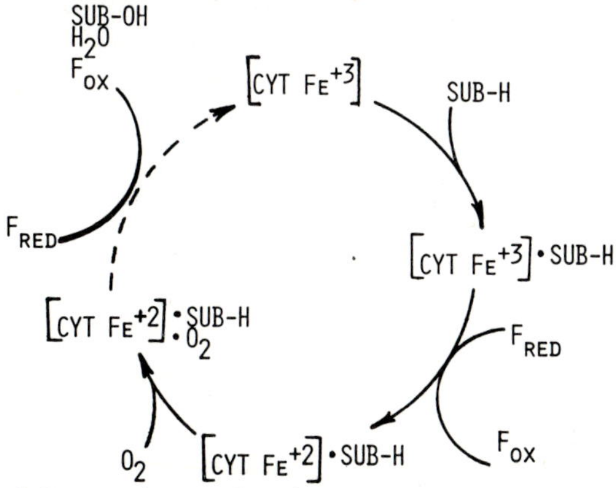

Fig. 1. *Stoichiometric mechanism for cyt P_{450} substrate hydroxylation system (after Sharrock* et al., *1976; Sligar and Gunsalus, 1976).*

suffice. In contrast, the intimate mechanism involves a further analysis of the elementary steps themselves and seeks to answer the question: How do the molecules interact on an atomic-molecular level to cause the particular elementary steps to occur Included are structural aspects of reactants, transition states and products, orbital symmetry and energy restrictions and free energy considerations within the elementary step. The intimate mechanism goes beyond stoichiometry, kinetics and thermodynamics and calls upon various forms of spectroscopy as well as quantum mechanics to seek an understanding of the chemical pathway within the step. Such phenomena as electrostatic interactions, hydrogen bonding, hydrophobic interactions and the details of conformational change are included within this mechanism.

At this point in time, much progress has been made in understanding the stoichiometric mechanisms of nitrogen fixation processes and more importantly, the means to establish these seem firmly in hand. In contrast, the intimate mechanism for which the chemist can draw a probable structural diagram, is quite unknown and the means for its determination are only now being developed. In consequence, speculations about the intimate mechanism of nitrogenase action are greatly influenced by results from model systems and coordination chemistry. The distinction between stoichiometric and intimate mechanism is beautifully illustrated by the papers in Volume 1 of the Proceedings of the First International Symposium on Nitrogen Fixation (Newton and Nyman, 1976). With the single exception of one figure in the paper by Burns, Stasny and Hardy (1976), none of the papers on enzymology display a chemical structural formula. In contrast, each of the chemical papers show proven or postulated structural formulas for reactants, intermediates and products.

In the present article, we discuss reductive nitrogen fixation systems for, although oxidative systems exist in both nature (Burns and Hardy, 1975) and technology (Leigh, 1971), they are currently dwarfed as far as quantity is concerned. Since all reaction mechanisms must have a thermodynamic underpinning, we first consider the energetics of the known or postulated reactions of nitrogenase. Some of the chemical fixation systems are then discussed with the idea of scrutinizing these processes in comparison with nitrogenase. As the role of molybdenum remains one of the great mysteries of nitrogenase action, we also briefly consider the coordination chemistry of molybdenum and the reactivity of other Mo-containing enzymes. Finally, in a clearly speculative disposition we discuss possible stoichiometric and intimate mechanisms for nitrogen fixation and the associated reactions of nitrogenase. It is our hope that the superposition of data from chemical and biological systems will in some way prove stimulatory in both fields.

THERMODYNAMIC CONSIDERATIONS

Thermodynamic considerations provide the limits within which all potentially valid mechanistic proposals *must* lie. Here, we discuss the thermodynamics of N_2 and its reduction products, and, in anticipation of later discussion, we briefly consider the reduction of other nitrogenase substrates and CO as well as the energetics of ATP hydrolysis. Potentially relevant data for Mo and Fe complexes are also given. Where appropriate we judiciously combine thermodynamic data to present estimated values for processes which have been hypothesized as part of nitrogenase action.

Dinitrogen and its Reduction Products

Thermodynamic parameters for the gas-phase reduction of N_2 by H_2 are shown in Table I. While the overall reduction to NH_3 is clearly exergonic, the formation of N_2H_2 or N_2H_4 are each both endothermic and endergonic. However, once having attained

Table I

Gas Phase Thermodynamics of Dihydrogen Reactions with N_2, N_2H_2 and N_2H_4[a]

Reaction	H°	G°
(1) $N_2 + 3H_2 \longrightarrow 2NH_3$	-22	-7.98
(2) $N_2 + H_2 \longrightarrow N_2H_2$	+36[b]	(+51)[c]
(3) $N_2 + 2H_2 \longrightarrow N_2H_4$	+22.7	+38
(4) $N_2H_2 + H_2 \longrightarrow N_2H_4$	-13.3	(-13)[d]
(5) $N_2H_2 + 2H_2 \longrightarrow 2NH_3$	-58	(-59)[d]
(6) $N_2H_4 + H_2 \longrightarrow 2NH_3$	-44.7	-46

[a] Values in Kcal/mole obtained from NBS Bulletin No. 500 unless otherwise noted.
[b] Value obtained from Willis *et al.*, 1976
[c] Calculated assuming $T\Delta S° = +15$ Kcal/mole, an estimate based upon similar $T\Delta S°$ values for N_2H_4 and $2NH_3$ formation.
[d] Based upon estimated $\Delta G°$ for reaction (2).

either N_2H_2 or N_2H_4, the additional reduction(s) required to reach NH_3 are again exergonic. It is clearly the production of the intermediate states between N_2 and NH_3 which constitute the energy-demanding steps in dinitrogen reduction schemes.

To eliminate consideration of H_2 as a specific reductant and to foster comparison with biologically available reductants, we have estimated E°' values for the relevant half-reactions (Table II). The potential for the overall conversion of N_2 to NH_4^+ at -280 mV is clearly accessible to nitrogenase (Mortenson *et al.*, 1976; Watt and Bulen, 1976; Evans and Albrecht, 1974).

Table II

Estimated E°' Values for Nitrogenase, Half-Reactions

Reaction	E°' (mV)
$N_2 + 6e^- + 8H^+ \longrightarrow 2NH_4^+(aq.)$	-280
$N_2 + 2e^- + 2H^+ \longrightarrow N_2H_2$	[-1050, -1450][b]
$N_2 + 4e^- + 5H^+ \longrightarrow N_2H_5^+$	-695
$N_2H_2 + 2e^- + 3H^+ \longrightarrow N_2H_5^+$	[-350, +60][c]
$N_2H_2 + 4e^- + 6H^+ \longrightarrow 2NH_4(aq.)$	[-50, 305][c]
$N_2H_5^+ + 2e^- + 3H^+ \longrightarrow 2NH_4^+(aq.)$	+ 550

[a] Values obtained from Lattimer (1952); Watt *et al.* (1976). calculated from free energy data in NBS Circular No. 500 as E° in acid solution under standard conditions and corrected to pH 7 using the Nernst Equation.

[b] Based on $G°_f$ as estimated in Table I. The first number is uncorrected for pH. If corrected to pH 7.0 by $E°' = E° - \frac{0.059}{n} \log \frac{1}{10^{-7n}}$ the second number obtains. Corrections for solvation are not made and could make E°' less negative by as much as 250 mV.

[c] Based on value for reaction 2[b].

However, this number is only valid for the overall six-electron process and there is substantial evidence (Bulen, 1976; Stiefel *et al.*, in press; Newton *et al.*, 1976; Newton *et al.*, these proceedings) that stepwise two-electron reactions are involved, *i.e.* that N_2H_2 and N_2H_4 are involved as enzyme-bound intermediates. Although the diimide values in Table II are only rough estimates, it is nonetheless clear that the production of either N_2H_2 or N_2H_4 from N_2 requires a reduction potential more negative than has been found for nitrogenase. The formation of N_2H_2 seems particularly difficult.

Despite these apparent limitations, we cannot definitely eliminate N_2H_2 as a possible bound intermediate on thermodynamic grounds. We first recall that the values calculated are either for free gas phase or free solution species. However, it is clear (Garcia-Rivera and Burris, 1967; Burris *et al.*, 1966) that free diimide or hydrazine is not produced during nitrogenase turnover and evidence supports *bound* intermediates (Bulen, 1976; Newton *et al.*, this volume). Therefore, each of the $E°'$ values must be modified by the difference in free energy of binding of the reactants and products. If the reactant (e.g. N_2) binds less strongly than the product (e.g. N_2H_2) then the $E°'$ for the reaction would be less negative, *i.e.* a tighter enzyme-product complex makes it easier to reduce the reactants.

A second potential consideration is the role of ATP in the redox process. Much evidence (Orme-Johnson, *et al.*, 1972; Mortenson *et al.*, 1976; Smith *et al.*, 1976) is consistent with the hydrolysis of ATP in a manner coupled with the transfer of electrons from the Fe protein(s) to the MoFe protein. It is also known that, in the presence of ATP, the reduction potential of the Fe protein becomes more negative. For example, in the *Clostridium pasteurianum* Fe protein, the presence of ATP causes a shift from -294 to -400 mV (Mortenson *et al.*, 1976). However, this shift is *not dependent upon ATP hydrolysis* and, in fact, can be brought about equally well by ADP. This modification may, however, be an indication that ATP hydrolysis could be coupled to electron transfer, *i.e.* the free energy of ATP hydrolysis being directly transduced into reducing power. If we make the assumption that all the free energy of ATP hydrolysis can be channelled into increased reducing power, then we can calculate the difference in potential which this may achieve. Taking the value for hydrolysis of MgATP at pH = 7 as ΔG = -8.8 Kcal/mole (Alberty, 1968; Phillips *et al.*, 1969) and assuming an ATP/$2e^-$ ratio of 4 as the probable minimum value attainable *in vitro* (Watt *et al.*, 1975; Burris and Orme-Johnson, 1976), the total G for ATP hydrolysis is -35 Kcal/mole, which, expressed as potential for a two-electron process, is ~750 mV. If this potential can be combined with the roughly -400 to -500 mV potential required for nitrogenase turnover (Watt and Bulen, 1976; Evans and Albrecht, 1974), then a potential more negative than -1000 mV is achievable. Coincidentally (?), this is the potential required to reduce N_2 to N_2H_2 and thus suggests the intriguing possibility that the source of the large ATP requirement in nitrogenase may be the large free energy requirement for the first step in dinitrogen reduction, namely $N_2 \longrightarrow N_2H_2$. Of course, the transduction of all the energy of ATP hydrolysis into reducing power for this one step is an efficiency of improbable closeness to perfection. However, binding and solvation considerations will modify the free energy requirements with respect to N_2 and N_2H_2 such that perfect coupling may not in fact be required. In any event, admitting the crudeness of the estimated values, thermodynamics provides us with this possible rationale for the ATP requirement of nitrogenase.

Nitrogenase Substrate and Inhibitor

Inspection of Tables III and IV reveals that reduction of other substrates of nitrogenase requires substantially less energy than does reduction of N_2. For example, the common assay reaction, acetylene reduction, is clearly more easily accomplished than is N_2 reduction. In fact, acetylene ($E^{\circ\prime}$ = + 320 mV) is a fairly good oxidizing agent, comparable to $Fe(CN)_6^{-3}$ and better than cyt_c. This fact must appear as a warning to those using acetylene reduction indiscriminately as a probe for nitrogen fixation in either biological or chemical systems, *i.e.* there are undoubtedly many systems which will reduce acetylene without at all affecting dinitrogen. The situation for N_2O and HN_3 is even more startling, each being a powerful oxidant ($E^{\circ\prime}$> + 1000 mV)

Table III

Thermodynamic Values of H_2 Reaction with Nitrogenase Substrates[a]

Reaction	ΔH°	G°
$H_2(g) + C_2H_2(g) \longrightarrow C_2H_4(g)$	-41.7	-33.7
$H_2(g) + N_2O(g) \longrightarrow N_2 + H_2O(aq.)$	-87.8	-81.5
$H_2(g) + HN_3(aq.) + H^+ \longrightarrow NH_4^+ + N_2$	-83.6	-96.3
$3H_2(g) + HCN(aq.) + H^+ \longrightarrow CH_4 + NH_4^+$	-75.1	-58.1
$CO + H_2 \longrightarrow H_2CO(g)$	- 1.3	+ 6.5
$CO(g) + 2H_2(g) \longrightarrow CH_3OH(g)$	-32	- 9.1
$CO(g) + 3H_2(g) \longrightarrow CH_4(g) + H_2O$ 1	-59.8	-36.0
$MgATP + H_2O \longrightarrow MgADP + P_i$		- 8.8[b]

[a] Values in Kcal/mole from NBS Bulletin No. 500 unless otherwise indicated.
[b] Value from Alberty (1968) and Phillips *et al.* (1969).

Table III also contains thermodynamic data for CO, a potent inhibitor of substrate reduction by nitrogenase. The 2-, 4- and 6-electron reduction products of CO (formaldehyde, methanol and methane plus water, respectively) are each more easily produced (from thermodynamic considerations) than are their analogues in the dinitrogen reduction process. This raises the important question as to why, in some N_2-reducing systems, CO is also reduced but in others it is not and in fact serves as an inhibitor. This point is discussed further below for individual systems.

Table IV

Estimated E°' Values for Half-Reactions of Nitrogenase Substrates[a]

Half-Reaction	E°' (mV)
$C_2H_2 + 2e^- + 2H^+ \longrightarrow C_2H_4$	+ 320
$N_2O + 2e^- + 2H^+ \longrightarrow N_2 + H_2O$	+1,350
$HN_3(aq.) + 2e^- + 3H^+ \longrightarrow N_2 + NH_4^+$	+1,470
$HCN + 6e^- + 7H^+ \longrightarrow NH_4^+ + CH_4$	- 60

[a] Values calculated from free energy data in NBS Circular No. 500 as E° in acid solution under standard conditions and corrected to pH 7 using the Nernst Equation.

Fe and Mo Complexes

Finally, we consider the ability of either Fe or Mo sites in nitrogenase to carry out the necessary reduction at the required potentials. Clearly, Fe-S type sites can achieve highly negative (reducing) potentials (Orme-Johnson, 1973; Holm, 1975, 1976). The coordination chemistry of ferredoxin model systems shows that, for example, the Fe_4S_4 cluster can achieve reduction potentials as low as -1.2 volts (Holm, 1975). Coupled with the existence of a super-reduced form of HiPiP (Cammack, 1973) whose potential may be more negative than -600 mV, we conclude that an Fe-S site can have sufficient potential to serve either as the direct reductant of substrate and/or of the molybdenum site.

As little is known about the coordination or oxidation state of Mo in nitrogenase, we are clearly at a loss when we try to choose a model. Nonetheless, the scope of oxidation states of Mo (Stiefel, 1976) and the various potentials required for their interconversion would seem to provide sufficient versatility for the requisite potential to be achieved. For example, the conversion shown in equation (1) at pH 7 has E°' = -550 mV (Ott and Schultz, 1975) for a $4e^-$ couple. This potential is sufficient to effect all but the diimide formation reaction and is very pH sensitive rising to -700 mV at pH = 10. Since local effective pH at an active site can vary considerably,

$$Mo_2O_4(edta)^{-2} + 4e^- + 4H^+ \longrightarrow MoO_2(H_2O)_2(edta)^{-2} \quad (1)$$

it is possible that the enzyme exerts control of the redox potential in this manner.

Recently, Watt *et al.* (1976) have evaluated the thermochemistry of the dithiocarbamate complexes of Mo(IV), Mo(V) and Mo(VI). It was calculated that reactions (2) and (3) are endothermic to the extent of + 30.1 and + 51.7 Kcal/mole, respectively, and are therefore likely to be highly endergonic. Thus,

$$MoO(R_2dtc)_2 + H_2O \longrightarrow H_2 + MoO_2(R_2dtc) \quad (2)$$

$$3H_2O + 3MoO(R_2dtc)_2 + N_2 \longrightarrow 3MoO_2(Rdtc)_2 + 2NH_3 \quad (3)$$

these reactions are not viable as models for N_2-reducing or H_2-evolving systems. However, it is well known that the redox potentials for Mo complexes are very ligand sensitive (Howie and Sawyer, 1976) and may change drastically with donor atom set. The studies of Chatt *et al.* (1976) for example, clearly reveal the importance of O donors in the reduction of bound dinitrogen. Thus, thermodynamic results do not necessarily eliminate the possibility of a Mo(IV)/Mo(VI) couple in nitrogenase although lower oxidation states remain viable as alternatives.

ABIOLOGICAL NITROGEN FIXING SYSTEMS

The Haber Process

The commercially enormous Haber-Bosch process involves a heterogeneous reaction wherein H_2 is used as the reductant for N_2. The process works at high temperatures (300 to 600°C) and high pressures (200 to 800 atm) even though the reaction catalyzed, $N_2 \longrightarrow NH_3$, is exergonic (Table I). The high temperature is required for the reaction to proceed at an economically desirable rate. The high pressure drives the reaction to the right (Le Chatalier, 1884) and compensates in part for the fact that the high temperature moves the exothermic reaction to the left. Under operating conditions, roughly 20 to 40% of the gas mixture is NH_3 and, as this agrees well with yields calculated from thermodynamics, it is clear that a true equilibrium is established. This situation is further illustrated by addition of ^{15}N to the reaction mixture (as $^{30}N_2$). The ^{15}N is soon found not only in $^{15}NH_3$ but also in ^{15}N-^{14}N and the rate of $^{29}N_2$ production is comparable to the rate of N_3 synthesis (Vancini, 1971).

Numerous catalysts have found use in the process and it is of some interest that mixed iron-molybdenum catalysts (Larson and Brooks, 1926) are quite effective (but unfortunately do not survive sufficiently long in the reaction milieu). The catalyst of choice remains iron, produced by reduction of magnetite ($Fe_3O_4 + H_2 \longrightarrow Fe + H_2O$). The magnetite is initially brought into solid solution with smaller amounts of Al_2O_3 and K_2O which are known as promotors. The promotors prolong catalyst life apparently by preventing sintering, the process whereby small crystallites of Fe fuse to produce larger crystals with smaller surface area and lower activity. The catalyst is readily poisoned and an interesting poison is CO which is reduced by the system to CH_4 and H_2O with such exothermicity and rapidity that local heating occurs causing catalyst destruction. This reaction accentuates the fact that CO (at least thermodynamically) is more easily reduced than is N_2 (see Section II). The inability of CO to be reduced by nitrogenase must be viewed in this perspective.

Homogeneous Inorganic Nitrogen-Fixing Systems

Non-Aqueous Systems The reports of Volpin and Shur (1964, 1965) commenced the extensive activity in this field and several reviews cover recent aspects (Brintzinger, 1976; van Tamelen, 1971; Shilov, 1974). Typically, an early transition metal, such as Ti, V, Zr, or Nb, is involved as either an alkoxide, halide or cyclopentadienyl complex. The reductant is usually derived from an alkali metal, an alkaline earth metal or from aluminum. The most thoroughly studied of these systems involve bicyclopentadienyl-Ti and more recently bicyclopentadienyl-Zr complexes (Brintzinger, 1976; Bercaw, this volume). In these cases, intermediate species have been isolated which are convertible to product by treatment with acid. Shilov (1972), van Tamelen (1971) and Teuben (1973) have each isolated dinitrogen derivatives of the titanocene complex at low temperature and the recent work of Manriquez and Bercaw (1974) and Manriquez *et al.* (1976) has led to structural characterization of the dinitrogen complex in the permethylzirconocene system. The dinitrogen ligands in the dinuclear complex $(Me_5C_5)_2(N_2)ZrN_2Zr(N_2)(C_5Me_5)_2$ are similar in all respects to the linear terminal or linear bridging dinitrogen ligands which have been found for numerous transition metal dinitrogen complexes (Sellmann, 1974). As only very few of these have been reduced to N_2H_4 or NH_3, the structural result itself gives no real clue. Only one of the three N_2 molecules in the dinuclear complex is reduced by treatment with HCl [reaction (4)] and isotopic-labelling studies indicate a symmetrical intermediate since the bridging and terminal nitrogens are scrambled in the N_2H_4 product (Manriquez *et al.*, 1976).

$$(Me_5C_5)_2Zr(N_2)N_2Zr(N_2)(C_5Me_5)_2 + 4HCl \longrightarrow 2(C_5Me_5)_2ZrCl_2 + N_2H_4 + 2N_2 \tag{4}$$

Mechanistically, the role of both the proton and the anion of the acid (Cl^-) must be scrutinized. Coupled with protonation of bound nitrogen and internal electron transfer, it seems likely that the anion of the acid (Cl^-) enters the coordination sphere of at least one of the Zr atoms. The probable displacement of one of the terminal Me ligands involves replacement of the -acceptor dinitrogen with the -donor chloride ligand which raises the energy of the d-levels, thereby stabilizing the higher oxidation state [Zr(IV)], and could be important in channelling the redox flow into a single Zr atom. From this state, additional protonation and ligand substitution would allow all four electrons to enter one N_2 molecule to produce hydrazine.

In many respects, the Bercaw system is similar to that of Chatt and co-workers (Chatt *et al.*, 1972; Chatt, 1975; Chatt *et al.*, 1976). In both cases, there are two dinitrogen molecules present in the coordination sphere of each metal with only one N_2 in each complex being reduced. In both cases, the reduction of N_2 is triggered by mineral acid whose protons and counterions both play key roles. In the zirconocene system, only four electrons are available and only hydrazine is formed. In the molybdenum or tungsten phosphine systems, six electrons are available and these all flow to a single coordinated dinitrogen. In each case, the reducing agent used to produce the low valent dinitrogen complex is Na(Hg). Thus, while the NH_3- or N_2H_4-forming reactions are conducted under mild conditions, the reducing power of Na metal is already built into the active complex. It remains to be seen if significantly milder reductants can affect the syntheses of these or similar species. As discussed previously, even if all the energy of ATP hydrolysis can be transduced into reducing potential, nitrogenase can still only achieve a potential of -1.2 volts. It does not seem likely that lower valence states such as Mo(0) could be produced in nitrogenase.

While clearly not chemically similar to the active site of nitrogenase, these non-aqueous N_2-reducing systems may nonetheless teach us something in principle about the activation of N_2 toward reduction and the induction of internal redox reactions. The first step in N_2 reduction by nitrogenase undoubtedly involves the binding of N_2 to a transition metal site on the enzyme. There is nothing in the known enzymology which precludes either changes in metal coordination sphere or protonation of bound N_2 (or both) as initiators of the redox process.

A change in the coordination geometry and donor atom set might be part of the oft quoted conformational change (Silverstein and Bulen, 1970; Walker and Mortenson, 1974) which may accompany the ATP reaction in nitrogenase. Moreover, ATP hydrolysis, which delivers protons above pH 6, could be the intimate or indirect proton donor. It is also possible that protons on protein ligands coordinated to Mo (or Fe) in the enzyme have pKa values which allow them to be transferred to the bound dinitrogen in conjunction with electron flow (Stiefel, 1973; Stiefel and Gardner, 1974). As the reducing site of nitrogenase may be hydrophobic in nature, its *mechanistic* resemblance to the inorganic systems discussed above remains a distinct possibility.

Aqueous Systems (a) Molybdothiol System. This system was developed and extensively studied by Schrauzer and co-workers and has been thoroughly reviewed elsewhere (Schrauzer, 1976). Here, some of the newer developments are considered, stressing reports other than those of Professor Schrauzer (the contribution of Professor Schrauzer is included herein).

The molybdothiol catalyst system consists of a molybdenum salt, thiol ligand, buffer solution (usually pH 9-10) and reductant (usually BH_4^-). The system effects the reduction of all tested nitrogenase substrates, although the rates of some of the reactions are very slow and dinitrogen reduction is not catalytic, occurring only to the extent of 0.1% of the total molybdenum present. An interesting feature of the system is its stimulation by ATP which is said to parallel the stimulation found in nitrogenase.

Recently, Corbin *et al.* (1976), although confirming that molybdothiol catalysts effect the BH_4^- reduction of C_2H_2 to C_2H_4 and C_2H_6, find that under many conditions, butadiene (C_4H_6) is the major reaction product. For example, in borate buffer at pH 9.6 using BH_4^- as reductant and $Mo_2O_4(cys)_2^{-2}$ as the molybdothiol, roughly 80% of the C_2H_2 reduced ends up as butadiene (100% C_2H_2 as initial gas phase). The absence of reported butadiene in earlier work under virtually identical conditions may be attributable to the long retention time of this C_4 hydrocarbon under the chromatographic conditions normally used to assay for C_2H_2, C_2H_4 and C_2H_6. The observation of butadiene in this model system illustrates that care must be taken in the investigation of chemical systems to examine their full range of behaviour so as to be certain that the enzyme-like reaction is indeed the major pathway in the "model" system. We have additionally found (Corbin *et al.*, 1976) that, even at 100% C_2H_2, nitrogenase does not produce butadiene although high concentrations of C_2H_2 do lead to substrate inhibition and lower levels of C_2H_4 production (Hwang and Burris, 1972). Recently, acetylene has been found to inhibit hydrogenase (Smith, *et al.*,

1976b) and methane monooxygenase (Dalton and Whittenbury, 1976), each of which plays an important role in the metabolism of certain nitrogen-fixing organisms. In view of the results from both inorganic and biological systems and in the light of the relative thermodynamic ease with which C_2H_2 is reduced (Section II) or oxidized (by O_2), it is clear that extreme caution must be used when applying C_2H_2 reduction as the *sole* probe for nitrogen-fixing activity.

Werner *et al.* (1973) have shown that a complex of Mo and the tripetide glutathione is active in BH_4^- reduction of C_2H_2 to C_2H_4 and N_2H_4 to NH_3 but apparently not in the conversion of N_2 to NH_3. ATP was reported by these authors to stimulate the reduction of acetylene. Corbin *et al.* (1976) have found that a glutathione/molybdate catalyst also produces C_4H_6 as a major reaction product. Shilov and co-workers (Krushch *et al.*, 1975) investigated the ATP effect in the glutathione systems and concluded that the function of ATP was to act as an acid, *i.e.* the "ATP effect" could be achieved using sulfuric acid. Schrauzer *et al.* (1975), on the other hand, utilizing somewhat different procedures, claims ATP hydrolysis and stimulation in reaction systems using $Mo_2O_4(cys)_2^{-2}$. Both groups agree that ATP may produce acid which activates the Mo complex.

Chemical and electrochemical techniques have been applied to $Mo_2O_4(cys)_2^{-2}$ to elucidate the oxidation states and reaction intermediates involved. Ichikawa and Meshitsuka (1973) showed that ~ -1.0V (*vs.* s.h.e.) was required to maximize catalytic activity by using a set of charge carriers (redox mediators) to set the potential. These polarographic studies showed a reduction wave at -1.03V which was attributed to the Mo(V)⟶Mo(IV) couple. More recently, Schultz (1975) and Ledwith and Schultz (1976) have carefully studied the $Mo_2O_4(cys)_2^{-2}$ system electrochemically and, based upon coulometric determination, conclude that the reduction at ~ -1000 mV is a $4e^-$, $4H^+$ process leading to the production of Mo(III). Using the electrode as the source of electrons, they found catalysis of acetylene reduction only after formation of the Mo(III) complex. The identification of the Mo(III) state by Schultz shows that the molybdothiol systems may be closer to the Mo(III) systems devised by Shilov (discussed below) than had been previously surmised.

A second observation by Schultz on the electrochemical system may be both mechanistically relevant and analogous to nitrogenase. If a solution of $Mo_2O_4(cys)_2^{-2}$ is electrochemically prereduced to the Mo(III) stage and the potential is cut off prior to C_2H_2 gassing, it is found that C_2H_2 is not reduced. Reapplication of the potential leads to immediate reduction of C_2H_2. It is clear that catalyst, reduced in the absence of substrate, cannot effect substrate reduction but requires additional reductant, *i.e.* the presence of Mo(III) seems necessary but not sufficient to reduce substrate. This result is closely analogous to the observation of Watt (unpublished) that the fully reduced 1:1 complex from *A. vinelandii* nitrogenase in the absence of external reductant fails to turn over, even in the presence of ATP. Addition of further reductant causes immediate turnover. The juxtaposition of inorganic and biochemical results implicate the presence of a reduced state which is nonetheless incompetent in substrate reduction.

There has been little systematic study in the variation of ligand structure in molybdothiol systems. That such studies could yield fruitful results is clear from the results of Suzuki *et al.* (1975), who have linked cysteine to a dextran polymer and synthesized a polymer-bound Mo complex (Fig. 2). They find a

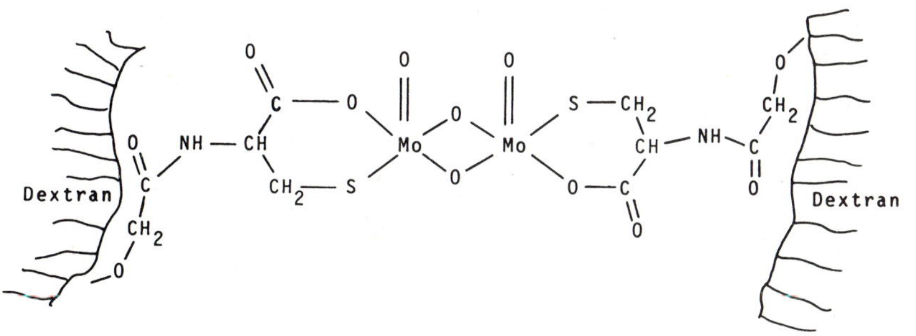

Fig. 2. *Proposed structure for dextran-bound molybdocysteine complex active in catalysis of BH_4^- reduction of C_2H_2 (Suzuki* et al., *1975).*

27-fold enhancement of rate for the BH_4^- reduction of acetylene by the dextran bound Mo-cys complex as compared to $Mo_2O_4(cys)_2^{-2}$. This result may mean that the monomeric form of the catalyst, the presumed active species, is preferentially stabilized on the polymer support. Furthermore, a relatively large ethane/ethylene ratio was reported, indicating the possibility that the two monomeric centres, which form upon dissociation, may remain in sufficiently close proximity to affect the four-electron reduction of acetylene.

Recently, Corbin *et al.* (1976) have demonstrated additional interesting ligand effects. The complex $Mo_2O_4(pen)_2^{-2}$, [*1*; X = CH_3, Y = CH_3], was catalytically inactive in C_2H_2 reduction, although the ligand penicilamine differs from cysteine only in having methyl groups substituted for hydrogens in the β-position. The effect was probed by investigating complexes of the two isomeric β-methyl-cysteine ligands

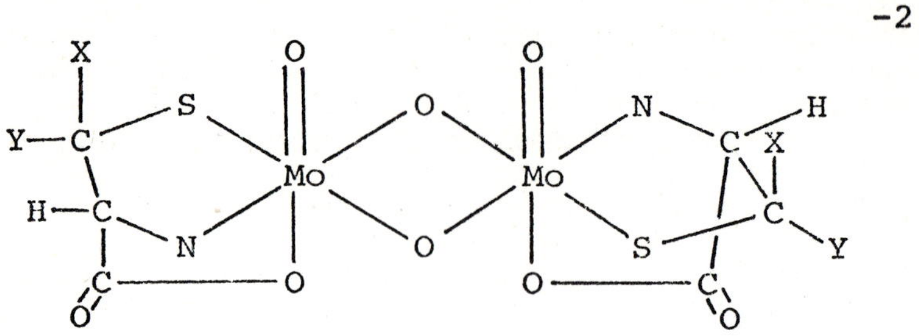

1

(designated erythro and threo). Remarkably, although these isomeric complexes display nearly identical spectroscopic properties, there is a gross difference in reactivity. The threo complex [*1*; X = CH_3, Y = H] has activity similar to the cysteine complex but the erythro complex [*1*; X = H, Y = CH_3] is inactive just like the penicillamine complex. The great differences in reactivity caused by small ligand variations implicate subtle steric effects as determining whether or not an active species can be achieved.

Another structural variation pursued by Corbin *et al.* (1976) involved combining the two cysteine ligands through an ethylenediamine bridge such that they are constrained to bind to adjacent Mo centres (proposed structure *2*) in a similar manner

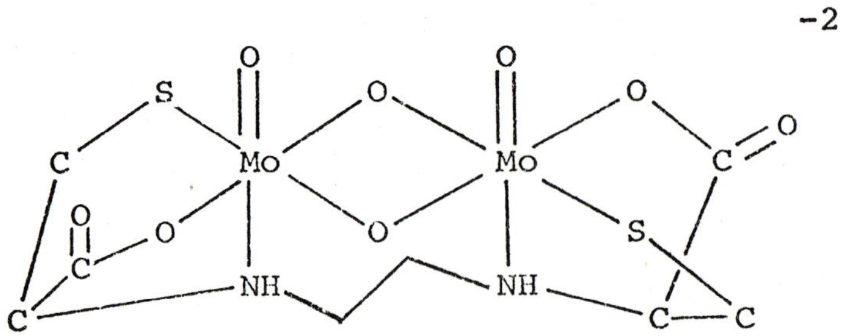

2

to $Mo_2O_4(edta)^{-2}$. The resulting complex is extremely unreactive in the reduction of C_2H_2. Significantly, of the small amount of C_2 reduction product formed, a much higher percentage (30%) compared to the cysteine case (1%) is in the form of ethane. These results support a mononuclear Mo site as the active species with the ethylenediamine bridge both enforcing the lack of reactivity of the complex and enhancing its ability to engage in $4e^-$, $4H^+$ transfer processes.

Despite significant differences between the molybdothiol system and nitrogenase, much similarity is present (Schrauzer, 1976). The similarity provides presumptive evidence for the presence of Mo at the active site of nitrogenase. The implication of mononuclear catalytic entities, the identification of the Mo(III) oxidation state and the stimulation of the Mo catalyst by ATP in the model system are particularly interesting. As more structural and/or mechanistic data becomes available from both enzyme and model systems, additional similarities and differences will surely be articulated.

(b) Other Aqueous Models. N_2-fixing systems other than the molybdothiol systems have been studied by Shilov and his coworkers (Shilov, 1974, 1976). Either Mo or V is required to produce an active system. Although there is a lack of consensus (Benneman *et al.*, 1972), V may be the one element capable of substituting for Mo in the Mo-Fe protein of nitrogenase (Burns *et al.*, 1976). The Shilov Mo system utilizes Cr^{+2} or Ti^{+3} as reductant and presumably, Mo^{+3} as catalyst in strongly basic media with no additional ligand present (*i.e.* metals coordinated only by O, OH or OH_2). The homogeneous nature of these preparations has been questioned (Shilov, 1976; Schrauzer, 1976) and these are most likely examples of heterogeneous catalytic systems. However, the very active V system, containing catecholate ligands (Nikonova *et al.*, 1974) is homogeneous and clearly reduces N_2 to NH_3 while the V(II) is oxidized to V(III) in a pH-sensitive reaction. These systems evolve hydrogen in the absence of N_2 and are capable of reducing acetylene. Interestingly, CO is found to be an effective inhibitor. Further development of these systems to a point where intermediates are detected and/or mechanistic pathways delineated, would be most welcome.

MOLYBDENUM COORDINATION CHEMISTRY

Compared to other transition metals present in biological redox systems, the oxidation states likely to be used by Mo are very high (Stiefel, 1976). The IV, V and VI states are a likely set of participants in Mo oxidases and, while the II and III states remain viable for nitrogenase, it is possible that higher oxidation states will be found here as well. Indeed, the apparent substitution of W for Mo in nitrogenase (Benneman *et al.*, 1973) indicates this likelihood as it is far more difficult to obtain the lower oxidation states of W (Kepert, 1972).

Examination of the coordination chemistry of Mo reveals that the higher oxidation states are always found coordinated by deprotonated ligands. In the most common case, these ligands are waters, which when fully deprotonated become the common oxo groups. The compounds with dithiocarbamate ligands, $(MoO_2(R_2dtc)_2$, $Mo_2O_3(R_2dtc)_4$ and $MoO(R_2dtc)_2$ reveal the decrease in the number of oxo groups in going from Mo(VI) to Mo (V) (Stiefel, 1976; Newton *et al.*, 1974).

Another aspect of the deprotonated (acidic) ligand effect is revealed when the oxo groups are removed from the coordination sphere. For example, the reaction of MoO_4^{-2} with o-aminobenzenethiol (Stiefel and Gardner, 1974) proceeds according to equation (5), where the product is formally a Mo(VI) complex

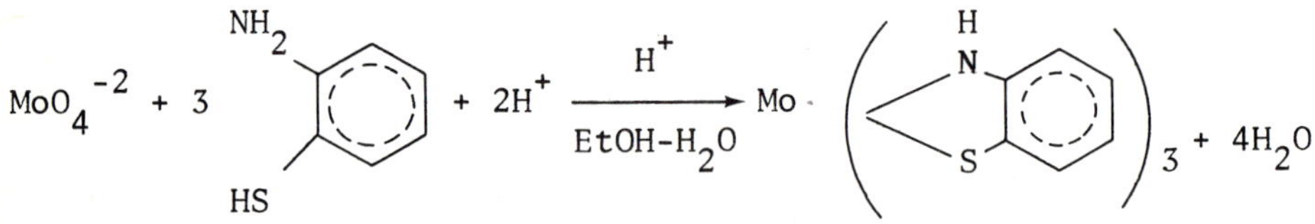

(5)

coordinated by a deprotonated amine ligand. The formation of the deprotonated products results from the acidity-enhancing properties of Mo(VI) (Stiefel, 1973; Stiefel and Gardner, 1974). When the more acidic aquo ligands (oxos) are removed from the coordination sphere, the acidity manifests itself in the ionization of a coordinated amine ligand which ordinarily would not be considered as a potentially ionizable grouping. This effect of oxidation state on ligand acidity is further illustrated in Mo chemistry by the aquo ions (Stiefel, 1976). In strong acid solutions, the principal species of Mo(VI) is thought to be $[MoO_2(H_2O)_4]^{+2}$ while for Mo(III), the ion $[Mo(H_2O)_6]^{+3}$ is present. These results show that ligands coordinated to Mo have the ability to engage in proton transfer reactions which, through the effect of oxidation state on pKa, can be coupled to electron transfer reactions.

MOLYBDOENZYMES

Nitrogenase is one of a group of Mo-enzymes involved in redox reactions (Bray, 1975 Bray and Swann, 1972). For each of the other molybdenum enzymes (xanthine oxidase, sulfite oxidase, aldehyde oxidase and nitrate reductase), the reduction or oxidation of substrate is a $2e^-$ step, accompanied by transfer of one or two protons (Stiefel, 1973). Like nitrogenase, each of the other Mo-enzymes has additional prosthetic groups which are capable of redox activity. However, in each case except nitrogenase, there is evidence that the Mo site of the enzyme interacts directly with substrate (*i.e.* effects the $2e^-$ change). For nitrogenase, much evidence is also consistent with substrate reduction by a series of $2e^-$, $2H^+$ steps (Burns and Hardy, 1975; Bulen, 1976; Newton *et al.*, this volume). In view of this and the concept of common molybdenum-cofactor (Pan *et al.*, 1976) it is possible that the Mo site in nitrogenase bears similarity in both structure and mechanism to that in other Mo-enzymes. Therefore, some mechanistic details for xanthine oxidase, the most thoroughly studied of these Mo enzymes are considered. Its stoichiometric mechanism is well established and there is considerable information, mostly from EPR studies, relating to the intimate mechanism of xanthine oxidation.

Stoichiometric Mechanism for Xanthine Oxidase

Xanthine oxidase has a molecular weight of 300,000 and contains 2 Mo, 8Fe, $8S^{-2}$ and 2FAD moieties per molecule. The evidence (Massey *et al.*, 1970; Bray, 1975 strongly favours the organization of these prosthetic groups into two catalytically independent groupings (presumably in different subunit sets), each containing 1Mo, 1FAD and $2Fe_2S_2$ redox clusters. The work of several groups (Olsen *et al.*, 1975; Massey, 1973; Bray, 1975; Cammack *et al.*, 1976) has established that Mo is the site of substrate oxidation (electron ingress) whereas Fe_2S_2 or more likely flavin is the site of electron egress. Both redox titration and kinetic studies are consistent with the distribution of electrons within the enzyme being determined solely by the potentials of the redox components (appropriately modified by substrate binding in the case of Mo).

The general picture of activity with respect to the enzyme is as follows. The resting oxidized enzyme has Mo in the VI oxidation state and flavin and FeS clusters in their oxidized forms. Xanthine binds to the Mo site and transfers two electrons and one proton to that site forming Mo(IV) with a tightly associated proton. The Mo(IV) component then transfers its electrons one at a time to flavin and Fe_2S_2, leading to the formation of an intermediate Mo(V) state which is EPR active and displays proton superhyperfine splitting from a single H atom. This hydrogen has been shown by deuterium labelling to originate in the substrate xanthine. In all, three molecules of substrate can be oxidized by each catalytic unit with the six electrons being accommodated in the flavin (2), $2Fe_2S_2$(2) and Mo(2) centres. Oxidation by O_2 can occur at any stage of reduction of the enzyme, *i.e.* when either 2, 4 or 6 electrons are present per catalytic unit. It has been suggested that the function of the Fe_2S_2 centres is to keep the Mo site in its fully oxidized form and, therefore, ready to accept substrate while simultaneously keeping the flavin site in its reduced form and so ready to react with oxidant (Olson, 1974). In view of the rapidity of the internal electron transfer, the enzyme can therefore be ready for redox action even if it has already accepted electrons from two molecules of substrate without oxidant action. This clearly efficient mode of action provides an interesting rationale for the presence of three redox active centres.

Intimate Mechanism of Xanthine Oxidation

Here, those aspects which are relevant to the Mo site and its interaction with substrate are considered. The simplified picture in Fig. 3 is one which fits current experimental data.

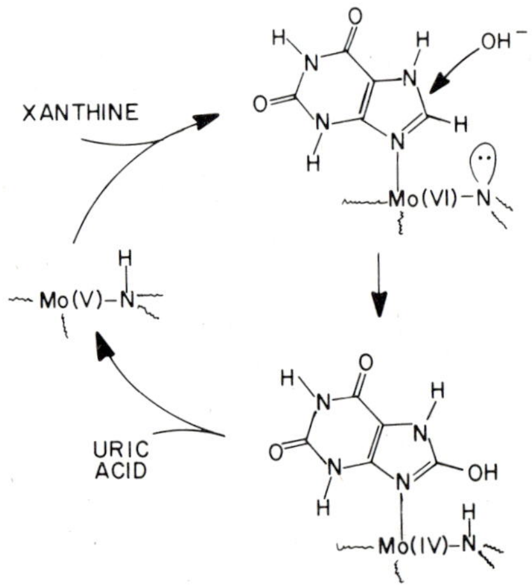

Fig. 3. *Proposed mechanism for xanthine oxidation at the Mo site in xanthine oxidase: coupled proton-electron transfer (Stiefel, 1973).*

The upper right of the diagram shows the resting Mo(VI) site which is assumed to have a deprotonated N-containing ligand (see discussion in section IV and Stiefel, 1973; Stiefel *et al.*, in press). Coordination of xanthine to Mo(VI) polarizes the N=C bond and makes the carbon susceptible to nucleophilic attack either from OH^- or from an enzyme-bound persulfide as suggested by Edmonson *et al.* (1973 In conjunction with this attack, a coupled one-proton/two-electron transfer can occur from xanthine to the Mo site, producing the protonated Mo(IV) state. The rationale for proton transfer accompanying electron transfer is the change in pKa of coordinated ligands which accompanies the change in oxidation state (Stiefel, 1973; Stiefel and Gardner, 1974). Two internal one-electron transfers follow producing sequentially the EPR-active Mo(V) state and the reactivated deprotonated Mo(VI) state. A similar coupled proton-electron transfer has been suggested to occur in all other Mo enzymes including nitrogenase and is discussed at greater length below.

Recent electron spin resonance studies on inorganic Mo(V) complexes (Pariyadath *et al.*, 1976) have revealed proton super-hyperfine splitting of a magnitude comparable with that observed in molybdoenzymes. The location of the protons on coordinated ligands in these Mo complexes adds further weight to the suggestion that similar species are responsible for the splittings observed in Mo enzymes.

NITROGENASE - THE STOICHIOMETRIC MECHANISM

General Considerations

The Mo-Fe protein (designated [Mo-Fe]), the Fe protein (designated [Fe]), ATP, Mg^{+2}, reductant and protons each play roles in the stoichiometric mechanism. Evidence (Thorneley and Willison, 1974) shows that the 1:1 Mg:ATP complex is acted upon by nitrogenase with two molecules of MgATP binding to each [Fe] (Tso and Burris, 1973; Zumft *et al.*, 1973). Each [Fe] is capable of delivering a single electron going from an EPR-active reduced form to an EPR-inactive oxidized form (Burris and Orme-Johnson, 1976; Mortenson *et al.*, 1976). If two Fe proteins (each complexed with 2 MgATP molecules) interact with one [Mo-Fe] (or site) and transfer electrons in concert with ATP hydrolysis, then an ATP/2e$^-$ ratio of 4 is predicted in agreement with experiment (see below). [Mo-Fe], acted upon by the [Fe]·2MgATP complexes is initially in an EPR-active (g = 4.3, 3.7, 2.01) state but during turnover, an EPR-inactive state is formed which is reduced with respect to the EPR-active form (Orme-Johnson *et al.*, 1972; Smith *et al.*, 1973; Mortenson *et al.*, 1976). This ill-defined super-reduced state may be responsible for hydrogen evolution and/or substrate reduction reactions, although the enzyme states responsible for H_2 evolution, C_2H_2 reduction and nitrogen fixation may not be identical (Smith *et al.*, 1976).

The total number of electron pairs transferred per unit time, as measured either by dithionite ($S_2O_4{}^{2-}$) utilization or by combined substrate reduction and hydrogen evolution, is dependent upon enzyme, ATP and reductant levels but not upon substrate concentration (but see below for exceptions). This situation indicates that the rate-determining step for nitrogenase turnover occurs independently of the substrate reduction and/or hydrogen evolution processes and is distinct from the product-determining step(s). It is, therefore, convenient to separately discuss the enzyme-activating reaction, which is rate-determining and the substrate binding and reducing reactions, which are product-determining.

ATP/2e Ratio A minimum of 4 to 5 moles of ATP are hydrolyzed to ADP and phosphate for each pair of electrons which travels through nitrogenase (Watt *et al.*, 1976). The ATP/2e$^-$ ratio becomes larger as the ratio of [Mo-Fe] to [Fe] increases (Burris and Orme-Johnson, 1976; Mortenson, 1976). The reason for this seemingly wasteful phenomenon (which may be analogous to uncoupling) is not clear but its existence implies the presence of non-productive ATP hydrolyses and causes the lower range of ATP/2e$^-$ values to be taken most seriously as representative of effective turnover.

Interaction between Fe and Fe-Mo Proteins The Sussex group (Eady, 1973; Thorneley *et al.*, 1975; Smith *et al.*, 1976) has reported that [Fe] and [Mo-Fe] form a 1:1 complex as the only identifiable aggregated species in the ultracentrifuge. However, the observation of complex formation requires the absence of dithionite (always present when enzyme is turning over) and the mechanistic relevance of the observed complex is therefore in some doubt.

Watt and Bulen (unpublished) have characterised the isolated nitrogenase complex from *Azotobacter vinelandii* as a roughly 1:1 complex of [Fe] and [Mo-Fe]. While it is possible that this unfractionated system represents the true cellular proportions of [Fe] and [Mo-Fe], it is also possible that the far more labile [Fe] is lost during purification to an extent far greater than [Mo-Fe]. Electron microscopy studies by the DuPont group (Stasny *et al.*, 1974; Burns *et al.*, 1976) have tentatively identified both 1:1 and 2:1 complexes of [Fe] and [Mo-Fe] during presumed catalytic conditions. While the formation of a 1:1 complex is probable, the formation of a 2:1 complex is not disallowed by the evidence at hand. In fact, the sigmoidal plot of activity versus [Fe] concentration (Burns and Hardy, 1975) and the stoichiometry of the reaction would appear to demand the interaction of [Mo-Fe] with at least two [Fe] proteins although not necessarily at the same time.

Electron Transfer There is little doubt now that electrons flow from reductant to [Fe] to [Mo-Fe] and then to substrate. The ATP utilization reaction appears to occur at the stage where electrons transfer from [Fe] to [Mo-Fe] although the presence of substrate at that point is not precluded. Although containing two identical subunits, [Fe] appears to possess a single Fe_4S_4 cluster (Orme-Johnson *et al.*, 1976) and is capable of delivering a single electron to [Mo-Fe]. It is reasonable to assume that both molecules of bound ATP are hydrolyzed concomitant with the electron transfer from [Fe] to [Mo-Fe]. Whether both electrons required for a single turnover are simultaneously or sequentially delivered to [Mo-Fe] is unknown.

The sequential delivery is, however, favoured by some of the experimental results. Assuming that there are two independent sites on [Mo-Fe] (as strongly favoured in other molybdoenzymes), then in the reduced state, the EPR signal (g = 4.3, 3.7) from each site is observed (Orme-Johnson *et al.*, 1972). Although the exact origin of this EPR signal is uncertain, it is definitely attributable to an *odd-electron* species. The EPR signal disappears upon further reduction during the steady state which is most likely due to generation of an even-electron system (likely to be EPR-inactive) at each site. This scheme implies a one-electron reduction of the reduced state. However, unless [Mo-Fe] can achieve a form more oxidized than the EPR active state during turnover, this state should not be competent in substrate reduction (although it might be weakly hydrogen-evolving). Thus, a second reduction by [Fe] would be required to bring the enzyme to a substrate-reducing level or to trigger the electron transfer to substrate. The notion of a super-reduced state, which is unable to reduce substrate, has a distinct analogy in the observations of Schultz on the molybdothiol system (section III) and may be supported by work of Watt (unpublished) on the 1:1 nitrogenase complex of *A. vinelandii*.

The proposed sequential model is also analogous to the stoichiometric mechanism for the cyt P_{450} hydroxylation system (Sharrock *et al.*, 1976; Sligar and Gunsalus, 1976) as shown in Fig. 1. Here, after substrate binding, a molecule of ferredoxin (e.g. adrenodoxin, putidaredoxin) primes the cytochrome to bind O_2 whereupon an additional molecule of ferredoxin triggers the completion of the reaction. Similarly, in nitrogenase, one [Fe] may prime the site but the second [Fe] may be required to trigger electron transfer to substrate and completion of the reaction.

In nitrogenase, there is a nonproductive ATPase activity which increases in prominence when [Mo-Fe] is in excess (*i.e.* at low levels of [Fe]). This process may be a back electron transfer (Orme-Johnson and Burris, 1976) from the one-electron reduced [Mo-Fe] site to oxidized [Fe]. At low [Fe], the second electron may not be available to trigger the final reaction, thereby allowing the one-electron super-reduced state to live sufficiently long to back-transfer to [Fe], leading to net unproductive ATP hydrolysis. This second reduced [Fe] may represent a distinct second molecule which enters into binding with [Mo-Fe] or could involve the same [Fe] which supplied the first electron. In the latter case, rapid reduction of oxidized [Fe] would be required, as has been demonstrated (Thorneley *et al.*, 1976), and replacement of MgADP by MgATP would also have to be rapid. If such rapid reactivation does not occur, then the unproductive back electron transfer could proceed.

Substrate Manipulation

Hydrogen Evolution versus Substrate Reduction The rate of appearance of reducing equivalents in product is specified by the levels of ATP, Mg^{+2}, [Fe], [Mo-Fe] and reductant as well as the pH and temperature. In the absence of reducible substrate, H_2 evolution (termed ATP-dependent) accounts for the entire electron flow. As reducible substrate is added to the system, the quantity of hydrogen evolved decreases commensurate with the diversion of reducing equivalents to substrate. Only one substrate, C_2H_2, seems to be capable of totally suppressing H_2 evolution (Burns and Hardy, 1975). For all others, even under saturating conditions, there is some residual H_2 evolution.

Burns and Hardy (1975) have introduced the electron allocation coefficient (EAC), defined as

$$\text{EAC} = \frac{\text{Electrons flowing to substrate/unit time}}{\text{Total electron flow/unit time}}$$

Thus, for acetylene reduction EAC = 1.0 (all electrons go to C_2H_4 production). For N_2 reduction, however, EAC only approaches 0.75 (Hadfield, 1970; Rivera-Ortiz and Burris, 1975), which means that only three out of four electrons go into NH_3 formation, the remaining going to H_2 production. Stoichiometrically, since reduction of N_2 requires 6 electrons, one mole of H_2 is evolved for each mole of dinitrogen reduced under saturating conditions. Relative and absolute EAC values are useful quantities for both *in vitro* (Burns and Hardy, 1975) and *in vivo* studies (Peterson and Burris, 1976; Schubert and Evans, 1976). *In vivo* and in some crude cell-free extracts, the presence of the enzyme hydrogenase allows the recycling of hydrogen produced to generate more reducing power and/or ATP (Dixon, 1972; Schubert and Evans, 1976; Hill *et al.*, 1976). Coupled with the recent observation that C_2H_2 inhibits H_2 uptake by hydrogenase (Hill *et al.*, 1976), this conservation mechanism indicates that relative and absolute EAC values obtained *in vitro* cannot be applied indiscriminately to *in vivo* systems. Rather, *differences* in EAC in the two cases become significant insofar as they indicate the probable involvement of additional enzyme systems *in vivo*.

Inhibition or Redirection With nitrogenase, great care must be taken in defining what is meant by inhibition. For example, C_2H_2 is often referred to as an inhibitor of nitrogen fixation but C_2H_2 is not (at least at usual levels) an inhibitor of nitrogenase. It is, rather an alternative substrate which redirects electrons from dinitrogen reduction to the reduction of acetylene. Other substrates, N_3^-, N_2O, RCN, RNC, are also inhibitors of nitrogen fixation in this sense (Burns and Hardy, 1975). They are more properly called redirectors as they do not change the basic rate of turnover of the enzyme (with the possible exceptions of N_3^-, CN^- and very high levels of C_2H_2). CO, although not a reducible substrate also serves as a redirector, shunting electrons away from substrate reduction and towards hydrogen evolution without affecting the rate (per electron pair transferred).

The inhibition of nitrogen fixation by H_2 is yet another case of electron reallocation. Thus, H_2 does not effect the rate of electron transfer, but causes the rate of dinitrogen reduction to decrease concurrently with an increase in the rate of H_2 evolution. Significantly, of all substrate reactions, H_2 only inhibits the reduction of N_2 to NH_3 (Newton *et al.*, this volume).

Experiments under D_2 provide clear data to assess the stoichiometry of this inhibition. Significantly, the rate of H_2 evolution under D_2 is the same as its rate in the absence of D_2, *whether N_2 is present or not* (Stiefel *et al.*, 1977; Newton *et al.*, this volume). Thus, H_2 redirects electrons from N_2 reduction to additional H_2 production, but D_2, while likewise decreasing NH_3 production, does not produce any additional H_2. Because D_2 does not inhibit the turnover rate of nitrogenase, the question arises as to where the redirected electrons go. The answer is that the production of HD under D_2 and N_2 is an electron-requiring process, *i.e.* the so-called HD exchange is not an exchange at all, but requires one electron for the appearance of each HD molecule (Bulen, 1976). Thus, D_2 inhibition, like H_2 inhibition, is an example of redirection with electrons being diverted from N_2 reduction into HD formation. To agree with electron balance studies, D_2(as the only source of D) must react with $2e^-$ and $2H^+$ to produce two molecules of HD. Since N_2 is an absolute requirement for HD production, a product of N_2 reduction by $2e^-$ and $2H^+$ is implicated. The H_2 inhibition and HD production reactions are expressed by equations (6) and (7).

$$E(N_2 + 2e^- + 2H^+) + H_2 \longrightarrow E + N_2 + 2H_2 \qquad (6)$$

$$E(N_2 + 2e^- + 2H^+) + D_2 \longrightarrow E + N_2 + 2HD \quad (7)$$

These equations emphasize that, stoichiometrically, the H_2 inhibition reaction and HD *production* reactions are one and the same process. [It is clearly tempting to represent ($N_2 + 2e^- + 2H^+$) as N_2H_2, a bound diimide-level species.] The above stoichiometric mechanism predicts that, during N_2 fixation by nitrogenase under D_2, no HDO will be formed in the solution phase. Experimental confirmation of this point would be difficult but is both possible and desirable.

NITROGENASE - THE INTIMATE (CHEMICAL) MECHANISM

All discussions of the intimate mechanism of nitrogenase action must be considered highly speculative and useful insofar as they suggest further experimentation on either enzyme or model systems.

It is assumed throughout that the Mo-Fe protein is the site of substrate reduction although the evidence for this is not unequivocal. Model system studies reveal (Schrauzer, 1976; Shilov, 1976) that Mo-containing systems mimic the substrate reactions of nitrogenase more closely than do those containing other metals with the possible exception of V in some of Shilov's systems. Therefore, not only do we assume that the Mo-Fe protein contains the site of substrate binding and reduction but also that Mo atoms are key components of that site.

ATP Reaction

ATP (as the MgATP complex) binds to the Fe protein. Because CTP, UTP and α,β-CH_2-ATP do not bind (Zumft *et al.*, 1973) but ADP does bind (again presumably as the MgADP complex), it seems likely that the adenosine-diphosphate portion of the ATP molecule binds to the Fe protein leaving the terminal phosphoryl groups exposed. It is this terminal phosphoryl group which may contact the Mo-Fe protein and be involved in the interaction of these two proteins which facilitates electron transfer between them.

Among the many suggestions for the role of ATP are those involving ATP in a conformational change of the Fe and/or Mo-Fe proteins (Walker and Mortenson, 1974). While this function of ATP may well be important, only suggestions involving specific chemical interactions between ATP and the Mo site are discussed. Such suggestions (which are not preclusive of a concurrent conformational change) involve ATP in the generation of an open coordination site on Mo.

As previously noted, Mo-O linkages pervade the chemistry of Mo in high oxidation states and ATP, by its ability to act as either a phosphorylating agent or a proton source, could facilitate the oxo removal reaction. One possibility is that ATP couples phosphorylation of the Mo-O group with reduction of the site by two electrons. The rationale for the coupled electron-phosphoryl group transfer is similar to that for the coupled electron-proton transfer (Stiefel, 1973) with the electrophilic phosphoryl group serving in place of the proton. Thus, Mo-O could, upon accepting electrons, become more nucleophilic and attack the terminal phosphoryl group of ADP (Fig. 4). Transfer of this group to Mo would produce ATP and a molybdenum phosphate. The phosphate might then dissociate (perhaps assisted

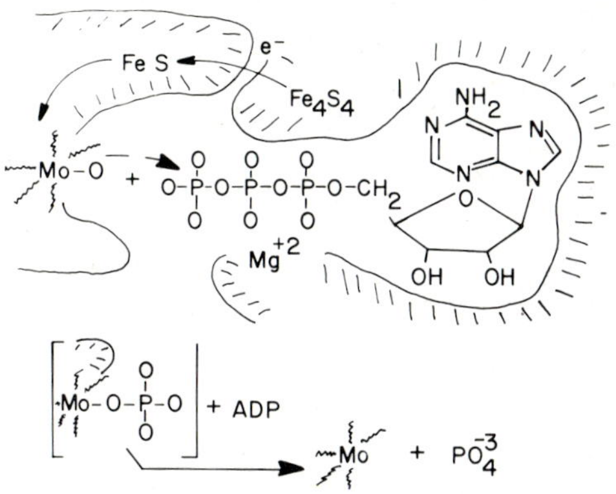

Fig. 4. *Possible mechanism for ATP action in nitrogenase: phosphorylation of Mo-O coupled with electron transfer.*

by acid formed from hydrolysis of additional ATP) to form the open active Mo site.

Alternatively, it is well known (Stiefel, 1976) that oxo group removal coupled with reduction is facilitated by the presence of acid. Protonation of oxo groups produces hydroxo or aquo ligands which are superior leaving groups. The hydrolysis of ATP near pH 7 is known to produce protons and this hydrolysis may be enzymically controlled to produce these protons in the vicinity of the oxo group facilitating the coupled reduction and oxo removal on the Mo site (Shilov, 1974; Newton *et al.*, 1976). In either case, the role of ATP would appear to be the production of an open site on Mo.

The coupling of ATP hydrolysis with electron transfer is, in effect, a reductive dephosphorylation of ATP. The popular chemiosmotic hypothesis of oxidative phosphorylation (Mitchell, 1976) involves electron transport producing a proton gradient which in turn, effects the synthesis of ATP. For nitrogenase, the above suggestion involves the hydrolysis of ATP to produce a proton gradient which in turn induces electron transport. Should these speculations on ATP utilization by nitrogenase achieve experimental support then there may be much to learn about both oxidative phosphorylation and ATP utilization in nitrogenase by their judicious comparison.

Nature of Reducible Substrate

Nitrogenase catalyzes the reduction of a number of small molecules most of which contain triple bonds (RC≡C-H, RC≡N, RN C, HC≡N) or potential triple bonds (N=N=O, HN=N=N). The triple bond must be terminal as, e.g. 2-butyne is not reduced although 1-butyne is (Burns and Hardy, 1975). Even for terminal triple bonds, there are stringent steric requirements for the remainder of the molecule as if a near linear cavity on the enzyme excluded all but long and thin molecules (Burns and Hardy, 1975; Hardy *et al.*, 1971). Recently, a number of molecules which do not contain triple bonds have been reported to be reducible substrates.

The reduction of cyclopropene by nitrogenase was reported by McKenna *et al.* (1976). It yields cyclopropane and propylene in a 1:2 ratio (equation 8) perhaps reflecting random

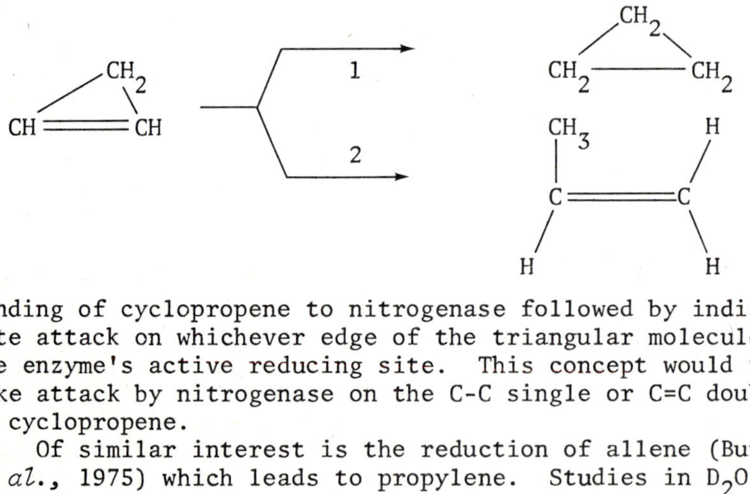

(8)

binding of cyclopropene to nitrogenase followed by indiscriminate attack on whichever edge of the triangular molecule faces the enzyme's active reducing site. This concept would then invoke attack by nitrogenase on the C-C single or C=C double bonds of cyclopropene.

Of similar interest is the reduction of allene (Burns *et al.*, 1975) which leads to propylene. Studies in D_2O reveal 2,3-dideuteropropylene to be the sole product (equation 9), eliminating a mechanism involving prior rearrangement of allene

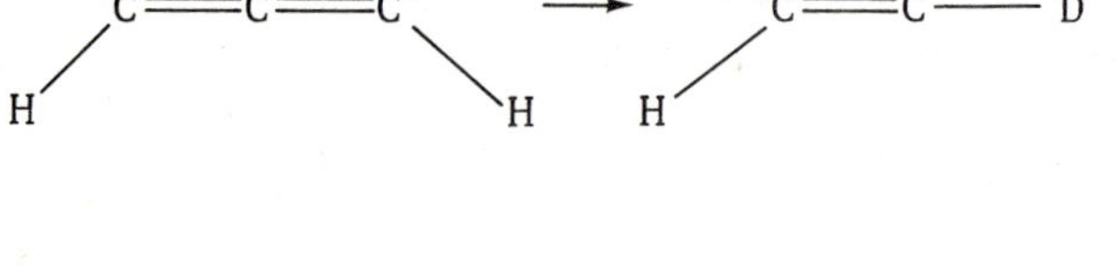

(9)

to the isomeric propyne because nitrogenase-catalyzed reduction of propyne should exclusively form 1,2-dideuteropropylene.

Taken together, the reduction of allene, cyclopropene and hydrazine (Bulen, 1976) show that in addition to triple bonds, nitrogenase also attacks highly reactive double and single bonds *in small molecules*. These observations provide circumstantially supportive evidence that multielectron reductions of triple bonds by nitrogenase can proceed by two-electron steps, *i.e.* that the doubly and singly bonded intermediates formed are likely to be reactive towards additional reduction. We adopt two-electron reduction steps in the discussions which follow.

Hydrogen Evolution and C_2H_2 versus N_2 Reduction

The activated state of N_2ase must be able to deliver two electrons and two protons (or some multiple thereof) to substrate. In view of the reduction potential of which N_2ase operates, it is not surprising that, in the absence of reducible substrate, H_2 is evolved. It is suggested that this H_2 evolution is associated with the substrate-reducing site, assumed for the purpose of this discussion to contain Mo. The intimate nature of the H_2 evolution reaction is unknown and although water is the ultimate source of the protons, metal hydrides, metal dihydrides, ATP, coordinated ligands or water are all possibilities as the intimate source.

Some substrates (like acetylene) completely curtail H_2 evolution while others (such as N_2) are unable to do so. Most data indicate the evolution of one mole of H_2 per mole of N_2 reduced at high levels of N_2 (Hadfield, 1970; Rivera-Ortiz and Burris, 1975). A related problem involves the fact that acetylene is a non-competitive inhibitor of nitrogen fixation, while N_2 is a competitive inhibitor of acetylene reduction (Rivera-Ortiz and Burris, 1975). Several hypotheses have been advanced to explain these facts (Burris and Orme-Johnson, 1976; Davis *et al.*, 1975; Shilov, 1974; Burns *et al.*, 1971; Newton *et al.*, 1976) and these have been enumerated previously (Stiefel *et al.*, 1977). What follows is an attempt to unify these proposals in terms of a single mechanistic scheme. While this scheme is likely to be far removed from intimate reality, it nevertheless correlates many of the existing observations and will hopefully elicit further thought and experimentation.

As shown in Fig. 5, C_2H_2 is assumed to be reduced at the Mo site by a two-electron, two-proton process. The production of *cis*-dideuteroethylene in D_2O (Dilworth, 1967) indicates the stereospecific nature of the hydrogenation reaction. As C_2H_2 reacts solely at this H_2-evolving site, it can at high concentrations totally eliminate H_2 evolution. Following Hardy *et al.*, (1971), we postulate a second metal in the active site (assumed to be Fe) which, although not necessary for C_2H_2 reduction, is essential for dinitrogen fixation.

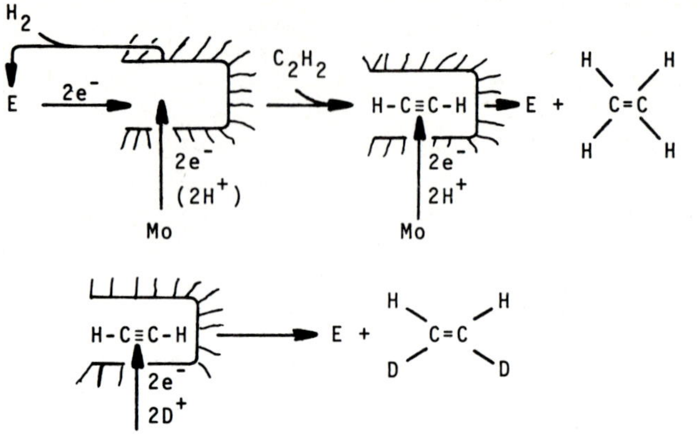

Fig. 5. *Acetylene reduction by nitrogenase illustrating formation of cis-$C_2H_2D_2$ in D_2O.*

While two-electrons (and perhaps two-protons) suffice to activate the Mo site, an additional two electrons and two protons may be required to activate the Fe site. Following Jackson *et al.* (1968), we designate this Fe site as dihydride. Fig. 6 shows the reaction of N_2 with the dihydride site leading to

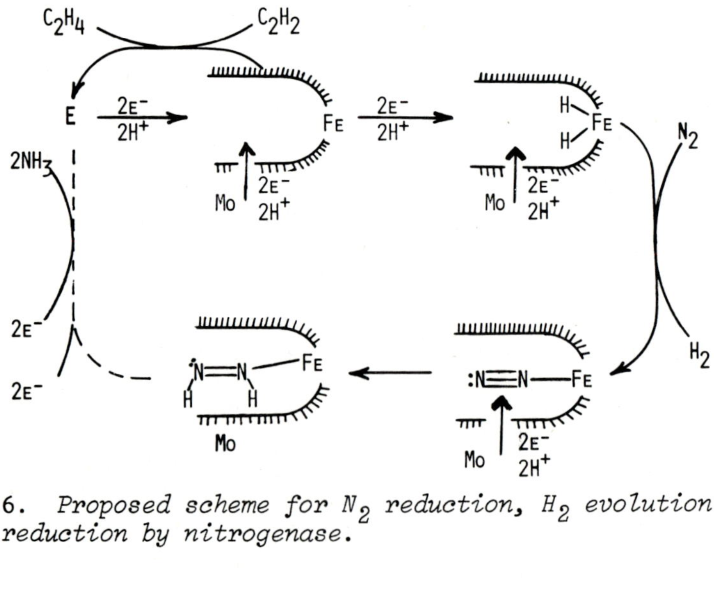

Fig. 6. *Proposed scheme for N_2 reduction, H_2 evolution and C_2H_2 reduction by nitrogenase.*

leading to release of H_2 and end-on binding of the dinitrogen. Such a reaction has direct counterparts in inorganic systems for both Mo and Fe (Stiefel, 1976) and its requirement as an intimate part of nitrogen fixation precisely explains the observed 1:1 stoichiometry of N_2 fixation and H_2 evolution. [The fact that D_2 does not alter H_2 evolution while inhibiting NH_3 production, indicates that H_2 evolution (as part of nitrogen fixation) occurs, as postulated, at a stage of the reaction prior to ammonia formation and prior to or in conjunction with D_2 or H_2 inhibition].

The fact that the postulated initial N_2-reducing site is in effect a $4e^-$ site agrees well with the apparently more stringent demands of N_2 fixation compared to C_2H_2 reduction for [Fe] and/or MgATP (Davis *et al.*, 1975; Smith *et al.*,1976). Furthermore, the nonreciprocity in inhibition type for C_2H_2 reduction is clearly explained. Fig. 6 shows that C_2H_2 reduction only requires the two-electron reduced state and therefore reacts with an earlier state of the enzyme than does N_2. If the Mo site is activated prior to the Fe site, then high concentrations of C_2H_2 could prevent the activation of the Fe site by continually withdrawing electrons from the Mo site. Thus, the $4e^-$ state required to bind and activate N_2 would never be formed. High concentrations of N_2 could not prevent this occurrence and so acetylene is a non-competitive inhibitor of N_2 fixation (*i.e.* V_{max} for N_2 reduction could not be reached in the presence of C_2H_2). On the other hand, high acetylene concentrations could clearly eliminate N_2 inhibition by rapidly emptying the $2e^-$ state, so preventing the N_2-binding site from ever being activated (*i.e.* V_{max} for acetylene reduction could be reached in the presence of N_2) and thus, N_2 is only a competitive inhibitor of C_2H_2 reduction.

The two-metal site hypothesis for N_2 reduction allows the Fe to hold the intermediate N_2H_2 and N_2H_4 complexes while the site is reactivated for the additional two-electron steps required to produce NH_3. Similarly, the CO inhibition of substrate reduction is nicely accommodated by this model.

CO Inhibition

CO inhibits all substrate reactions while allowing H_2 evolution to proceed unimpeded. The fact that the isoelectronic molecules N_2 and CO bind similarly to a variety of metal centres (including Mo and Fe) suggests that CO inhibits by binding to and blocking the N_2-binding site. This site is likely to be an end-on site and therefore, other substrates which have conventional lone pairs (N_3^-, RCN, RNC, N_2O, HCN) could also be denied access to that site (the Fe site in Fig. 6). The substrates which cannot achieve end-on binding (*i.e.* HCCH, allene, cyclopropene) could nevertheless be physically blocked from access to the reducing site. The reducing site, being activated independently of coordination at Fe could then simply discharge its two protons and two electrons to produce H_2, as observed. There is only one problem with this explanation. Thermodynamically, CO is more easily reduced than is N_2 (section II). If it indeed binds to the N_2-binding site, then, why is CO not reduced. One answer might involve the stability of the resultant CO complex. Thus, CO could produce a complex so stable as to resist reduction and/or prevent the conformational change required to bring it to the reducing site. Alternatively, the possible preference of the reducing site of nitrogenase for cis side-on addition to the triple bond (as demonstrated for C_2H_2 reduction, Fig. 5) would produce HCOH from CO reduction rather than the stable isomer formaldehyde (H_2CO). The instability of this hypothetical species may preclude its formation and cause the two-electron reduction of CO to be kinetically unfavourable.

H_2 Inhibition and HD Production

A proposed intimate mechanism for these phenomena is shown in Fig. 7. Electron balance and reaction studies indicate that

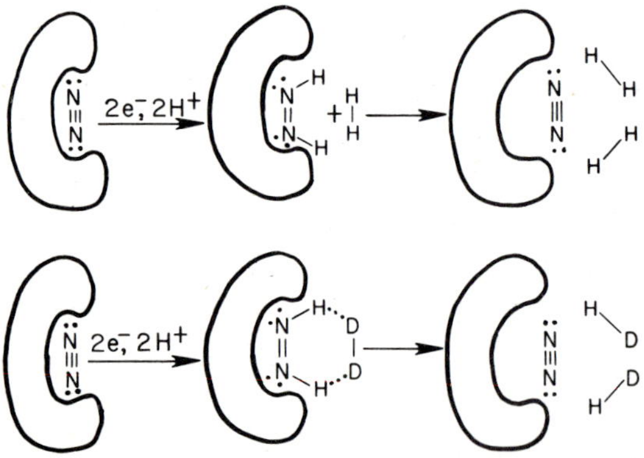

Fig. 7. *Proposed intimate mechanism for H_2 inhibition and HD formation by nitrogenase (Bulen, 1976; Newton* et al., *1976; Stiefel* et al., *in press).*

N_2, $2e^-$ and $2H^+$ are required to produce 2HD molecules ($1e^-$/HD) and therefore a diimide level intermediate has been proposed (Bulen, 1976; Newton *et al.*, 1976; Stiefel *et al.*, in press). In analogy with the production of cis-dideuteroethylene from acetylene, a cis-diimide species is suggested. Such a species could react with H_2 to generate 2 moles of H_2 and reform N_2

thereby inhibiting N_2 reduction and diverting electrons from NH_3 formation to H_2 production in agreement with the stoichiometric mechanism. Similarly, the reaction with D_2 diverts electrons from NH_3 formation to the production of HD with one electron required per HD formed in agreement with the electron balance studies. The reaction pictured is, in effect, the H_2-catalyzed decomposition of bound diimide to the elements, which is known to be exergonic (-50 Kcal/mole) for free diimide and is unlikely to be modified significantly by binding. The reaction shown is also allowed by orbital symmetry considerations (Stiefel, unpublished).

The implication of bound-N_2H_2 as a likely intermediate and additional evidence concerning hydrazine reduction (Bulen, 1976; Newton *et al.*, these proceedings) indicate that nitrogenase functions by a series of two-electron, two-proton steps, which encumbers the nitrogenase active site with the same stoichiometric reaction that the Mo-site is known to catalyze in other Mo enzymes.

The Activation and Reduction of Substrate

A number of proposals have recently been made for the mechanism of substrate activation by nitrogenase. In each case, although the proposal is based upon considerations of known coordination chemistry, there is no data which either confirm or deny its relevance with respect to nitrogenase.

Activation by Oxidative Addition Oxidative addition reactions of multiply bonded molecules (related to nitrogenase substrates and/or postulated intermediates) have been demonstrated on both Mo(II) and Mo(IV) centres. For example, acetylene was found to react with a Mo(II)-dithiocarbamate complex (McDonald *et al.*, 1975) (equation (10)) or with molybdocene generated photochemically from $Mo(C_5H_5)_2CO$ (Wong *et al.*, 1974), according to equation (11).

$$Mo(CO)_2(R_2dtc)_2 + C_2H_2 \longrightarrow Mo(C_2H_2)(CO)(R_2dtc)_2 + CO \quad (10)$$

$$Mo(C_5H_5)_2 + C_2H_2 \longrightarrow Mo(C_5H_5)_2C_2H_2 \quad (11)$$

In Mo(IV) systems, the complex, $MoO(R_2dtc)_2$, oxidatively adds activated acetylenes (e.g. acetylene dicarboxylate esters) and substituted diazenes (e.g. dimethyl azodicarboxylate). The latter upon treatment with proton donors yield the corresponding hydrazines (Schneider *et al.*, 1972). In a related study, it was shown (Newton *et al.*, 1974) that the Mo(IV) monomer, $MoO(R_2dtc)_2$, obtains upon disproportionation of a Mo(V) dimer, $Mo_2O_3(R_2dtc)_4$, thus providing a potential mechanism for a dimeric Mo site in the enzyme to channel reducing power to a single Mo atom.

For each of the above complexes, there is convincing evidence for side-on binding of the substrate (Stiefel, 1976). Extrapolation of these findings to nitrogenase invokes side-on binding of substrate followed by protonation and dissociation (Newton *et al.*, 1976; Schrauzer, 1976).

Stereochemical Control of Valence Enemark and Feltham (1972) used structural results on two nitrosyl complexes as a basis for the idea that an internal redox could be triggered by changing the other ligands in the coordination sphere. Specifically, reaction (12) occurs wherein coordination by thiocyanate causes bound NO to shift from linear to bent coordination. Since linear coordination is associated with NO^+ while bent coordination implicates bound NO^-, simple ligand addition and stereochemical change have caused an internal two-electron transfer from metal to ligand with cobalt changing from the I to the III

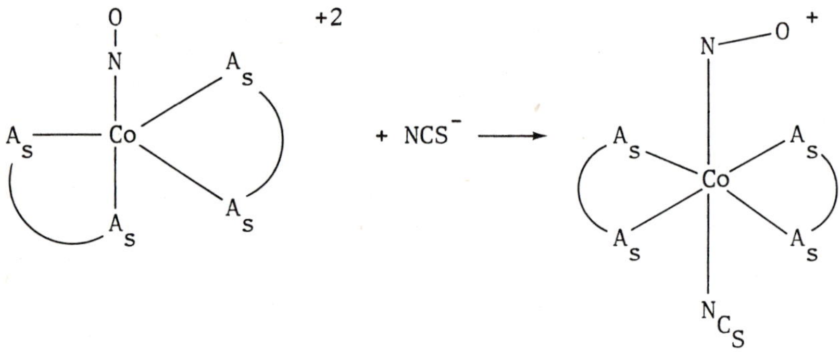

(12)

oxidation state. The additional ligand and stereochemical change are thought to raise the average d energy level and so facilitate electron transfer from the metal d level to the ligand π^* level of NO^+.

Because N_2 is isoelectronic to NO^+ and N_2^{-2} (or N_2H_2) is isoelectronic to NO^-, Enemark and Feltham (1972) suggested that so-called stereochemical control of valence could likewise play a role in dinitrogen reduction. Unlike the nitrosyl case, where the π^* level is similar in energy to the metal d-orbitals, in N_2, the π^* level lies significantly above the d-levels such that simple electron transfer is not possible. Rather, protonation of N_2 is required to accompany the electron transfer in order to stabilize the reduced nitrogen product. This effect is illustrated by reaction (13), discovered by Chatt *et al.*,

1972, which shows that replacement of a π-accepting N_2 ligand by two π-donating bromides facilitates internal electron transfer from Mo(0) to N_2 in conjunction with protonation. While

$$\text{trans-Mo}(N_2)_2(\text{dppe})_2 + 2\text{HBr} \longrightarrow \text{Mo}(N_2H_2)\text{Br}_2(\text{dppe})_2 + N_2 \qquad (13)$$

Feltham and Enemark stressed the importance of the stereochemical change, the subsequent work of Chatt *et al.* (1975, 1976) showed that of equal or greater importance is the change in donor atom set.

The Donor-Atom Set Chatt and co-workers (Chatt, 1975; Chatt *et al.*, 1975, 1976) have shown that N_2 bound to Mo(0) or W(0) complexes can be reduced all the way to NH_3 in conjunction with replacement of phosphine ligands by O-donor ligands in the metal coordination sphere. Clearly, lower valent complexes with O ligands have a more strongly reducing nature than those complexes having phosphine donors. The π-donating O atoms raise the d energy levels to the point where in conjunction with N_2 protonation they can donate electrons into the N_2-π^* levels. Chatt *et al.* (1976) therefore stress the importance of O donors in all N_2-reducing systems and opine that O-donors could be ligands for Mo in nitrogenase. It is clear from thermodynamic data that complexes containing Mo-O donors in a given oxidation state are more powerful reductants than similar complexes containing Mo-S. Nonetheless, recent x-ray absorption spectroscopy studies reveal a significant number of S atoms (3 or 4) as being bound to Mo in nitrogenase. If these results are confirmed, it will be interesting to see whether they might be reconciled with the importance of O atoms as discussed by Chatt. An interesting possibility would involve a ligand substitution in nitrogenase which changes the relative redox potentials of the metal and its nitrogen ligand to affect electron transfer. It seems possible that such a process could occur in nitrogenase where replacement of sulfur by oxygen ligands could occur. This transformation presents yet another possibility for the role of ATP in nitrogenase.

Coupled Proton-Electron Transfer Mechanism The effect of metal oxidation state on the pKa of its coordinated ligands has been discussed previously (Stiefel, 1973; Stiefel and Gardner, 1974). While the oxidation state of Mo in nitrogenase is unknown at present it is nevertheless likely that this state changes during turnover. For the sake of discussion, we assume a single Mo at the active site capable of existing in the VI and/or IV oxidation states. These states are chosen for their EPR-inactive nature but a Mo(II)-Mo(IV) couple would fit equally well and would not affect the arguments presented below.

In the higher oxidation state of Mo, coordinated ligands will be deprotonated, while these same ligands may be protonated in the lower oxidation state (other features of the coordination sphere remaining unchanged). The activation of nitrogenase towards turnover then involves the generation of the reduced-protonated form. Since the active site has sufficient reducing potential to produce H_2 from protons and as protons are already present in the active site, it is not surprising that this site produces H_2 when other reducible substrate is not present.

If we envision trans protonated donor atoms on Mo, then a reaction sequence as shown in Fig. 8 could be involved in the

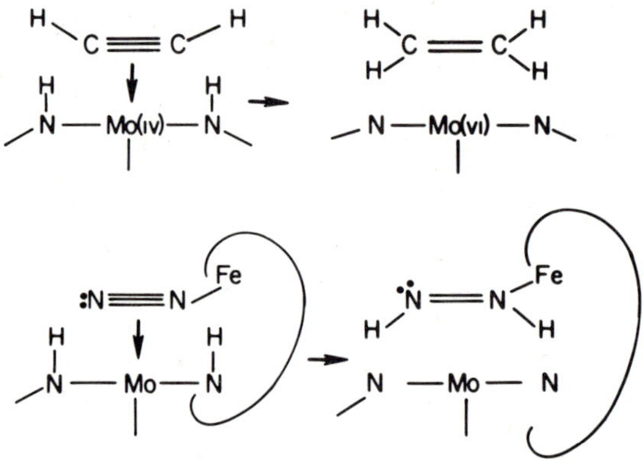

Fig. 8. *Proposed coupled proton-electron transfer mechanisms for a) C_2H_2 and b) N_2 reduction by nitrogenase (Stiefel, 1973; Stiefel and Gardner, 1947).*

intimate mechanism of C_2H_2 reduction by N_2ase. As two electrons transfer from Mo to C_2H_2, its oxidation state increases from IV to VI, causing the protons on the coordinated ligands to become highly acidic and transfer as well. The known cis stereochemistry of the transfer is assured by the location of the proton source in defined close juxtaposition to the electron source. The ability of the enzyme to control proton flow is clearly in concert with the ability of numerous hydrolytic enzymes to manipulate protons.

The two-site proposal discussed above is invoked for N_2 reduction with a postulated Fe atom as the N_2-binding site. This site would be spacially related to the Mo site in a manner which allows end-on binding at the Fe to cause side-on (or near side-on) binding to the Mo site. Then, a single turnover (a two-electron, two-proton transfer) can occur analogously to acetylene reduction to produce bound diimide. Dissociation of diimide from the Mo site could occur allowing that site to be reactivated while N_2H_2 remains coordinated to the Fe site. After reactivation of the Mo site, the bound diimide could be reduced to hydrazine by the coupled proton-electron transfer. Finally, with N_2H_4 remaining bound to Fe, a final reactivation occurs allowing the N_2H_4 to NH_3 conversion to take place. The dissociation and reassociation of the Mo-N_2 and Mo-intermediate complexes may require small or large conformational changes which could involve ATP participation. The coupled electron-proton transfer mechanism has the appealing characteristic of being similar to that suggested for xanthine oxidase and other Mo enzymes.

Which, if any, of these mechanisms is involved in nitrogenase action is at present a matter of conjecture. In fact, the proposals discussed above overlap to an extent and it is possible that each has grasped some part of the truth. It is hoped that further cooperation among inorganic chemists and biochemists will allow for elucidation of the intimate chemical mechanism of biological nitrogen fixation and encourage its ever closer duplication by simpler chemical systems.

ACKNOWLEDGEMENTS

I am grateful to Drs J. L. Corbin, J. W. McDonald, W. E. Newton, N. Pariyadath and G. D. Watt for many helpful discussions.

REFERENCES

Alberty, R.A. (1968). *J.Biol.Chem.*, 243: 1337-1343.

Benneman, J.R., McKenna, C.E., Lie, R.F., Traylor, T.G. and Kamen, M.D. (1972). *Biochim.Biophys.Acta.*, 264: 25-38.

Benneman, J.R., Smith, G.M., Kostel, R.J. and McKenna, C.E. (1973). *FEBS Letters*, 29: 219-221.

Borodko, Y.G., Ivleva, I.N., Kachapina, Salienko, S.I., Shilova, A.K. and Shilov, A.E. (1972). *Chem.Commun.*, 1178.

Bray, R.C. and Swann, J.C. (1972). *Structure and Bonding*, 14: 107-144.

Bray, R.C. (1975). *In:* The Enzymes (P.D. Boyer, ed.), 3rd ed., Academic Press, New York.

Brintzinger, H.H. (1976). *In:* Proceedings of the First International Conference on Nitrogen Fixation (W.E. Newton and C.J. Nyman, eds), pp. 33-41, Washington State University Press, Pullman, Washington.

Bulen, W.E. (1976). *In:* Proceedings of the First International Conference on Nitrogen Fixation (W.E. Newton and C.J.

Nyman, eds), pp. 177-186. Washington State University Press, Pullman, Washington.
Burns, R.C., Hardy, R.W.F. and Phillips, W.D. (1975). *In:* Nitrogen Fixation by Free-Living Micro-organisms (W.D.P. Stewart, ed.), pp. 447-452. Cambridge University Press.
Burns, R.C. and Hardy, R.W.F. (1975). *In:* Nitrogen Fixation in Bacteria and Higher Plants. Springer Verlag, Berlin.
Burns, R.C., Stasny, J.T. and Hardy, R.W.F. (1976). *In:* Proceedings of the First International Conference on Nitrogen Fixation (W.E. Newton and C.J. Nyman, eds), pp. 187-195. Washington State University Press, Pullman, Washington.
Burns, R.C., Stasny, J.T. and Hardy, R.W.F. (1976). *In:* Proceedings of the First International Conference on Nitrogen Fixation (W.E. Newton and C.J. Nyman, eds), pp. 196-207. Washington State University Press, Pullman, Washington.
Burris, R.H. and Orme-Johnson, W.H. (1976). *In:* Proceedings of the First International Conference on Nitrogen Fixation (W.E. Newton and C.J. Nyman, eds), pp. 208-233. Washington State University Press, Pullman, Washington.
Cammack, R. (1973). *Biochem.Biophys.Res.Commun.*, 54: 548-554.
Cammack, R., Barber, M.J. and Bray, R.C. (1976). *Biochem.J.*, 157: 469-478.
Chatt, J., Heath, G.A. and Richards, R.L. (1972). *J.Chem.Soc. D.*, 1010-1011.
Chatt, J. (1975). *J.Organometal.Chem.*, 100: 17-28.
Chatt, J., Pearman, A.J. and Richards, R.L. (1975). *Nature*, 253: 39-40.
Chatt, J., Pearman, A.J. and Richards, R.L. (1976). *Nature*, 259: 204.
Corbin, J.L., Pariyadath, N. and Stiefel, E.I. (1976). *J.Am. Chem.Soc.*, submitted for publication.
Cramer, S.P., Eccles, T.K. Kutzler, F., Hodgson, K.O. and Mortenson, L.E. (1976). *J.Amer.Chem.Soc.*, 98: 1287-1288.
Dalton, H. and Whittenbury, R. (1976). *Arch.Microbiol.*, 109: 147-151.
Davis, L.C., Shah, V.K. and Brill, W.J. (1975). *Biochim.Biophys.Acta*, 403: 67-78.
Dilworth, M.J. (1966). *Biochim.Biophys.Acta*, 127: 285-294.
Dixon, R.O.D. (1972). *Arch.Mikrobiol.*, 85: 193-201.
Eady, R.R. (1973). *Biochem.J.*, 135: 531-536.
Edmondson, D., Ballou, D., van Heuvelen, A., Palmer, G. and Massey, V. (1973). *J.Biol.Chem.*, 248: 6135-6144.
Enemark, J.H. and Feltham, R.D. (1972). *Proc.Nat.Acad.Sci. USA*, 69: 3534-3536.
Evans, M.C.W. and Albrecht, S.L. (1973). *Biochem.Biophys.Res. Commun.*, 61: 1187-1192.
Garcia-Rivera, J. and Burris, R.H. (1967). *Arch.Biochem.Biophys.*, 119: 167-172.

Hadfield, L. (1970). Ph.D. Thesis, Brigham Young University.
Hardy, R.W.F., Burns, R.C. and Parshall, G.W. (1971). *In:* Bioinorganic Chemistry, Advances in Chemistry series 100, pp. 219-247. American Chemical Society Publications, Washington, D.C.
Holm, R.H. (1975). *Endeavour*, 34: 38-42.
Holm, R.H. (1976). *In:* Proceedings of the First International Conference on Nitrogen Fixation (W.E. Newton and C.J. Nyman, eds), pp. 75-78. Washington State University Press, Pullman, Washington.
Howie, J.K. and Sawyer, D.T. (1976). *Inorg.Chem.*, 15: 1892-1899.
Hwang, J.C. and Burris, R.H. (1972). *Biochim.Biophys.Acta*, 283: 339-350.
Ichikawa, M. and Meshitsuka, S. (1973). *J.Amer.Chem.Soc.*, 95: 3411-3412.
Jackson, E.K., Parshall, G.W. and Hardy, R.W.F. (1968). *J.Biol. Chem.*, 243: 4952-4958.
Kepert, D.L. (1972). Early Transition Metals. Academic Press, New York.
Kloubek, J. and Podlaha, J. (1971). *J.Inorg.Nucl.Chem.*, 33: 2981-2988.
Krushch, A.P., Shilov, A.E. and Vorontsova, T.A. (1975). *J. Amer.Chem.Soc.*, 96: 4987-4989.
Larson, A.T. and Brooks, A.P. (1926). *Ind.Eng.Chem.*, 18: 1305-1307.
Lattimer, W. (1952). Oxidation Potentials, 2nd ed. Prentice Hall, Inc., Englewood Cliffs, New Jersey.
Ledwith, D.A. and Schultz, F.A. (1975). *J.Amer.Chem.Soc.*, 97: 6591-6593.
Leigh, G.J. (1971). *In:* The Chemistry and Biochemistry of Nitrogen Fixation (J.R. Postgate, ed.), pp. 19-56, Plenum Press, London.
Manriquez, J.M. and Bercaw, J.E. (1974). *J.Amer.Chem.Soc.*, 96: 6229-6230.
Manriquez, J.M., Sanner, R.D., Marsh, R.E. and Bercaw, J.E. (1976). *J.Amer.Chem.Soc.*, 98: 3042-3044.
Massey, V., Komai, G., Palmer, G. and Elion, G.B. (1970). *J. Biol.Chem.*, 245: 2837-2844.
Massey, V. (1973). *In:* Iron-Sulfur Proteins (W. Lovenberg, ed.) Vol. I. Academic Press, New York.
McDonald, J.W., Corbin, J.L. and Newton, W.E. (1975). *J.Amer. Chem.Soc.*, 96: 1970-1971.
McKenna, C.E., McKenna, M.C. and Higa, M.T. (1976). *J.Amer. Chem.Soc.*, 98: 4657-4659.
Mitchell, P. (1976). *Biochem.Soc.Trans.*, 4: 399-430.
Mortenson, L.E., Walker, M.N. and Walker, G.A. (1976). *In:* Proceedings of the First International Conference on Nitrogen Fixation (W.E. Newton and C.J. Nyman, eds), pp. 117-149.

Washington State University Press, Pullman, Washington.
Newton, W.E., Corbin, J.L., Bravard, D.C., Searles, J.C. and McDonald, J.W. (1974). *Inorg.Chem.*, 13: 1100-1104.
Newton, W.E., Corbin, J.L. and McDonald, J.W. (1976). *In:* Proceedings of the First International Conference on Nitrogen Fixation (W.E. Newton and C.J. Nyman, eds), pp. 53-74. Washington University Press, Pullman, Washington.
Newton, W.E. and Nyman, C.J. (1976). *In:* Proceedings of the First International Conference on Nitrogen Fixation. Washington State University Press, Pullman, Washington.
Nikonova, L.A., Pershikova, N.I., Bodeiko, M.V., Olijnik, L.G., Sokolov, D.N. and Shilov, A.E. (1974). *Dokl.Akad.Nauk SSSR*, 216: 140-141.
Olson, J.S., Ballou, D.P., Palmer, G. and Massey, V. (1974). *J.Biol.Chem.*, 249: 4363-4382.
Orme-Johnson, W.H., Hamilton, W.D., Ljones, T., Tso, M.-Y., Burris, R.H., Shah, V.K. and Brill, W.J. (1972). *Proc. Nat.Acad.Sci. USA*, 69: 3142-3145.
Orme-Johnson, W.H. (1973). *Ann.Rev.Biochem.*, 42: 159-204.
Orme-Johnson, W.H., Jacob, G., Menzl, M. and Averill, B.A. (1976). Presented at 100th National Meeting of American Chemical Society, New York. *In:* Recent Advances in Bioinorganic Chemistry (K.N. Raymond, ed.), in press. ACS Advances in Chemistry Series. American Chemical Society Publication, Washington, D.C.
Ott, V.R. and Schultz, F.A. (1975). *J.Electroanal.Chem.*, 61: 81-98.
Pan, S.-S., Erickson, R.H., Lee, K.-Y and Nason, A. (1976). *In:* Proceedings of the First International Conference on Nitrogen Fixation (W.E. Newton and C.J. Nyman, eds), pp. 293-311. Washington State University Press, Pullman, Washington.
Pariyadath, N., Newton, W.E. and Stiefel, E.I. (1976). *J.Amer. Chem.Soc.*, 98: 5388-5390.
Peterson, R.B. and Burris, R.H. (1976). *Anal.Biochem.*, 73: 404-410.
Phillips, R.C., George, P. and Rutman, R.J. (1969). *J.Biol. Chem.*, 244: 3330-3342.
Rivera-Ortiz, J.M. and Burris, R.H. (1975). *J.Bacteriol.*, 123: 537-545.
Schneider, P.W., Bravard, D.C., McDonald, J.W. and Newton, W. E. (1972). *J.Amer.Chem.Soc.*, 94: 8640-8641.
Schrauzer, G.N., Kiefer, G.W., Tano, K. and Robinson, P.R. (1975). *J.Amer.Chem.Soc.*, 97: 6088-6094.
Schrauzer, G.N. (1976). *In:* Proceedings of the First International Conference on Nitrogen Fixation (W.E. Newton and C.J. Nyman, eds), pp. 79-116, Washington State University, Press, Pullman, Washington.

Schubert, K. and Evans, H.L. (1976). *Proc.Nat.Acad.Sci.USA*, 73: 1207-1211.
Sellmann, D. (1974). *Angew.Chem.internat.Edit.*, 13: 639-649.
Sharrock, M., Debrunner, P.G., Schulz, C., Lipscomb, J.D., Marshall, V. and Gunsalus, I.C. (1976). *Biochim.Biophys. Acta*, 420: 8-26.
Shilov, A.E. (1974). *Russ.Chem.Rev.*, 43: 378-398.
Shilov, A.E. (1976). *In:* Proceedings of the First International Conference on Nitrogen Fixation (W.E. Newton and C.J. Nyman, eds), pp. 42-52. Washington State University Press, Pullman, Washington.
Silverstein, R. and Bulen, W.A. (1970). *Biochemistry*, 9: 3809-3815.
Sligar, S.G. and Gunsalus, I.C. (1976). *Proc.Nat.Acad.Sci.USA*, 73: 1078-1082.
Smith, B.E., Lowe, D.J. and Bray, R.C. (1973). *Biochem.J.*, 135: 331-341.
Smith, B.E., Thorneley, R.N.F., Eady, R.R. and Mortenson, L.E. (1976). *Biochem.J.*, 157: 439-447.
Smith, B.E., Thorneley, R.N.F., Yates, M.G., Eady, R.R. and Postgate, J.R. (1976). *In:* Proceedings of the First International Conference on Nitrogen Fixation (W.E. Newton and C.J. Nyman, eds), pp. 150-176. Washington State University Press, Pullman, Washington.
Smith, L.A., Hill, S. and Yates, M.G. (1976). *Nature*, 262: 209-211.
Stasny, J.T., Burns, R.C., Korant, B.D. and Hardy, R.W.F. (1974). *J.Cell Biol.*, 60: 311-316.
Stiefel, E.I. (1973). *Proc.Nat.Acad.Sci.USA*, 70: 988-992.
Stiefel, E.I. and Gardner, J.K. (1974). *In:* Proceedings of the First International Conference on Chemistry and Uses of Mo (P.C.H. Mitchell, ed.), pp. 272-277. Climax Molybdenum Corporation, London. *J.Less Common Metal*, 36: 521-533.
Stiefel, E.I. (1976). *Progr.Inorg.Chem.*, 22: 1-223.
Stiefel, E.I., Newton, W.E., Watt, G.D., Hadifield, K.L. and Bulen, W.A. (1977). *In:* Recent Advances in Bioinorganic Chemistry (K.N. Raymond, ed.), in press. ACS Advances in Chemistry Series, American Chemical Society Publications, Washington, D.C.
Suzuki, H., Meshitsuka, S., Takashima, T. and Ichikawa, M. (1975). *Chem.Letters*, 285-288.
Teuben, J.H. (1973). *J.Organometal.Chem.*, 57: 159-167.
Thorneley, R.N.F. and Willison, K.R. (1974). *Biochem.J.*, 139: 211-214.
Thorneley, R.N.F., Eady, R.R. and Yates, M.G. (1975). *Biochim. Biophys.Acta*, 403: 269-284.
Thorneley, R.N.F., Yates, M.G. and Lowe, D.J. (1976). *Biochem. J.*, 155: 137-144.

Tso, M.-Y. and Burris, R.H. (1973). *Biochim.Biophys.Acta,* 309: 263-270.

Vancini, C.A. (1971). Synthesis of Ammonia. MacMillan Press Ltd., London.

Van Tamelen, E.E., Seely, D., Schneller, S., Rudler, H. and Cretney, W. (1970). *J.Amer.Chem.Soc.,* 92: 3231-3233.

Van Tamelen, E.E. (1971). *In:* Bioinorganic Chemistry (R.F. Gould, ed.), ACS Advances in Chemistry Series, 100, pp. 95-110. American Chemical Society Publications, Washington, D.C.

Volpin, M.E. and Shur, V.B. (1964). *Dokl.Acad.Nauk SSSR,* 156: 1102-1104.

Walker, G.A. and Mortenson, L.E. (1974). *Biochemistry,* 13: 2382-2388.

Watt, G.D., Bulen, W.A., Burns, A. and Hadfield, K.L. (1975). *Biochemistry,* 14: 4266-4272.

Watt, G.D. and Bulen, W.A. (1976). *In:* Proceedings of the First International Conference on Nitrogen Fixation (W.E. Newton and C.J. Nyman, eds), pp. 248-256. Washington State University Press, Pullman, Washington.

Watt, G.D., McDonald, J.W. and Newton, W.E. (1976). Presented at Second International Conference on Chemistry and Uses of Molybdenum, Oxford, England.

Watt, G.D. and Burns, A. (1976). *Biochemistry,* in press.

Werner, D., Russell, S.A. and Evans, H.J. (1973). *Proc.Nat.Acad. Sci.USA,* 70: 339-342.

Wong, K.L.T., Thomas, J.L. and Brintzinger, H.H. (1974). *J. Amer.Chem.Soc.,* 96: 3694-3695.

Willis, C., Lossing, F.P. and Back, R.A. (1976). *Can.J.Chem.,* 34: 1-3.

Zumft, W.G., Palmer, G. and Mortenson, L.E. (1973). *Biochim. Biophys.Acta,* 292: 413-421.

NITROGENASE MODEL SYSTEMS AND THE MECHANISM OF BIOLOGICAL NITROGEN REDUCTION: ADVANCES SINCE 1974

G. N. SCHRAUZER

Department of Chemistry, University of California at San Diego, Revelle College, La Jolla, California, U.S.A.

INTRODUCTION

At the First Symposium, plausible mechanisms for the reduction of all nitrogenase substrates were formulated on the basis of similar reactions observed in non-enzymic model systems. The reduction of the substrates was shown to occur in reactions catalyzed by mononuclear thiol complexes of oxomolybdate. The non-haem iron components in the enzyme were suggested to act as catalysts of electron transfer and nitrogen was shown to be reduced to ammonia *via* diimide (N_2H_2) and hydrazine. It was observed, however, that diimide as such is not reduced and that it disproportionates and decomposes under the reaction conditions. The stoichiometry of N_2 reduction by nitrogenase was expressed in terms of reaction (1) (Schrauzer, 1974, 1975, 1976; Schrauzer *et al.*, 1976a). According to reaction (1), H_2 is an

$$N_2 + 8e^- + 8H^+ \longrightarrow 2NH_3 + H_2 \qquad (1)$$

obligatory by-product of the reduction of N_2 to ammonia and nitrogenase utilizes only 75% of the electrons for N_2 reduction, the remaining 25% are apparently diverted to produce H_2. Because H_2 is formed even at saturating pressures of N_2 but not with acetylene as the substrate, this phenomenon is difficult to explain unless diimide is assumed to be an intermediate of N_2 reduction. The same is true for the interpretation of the H_2-D^+ exchange reaction by nitrogenase, which is promoted only if molecular nitrogen is present. Although the details of this process are not yet fully understood, we believe that diimide decomposition into the elements during the enzymic reaction is the underlying cause. It is also noteworthy that the stimulatory effect of N_2 on H_2-D^+ exchange has been observed in studies with the non-enzymic model systems (Schrauzer, 1976; Schrauzer *et al.*, 1976a).

The kinetic evidence which has been accumulated over the years indicates that the catalytically active species in the molybdothiol model systems are mononuclear complexes of oxomolybdates. Since the oxidation state of molybdenum in the reduced forms of the catalyst was not quite certain, we first decided to designate them simply as "Mo^{red}". However, later studies indicated that the active reduced forms showed a behaviour consistent with Mo(IV) with the structures A or B. Oxidized forms of the catalytic species in the molybdothiol model systems were referred to as "Mo^{ox}", a plausible structure for such a complex with cysteine as the ligand is depicted in formula C.

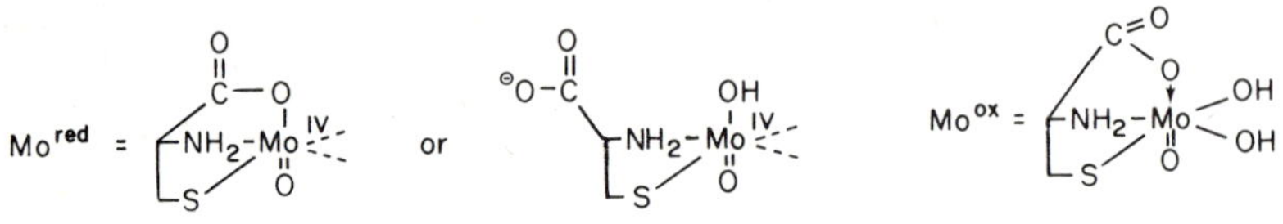

A B C

The assumed structures, A and B, for Mo^{red} in the molybdocysteine systems is supported by the fact that cysteine acts as a tridentate ligand in the complexes with oxomolybdate (V). Systematic studies revealed that catalytic activity in the reduction of typical substrates reaches a maximum at molybdenum: thiol ratios of 1:1, indicating that the catalytically active complexes contain one molecule of thiol ligand per molybdenum. Mo^{red} must contain at least two *cis*-positions available for interaction with the substrates. From the observed stereochemical course of the reduction of alkynes, it was concluded that these substrates interact with Mo^{red} in the side-on fashion. The reduction of acetylene to ethylene was accordingly formulated as shown in eq. (2). By analogy, nitrogen was assumed to

$$Mo^{red} + C_2H_2 \longrightarrow Mo\begin{smallmatrix}CH\\ \| \\ CH\end{smallmatrix} \xrightarrow[(H_2O)]{+2H^+} Mo^{ox} + C_2H_4 \qquad (2)$$

react with Mo^{red} in the side-on fashion as well and since Mo^{red} showed the behaviour typical of two-electron reductant, *cis*-diimide was postulated to be the first product of nitrogen reduction (eq. (3)). Other substrates which appear to interact

$$Mo^{red} + N_2 \longrightarrow Mo\begin{smallmatrix}N\\ \| \\ N\end{smallmatrix} \xrightarrow[(H_2O)]{+2H^+} Mo^{ox} + N_2H_2 \qquad (3)$$

with Mo^{red} in the side-on fashion include saturated and unsaturated nitriles, but CN^-, N_3^-, RNC (= isonitriles), N_2O and N_2H_4 were assumed to form intermediate "end-on" complexes. The reduction of these substrates need not be described at this point (for details see Schrauzer, 1974, 1975, and original papers cited therein).

During the past two years, we have directed our attention to a number of, as yet, unresolved problems. A brief progress report of this work follows.

THE ATP EFFECT

The addition of ATP and of other nucleoside phosphates to functional molybdothiol systems causes a stimulation of the reduction of most substrates. The effect was investigated in greater detail in the reduction of CN (Schrauzer *et al.*, 1973). We concluded that ATP forms a 2:1 complex with Mo^{ox} on the basis of the observed dependence of the rate of substrate reduction on $[ATP]^2$ and that the resulting complex is protonated. Other authors in turn suggested that ATP has no specific effect in our systems and that the stimulation of substrate reduction is merely due to the pH change which occurs upon the addition of ATP to the reaction mixture Krusch *et al.*, 1974). To settle this argument, we therefore performed a study in which the effects of pH and of 5 different acid anions on the reduction of isonitriles, cyanide, acrylonitrile and acetylene were carefully separated (Schrauzer *et al.*, 1975a). In this manner, we were able to demonstrate that the reduction of the molybdenum catalysts is clearly and significantly *anion assisted.* Of the acids studied, phosphoric acid showed the strongest stimulatory effects which were attributed to the high affinity of phosphate for oxomolybdate. ATP was most effective of all the nucleoside phosphates and acids studied. We noted, however, that only one rather than two ATP molecules interacts with the molybdenum catalyst because the rates of substrate reduction showed a linear dependence on $[ATP].[H^+]$. In the course of our work, we also rediscovered the important fact that *molybdate is a specific catalyst of ATP hydrolysis to ADP and inorganic phosphate.* The acceleration of ATP hydrolysis by traces of molybdate has actually been known for years and was studied particularly by Weil-Malherbe and Green (1951). We accordingly postulated that ATP forms a protonated complex with the oxidized molybdenum catalyst *via* the terminal phosphate group, leading to ATP hydrolysis and the generation of a reactive molybdenum species which is more rapidly reduced than the parent complex. The process is schematically represented in eq. (4). Our views on the nature

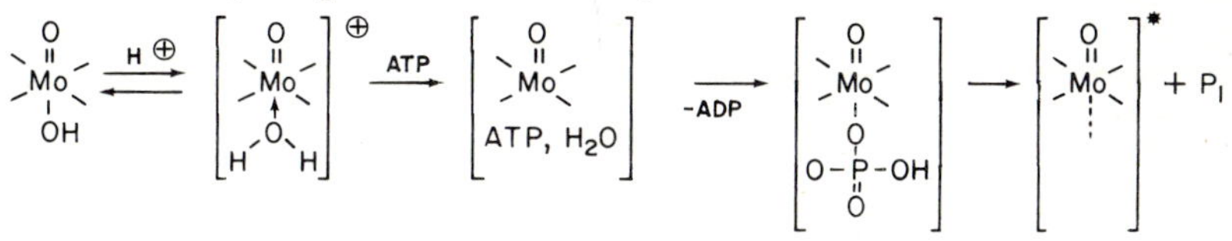

(4)

of the ATP effect are thus similar to the original proposal by Hardy and Knight (1967) and Parshall (1967). The ATP consumption by functional nitrogenase holoenzyme is therefore linked to the molybdenum active site. Last, but not least, it may be assumed that some of the energy released during ATP hydrolysis is transferred to the molybdenum active site.

ACETYLENE REDUCTION IN THE PRESENCE OF FERREDOXIN MODEL COMPOUNDS

Ferredoxin model compounds, *i.e.* anions of the type $[Fe_4S_4(SR)_4]^{-2}$ are efficient catalysts of the transfer of electrons from external reductant to molybdenum. Clear evidence for stimulation of the molybdenum-dependent reduction of acetylene to ethylene was obtained when catalytic amounts of the salt $[\overset{+}{N}\text{-}(nC_4H_9)_4]_2\ [Fe_4S_4(S_4\text{-}nC_3H_7)_4]^{=}$ were added to a solution of Complex I (= the binuclear oxobridged complex of L(+)-cysteine with Mo(V)) in the presence of $S_2O_4^{=}$ as the reducing agent (eq. (5)). Reduced ferredoxin model compounds have also been em-

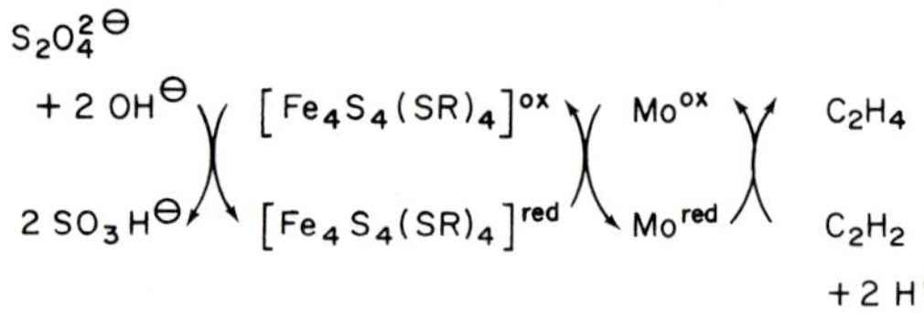

(5)

ployed as stoichiometric reductants in molybdothiol catalyst systems. The new result confirms the concept that the non-haem iron constituents of nitrogenase represent an electron-storage and -transfer system. The effects of the components of the ferredoxin model compounds, *i.e.* of $S^{=}$, Fe^{+2} and RS^{-}, were also investigated individually. It was observed that the maximum rates of C_2H_2 reduction occur at the Fe:S:RS^{-} ratio of 1:1:1 (Tano and Schrauzer, 1975).

HYDROGEN EVOLUTION

Solutions of Complex I evolve H_2 upon addition of stoichiometric amounts of reduced ferredoxin model compounds. We assume that H_2 is produced *via* hydridomolybdenum species according to pathways outlined in eq. (6). ATP stimulation of H_2 evolution

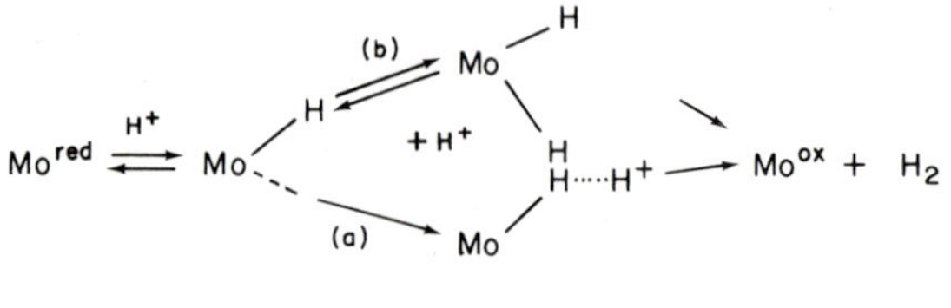

(6)

has not yet been clearly demonstrated in the model systems, presumably due to the instability of the cluster anions in aqueous solutions (Tano and Schrauzer, 1975).

REDUCTION OF COORDINATED CN^-

Functional nitrogenase holoenzyme reduces caynide to NH_3, CH_4, C_2H_4, C_2H_6 and CH_3NH_2. In a previous investigation of the reduction of this substrate in the molybdothiol model system (Schrauzer *et al.*, 1973), the formation of the end-on complexes of Mo^{red} with anions of cyanide was assumed to occur in the initial steps of the reaction. The well-defined cyanocomplexes of oxomolybdate (IV), whose structures have been determined (Day and Hoard, 1968; Robinson *et al.*, 1975), may be regarded as models of the reactive intermediates in the molybdenum-catalyzed reduction of cyanide. The three known complex anions D-F are connected by the protonation-deprotonation equilibria shown in eq. (7). Of the three complex anions, F reacts with reducing

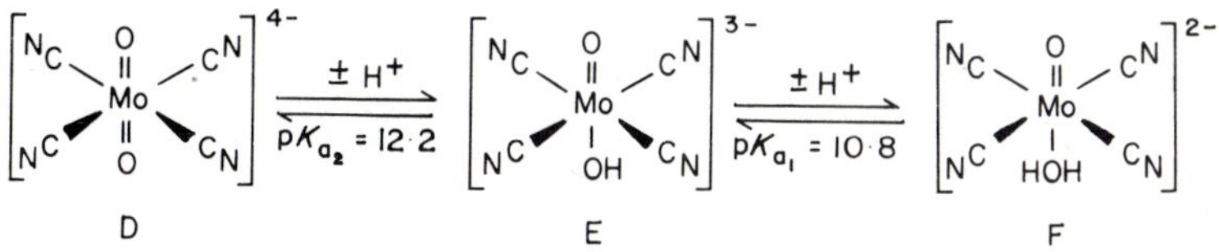

(7)

agents to produce NH_3, CH_4, C_2H_4 and C_2H_6 as well as traces of CH_3NH_2. The observed hydrocarbon ratios are similar to those found in the reduction of cyanide with molybdothiol catalysts. Evidence for significant ATP stimulation and inhibition by phosphate was obtained. The mechanism of methane production is summarized in eq. (8) (Schrauzer *et al.*, 1975b). Ethylene and

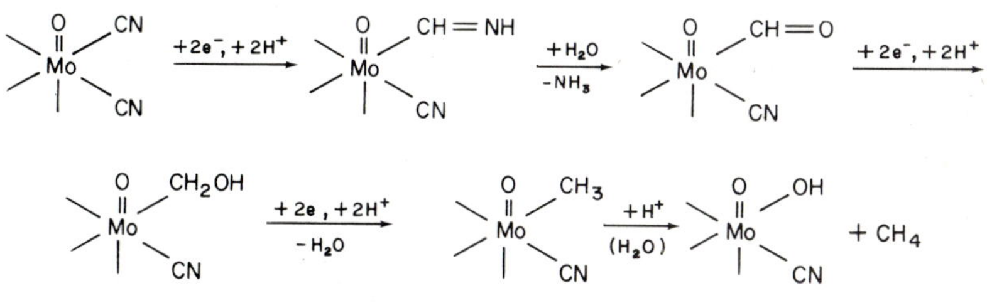

(8)

ethane are formed by way of CN^- insertion reactions into the Mo-C bond of organomolybdenum intermediates or, alternatively, by intra-molecular alkylation of coordinated cyanide. Only one reaction, *i.e.* the formation of ethane, is shown (eq. (9)). The study of the behaviour of anions D-F thus demonstrates that mononuclear complexes of oxomolybdate (IV) possess all the predicted properties under conditions of substrate reduction. All

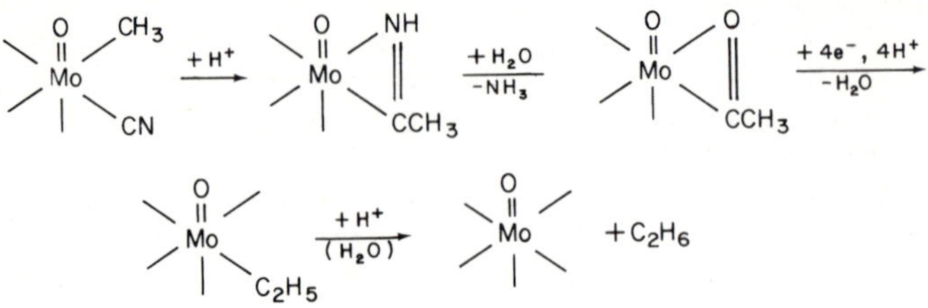

(9)

previous conclusions concerning the mechanism of CN^- reduction in the molybdothiol model systems are therefore substantiated. We also observed (Schrauzer *et al.*, 1975b) that reduced ferredoxin model compounds act as reductants in the conversion of coordinated cyanide to ammonia and hydrocarbons as well.

"MOLYBDO-CYANO" MODEL SYSTEMS

Investigation of the behaviour of salts containing the anion $Mo(O)(OH_2)(CN)_4^{-2}$ as potential catalyst for the reduction of substrates other than cyanide produced "molybdo-cyano" model systems of nitrogenase. The anion F becomes a catalyst of the reduction of nitrogenase substrates after coordinated CN^- is reduced off. The catalytically active species in this system are coordinatively unsaturated cyanide complexes of oxomolybdate (IV), such as H and I. The same reactive complex fragments may

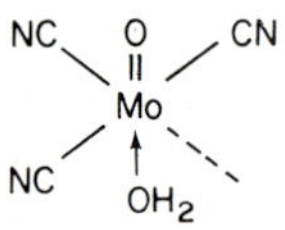

G

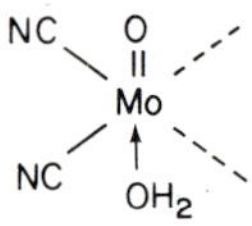

H

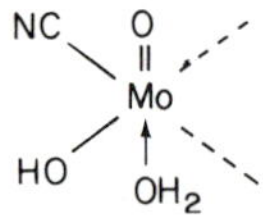

I

be generated by combining MoO_4 and CN^- in the molar ratios of 1:1 and 1:2 under reducing conditions. Reduction of acetylene in these systems yields ethylene efficiently. In D_2O as the solvent, *cis*-1,2-dideuterioethylene is formed while the reduction of 2-butene affords *cis*-2-butene. Both observations are consistent with a side-on interaction of the acetylenic substrates with the catalysts (eq. (10)).

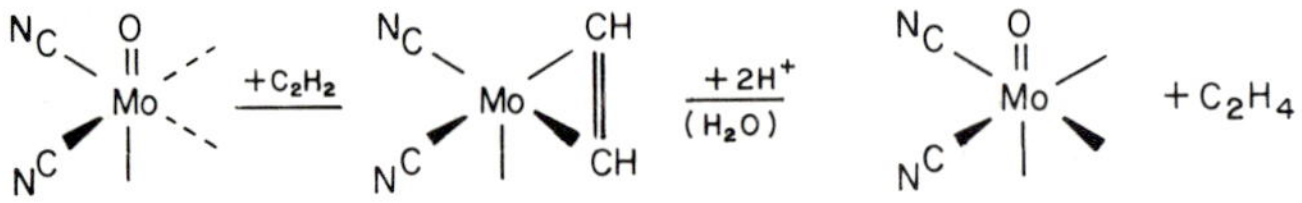

(10)

The reduction of molecular nitrogen in the molybdo-cyano model systems was also demonstrated. The yields of ammonia are actually higher than in the reduction of N_2 with molybdothiol catalysts, presumably because the catalytic decomposition of diimide is inhibited by cyanide. Evidence for diimide as the first product of N_2 reduction was obtained by conducting the reactions in the presence of olefinic substrates, such as fumarate. The overall reduction of N_2 to ammonia was accordingly formulated in terms of eq. (11) (Schrauzer *et al.*, 1975b, 1976b; Moorehead *et al.*, 1976).

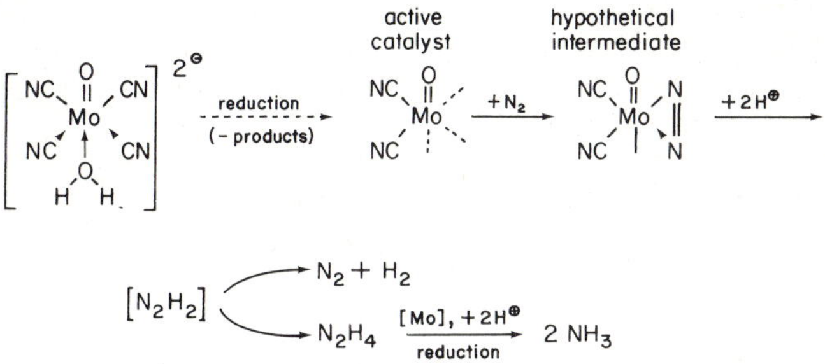

(11)

STOICHIOMETRIC REDUCTION OF NITROGENASE SUBSTRATES

We previously postulated that Mo^{red} in the molybdothiol systems are mononuclear thiolates of oxomolybdate (IV). The reduction of acetylenes was accordingly assumed to involve an oxidation of molybdenum from the +4 to the +6 state. The recent suggestion (Ledwith and Schultz, 1975) that the active reduced forms of the catalysts are derivatives of Mo(III) prompted us to investigate the stoichiometric reduction of acetylenes and N_2 with mononuclear cyano complexes of oxomolybdate (IV). The complex anion F decomposes in acidic solution with release of HCN. Although defined intermediates have not yet been characterized, it appeared plausible to expect that species such as G, H or I could be formed under conditions of protolysis. When the protolysis of F was performed in the presence of acetylene at low pressures (0.3 atm.), ethylene and traces of ethane and hydrogen were formed. The main reaction, *i.e.* the reduction of acetylene to ethylene, occurs according to eq. (12). The Mo(VI) produced undergoes a secondary reaction with remaining Mo(IV) to yield Mo(V)(eq. (13)). The idealized reaction stoichiometry thus is as shown in eq. (14) (Moorehead *et al.*, 1976). At higher C_2H_2 pressures (*e.g.* 1 atm.), C_4-hydrocarbons such as 1,3-butadiene and 1-butene are formed in addition to C_2H_4 and C_2H_6. Butadiene has also been observed (Stiefel, this volume and personal communication) in the reduction of C_2H_2 in

$$Mo^{+4} + C_2H_2 + 2\,H^+ \longrightarrow Mo^{+6} + C_2H_4 \qquad (12)$$

$$Mo^{+4} + Mo^{+6} \longrightarrow 2Mo^{+5} \qquad (13)$$

$$2Mo^{+4} + C_2H_2 + 2\,H^+ \longrightarrow 2Mo^{+5} + C_2H_4 \qquad (14)$$

the molybdothiol systems. The formation of C_4-hydrocarbons can be effectively suppressed, *e.g.* by lowering the C_2H_2 pressure or addition of a small excess of thiol ligand (Schrauzer *et al.*, unpublished observations).

For simplicity, the reactive fragments of F are designated $Mo^{+}4$ and the Mo(IV)-consuming side-reaction (eq. (13)) can be prevented by increasing the concentration of acetylenic substrate. Using the water-soluble propargyl alcohol in concentrated solution, high yields of allyl alcohol and quantitative oxidation of Mo(IV) to Mo(VI) were observed (eq. (15)). With

$$Mo^{+4} + HC{\equiv}C\text{-}CH_2OH + 2H^+ \longrightarrow Mo^{+}6 + CH_2{=}CH\text{-}CH_2OH \qquad (15)$$

N_2, NH_3 is formed *via* diimide and hydrazine as the intermediates. Hydrazine accumulates during initial periods of the reaction and is subsequently reduced to ammonia; the reduction of hydrazine was independently demonstrated. Evidence for diimide was obtained by experiments in the presence of ^{14}C-labelled fumarate. A reduction of the latter to ^{14}C-labelled succinate was observed only in the presence of molecular nitrogen. Because diimide is known to reduce olefinic substrates according to eq. (16), this experiment conclusively demonstrates the intermediate formation of *cis*-diimide in the reduction of N_2 to ammonia. We also observed an inhibitory effect of molecular hydrogen on the reduction of N_2 in this system (Moorehead *et al.*, 1976).

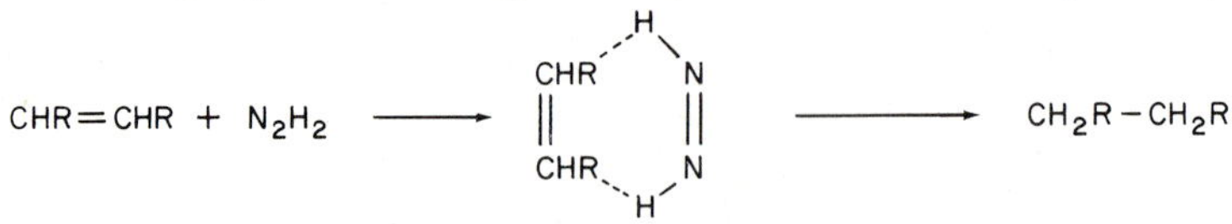

(16)

RELEVANCY OF THE RESULTS TO THE MECHANISM OF BIOLOGICAL NITROGEN FIXATION

Combining the available evidence from enzymological studies and the model experiments, the mechanism of biological nitrogen fixation was schematically represented in terms of SCHEME 1

(Schrauzer, 1975). We believe that this scheme provides a valid description of the mode of action of functional nitrogenase holoenzyme. Our new studies have furthermore confirmed and extended virtually all of our earlier conclusions and yielded deeper insights into the mechanisms of reactions catalyzed by complexes of oxomolybdate. Further details still remain to be elucidated, of course, but there can by now be little doubt in the validity of at least the key mechanistic interpretations.

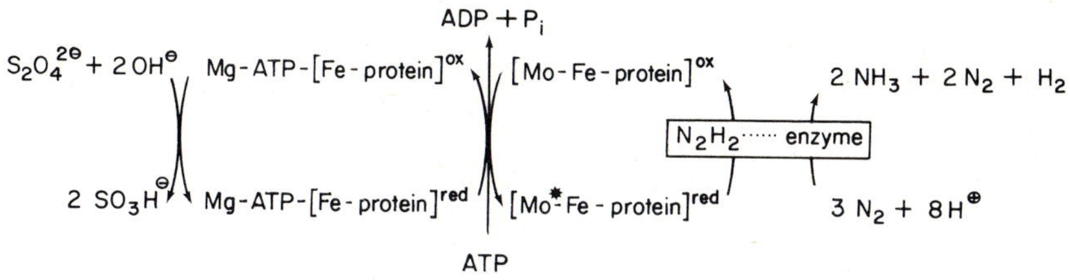

SCHEME 1
Proposed mechanism of action of nitrogenase (from Ref. 3)

As to the nature of the molybdenum active site in nitrogenase, the model studies suggest that it consists of a mononuclear oxomolybdate ion attached to the apoprotein.

ACKNOWLEDGEMENTS

The work described in the present review was supported by Grants GP 28485X and CHE 76-10890 of the National Science Foundation. Support by Climax Molybdenum Corporation is also gratefully acknowledged. I am deeply indebted to P. R. Robinson, E. L. Moorehead, Dr. T. M. Vickrey and K. Tano for their skilful experimental assistance and enthusiasm.

REFERENCES

Day, V.W. and Hoard, J.L. (1968). *J.Am.Chem.Soc.*, 90: 3774.
Hardy, R.W.F. and Knight, E.Jr. (1967). *Bacteriol.Proc.*, 112.
Krushch, A.P., Shilov, A.E. and Vorontsova, T.A. (1974). *J.Am. Chem.Soc.*, 96: 4987.
Ledwith, D.A. and Schultz,F.A. (1975). *J.Am.Chem.Soc.*, 97: 6591.
Moorehead, E.L., Robinson, P.R., Vickrey, T.M. and Schrauzer, G.N. (1976). *J.Am.Chem.Soc.*, 98: in press.
Parshall, G.W. (1967). *J.Am.Chem.Soc.*, 89: 1822.
Robinson, P.R., Schlemper, E.O. and Murmann, R.K. (1975). *Inorg. Chem.*, 14: 2035.
Schrauzer, G.N. (1974). *J.Less-Common Metals*, 36: 475.

Schrauzer, G.N. (1975). *Angew.Chem.International Ed.Engl.*, 14: 514.
Schrauzer, G.N. Kiefer, G.W., Doemeny, P.A. and Kisch, H. (1973). *J.Am.Chem.Soc.*, 95: 5582.
Schrauzer, G.N., Kiefer, G.W., Tano, K. and Robinson, P.R. (1975a). *J.Am.Chem.Soc.*, 97: 6088.
Schrauzer, G.N., Robinson, P.R., Moorehead, E.L. and Vickrey, T.M. (1975b). *J.Am.Chem.Soc.*, 97: 7069.
Schrauzer, G.N., Kiefer, G.W., Tano, K. and Doemeny, P.D. (1976a). *J.Am.Chem.Soc.*, 96: 641.
Schrauzer, G.N., Robinson, P.R., Moorehead, E.L. and Vickrey, T.M. (1976b). *J.Am.Chem.Soc.*, 98: 2815.
Schrauzer, G.N. (1976). *In:* Proc. of the 1st International Symposium on Nitrogen Fixation (W.E. Newton and C.J. Nyman, eds), pp. 79-116. Washington State University Press, Pullman, Washington.
Tano, K. and Schrauzer, G.N. (1975). *J.Am.Chem.Soc.*, 97: 5404.
Weil-Malherbe, H. and Green, R.H. (1951). *J.Biol.Chem.*, 49: 286.

HD FORMATION AS A PROBE FOR INTERMEDIATES IN N_2 REDUCTION

W.E. NEWTON, the late W.A. BULEN, K.L. HADFIELD, E.I. STIEFEL and G.D. WATT

Contribution No. 572 from Charles F. Kettering Research Laboratory, Yellow Springs, Ohio 45387, U.S.A.

INTRODUCTION

Inhibition by H_2 of the nitrogenase-catalyzed reduction of N_2 has been known for some time. The apparent competitive nature of this inhibition was first demonstrated in red clover plants (Wilson and Umbreit, 1937) and later in other organisms including *Azotobacter vinelandii* (Strandberg and Wilson, 1967; Hadfield and Bulen, 1969; Hwang *et al.*, 1973). In other experiments, Hoch *et al.* (1960) demonstrated that both H_2 evolution and HD formation occur in soybean nodules under a D_2/N_2 atmosphere. They found that HD formation was inhibited by carbon monoxide (an inhibitor of N_2 reduction) and nitrous oxide (an alternative substrate) but stimulated by N_2 and suggested a reversible exchange of D_2 with an enzyme bound diimide moiety. They also found that H_2 evolution was unaffected by carbon monoxide but inhibited by both N_2 and nitrous oxide. These observations have since been tested with cell-free preparations from *Azotobacter vinelandii* and other organisms (Burns and Bulen, 1965; Jackson *et al.*, 1968; Turner and Bergersen, 1969; Kelly, 1968, 1969; Hwang *et al.*, 1973) with only the dependency of HD formation on the presence of N_2 being in dispute (Kelly, 1968, 1969).

We sought to confirm N_2 as the only substrate of nitrogenase sensitive to D_2 or H_2 in the atmosphere over the reaction mixture (Jackson *et al.*, 1968; Hwang *et al.*, 1973) because these interactions, if they could be rationalized, might provide valuable insights into the mechanism of N_2 reduction. Previously, Bulen (1976) reported our preliminary data on HD formation under varying N_2 pressures at effectively only one D_2 pressure. Those data indicated that N_2 is required and may act as a catalyst for HD formation and that the HD accounts for a portion of the electrons consumed. An hypothesis which attempts to explain these observations and suggests that H_2 inhibition and HD formation are manifestations of the same molecular phenomenon has also been reported (Newton *et al.*,1976). Here, we report further experiments with purified nitrogenase complex from *A. vinelandii* (Hadfield and Bulen, 1969; Bulen and LeComte, 1972) on this subject,using enzyme preparations with activities of 350 nmoles H_2 evolved/min/mg protein. HD formation at two additional D_2 pressures and similar experiments using hydrazine as the reducible substrate (Bulen, 1976), are considered. These new data are compatible with our previous conclusions (Bulen, 1976; Newton *et al.*, 1976; Stiefel *et al.*, 1977) implicating a bound diimide-level species as the intermediate which reacts with D_2 and H_2 during the reduction of N_2 to ammonia. In addition, some preliminary experiments with molybdenum complexes containing N_2, N_2H_2, N_2H_4, CO and/or H^- as ligands are reported in an attempt to discover inorganic analogues of the biochemical reactions.

NITROGENASE-CATALYZED HD FORMATION

Effect of Reducible Substrates and Inhibitors

Jackson *et al.*(1968) reported that HD formation by nitrogenase preparations from *A. vinelandii* depended on N_2, ATP and reductant, was completely inhibited by CO and did not occur when azide, nitrous oxide, acetylene, cyanide, methyl isonitrile or acrylonitrile was present as reducible substrate. Kelly (1968) observed inhibition of HD formation to varying degrees with extracts of *A. vinelandii* and nitrogenase from *A. chroococcum* when acetylene, methyl isonitrile, cyanide or carbon monoxide were introduced into active reaction mixtures. In addition, he reported slight inhibition in the presence of N_2. Our previous results indicate that N_2 is a requirement, not an inhibitor (Bulen, 1976), and some of our more recent confirmatory findings are shown in Table I. Only under a D_2/N_2 atmosphere is HD produced. Levels of HD comparable to controls lacking enzyme, ATP or sodium dithionite were found when argon replaced N_2. Carbon monoxide and acetylene were both strongly inhibitory toward HD formation when added to reaction mixtures containing D_2/N_2. These results confirm that neither of these compounds promote HD formation and, because both are known to be potent inhibitors of N_2 reduction (Hwang *et al.*, 1973; Rivera-Ortiz and Burris, 1975), suggest that they operate by eliminating N_2 reduction which is necessary for HD to be produced. The alternative substrate, cyanide, also did not support HD formation; in fact, it eliminated even the "background" HD formation which is attributable to N_2 in the Ar used.

Hydrazine is a substrate for our preparations and shows the normal requirements for ATP, Mg^{2+} and reductant. It is

Table I

HD Formation as a Function of Substrate or Inhibitor

Atmosphere[a]		HD formed
Gases	Ratio	(moles)
$D_2:N_2$	1:1	7.0
$D_2:Ar$	1:1	1.3
$D_2:N_2:CO$	5:4:1	0
$D_2:N_2:C_2H_2$	5:4:1	0.1
$D_2:Ar(+CN^-)$[b]	1:1	0.1
$D_2:Ar(+N_2H_4)$[c]	9:1	1.9

[a]Total pressure of 1 atmosphere. Nitrogenase from *Azotobacter vinelandii* (see text) incubated at pH 7.25 at 30°C for 30 min, except those containing $N_2H_4 \cdot HCl$ which were at pH 8 for 20 min.

[b]5 mM CN^- added.

[c]30 mM $N_2H_4 \cdot HCl$ added.

believed to react with the substrate-reducing site, though an oblique reaction whereby the H_2-evolving function of nitrogenase leads to hydrazine reduction *via* contaminating conventional hydrogenase has not yet been excluded. Assuming a direct interaction, we find that hydrazine does not eliminate the "background" HD formation, thus confirming that it is a poor substrate for nitrogenase (Bulen, 1976) which allows the residual N_2 to compete successfully for electrons and produce HD. The slightly higher level of HD observed with hydrazine can be attributed to the higher level of D_2 used in the experiments or to N_2 produced in side reactions. As HD formation is not enhanced by N_2H_4, it probably occurs before the hydrazine-level is reached in the reduction of N_2.

Electron Distribution Studies

Watt and Burns (1976) have found that the rate of turnover of the enzyme, measured as the rate of utilization of either sodium dithionite or ATP, is independent of the nature of the substrate(s) being reduced (at least for N_2 and C_2H_2 reduction and H_2 evolution at the usual assay levels). This observation is valid even in a N_2-fixing system that is inhibited by H_2. It therefore appears that the rate-determining step for enzyme turnover occurs prior to substrate reduction, indicating that H_2 inhibition and D_2 inhibition (HD formation) affect only the distribution of electrons among the products. To test this hypothesis further, careful electron balance studies were carried out and representative data are listed in Table II.

Table II

Electron Distribution as a Function of Partial Pressure of N_2 and D_2

Gases[a]		Products[b]			
pN_2	pD_2	H_2	HD	NH_3	Total
	75	14.8(14.8)	1.1(0.55)	0.6(0.90)	(16.25)
10	150	15.1(15.1)	1.5(0.75)	0.5(0.75)	(16.60)
	350	15.4(15.4)	2.5(1.25)	0.4(0.60)	(17.25)
	75	13.3(13.3)	2.2(1.10)	1.3(1.95)	(16.35)
50	150	14.0(14.0)	2.9(1.45)	0.9(1.35)	(16.80)
	350	13.1(13.1)	5.0(2.50)	0.7(1.05)	(16.65)
	75	10.0(10.0)	2.4(1.20)	2.8(4.20)	(15.40)
150	150	10.4(10.4)	3.5(1.75)	2.2(3.30)	(15.40)
	350	10.8(10.8)	6.5(3.25)	1.8(2.70)	(16.75)
	75	8.8(8.8)	2.4(1.20)	4.0(6.00)	(16.00)
300	150	8.8(8.8)	3.6(1.80)	3.6(5.40)	(16.00)
	350	9.3(9.3)	6.7(3.35)	2.6(3.90)	(16.55)

[a] Reactant gases (in torr) contained in 1 atmosphere together with 75 torr Ar and He is diluent.

[b] Expressed as μmoles/incubation. The values in parentheses represent the numbers of electron pairs transferred to substrate, calculated by multiplying the number of product moles by 1 for H_2, 1.5 for NH_3 and 0.5 for HD. Total electron pairs transferred equals the sum of the numbers in parentheses in the previous three columns.

As expected, H_2 evolution decreased and NH_3 formation increased with increasing pN_2. However, at each pN_2 studied, the H_2 evolved remained effectively constant (varying by $\pm$ 0.5 electron pairs or less) even though pD_2 was varied over a 4-fold range. In contrast, a decrease in NH_3 production and an increase in HD formation was observed with increasing pD_2. These data show that the presence of D_2 (and by inference H_2 also, when H_2 replaces D_2) causes redirection of electrons away from NH_3 production and into HD (or H_2) formation, while ATP-dependent H_2 evolution remains invariant. As H_2 and D_2 affect only N_2 reduction, these species probably take part in a chemical reaction with an intermediate of N_2 reduction resulting in its decomposition to products other than NH_3 (Bulen, 1976; Newton *et al.*, 1976) as illustrated in Fig. 1.

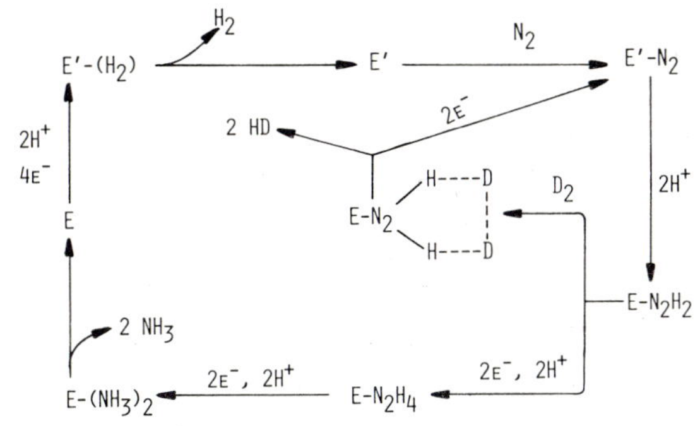

Fig. 1. *Suggested relationship of HD formation (H_2 inhibition) to N_2 reduction and H_2 evolution (E = oxidized enzyme; E' = enzyme reduced by two electrons and capable of substrate binding and reduction).*

The data in Table II further confirm our previous observation (Bulen, 1976) that unless the quantity of HD formed is taken into account at one electron (0.5 electron pair)/mole HD, a reasonable electron balance and constant electron flow (unaffected by gas atmosphere composition) is *not* maintained, thereby contradicting the results from ATP and $Na_2S_2O_4$ utilization studies (Watt and Burns, 1976). We find that the variation in the total products formed (Table II) more than doubles if electron consumption by HD formation is ignored. In addition, these results show that, at a given pD_2, the quantity of HD produced doubles on increasing pN_2 from 10 to 50 torr, increases only slightly when pN_2 is further increased to 150 torr and then remains unchanged on an additional increase to 300 torr N_2. Therefore, the major effect of pN_2 on HD formation is almost complete at 50 torr although NH_3 production continues to increase while H_2 evolution continues to decrease as pN_2 is raised above this value. Thus, we find that HD formation increases at a much more rapid rate and saturates at a much lower pN_2 than N_2 reduction which contrasts to the constant $HD:NH_3$ ratio reported previously (Jackson *et al.*, 1968).

The reduction of hydrazine is insensitive to the presence of H_2 in the atmosphere over the reaction mixture, consistent with the observation that no HD is formed during hydrazine reduction. Assuming that the reduction of free hydrazine proceeds similarly to that of hydrazine produced as a bound intermediate during N_2 reduction, these observations give added credence to the suggestion (Newton *et al.*, 1976; Bulen, 1976) that a bound intermediate at the diimide level is the reactive entity responsible for HD formation and H_2 inhibition.

Kinetics of HD Formation

Lineweaver-Burk plots of our initial HD formation data as a function of pD_2 at a variety of partial pressures of N_2 appear curved, but become linear when 1/HD is plotted against $1/pD_2^{1/2}$, with the lines intersecting at (or very near) the origin. The apparent K_m for D_2 in HD formation is thus extremely large (approaching ∞). The slopes of these lines, but not their intercepts, are obvious functions of pN_2. The intersection of all lines at zero indicates that nitrogenase never becomes saturated with D_2, even at infinitely high D_2 pressures (i.e. D_2 is reacting "chemically" without direct interaction with the N_2-binding site on the enzyme). The half-order dependence of the rate of HD formation on pD_2 may reflect the formation of 2 moles of HD from each D_2 molecule utilized. Plots of the reciprocal rate of HD formation versus $1/pN_2$ at various pD_2 are linear, with varying non-zero intercepts on the 1/HD axis. Both slope and intercept show dependence on pD_2. At these intercepts, pN_2 is infinitely high and is completely saturating the enzyme with N_2, thus making the rate of HD formation dependent upon pD_2 only. These data [using Cleland's rules (1963) for steady state analysis] suggest that N_2 and nitrogenase interact initially to form a "complex" which then reacts with D_2. Changes in pN_2 therefore alter the rate of HD formation by changing the amount of "complex" available to react with D_2.

The apparent K_m for N_2 [$K_m(N_2)$] in N_2 reduction is reported as ~0.15 atmosphere (Strandberg and Wilson, 1967; Hadfield and Bulen, 1969; Turner and Bergersen, 1969; Hwang *et al.*, 1973). For HD formation, we find the apparent $K_m(N_2)$ to be 0.015 atmosphere, i.e. ten-times smaller and similar to that reported previously (Turner and Bergersen, 1969; Burris and Orme-Johnson, 1976). The difference in apparent $K_m(N_2)$ in N_2 reduction and HD formation is puzzling if both reactions involve the same intermediate(s) (see Fig. 1). As the pN_2 required to give half-maximal rate of N_2 reduction is significantly higher than for HD formation, an additional interaction with N_2 might be required for the reduction to ammonia to proceed. This requirement might be related to the reduction of bound N_2 in purely chemical systems, where it occurs only in complexes containing more than one N_2 molecule (Chatt *et al.*, 1975; Manriquez *et al.*, 1976; Enemark and Feltham, 1972). Alternatively, the binding of D_2 close to (or at) the N_2-binding site could create an effect which preferentially attracts N_2 to such sites to produce HD rather than to sites without D_2 which would produce NH_3. This latter explanation is somewhat unsatisfactory as our kinetic analysis indicates that N_2 binds *before* D_2 reacts. However, it is very clear that these apparent $K_m(N_2)$'s are complex functions of many individual kinetic parameters, the detailed analysis of which (currently in progress) will hopefully lead to a satisfactory explanation of these divergent values.

Whatever the actual cause, the following correlation appears instructive. The concentration of nitrogenase used in these experiments is 2 to 3 mg protein per ml of the complex (~1:1 ratio of Mo-Fe protein to Fe protein) of molecular weight ~300,000, i.e. ~10^{-5}M. A comparable concentration of N_2 in solution results at pN_2 of 0.015 atmosphere. This pN_2 coincides with the apparent $K_m(N_2)$ for HD formation and suggests the possibility that N_2 is strongly attracted and firmly bound to the N_2-reducing (HD-evolving) site on the enzyme at this low pN_2. This initial interaction could involve the two-electron, two-proton reduction to the diimide-level and under the conditions of relatively high pD_2 and low pN_2, this intermediate could primarily function to channel electrons into HD, with the N_2 molecule remaining bound to the enzyme ready to catalyze further HD formation (see Fig. 1). Thus, comparable quantities of N_2 and N_2-binding sites on the enzyme are likely to produce maximal HD formation, which correlates well with our experimental observations. Thus, in the system we are studying, where the rate of enzyme turnover is limited by electron and/or proton availability (i.e. $S_2O_4^{2-}$ and ATP utilization), it may be that the diimide-level is rapidly attained on N_2 binding but that this species has various options for further reaction while awaiting the second electron/proton pairs, one of which is reaction with D_2 (or H_2), if available, to produce HD (or more H_2.

According to the above mechanism, H_2 inhibition (and HD formation) should not demonstrate pure competitive inhibition kinetics. Our experimental results indicate that competitive inhibition cannot occur if the electrons consumed in HD formation are to be accounted for, unless the rate of enzyme turnover increases at high pressures of N_2 and/or D_2 (rather unlikely). However, the number of electrons going to HD (the maximum observed is ~20% of the total number of electrons consumed, see Table II) is relatively small and so should cause only a small (in most instances, not *easily* discernible) shift in the intercept on the 1/v axis giving an apparent competitive inhibition pattern.

Our proposed mechanism of HD formation differs significantly from those proposed previously (Hoch *et al.*, 1960; Jackson *et al.*, 1968; Turner and Bergersen, 1969). We suggest that HD formation is not a simple exchange, but is both ATP- and reductant-dependent, consuming electrons which would otherwise have appeared in NH_3, with the diimide-level only (and *not* the hydrazine-level) pinpointed as the responsible intermediate. The simple exchange of D_2 on $CoH(N_2)L_2$,

$$[(p\text{-}FC_6H_4N{=}NH)PtClL_2]^+ \text{ and } [(p\text{-}FC_6H_4NH\text{-}NH_2)PtClL_2]^+$$

(with L = tertiary phosphine) reported previously (Parshall, 1967, 1968; Jackson *et al.*, 1968) and the demonstrated ability of the Pt complexes to catalyze exchange with protonic solvents are not good models therefore. Below we discuss our recent attempts to support this hypothesis of HD formation by nitrogenase using various molybdenum complexes.

HD FORMATION WITH INORGANIC COMPLEXES

Based on the hypothesis that bound diimide is responsible for HD formation with nitrogenase, the possibility of producing HD from inorganic $Mo\text{-}N_2H_2$ complexes was explored. Using $Mo(N_2H_2)Br_2dpe_2$ [dpe = 1,2-bis(diphenylphosphino)ethane] prepared by the method of Chatt *et al.* (1974), we attempted to determine its requirements for interaction with D_2 to produce HD. Any reasonable model for the enzyme should demonstrate similar requirements for HD formation and so reaction mixtures with and without possible sources of electrons and protons and under N_2 or argon were set up. The somewhat limited solubility of the complex dictated the use of dichloromethane, although this restricted the reductants available to us. For example, we could not use Na/Hg (used in the preparation of the precursor N_2 complex) due to possible reaction with the solvent. Metallic zinc was chosen as a starting point, but this eliminated the use of HBr as the proton source (as used in the preparation of the N_2H_2 complex). Under these circumstances, zinc and thiophenol were used as a possible reductant/proton source combination. Our initial results are reported in Table III.

Significant amounts of HD were only formed from D_2 when $Mo(N_2H_2)Br_2dpe_2$ was incubated with both zinc and thiophenol in dichloromethane solution. When any of these three ingredients was omitted, then HD formation was eliminated. When HBr replaced both zinc and thiophenol, again no significant quantity of HD was produced. We found that 0.05 mmol of complex converted 0.03 mmol D_2 to HD, i.e. a 60% yield, assuming that the hydrogen atom of the HD molecule originated from the bound diimide. This system therefore approaches a stoichiometric production of HD analagous to that presumed to occur in a single turnover of nitrogenase. Replacement of N_2 by argon does not significantly affect the amount and/or distribution of the products.

Table III

Requirements for HD formation by $Mo(N_2H_2)Br_2dpe_2$

System[a]	Products[b]		
	H_2	HD	D_2
Complete	2.4	15.3	82.3
Complete[c]	1.7	13.9	84.4
Minus Zn	1.0	1.0	98.0
Minus C_6H_5SH	1.2	1.8	97.0
Minus Complex	1.0	1.0	98.0
Minus Zn and C_6H_5SH plus HBr[d]	6.6	1.9	91.5
Minus Complex and C_6H_5SH	1.0	1.0	98.0
D_2 and N_2 only	1.3	1.2	97.5

[a]Reaction mixtures were contained in 20 ml bottles with complex (50mg), Zn (100mg), C_6H_5SH (0.1ml), D_2 (10cc), CH_2Cl_2 (5ml) and N_2 to 1 atmosphere and were incubated at 30°C for 18 h.

[b]Expressed as % of total of masses 2, 3 and 4 determined mass spectrometically.

[c]Argon to 1 atmosphere replaced N_2.

[d]Gaseous HBr (2cc) added.

Table IV

HD Formation from D_2 by Various Molybdenum Complexes[a]

Complex	Products		
	H_2	HD	D_2
$MoBr_2(N_2H_2)dpe_2$	1.9	17.6	80.5
$Mo(N_2)_2dpe_2$	0.6	1.7	97.7
$Mo_2O_4(S_2CNEt_2)_2$	0.8	1.2	98.0
$Mo(CO)_2(P\emptyset_3)_2Cl_2$	16.2	2.0	81.8
cis-$Mo(CO)_2dpe_2$	0.9	1.4	97.7
$[Mo(CO)_2(S_2CNEt_2)_2]_2 \cdot N_2H_4$	29.1	1.6	69.3
$MoH_2Cl_2dpe_2$	5.3	20.9	73.8
MoH_4dpe_2	3.7	29.8	66.5
None	0.8	1.2	98.0

[a]Reactions mixtures and expression of product composition are as in Table III.

Other molybdenum complexes were also tested, but no HD was observed with $Mo(N_2)_2dpe_2$, $Mo_2O_4(S_2CNEt_2)_2$, $Mo(CO)_2dpe_2$, $[Mo(CO)_2(S_2CNEt_2)_2]_2 \cdot N_2H_4$ or $Mo(CO)_2(PPh_3)_2Cl_2$, although the last two complexes gave significant quantities of H_2 (see Table IV). However, MoH_4dpe_2 and $MoH_2Cl_2dpe_2$ did produce HD in comparable quantities to $Mo(N_2H_2)Br_2dpe_2$, but this formation was also found to occur (to the extent of ~60%) in the absence of zinc and/or thiophenol. In both these systems, roughly statistical exchange occurs between D_2 and the molybdenum-hydrides, a situation commonly found for transition metal hydrides in general. These molybdenum-hydride complexes, like the aryldiimide and arylhydrazine complexes (Parshall, 1967, 1968; Jackson *et al.*, 1968), do not therefore satisfy the criteria for models of HD formation by nitrogenase.

Assuming $Mo(N_2H_2)$ survives intact under these reaction conditions, it is unique (so far) in producing HD from D_2 in a zinc/thiophenol-dependent reaction. This result is supportive of the involvement of a diimide-level intermediate in the formation of HD by nitrogenase. Further experiments are planned which will extend this investigation to other metal complexes and reductant/proton sources and which hopefully will elucidate the roles of the various reactants and the nature of both the metal-containing and other gaseous products.

CONCLUSIONS

HD formation during the enzymic reduction of N_2 offers a valuable insight into both the mechanism of biological N_2 fixation and the nature of the intermediates involved in the process. The work described above strongly suggests that nitrogenase operates in steps of not more than two-electron, two-proton transfers, very much like the other known molybdoenzymes (Stiefel *et al.*, 1977), with bound intermediates at the diimide and hydrazine levels. The former species is suggested to be the level at which H_2 inhibition and HD formation during N_2 reduction occur *via* the same mechanism. We stress that HD formation under N_2/D_2 mixtures does *not* represent a reversible exchange process. Thus, unlike other proposed mechanisms, our model predicts that HDO will *not* be formed in the aqueous phase concurrently with the formation of gaseous HD.

ACKNOWLEDGEMENTS

We thank Drs J.W. McDonald and J.L. Corbin for helpful discussions and J.R. LeComte, S. Lough, R. Land, Ms. D. Jacobs, D.C. Bravard and W. Braunlin for skilled technical assistance.

REFERENCES

Bulen, W.A. (1976). *In:* Proc.1st Int.Symp. on Nitrogen Fixation (W.E. Newton and C.J. Nyman, eds), pp. 177-186. Washington State University Press, Pullman, Wash.

Bulen, W.A. and LeComte, J.R. (1972). *In:* Methods in Enzymology (A. San Pietro, ed.), Vol. XXIV, Part B. Academic Press, New York and London.

Burns, R.C. and Bulen, W.A. (1965). *Biochim.Biophys.Acta,* 105: 437-445.

Burris, R.H. and Orme-Johnson, W.H. (1976). *In:* Proc.1st Int. Symp. on Nitrogen Fixation (W.E. Newton and C.J. Nyman, eds), pp. 217-219. Washington State University Press, Pullman, Wash.

Chatt, J., Heath, G.A. and Richards, R.L. (1974). *J.Chem.Soc., Dalton Trans.,* 2074-2082.

Chatt, J., Pearman, A.J. and Richards, R.L. (1975). *Nature,* 253: 39-40.

Cleland, W.W. (1963). *Biochim.Biophys.Acta,* 67: 104-137; 173-187; 188-196.

Enemark, J.H. and Feltham, R.D. (1972). *Proc.Nat.Acad.Sci., USA,* 69: 3534-3536.

Hadfield, K.L. and Bulen, W.A. (1969). *Biochemistry,* 8: 5103-5108.

Hoch, G.E., Schneider, K.C. and Burris, R.H. (1960). *Biochim. Biophys.Acta,* 37: 273-279.

Hwang, J.C., Chen, C.H. and Burris, R.H. (1973). *Biochim.Biophys.Acta,* 292: 256-270.

Jackson, E.K., Parshall, G.W. and Hardy, R.W.F. (1968). *J.Biol. Chem.,* 243: 4952-4958.

Kelly, M. (1968). *Biochem.J.,* 109: 322-324.

Kelly, M. (1969). *Biochim.Biophys.Acta,* 191: 527-540.

Manriquez, J.M., Sanner, R.D., Marsh, R.E. and Bercaw, J.E. (1976). *J.Am.Chem.Soc.,* 98: 3042-3044.

Newton, W.E., Corbin, J.L. and McDonald, J.W. (1976). *In:* Proc. 1st Int.Symp. on Nitrogen Fixation (W.E. Newton and C.J. Nyman, eds), Vol. 1, pp. 53-74. Washington State University Press, Pullman, Wash.

Parshall, G.W. (1967). *J.Am.Chem.Soc.,* 89: 1822-1826.

Parshall, G.W. (1968). *J.Am.Chem.Soc.,* 90: 1669-1670.

Rivera-Ortiz, J.M. and Burris, R.H. (1975). *J.Bacteriol.,* 123: 537-545.

Stiefel, E.I., Newton, W.E., Watt, G.D., Hadfield, K.L. and Bulen, W.A. (1977). *In:* Recent Advances in Bio-inorganic Chemistry (K.N. Raymond, ed.), in press. Advances in Chemistry Series, American Chemical Society Publications, Washington, D.C.

Strandberg, G.W. and Wilson, P.W. (1967). *Proc.Nat.Acad.Sci., USA,* 58: 1404-1409.
Turner, G.L. and Bergersen, F.J. (1969). *Biochem.J.,* 115: 529-535.
Watt, G.D. and Burns, A. (1976). *Biochemistry,* in press.
Wilson, P.W. and Umbreit, W.W. (1937). *Arch.Mikrobiol.,* 8: 440-457.

COMPONENTS AND PATHWAYS IN BIOLOGICAL NITROGEN FIXATION

W.H. ORME-JOHNSON[1], L.C. DAVIS[1,2], M.T. HENZL[1], B.A. AVERILL[1,3], N.R. ORME-JOHNSON[1], E. MUNCK[4] and R. ZIMMERMAN[4]

I. INTRODUCTION

The metabolism of inorganic forms of nitrogen, as well as the catabolism of amino acids and purines, affords many interesting prospects for those who are concerned with the molecular details of biological oxidations. This is nowhere more so than in the area of the reduction of dinitrogen to ammonia, catalyzed by nitrogenase. The literature of this sub-field of nitrogen fixation research is growing rapidly. It would not be useful to repeat here in detail summaries of earlier work which can be found in numerous reviews (*viz*, Orme-Johnson and Davis, 1976; Winter and Burris, 1976; Zumft and Mortenson, 1975; Burris and Orme-Johnson, 1974; Eady and Postgate, 1974) and in the recently published Proceedings of the 1974 Symposium (Newton and Nyman, 1976). Here, we propose to examine the nitrogenase system mainly from the viewpoints of protein chemistry and metallobiochemistry, concentrating on selected published reports of the last three years or so. We take it as our task to synthesize an overview of the present state of knowledge of nitrogenase and to identify areas of anomaly or plain ignorance. We also take it as our task to speculate about some of the less

[1]Department of Biochemistry and Center for the Study of N_2 Fixation, College of Agriculture and Life Sciences, The University of Wisconsin-Madison, Madison, Wisconsin 53706
[2]Present address: Department of Biochemistry, Kansas State University, Manhattan, Kansas 66044
[3]Present address: Department of Chemistry, Michigan State University, East Lansing, Michigan 48823
[4]Freshwater Biological Research Institute, The University of Minnesota, Navarre, Minnesota 55455

clear aspects of this fascinating enzyme. These speculations will surely be found provocative and, we hope, be a stimulus to further work.

Some overall features of the nitrogenase system are generally agreed upon. As depicted in Figure 1, electrons pass through ferredoxins and other carriers to the iron protein (MW ∿ 60,000, ∿ 4Fe and 4 labile S per molecule) which binds MgATP and reduces the molybdenum-iron protein (MW ∿ 220,000; 1-2 Mo/molecule, ∿ 24 Fe and 24 labile S per molecule). Subsequently, MgADP is released and either N_2 or H^+ are reduced. The presence of both proteins is required for catalysis and the sum of electron fluxes to N_2 and H^+ remains approximately constant if pN_2 is varied. The amount of MgATP converted to MgADP and Pi depends on the ratio of Fe to MoFe protein, temperature, and pH; the minimum figure is about 2 MgATP hydrolyzed per electron transferred. MgADP and O_2 are inhibitors of the system; soluble forms of the nitrogenase proteins are rapidly (and apparently irreversibly) inactivated by O_2.

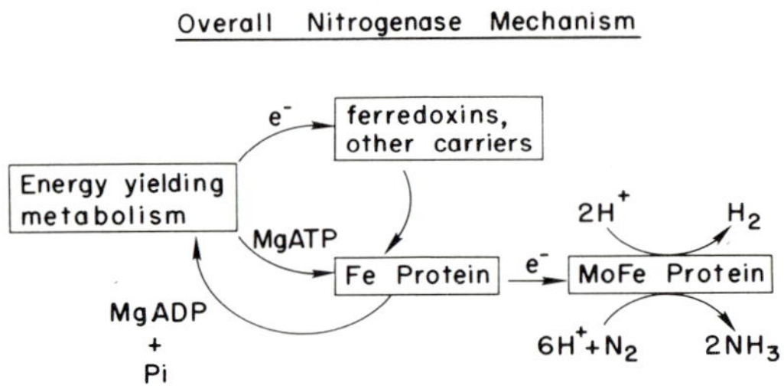

Fig. 1 *General features of the nitrogenase system.*

In the following sections, we examine evidence bearing on the molecular details of the nitrogen reduction process, particularly as to the number and kinds of prosthetic groups in the component proteins, their function, and the interaction of the proteins with each other and with the substrates. Finally, we review hypotheses that seek to explain the enzymic peculiarities of the system and discuss the meaning of these constraints for the integration into cellular metabolism.

II. NOMENCLATURE

There are several systems of nomenclature in use for the protein components of nitrogenase. Unfortunately, the IUB-IUPAC commission on enzyme nomenclature has not seen fit to name the protein components of the enzyme, so that it is up to the present workers in the field to establish a reasonable system. Since the established investigators all know the various systems in use, the sole purpose of this is to provide some consistency in the literature so as to lessen the confusion experienced by abstractors, students, and others less familiar with the history of work in this area. For this laudable purpose, we would like to propose that all systems which are non-informative, such as "Component I", "Component II", "azofermo", and "azofer", or may convey mininformation, such as "azoferredoxin" and "molydoferredoxin" (which incorrectly imply that the nitrogenase components are ferredoxins) are abandoned. That which we do know and which can be conveyed concisely is that the nitrogenase components are iron-sulfur proteins, one of which also contains molybdenum. A satisfactory nomenclature can be assembled from some systems currently in use and we propose that the proteins be called "the iron-protein of nitrogenase", abbreviated "Fe protein", and "the molybdenum-iron protein of nitrogenase", abbreviated "the MoFe protein". Furthermore, we propose that the suggestion of the Sussex workers (namely, that the component proteins of nitrogenase be designated by a capital letter for the genus and a lower case letter for the species, followed by the numeral 1 for the MoFe protein or the numeral 2 for the Fe protein) be used after a definition in terms of the previous nomenclature is given. The Sussex system is convenient and unambiguous if used in this way, but care should be taken *not* to use the nomenclature in titles or in abstracts unless a definition is provided. Here, we will utilize the Sussex abbreviations when specific proteins[1] are discussed. We think that nothing of value would be lost and much would be gained if workers in the area would adhere to this nomenclature, at least until subsequent advances in our knowledge of nitrogenase make a more informative nomenclature possible and desirable.

III. THE Fe PROTEIN

A. *Structural studies*

Where they have been examined, the Fe protein component of nitrogenase is composed of two subunits, indistinguishable by SDS-polyacrylamide gel electrophoresis. The published data include (holoprotein MW; subunit MW; reference): Kp2 (66,800, 34,600; Eady, *et al.*, 1972); Cp2 (56,000; 27,500; Tso,*et al.*, 1972); Av2 (64,000; 33,000- Kleiner and Chen, 1974); Ac2 (64,000; 30,800; Yates and Planque, 1975). Sequence studies of Cp2 are well advanced, and it appears that the subunits of this protein are indeed identical (Mortenson, personal communication). As isolated, the proteins generally contain 3 to 4 atoms of Fe and labile S per molecule, and it is thought that native active molecules contain 4 of each component. The following experiment (Averill, *et al.*; quoted in Orme-Johnson and

Davis, 1976; see also N.R. Orme-Johnson, *et al.*, Abstracts of this Symposium) strongly suggests that the inorganic constituents are present as a single iron-sulfur cluster. If the protein is dissolved in 80% hexamethylphosphoramide[2] - 20% 0.1 M Tris chloride, pH 8, to which a 100-fold (over protein) molar excess of thiophenol is added, the iron-sulfur cluster is displaced from the protein and can be identified (Fig. 2) by absorption spectroscopy as $Fe_4S_4\ (\text{ØS})_4^{2-}$, *i.e.*, the analog of the FeS cluster in bacterial ferredoxins (Orme-Johnson, 1973; Holm and Ibers, 1976). All of the iron (>95%) in the preparation is accounted for in this way. Control experiments with low MW ferredoxins with known cluster types (Fig. 2) show that, under these conditions, the integrity of the cluster is maintained. A similar result is obtained if 80% N-methylformamide plus *o*-xylyldithiol is used as the displacing system. Thus, the iron-sulfur cluster in the Fe protein of nitrogenase appears to be of the same type as that found in bacterial ferredoxins. Supporting evidence for this conclusion comes from the Mössbauer spectrum of Kp2 (Smith and Lang, 1974), which bears a close resemblance in its form and temperature-dependent behaviour to that of the bacterial ferredoxins. Orme-Johnson and Münck (unpublished) have made similar observations on Av1. This evidence is not too compelling, however, because the electronic relaxation time of the unpaired electron in the Fe protein is short at the normally used temperatures ($\leq 4.2°$ K), so that the Mössbauer spectrum is a quadrupole-split doublet, without the magnetic features that were useful in the analysis of 2Fe-2S ferredoxin spectra (Dunham, *et al.*, 1971). Further evidence on this point may be derived from recent linear electric field effect (LEFE) experiments on Cp2 (see N.R. Orme-Johnson, *et al.*, Abstracts of this Symposium). The experiment consists of the measurement of the extent to which EPR g-values of a center are shifted when a strong (10^4 to 10^5 v/cm) electric field is applied to the sample. A shift occurs only when the ligand charge distribution is non-centrosymmetric. In principle, reduced 2Fe-2S clusters (having a center of symmetry) may be distinguished from reduced 4Fe-4S clusters (which have no center of symmetry). In practice (Peisach, *et al.*, submitted for publication), a study of various ferredoxins of both cluster types has revealed that 2-Fe ferredoxins possess a small LEFE (*i.e.*, the protein lowers the effective symmetry) and that the LEFE of 4-Fe ferredoxins is nearly an order of magnitude larger. When the LEFE vs. magnetic field curves are examined, the two cluster types are seen to yield distinctive LEFE dependencies on the field. Reduced Cp2 and the Cp2·MgATP complex gave LEFE entirely consistent with the presence of a 4Fe-4S cluster, as deduced from independent studies. All of the evidence supports the presence of a *single* cluster, and since the EPR of the system suggests that the cluster shuttles between the states corresponding to the $Fe_4S_4(RS)_4^{-2}$ (diamagnetic) and $Fe_4S_4(RS)_4^{-3}$ (paramagnetic) ferredoxin-like synthetic clusters (Holm and Ibers, 1976, and references therein), the most reasonable position at present is that the *paramagnetic center* of Cp2 is a $1e^-$ carrier during nitrogen fixation.

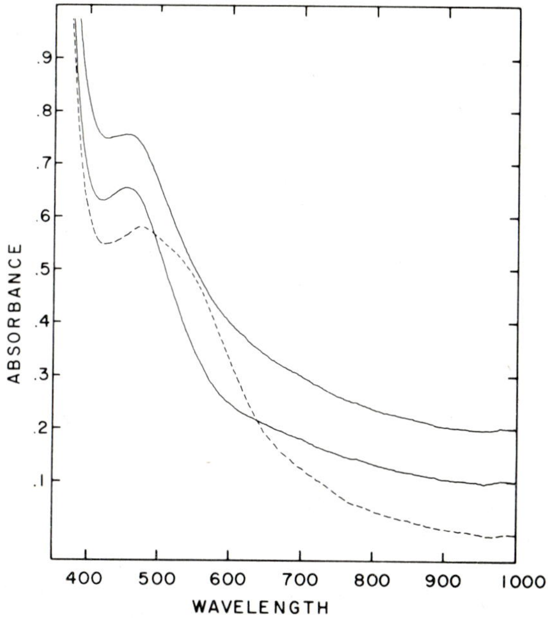

Fig. 2 *Absorbance spectra of* (top) *Cp2,* (middle) *Cp ferredoxin (4Fe4S cluster donor),* (bottom) *adrenodoxin (2Fe2S cluster donor) in 80% hexamethylphosphoramide-20% H_2O containing 0.1* M *tris chloride pH8 and 10 mM thiophenol. Each sample contained 120 nmoles Fe in 1.5 ml final volume. The samples were examined after 8 minutes incubation at 23°C and are displayed displaced vertically for clarity. The recovery of Fe from Cp2 as $Fe_4S_4(\phi S)_4^{2-}$ was better than 95% according to this experiment. From Averill,* et al. *(unpublished).*

Two aspects of the Fe protein worth considering here are: first, that the single cluster is bound to a protein containing two apparently identical subunits; and second, that the molecular weight of the Fe protein is 5 to 10 times that of the ferredoxins. As to the first point, because there are a minimum of four sites of attachment of a Fe_4S_4 cluster to a protein (the four exo cysteine ligands in the ferredoxins), it is possible that the cluster is bound equivalently to both subunits. However, it is equally possible that upon associating, the two subunits become non-equivalent so that only one of the subunits is covalently attached to the cluster. The elevated MW of the Fe protein, in contrast to that of the ferredoxins, is understandable if one considers that the protein has at least two binding sites for magnesium nucleotides (Tso and Burris, 1973) as well as binding sites for the MoFe protein and perhaps ferredoxins, and that the Fe protein may be the site of energy transduction, *i.e.*, conversion of MgATP hydrolysis energy into redox energy (*vide infra*)

A. *MgATP binding and redox potential*

Tso and Burris (1973), using a gel-equilibrium technique, showed that Cp2 binds 2 MgATP, with a Kd of about 17 μM. MgADP binds even more tightly, apparently at one of the MgATP sites. Zumft, *et al.* (1974) measured the height of the EPR signal associated with reduced Cp2 as a function of the redox potential in a cell containing the protein in buffer plus a variety of redox mediators. The potentials were determined at the platinum electrode and it was found that the measured potentials were insensitive to mediator concentration in the range of 0.2 to 1 times the concentration of Cp2, suggesting that mediator (Stombaugh, *et al.*, 1976) was not biasing the results. The potential of the protein was not dependent on pH (at 25°) in the range 6.5 to 8, so that transfer of protons during reduction is not required. The addition of saturating MgATP shifted the potential from -294 ± 20 mv to -402 ± 22 mv, which corresponds to a change in Kd for the MgATP complex from 17 μM with reduced Cp2 (Tso and Burris 1973) to *ca.* 0.3 M with oxidized Cp2. The value of n in Nernst's equation was found to be 1, indicating that independent single electron transfers cause the change in the EPR signal upon reduction. MgADP induces similar changes in the potential. Though details are not given by Zumft, *et al.*, it was suggested that the titration curve of Cp2 in the presence of MgADP was biphasic with potentials at -340 mv and -400 mv, which may correspond to forms of the protein with high (Tso and Burris, 1973) and low affinity for MgADP, or to catalytically active and inactive species in the preparation. Since the EPR signal is due to a single FeS center in the protein, this phenomenon cannot be due to the presence of two types of centers in the same protein molecule. NgATP does not appear to be hydrolyzed by the Fe protein in the absence of the MoFe protein and MgADP shifts the potential nearly as much as does MgATP, so that the potential shift induced by MgATP is not *per se* the energy transduction event involved in ATP-driven substrate reduction.

C. *EPR and chemical evidence for conformation changes on binding nucleotides*

The low temperature (<30° K) EPR spectrum of reduced active Fe protein is well known (Orme-Johnson, *et al.*, 1972; Smith, *et al.*, 1973; Evans, *et al.*, 1973; Zumft, *et al.*, 1972) to be a rhombic (g = 2.05, 1.94, 1.88) form similar to that in reduced ferredoxins (Orme-Johnson, 1973, and references therein). This signal undergoes a change toward a more axial form (g = 2.04, 1.92) when MgATP binds to the Fe protein. A similar response is elicited by MgGTP, Mg CH_2ATP, and MgATP (Zumft, *et al.*, 1973; 1974) at 10 mM but Mg salts of AMP, UTP, CTP and NAD gave no detectable change at this concentration. Tso (1974) used this change to quantitate the binding of MgATP to Cp2 and found a Kd of 20 μM, in good agreement with the direct measurement. Zumft, *et al.*, (1973, 1974) report the binding of 2 MgATP per Cp2 in experiments where the component concentrations were well above the Kd. Smith, *et al.*, (1973) titrated Kp2 with ATP in the presence of a large excess of Mg and found that two moles of nucleotide bind, but the Kd was 400 μM. Thorneley and Willison (1974) repeated these measurements with low Mg levels with the same results. These experiments have generally been interpreted as evidence that MgATP induces a conformational change upon binding. This interpretation has been reinforced by the observation that high concentrations of urea provoke similar changes (Mortenson, *et al.*, 1973) and that the reactivity of SH groups and Fe in the Fe protein are influenced by the presence of the nucleotides (Mortenson, *et al.*, 1976; Smith, *et al.*, 1976; and references therein). The EPR studies have been carried out at pH > 7.5. Recently, we have found that there is no change in the shape of the EPR of Cp2 upon addition of MgATP at pH 7, where we know ATP binds (from the work of Tso and Burris), though there is a consistent loss of 30% in integrated EPR upon binding in the pH range 6.5 to 9. Nitrogenase is, in general, more active at pH 7 than 8 (though this may be as much a function of the properties of the ATP generator as a property of nitrogenase itself (Davis and Orme-Johnson, 1976)), so that the conformational change detected by EPR and conveniently used as a measure of MgATP binding may be the result of a deprotonation leading to a less active form of the Fe protein. At the very least, the chemical modification phenomena should if possible be re-studied as a function of pH.

D. *Kinetic Studies*

Since the Fe protein of nitrogenase will undergo oxidation-reduction reactions in the absence of the MoFe protein, it is possible to examine at least this partial reaction of the nitrogenase electron transfer sequence. Such undertakings are complicated by the instability of oxidized Fe protein, but Yates, *et al.* (1975) and Thorneley, *et al.* (1976) have reported studies with Ac2, using an anaerobic stopped flow system devised by Thorneley (1974) as well as the freeze-quench EPR system of Bray (1961). The Fe proteins are normally maintained in buffer containing sodium dithionite, which was removed from solutions of Ac2 by an aerobic chromatography on Sephadex G-25, followed by oxidation of the protein with a ten-fold molar excess of phenazine methosulfate. The Ac2 was then rechromatographed on G-25 to remove phenazine methosulfate and reduction products. The authors state that this oxidation is accomplished with "marginal loss of enzymic activity". Oxidized Ac2 was reacted with sodium dithionite solutions, under various conditions, in the rapid reaction apparatus. The reduction of Ac2 and the oxidation of dithionite were estimated from absorbance changes at 425 and 315 nm, respectively. Several important conclusions are brought forth by Yates, *et al.* (1975) and Thorneley, *et al.* (1976). The reduction of Ac2 proceeds in four phases with estimated second order rate constants ($M^{-1}s^{-1}$ at 23°C and pH 7.4) of $>10^8$, 1.5×10^6, 9.1×10^4, and 2.3×10^4. In the initial rapid steps, *ca.* one electron per molecule of Ac2 is transferred and nearly one equivalent of electrons goes into Ac2 during the subsequent three steps. Oxygen-damaged Ac2 does not undergo the first step, but does slowly bleach in slow steps comparable to the last three. The first step is thought to be catalytically relevant and corresponds to the full development of EPR. Quantitation of the numbers of electrons transferred yields a ΔA_{425}/mole Ac2 of 2900 M^{-1} cm^{-1} for all four steps combined (2-e^- reduction) (calculated from Figure 4 of Thorneley, *et al.*, 1976). A second conclusion is that dithionite oxidation is first order in $[S_2O_4^{=}]^{1/2}$. In agreement with other studies of dithionite as a reductant of electron-transfer proteins (Lambeth and Palmer, 1973; Creutz and Sutin, 1973) the actual reductant is $SO_2^{\cdot -}$ which is in equilibrium with dithionite:

$$S_2O_4^{=} \rightleftarrows 2SO_2^{\cdot -}$$

K_{eq} for this reaction is *ca.* 10^{-9} M and the rate constant in the forward direction is *ca.* 1.7 s^{-1} near pH 7 (Lambeth and Palmer, 1973), so that under conditions where [Ac2] >> $[SO_2^{\cdot -}]$ the observed rate is that of dithionite dissociation, *i.e.*, zero-order in Ac2. This condition could be experimentally realized for the first phase of Ac2 reduction in the work of Thorneley, *et al.* (1976). Yates, *et al.* (1975) further found that the presence of MgADP inhibits the reduction of Ac2 by dithionite, whereas MgATP was without effect in this regard. The implication is that the MgADP complex of the Fe protein is difficult to reduce and accounts for the strongly inhibiting nature of MgADP in nitrogen reduction. Finally, Thorneley, *et al.* (1976) point out the following anomaly in that their EPR experiments suggest that the rate of transfer of electrons between the Fe protein and the MoFe protein is about 220 s^{-1} (which is slow compared to the rate of reduction by dithionite in the absence of the MoFe protein (>10^3 s^{-1} with 10 mM dithionite)), yet the EPR shows the Fe protein to be largely oxidized in the steady-state. They suggest that, in the N_2-fixing complex, re-reduction of the Fe protein has somehow been slowed. This could be due to a slow release of MgADP, of course, but it should be noted that in the original experiments of Orme-Johnson, *et al.* (1972), the Fe Protein remains mostly reduced in the steady-state with Cp and Av nitrogenase, though the shapes of the Fe protein signals in the steady-state suggest that, in these experiments, the Fe protein may be mostly the MgADP form (for further discussion, see Davis and Orme-Johnson (1976)). It is well to remember here that in addition to possible unevaluated experimental difficulties, these problems may be due to differences in relative rate constants in nitrogenases from various sources. It would be surprising if the overall *mechanism* or sequence of chemical steps were different from organism to organism.

With the foregoing in mind, a further difficulty with the kinetics experiments deserves exploration. The result (Thorneley, *et al.*, 1976) that 1-e^- is rapidly transferred (in a step in which the EPR-visible centre is affected) agrees nicely with the concept, from chemical and potentiometric studies showing the presence of a 1-e^--accepting 4Fe-4S centre, and the finding that a total of 2/e^- can be removed upon oxidation (Walker and Mortenson, 1973). However, Ljones (1973), who first carefully studied the oxidation of Cp2 by the natural oxidant system (Cp1 plus MgATP), found that the absorbance change (with Cp2 Cp2 of comparable activity) was 4500 M^{-1} cm^{-1} compared to the value of 2900 M^{-1} found by Thorneley, *et al.* for the rapid phase. Ljones points out that one actually expects a change of *ca.* 7000 M^{-1} cm^{-1} based on the properties of purified bacterial ferredoxins (Orme-Johnson, 1973, and references therein) and that only if this last value is used can he obtain the same rate of electron transfer during the steady state and during oxidation of excess Fe protein. Ljones quite logically (to our minds) attributes these differences to the presence of *inactive* forms of Cp2 in his preparations. This problem can be hardly less so with the experiments of Thorneley, *et al.* (1976) and

Walker and Mortenson (1973). In fact, Thorneley, *et al.* indicates that the specific activities of their preparation are lower by up to 40% after recovery from the stopped-flow experiment. In view of difficulties with quantitation of the EPR signal, it is hard to believe that one should expect such excellent agreement between electrons transferred and molecules of Fe protein present. The implication is that the Fe protein may transfer more than one electron rapidly, but what then of Ljones argument about ΔA? Does this imply the presence of a second electron-transfer center in *functional* Fe protein?

E. *Quantitation of EPR*

An ongoing problem with understanding the function of the Fe protein of nitrogenase has been the lack of any simple relationship between the specific activity, the iron content and the number of spins per molecule. We found a value of 0.79 spins per molecule for a preparation with an iron content of about 3.5 Fe per molecule and a specific activity[3] of 1200 (Orme-Johnson, *et al.*, 1972). More recently, we examined a preparation for which the specific activity was *ca.* 2400 and found 0.25 spins per molecule. Zumft, *et al.* (1973) reported a value of 0.2 spins per mole with a preparation of somewhat lower specific activity. Smith, *et al.* (1973) found a value of 0.45 ± 0.07 spins/molecule for Kp2 with preparations with specific activities varying from 600 to 1500, while Yates and Planque (1975) report that Ac2 with a specific activity of 2000 yielded an EPR integration value of 0.17 spins/molecule. As noted above, this type of Ac2 gave a value of 1 e^-/molecule transferred as determined from the kinetics of dithionite oxidation. No assumptions were made except that ΔA_{315} for Ac2 is zero during oxidation. According to Ljones's (1972) work with Cp2 ($\Delta A_{315}/\Delta A_{425} \sim 0.3$ during oxidation) this may yield an underestimate of perhaps 30% in the number of electron transferred in the phase during which the full extent of the EPR signal appears. Thorneley, *et al.* (1976) suggest that "damage to a more sensitive site" than the EPR center is responsible for a loss in catalytic activity without changing the dithionite reduction kinetics or EPR spin quantitation. This explanation sounds reasonable, but the questions remain. Why the low and, in the case of Kp2, reproducible spin quantitation with highly purified preparations? Why with Ac2 does the reduction of a center quantitated at 0.17 per molecule required 1 e^-? As discussed above, the potentiometric measurements of Zumft, *et al.* (1974) show that a 1 e^- change is responsible for the development of an EPR signal quantitating at 0.2 spins per molecule which leads to the same question. An independent measurement of the number of spins per molecule, by magnetic susceptometry, would be most useful.

The data suggest that, firstly, we do not understand how to integrate the EPR of the Fe protein, secondly the spins in some fraction of the molecules do not give rise to *observable* EPR or thirdly, the preparations are heterogenous in a miraculous way that allows a low value of the EPR intensity while preserving a 1 (or 2, see Walker and Mortenson, 1973) electron transfer character for the preparation as a whole. As to the first point, the EPR of the Fe protein appears to resemble that of the 4Fe-4S ferredoxin from *Bacillus polymyxa* (Stombaugh, *et al.*, 1973) in temperature sensitivity and general appearance. The recent advances made by Aasa and Vanngard (1975) in the theory of quantitation of such signals should not affect previously reported values, which were generally obtained by comparison to the EPR of Cu(II) salts, by more than about 10%. The integrated spin numbers are in fact *decreased* by the correction. It might be that a low-lying EPR-silent excited state may be thermally populated to the disadvantage of the EPR-active state. We have searched (Davis and Orme-Johnson, unpublished) unsuccessfully for such a state in the range of 6-20° K. At present, there is no reason to believe that the integrations are as inaccurate as the reported values indicate. On the second point, the work of Thorneley, *et al.* (1976) and Yates and Planque (1975) on Ac2 make one suspect that there may be a second 1-electron-accepting centre in the Fe protein, which becomes paramagnetic on reduction and in so doing, couples to the spin on the FeS center forming an even-spin state without EPR activity. The extreme form of this view suggests that the more intact the Fe protein, the smaller the EPR signal of the reduced form should be! A variant would be that we are seeing the results of either a conformational or a redox equilibrium, frozen out at the temperature of the EPR experiment (<20° K), such that only a fraction of the Fe-S center is uncoupled from the "other" paramagnet and therefore visible. Under such circumstances, either the "other" center is oxidized and diamagnetic or the EPR of the "other" center lies beneath the EPR thought to be due solely to the FeS center. It should be noted that there is no evidence for *weak* coupling of the kind responsible for the multiple EPR features in 2-center (8Fe8S) ferredoxins (Mathews, *et al.*, 1974). As to the third point, the extreme O_2 sensitivity of the Fe protein and the fact that O_2-oxidized (inactive) Fe proteins lack any EPR due to Fe-S centers makes it easy to accept that part of the problems discussed here may be due to O_2 inactivation during isolation and handling of the protein. This may be a factor, but matters are obviously not that simple.

F. *"Third Components" and Activation Phenomena*

Nitrogenase activity requires at least six distinct entities (Fe protein, MoFe protein, an electron source, protons, MgATP and dinitrogen). Perhaps it is not surprising to read in the earlier literature that additional protein factors have been reported as requirements for activity in Av (Kajiyama, *et al.*, 1969) and Cp (Taylor, 1969) nitrogenase systems. It is highly likely that current purified systems do not require components beyond those listed above, but one cannot now dismiss the possibility that *in vivo* certain ancillary proteins serve in various roles, either in coupling dinitrogen reduction directly (as opposed to ferredoxin mediation) to other cellular oxidations, or in structural roles, such as binding the nitrogenase components together. In this atmosphere of health skepticism about "third protein components," Ludden and Burris have provided evidence for a possible regulatory system, first identified in *Rhodospirillum rubrum* (Ludden and Burris, 1976) and later, in *Spirillum lipoferum* (Ludden, Okon and Burris, unpublished). There does not appear to be a comparable system in *C. pasteurianum*, *A. vinelandii*, *B. polymyxa* or *K. pneumoniae* according to these workers.

The facts are as follows. Rr2, as isolated, has only slight activity compared to Fe proteins from other organisms. A protein, obtained from *R. rubrum* chromatophores with a 0.5 *M* NaCl wash, will activate Rr2 in the presence of ATP and a divalent cation. The activating protein is both trypsin- and O_2-sensitive and the activation, which requires high levels of (10^{-2} M) Mg^{++}, is stimulated several-fold by the addition of 5×10^{-4} M Mn^{++}. Once Rr2 is activated, the activating system can be removed by chromatography without inactivating the Fe protein. The Fe protein is not adenylated nor is ^{32}P from γ-^{32}P-ATP incorporated during activation. The EPR of Rr2 is apparently not altered during activation. Analysis of purified Rr2 and S 2, in connection with the activation studies, revealed the presence of a large number of non-dialyzable phosphorus and aldopentose residues. Whether these unique features are related to the activation phenomenon is not yet known (Ludden, Okon, and Burris, unpublished). These recent findings suggest the presence of novel regulatory and structural properties in certain Fe proteins and will without doubt lead to further advances in our knowledge of nitrogenase enzymology. Diagnostic characteristics of the activating system noted by Ludden and Burris (1976) are (1) a lag phase (during which activation occurs) in the reduction of dinitrogen by crude extracts (supplemented by a MgATP generator), and (2) marked stimulation of activity by low levels of Mn^{++}.

IV. THE MoFe PROTEIN

A. *Structural studies*

1. *Subunit Structure.* Kennedy, *et al.* (1976) have examined the subunit structure of a Kp1, Av1, Ac1 and Rj1 by sodium dodecylsulfate (SDS)-acrylamide gel electrophoresis. Previous workers had concluded that the MoFe protein generally has a molecular weight in the range of 200,000 to 220,000 with four subunits (see Table I in Orme-Johnson and Davis, 1976, for a summary), but there has been some uncertainty as to whether the structure is, in all cases, of the α_4 or $\alpha_2\beta_2$ type. Kennedy, *et al.* find that, under appropriate conditions, all of the MoFe proteins they examined yielded two kinds of subunits of molecular weights near 50,000 and 60,000. In the case of Kp1, they further showed that tryptic peptide maps, prepared with subunits that had been resolved by electrophoresis, were significantly different. Kennedy, *et al.* made the disturbing observation that the resolution of the subunits, as well as the subunit molecular weight calculated by comparison to that of standard proteins, depended strikingly on the brand and age of the SDS employed. They found that the mobility of proteins is altered by the addition of n-dodecanol, a likely contaminant of SDS. Despite the rather damning implication of this last observation for the reliability of the SDS-electrophoresis method of subunit resolution, present evidence does not favour the $\alpha_2\beta_2$ type of structure for the MoFe protein. The disposition of the iron and molybdenum atoms among these subunits is not yet known.

Whether the hypothetical $\alpha_2\beta_2$ subunit structure suggests that the MoFe protein is a catalytic dimer with two active sites for dinitrogen reduction is not resolved and clearly depends, among other things, on whether the metal composition is divisible by 2. As discussed in more detail by Orme-Johnson and Davis (1976), the reported Fe compositions are useless in this regard, as they range between 12 and 32 Fe atoms per molecule. In no reported case could the distinction between odd and even numbers of Fe atoms be reasonably made, given the uncertainties of the analytical methods employed. The case of the Mo compositions is equally difficult at present, though in principle this should be easier to resolve. Nonetheless, reported Mo contents range from 1 to 2 per molecule with no obvious correlation between Mo content and specific activity. For example, Zumft and Mortenson (1973) separated partially purified Cp1 into an active fraction with 2 Mo per molecule and an inactive fraction with .5 to 1 Mo per molecule. On the other hand, Tso (1974) reported that her preparations of Cp1, with a specific activity at least as high as those of Zumft and Mortenson, contained 1 to 1.5 Mo per molecule. Obviously, one can obtain preparations with a full complement of metal but diminished catalytic activity. By definition, one cannot obtain preparations with a diminished content of required metals and still observe maximal activity. Perhaps none of the preparations qualifies as "fully active" in the sense that would allow us to know the required Mo content of the active species. However, since the organisms generally accumulate Mo, we may have in the MoFe protein preparations the unexpected case (given the low levels of Mo found in natural environments) of Mo contamination of the preparations, *i.e.*, the MoFe protein may *utilize* only 1 Mo per active molecule.

2. *Chemical Studies on the FeS Centers.* The approximate equality of the numbers of Fe and labile S atoms in MoFe protein preparations suggests the presence of numerous Fe-S centers of the types known in 2Fe and 4Fe ferredoxins (Orme-Johnson, 1973, and references therein). We have attempted to use the displacement procedure that was successful with the Fe protein (Averill, *et al.*, cited in Orme-Johnson and Davis, 1976) and with hydrogenase (Erbes, *et al.*, 1975), namely unfolding the protein in organic solvents containing thiols, but the results have been inconclusive. When Cp1 is dissolved in 80% hexamethylphosphoramide and 80 mM thiophenol of pH 8.5 (in the absence of oxygen), the absorbance spectra between 1000 and 300 nm change slowly and continuously for at least 24 hours. From the presence of a shoulder near 450 nm, characteristic of $[Fe_4S_4(\text{ØS})_4]^{2-}$, we deduce that *some* 4Fe - 4S clusters are present, but the spectrum, at no stage, resembles any mixture of 2Fe - 2S and 4Fe - 4S clusters. We concluded that FeS compounds (or other chromophores) of unknown types are present in addition to 4Fe - 4S clusters.

The problem of determining the prosthetic groups thus has two novel aspects that we had not encountered in our work on the "simpler" FeS proteins (ferredoxins) (Averill, Bale and Orme-Johnson, unpublished). First, the MoFe protein may unfold slowly due to a close-knit conformation, perhaps caused in part by disulfide bridging and by multiple attachments to Fe-S clusters and other prosthetic groups. We have begun to attack this problem with disulfide-cleaving reagents. Second, the protein may contain FeS clusters other than the recognized 2Fe-2S and 4Fe-4S types. To probe this problem, we devised a new technique for determining the cluster composition of proteins (Bale, Averill, and Orme-Johnson, unpublished), based on the discovery by Bale (Ph.D. Thesis, University of Wisconsin, 1975) that iron-sulfur clusters can be shifted from one protein to another. The concept is illustrated in Figure 3. The point is that specific iron-sulfur proteins accept only the cluster type for which they are "designed." Thus, the MoFe protein, the apo-proteins of 2Fe2S and 4Fe4S ferredoxins, an unfolding solvent and a low molecular weight thiol are mixed and subsequently diluted. They are then reduced with sodium dithionite and examined at 13° K by EPR spectroscopy. The dilution shifts the equilibrium from low-molecular weight complexes of the FeS clusters to holoproteins of the ferredoxins. When apoferredoxins prepared from *B. polymyxa* ferredoxin (Stombaugh, *et al.*, 1973) and adrenodoxin (Orme-Johnson, *et al.*, 1968) (both in 10-fold excess over the possible cluster content) are mixed with Cp1 in 80% DMSO/20% 0.2 *M* Tris-HCl, pH 8.5, plus 10 mM o-xylyldithiol, incubated anaerobically for 1 hour and then diluted with an equal volume of 0.1 *M* Tris HCl, pH 8.5, incubated a further hour, and reduced with sodium dithionite, the EPR spectrum of Figure 4 is obtained. By inspection, only 4Fe-4S clusters are detected, *i.e.*, <0.25 2Fe-2S centers per mole are present in the Cp1 by this test. Furthermore, quantitative estimations show that approximately half the Fe is recovered as 4Fe-4S clusters. If the composition (Huang, *et al.*, 1973) is 24 Fe per mole, then this indicates the presence of three 4Fe-4S clusters per molecule. Many more experiments of this kind must be done to conclusively establish the composition, but it is extremely significant that, in Figure 4, one can still observe EPR at g=4.2 and 3.6, which looks like a broadened version of the signal originally present in Cp1 (previously ascribed to *two* centers with 4Fe each for Av1 at least (Münck *et al.*, 1975) and which is *not* necessarily of the known 4Fe4S type.

Further evidence (from EPR and Mossbauer studies) about the occurrence of Fe-S centers in the MoFe protein is discussed in detail below.

3. *"The Molybdenum Cofactor"*. Pateman, *et al.* (1964) suggested that the existence of mutants of *Aspergillus nidulans* lacking both nitrate reductase and xanthine dehydrogenase implies the existence of a cofactor common to both enzymes. Nason, *et al.* (1971) observed that extracts of a nitrate reductaseless mutant (nit-1) of *Neurospora crassa* yielded nitrate reductase activity *in vitro* when supplemented with acid-treated extracts from a variety of molybdoenzymes including nitrogenase. They proposed the existence of a universal "molybdenum cofactor" and subsequently showed (Lee, *et al.*, 1974) that $^{99}MoO_4^=$, added to a mixture of acid treated molybdoprotein and nit-1 extract, was incorporated into nitrate reductase as resolved during sucrose density gradient fractionation. Though no well-characterized low molecular weight Mo compound has been shown to replace the Mo cofactor for reactivation, this latter experiment suggests that $MoO_4^=$ and Mo cofactor may be in equilibrium in solution. Presumably, the Mo

cofactor contains additional components, perhaps a peptide fragment or other organic moiety. Zumft (1975) has reported that Mo-containing fractions that will reactivate *nit-1* can be obtained by guanidine hydrochloride treatment and chromatography of Cp1.

Brill and his coworkers have developed a sensitive assay system for the isolation of the Mo cofactor. The mutant UW-45 of Av lacks the cofactor but when [NH_4^+] in the medium is low this mutant makes an apoAv1 that can be reactivated by acid

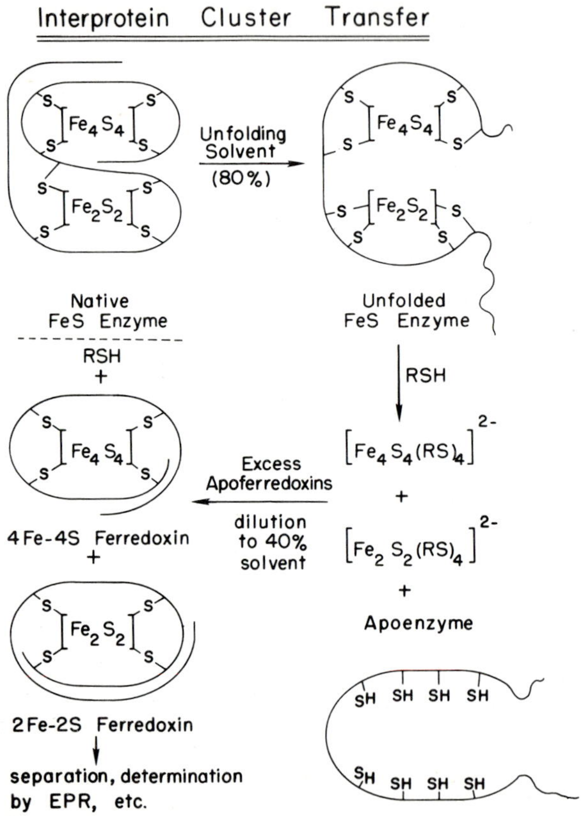

Fig. 3 *Principle of determination of the FeS cluster composition of an iron-sulfur enzyme by shifting the clusters onto apoferredoxins which are specific for 2Fe2S or 4Fe4S clusters. The holoferredoxins thus formed are stable and can be separated and recovered; more simply, after reduction by dithionite they are determined* in situ *by quantitation of their low temperature (∿ 13°K) EPR spectra.*

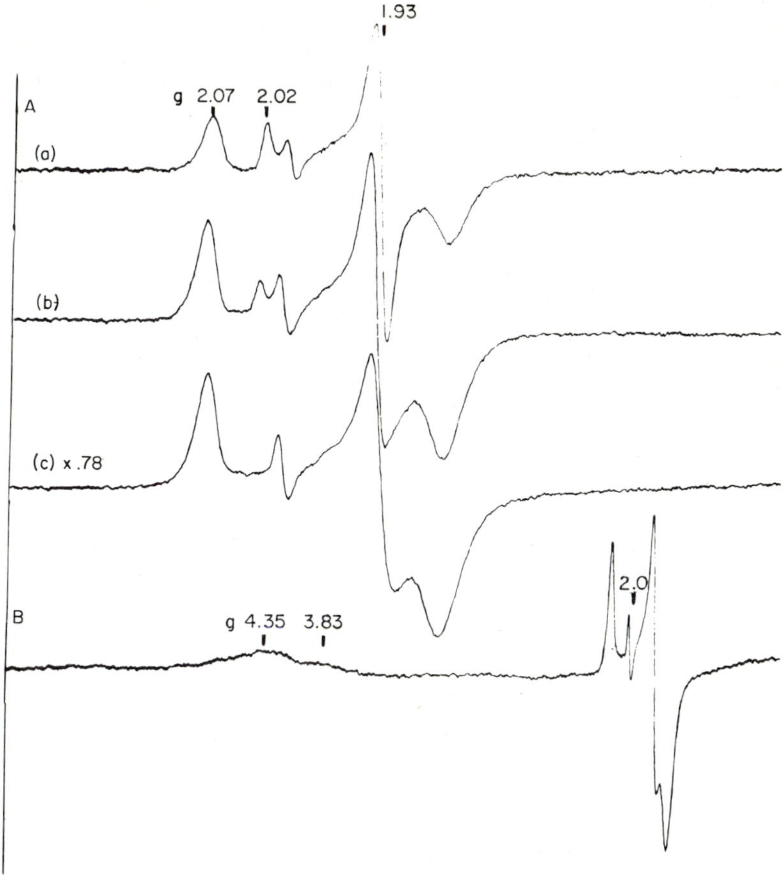

Fig. 4 (A) *Determination of the iron-sulfur clusters in Cp1 according to the experiment of Figure 3.* Top, *3:1 (molar) mixture of reduced ferredoxin from Bacillus polymyxa (4Fe4S) and adrenal cortical mitochondria (2Fe2S) (adrenodoxin);* middle, *reduced, diluted reaction mixture containing dimethyl sulphoxide and o-xylyldithiol as the carrier ligand (RSH of Figure 3) as well as excess apoadrenodoxin and apofd from* B. polymyxa; bottom, *as in middle spectrum, except that N-methylformamide and benzylthiol were used. All spectra at 13°K, microwave power 3 mw (Varian E-9 spectrometer, rectangular cavity), frequency 9.1 GHz, modulation 10 gauss at 100 KHz, time constant 0.3 s, sweep rate 500 gauss min^{-1}. The small amount of adrenodoxin (feature at g = 2.02) in the middle spectrum may be due to a*

extracts of several MoFe proteins (Pienkos, *et al.*, 1976). In this same laboratory, Nagatani, *et al.* (1974) showed that if tungsten replaces Mo in the growth medium of azotobacter and nitrogenase is derepressed (low[NH_4^+]), apoAvl is also formed. Whether the two forms of apoAvl are identical is not known but the systems are being used currently to isolate and study nitrogenase molybdenum cofactor. A very promising approach, using the organic solvent systems mentioned above, is under study by Brill and Shah.

B. *Oxidation-reduction Properties of the MoFe Protein*

If our present concept of the nitrogenase system is correct, the redox chemistry of the MoFe protein should divide itself naturally into two phases: (a) the addition of electrons up to (but not including) the state where substrate reduction can begin; and (b) the production of reduced forms of the MoFe protein that will themselves reduce H^+ or N_2. Redox reactions in the (a) region have been observed in several cases with isolated MoFe proteins but redox reactions of the (b) type seem to require the Fe protein plus a source of MgATP. The results of a number of studies are summarized in Table I. Together these studies suggest that there are two types of centers that titrate before the potential required for the reduction of substrates is reached. We have begun a combined EPR and Mössbauer study of the centers affected in the (a) region by examining spectra of thionine-oxidized Avl and Cpl. Excess dithionite is removed from Avl by incubating a 14 μM solution of the protein with 2 μM Cp ferredoxin and 0.5 μM Cp hydrogenase. The Avl is in the "resting" state (*i.e.*, as isolated in the presence of dithionite) according to both the size of the EPR signal at g=4.3, 3.65, 2.01 and the Mössbauer spectrum (Münck, *et al.*, 1975). When this Avl is titrated with thionine and the titration is followed by EPR at 13°K, the results are as shown in Figure 5. The titration proceeds in two distinct phases accounting for a total of *ca.*6 electrons. The presence of two classes of centers is indicated and these can be profitably studied by the Mössbauer measurements to be described below.

Fig. 4 (contd.) *small amount (less than 10%) of conversion of 4Fe-4S centers to 2Fe-2S centers in this solvent system. Control experiments suggest that ¾ or more of the 2Fe-2S centers observed occur in this way. We estimate that about half the iron (2-3 clusters per molecule) in Cpl appears as 4Fe-4S clusters in this determination.* (B) *Full 0-4000 gauss scan of sample of middle spectrum in (A). Note the broad features between g = 4.5 and 3.5. These appear to be a broadened version of the signal seen in native Cpl (Figure 9, below).*

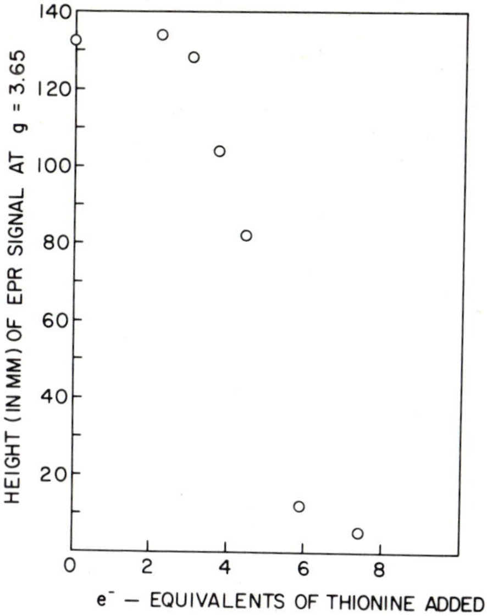

Fig. 5 *Thionine titration of Av1. Graded amounts of thionine were added to anaerobic EPR tubes (Orme-Johnson,* et al.*, 1972), containing 14 μM Av1 that had been freed of dithionite by incubation at 23°C with 2 μM Cp ferredoxin and 0.5 μM Cp hydrogenase. The samples were frozen (isopentane, -140°C) and observed in the EPR under conditions of Fig. 4. The height (at constant gain) of the feature near g = 3.65 is recorded as the ordinate.*

C. *Combined EPR and Mössbauer Spectroscopy of the MoFe Protein*

The MoFe protein as isolated yields EPR at low temperatures, with a nearly axial form. Two g-values are centered near g=4 and the third is found near g=2. This signal has been used frequently as a marker of the reduction state of the MoFe protein in static and kinetic investigations. As implied in previous sections, the signal disappears during oxidative titrations (Figure 5) (Smith, *et al.*, 1973) as well as during reduction of the MoFe protein by the Fe protein-MgATP complex (Smith, *et al.*, 1972, Orme-Johnson, *et al.*, 1972). The MoFe protein in the reduced state reduces H^+, N_2 and other substrates. Because of the reactions with H^+, it has been impossible to deal with the reduced MoFe system in the reversible manner required for thermodynamic analysis of titration data (see, for example, Table I). Thus, this state has been examined only in steady states of reduction in frozen samples.

Table I

Oxidation-reduction reactions of the MoFe protein

Species	E°',mv[2]	e^- added per molecule	Other substances present	Method	References
(a) region: no substrate reduction					
Cv1[1]	-60 -260	n=1[3] n=1	700 μM dyes	pH 7.5; thionine oxidation or dithionite reduction, followed by EPR (Dutton, 1971)	Albrecht and Evans, 1973
	-460			pH 8-10, dithionite reduction	Evans and Albrecht, 1974
Cp1	-70	3.6	dyes, concentrations not given	pH 7.5; thionine oxidation or dithionite reduction, followed by absorbance at 460 nm	M. Walker and Mortenson, 1973
Cp1	-33	n=1	140 μM dyes	pH 7.5; as for Cv1	Zumft, *et al.*, 1973
Av1	-320 -450	*ca.* 6 *ca.* 6, (n=2)	300 μM methyl viologen	pH 8; coulometry at fixed potentials	Watt and Bulen, 1976
(b) region: Fe protein + MgATP present; substrate reduction occurring					
Cv1	-465[4]	n=2	20 M dyes	pH 8; potential for half maximal reduction of acetylene	Evans and Albrecht, 1974

Cp1	n.d.	4		pH 7.5; difference titration with thionine; fixing *vs.* non-fixing system.	M. Walker and Mortenson, 1974
Av1	-450[4]	n.d.	300 μM methyl	pH 8; potential for half maximal reduction of protins	Watt and Bulen, 1976

[1]MoFe protein of *Chromatin vinosum*

[2]Values given versus Normal Hydrogen Electrode; to get Saturated Calomel Electrode values (Watt and Bulen, 1976), subtract 240 mv.

[3]When n-values are given this refers to the slope of the Nernst plot. The total number of electrons transferred per molecule of MoFe protein is an integral multiple of n, depending on the number of independent electron transferring centers.

[4]"Appearance potentials", *i.e.*, the potential at which half-maximal rates are obtained.

Smith and Lang (1974) and Münck, *et al.* (1975) have reported Mössbauer investigations of Kp1 and Av1 under various conditions of oxidation. Smith and Lang (1974) showed that the Mössbauer spectrum of Kp1 was dramativally altered upon oxidation with thionine. Since the titration of Av1 by thionine reveals two distinct events as monitored by EPR (previous section), we undertook to correlate this titration with Mössbauer spectra of the same samples.

Figure 6 shows a Mössbauer spectrum of the native Av1 protein taken at 4.2°K. Since a detailed analysis of this spectra was given by Münck, *et al.* (1975), we briefly quote the main results. The central doublet, labeled component D in Münck, *et al.*, contains 44% of all iron present and reflects iron atoms in complexes of integral or zero electronic spin. The two sharp lines at about -1 mm/s and about +2 mm/s belong to spectroscopic component Fe^{2+}, which contains 11% of the total iron. Finally, the broad absorption stretching from -2 mm/s to + 3 mm/s is the magnetic Mössbauer spectrum M_{EPR}. This last spectrum has been shown to be associated with two identical EPR active centers, each containing four iron atoms. Münck, *et al.* (1975) have shown that component M_{EPR} in Av1 is associated with two S=3/2 centers which give rise to the observed EPR signal at g=4.32, 3.67 and 2.01. Furthermore, under N_2-fixing conditions, the Mössbauer spectrum M_{EPR} disappears concurrently with the EPR signals, changing to a quadruple doublet in the Mössbauer spectrum. The data show that each EPR active center is reduced by one electron. (We have recently reanalysed our data taking into account effects due to the thickness of the samples. The relative amounts of each component quoted above therefore are slightly different from those given by Münck, *et al.*; see also Table II.)

Figure 7 shows a Mössbauer spectrum, taken at 4.2 K, of the MoFe protein after the sample was oxidized with thionine (6 e^- equivalents per molecule). Inspection of the data shows that all iron components have changed relative to the native enzyme. Figure 7 shows essentially two types of spectra, a quadruple doublet and a magnetic spectrum. The quadruple doublet in the centre of the spectrum is quite distinct from spectral component D in the native enzyme. Its center, corresponding to an isomeric shift δ=0.35 mm/s, is shifted by about 0.3 mm/s relative to the center of D (δ=0.64 mm/s). The remainder of the spectrum shown in Figure 7 is a magnetic Mössbauer spectrum (labelled M_D) reflecting iron atoms in an environment of half-integer electronic spin. At least four iron environments can be discerned in M_D. The Mössbauer spectrum shown in Figure 8 contains some of the information necessary to identify the components present in Figure 7. The spectrum in Figure 8 was

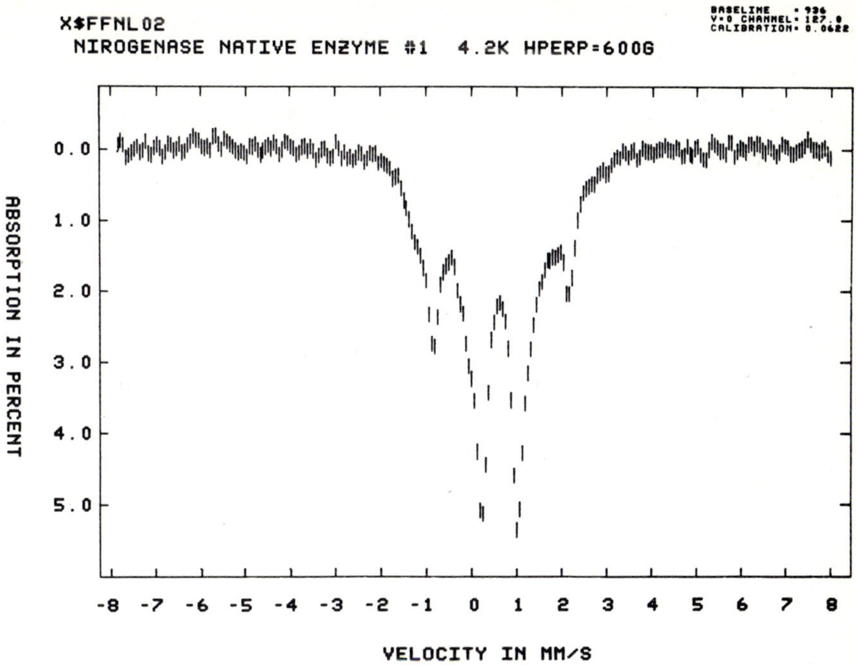

Fig. 6 *Mössbauer spectrum of native* ^{57}*Fe-enriched MoFe protein from* A. vinelandii. *The spectrum was taken at 4.1°K in a magnetic field of 600 gauss applied perpendicular to the Mössbauer radiation. The spectrum was analyzed and discussed in detail in Münck,* et al. *(1975).*

Table II

Mössbauer species in native Av1

Species	Mössbauer parameters		Fe[a]/molecule
	ΔE[b]	I.S.[c]	nearest integer
M_{EPR}	0.76	0.40	8
D	0.81	0.64	9
Fe^{2+}	3.02	0.69	3
S	1.4	0.6	1
M_D^d	*ca.* four species		10-12 (D + Fe^{2+})

[a]Recalculated from data of Münck, *et al.* (1975), assuming that the MW of Av1 is 220,000 (not 300,000) and that there are 1.6

Mo and 20 Fe per molecule. Under these assumptions and applying the correction suggested by Aasa and Vanngard (1975), M_{EPR} is due to 2.0 independent FeS clusters in the spin state S=3/2.

[b]Quadruple splitting, mm/sec.

[c]Isomer shift, mm/sec relative to metallic iron (300°K).

[d]Produced during the oxidation of Av1 with thionin.

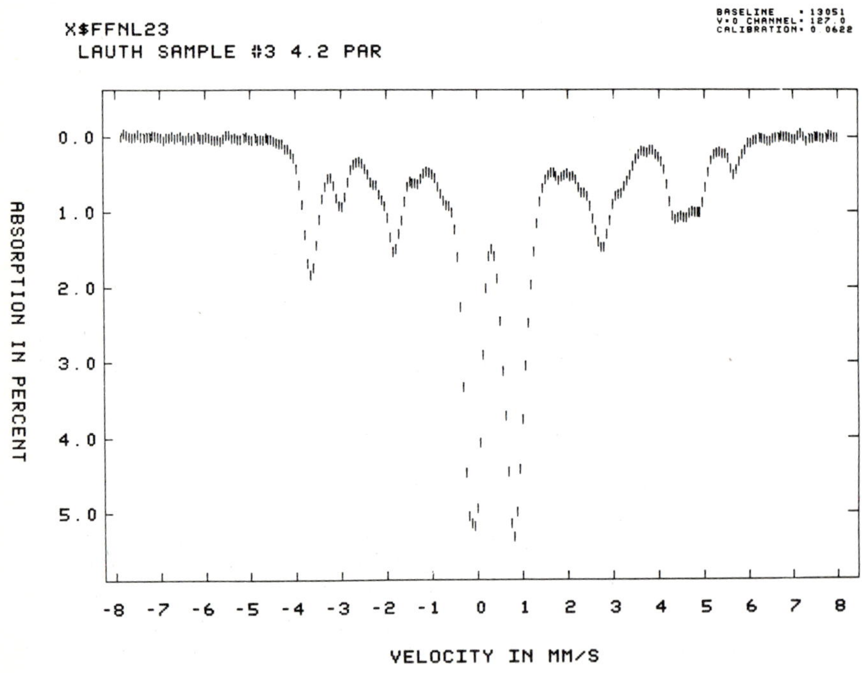

Fig. 7 *Mössbauer spectrum of the MoFe protein after the sample was oxidized with 6 e^--equivalents of thionine. The two strong lines in the center are due to component M_{EPR}^{+1}, an oxidized state of the S = 3/2 centers. The remainder is an intricate magnetic pattern (component M_D). M_D contains spectroscopic component D and very probably component Fe^{2+}. Three electrons are necessary to transform components D and Fe^{2+} into M_D. M_D reflects clusters which may be identical or different.*

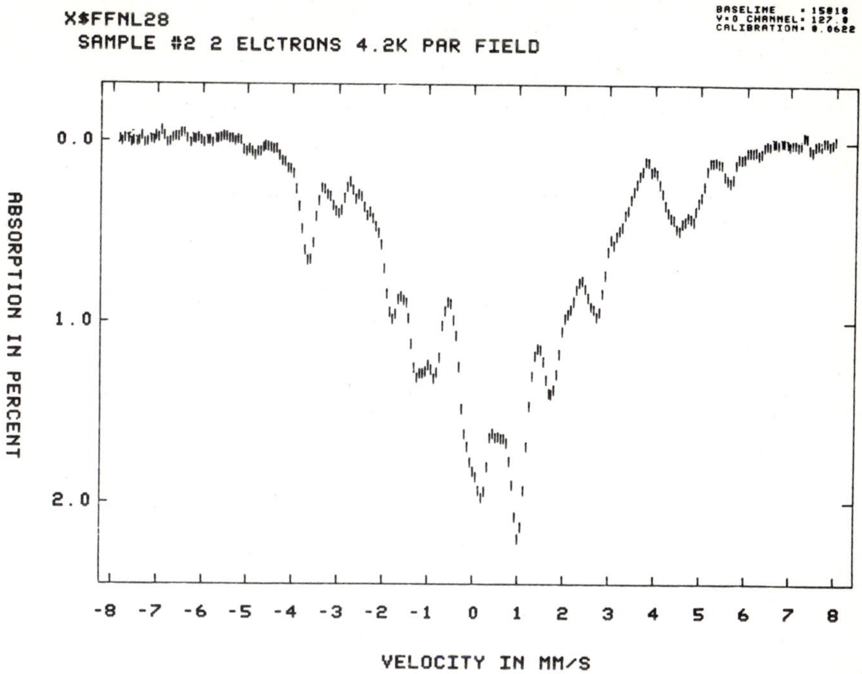

Fig. 8 *Mössbauer spectrum of the MoFe protein after the sample was oxidized with 2 e^--equivalents of thionine. Note the disappearance of the D and Fe^{2+} doublets and the appearance of compound M_D.*

obtained on a sample which had been oxidized by two electron-equivalents of thionine. Under these conditions, the S=3/2 centres are still in the same state as observed in the native enzyme. Comparison of Figures 6, 7, and 8 shows that a two-electron reduction causes the simultaneous disappearance of components D and Fe^{2+} and the appearance of component M_D. Spectral component M_{EPR} is not affected. Therefore we can draw the following conclusions. Firstly, the central doublet in Figure 7 is an oxidized form of spectral component M_{EPR} (the S=3/2 centres). Since this doublet is non-magnetic even at 1.5°K, the irons in the S=3/2 centres reside in a cluster with an even number of electrons. Since this cluster has an odd number of electrons in the native enzyme, we have removed *one* electron from the cluster (3 electrons are incompatible with the titration curve). Thus, the EPR active centres can exist in three distinct oxidation states (M°_{EPR} in the native form,

$M^{-1}{}_{EPR}$ in the fixing mixture, and $M^{+1}{}_{EPR}$ after oxidation with thionin). Secondly, quantitation of the spectra shows that 65 ± 3% of component M_D has appeared after a two-electron oxidation (Figure 7). Hence, approximately three electrons are required for the full appearance of M_D. Although the data analysis is still in progress, we can state that component M_D contains at least 52%, but less than 60%, of the total iron present. Making the reasonably assumption that the MoFe protein contains approximately 20 iron atoms, we see that a three-electron oxidation affects at least 10 iron atoms (probably 12). Thus, our data prove that M_D reflects an oxidized form of component D and, moreover, that the D-irons reside in spin-coupled clusters. What can be said about component Fe^{2+}? We are reasonably sure that Fe^{2+} is contained in M_D also and we have some evidence to support this conjecture.

We have mentioned above that component M_D shows iron atoms in at least four different environments. The Mössbauer spectrum M_D is a complex mixture of magnetic species and we are far from understanding the nature of M_D. However, a few experimental facts have emerged which we consider worthwhile to list here. (1) M_D contains at least four species reflecting iron atoms in different environments. These species appear as four distinct quadruple doublets at higher temperatures (170°K). (2) M_D reflects iron atoms in an environment of half-integer electronic spin. A spectrum of the M_D type has never been observed for any iron-sulfur protein. (3) A weak polarizing magnetic field of 600 gauss applied either parallel or perpendicular to Mössbauer radiation does not affect the intensities of M_D. This implies that M_D is associated with a Kramers doublet which has g-values such that $g_x \simeq g_y < g_z$. The data imply $g_z/g_x \geqslant 20$; for such a doublet, the EPR transition probability is vanishing small. Moreover, such doublets are generally substrates of a spin multiplet with electronic spin $S \geqslant 3/2$. Although such multiplets do often produce doublets which are EPR silent, in general at least one of the doublets gives rise to strong EPR absorption. Despite a thorough search, we have not been able to detect an EPR signal which can be uniquely associated with component M_D. There are two experimental clues as to why no EPR signal is observed. Firstly, the Mössbauer spectrum does not change between 1.5° K and 4.2° K, which implies that the first excited state doublet is separated from the ground doublet by more than 1°K; and secondly, the spectrum M_D broadens appreciably at 9°K, indicating (for the ground state doublet) an electronic spin relaxation time of about $\sim 10^{-9}$ sec. Thus at lower temperatures the EPR active state is not sufficiently populated and at temperatures >10°K the electronic relaxation rate becomes too fast.

Two other points are worth noting. The spectrum in Figure 7 can also be produced by air oxidation (Smith and Lang, 1974) and it is almost impossible to rereduce component M_D with dithionite, although Av2 plus MgATP returns the spectrum to normal as seen earlier. We do not know yet whether the iron atoms of M_D change their oxidation state during the reduction of substrates. It would be surprising if they did not. If the species M_D turns out to be a kinetically valid intermediate in nitrogen reduction, the lack of evidence for this species in fixing mixtures may be due to the rapid reduction of M_D compared to its rate of oxidation. Experiments on this point, using freeze-quenched Mössbauer spectroscopy, are in progress. With reference to the oxidation-reduction properties discussed above (Table I and Figure 5) it is obvious that oxidation of M_{EPR} gives rise to the potential just below 0 mv, while the (D + $Fe^{2+} \rightarrow M_D$ transitions may occur at even more *positive* potentials (assuming that the thionine couple is in equilibrium with this transition). EPR and Mössbauer correlates of the more negative potential species have not yet been found. These may, of course, be due to *Mo* species.

D. *CO-inhibited States of Nitrogenase*

Davis, *et al.* (unpublished) have recently examined the action of carbon monoxide as evidenced in the EPR of nitrogenase. No effect is seen on Fe protein or MoFe protein alone. Carbon monoxide inhibits the reduction of dinitrogen by purified nitrogenase from *Azotobacter vinelandii* and *Clostridium pasteurianum* in a noncompetitive manner (K_{ii} and K_{is} = 1.4 x 10^{-4} and 4.5 x 10^{-4}; 7 x 10^{-4} atm and 14 x 10^{-4} atm for the two enzymes, respectively. The onset of inhibition is within the turnover time of the enzyme, and CO does not affect the electron flux to the H_2-evolving site. The kinetics of CO inhibition of N_2 reduction are simple but CO inhibition of acetylene reduction is complicated by substrate inhibition effects. When low temperature (∿13°K) EPR spectra of CO-inhibited nitrogenase are examined, it is found that low concentrations of CO ([CO] = [Enzyme]) induce the appearance of a signal with g-values near 2.1, 1.98 and 1.92 with $t_{1/2} \approx 4$ sec, while higher concentrations of CO lead to the appearance of a signal with g-values near 2.17, 2.1, 2.05 with a similar time-course. The appearance of the signals is extremely sensitive to the temperature of incubation. Below 15° C, little of the intermediate is formed. The MoFe proteins from *Rhizobium japonicum* and *Rhodospirillum rubrum,* reduced with *Azotobacter* Fe protein in the presence of CO, give similar results. Under conditions which promote the accumulation of H_2 in the *absence* of CO, an additional EPR signal with g-values near 2.1, 2.0 and 1.98 is observed. The use of Azotobacter nitrogenase components enriched selectively with ^{57}Fe or ^{95}Mo, as well as the use of ^{13}CO, permitted the assignment of the centre(s) responsible for the induced signals. Only ^{57}Fe, when present in the MoFe protein, yielded broadened EPR signals. The signals are produced from steady state species, and not more than 0.5 spin per molecule has thus far been elicited. It was proposed that the CO-induced signals arise from states of the MoFe protein in which CO inhibits electron flow to the N_2-reducing site (but not to the H_2-evolving site) so that one or more iron-sulfur clusters achieve steady-state net charges of -1 (high CO complex) and -3 (low CO complex) in analogy to the normal paramagnetic states of high potential iron-sulfur proteins (hipip) and ferredoxins (fd), respectively. The "no-CO" signal may be either an additional center or the N_2-reducing site with H_2 bound competitively. These results make it clear that, although the MoFe protein does not normally yield easily detectable EPR that can be assigned to a known type of Fe-S cluster, in fact at least one such cluster is present per molecule. Our finding, described above, that two to three 4Fe-4S centers can be displaced by thiols, reinforces the idea that the low level seen in the CO experiments results from the observation of a steady state of the system. These results also indicate that there are *two* binding sites for CO, though the occupation of the first site is all that is required to stop the reduction of dinitrogen. Since CO can bind to 4Fe-4S clusters (Erbes, *et al.*, 1975), one may postulate that CO inhibits by complexing Fe-S centres not seen in either the native (g=3.65 signal) or CO inhibited (oxidized hipip and reduced fd signals) states.

E. *Solvent Perturbation Spectroscopy*

Cammack (1975) has reported that 2Fe-2S and 4Fe-4S ferredoxins give distinctive EPR in the reduced state in the presence of denaturing solvents such as 80% DMSO. When this technique is applied to Cp1, EPR spectra such as those of Figure 9 are obtained. Several important points should be noted. Firstly, a signal characteristic of 4Fe-4S centres is seen near g=2, which we estimate as representing *two* such clusters per molecule of Cp. Secondly, the g=3.65 signal due to the M_{EPR} centre is broadened *but not absent*. The broadening is due to relaxation as can be seen in the spectrum obtained at 4.2° K, where relaxation times are considerably longer than those at 13° K. Thirdly, treatment of Cp1 with 4 *M* imidazole in 3.6 *M* hexamethylphosphoramide abolishes the signal near g=2, because, as we have learned from experiments with ferredoxins, this solvent rapidly destroys FeS clusters of the normal types. The signal near g=3.65 is still present, reinforcing our conclusion that it is not due to a normal Fe-S cluster.

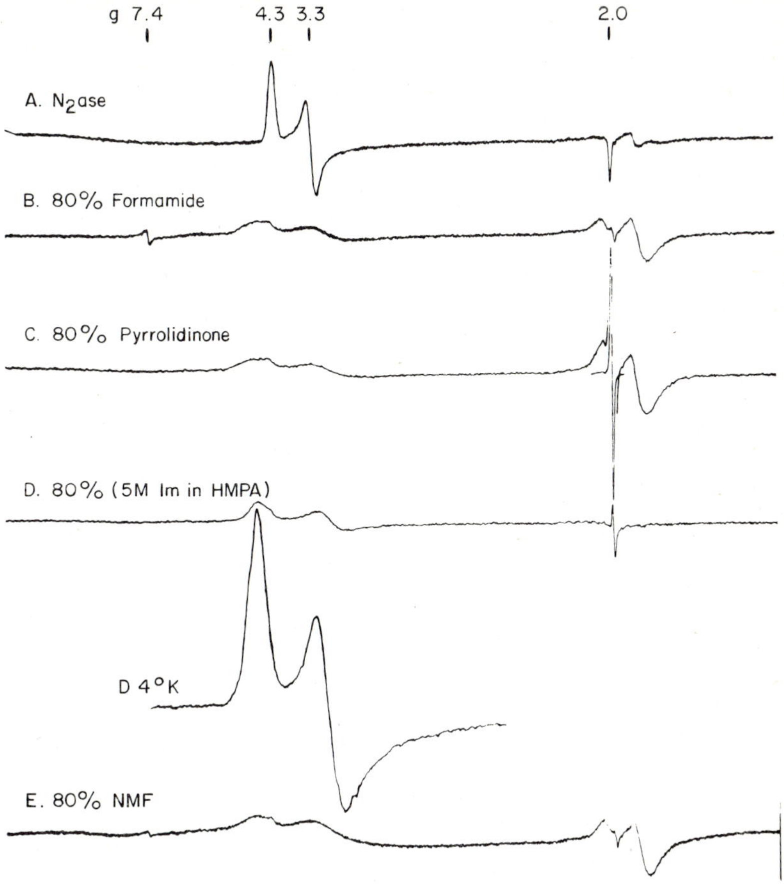

Fig. 9 *Solvent perturbation spectra of Cp1. The protein was dissolved in the solvents indicated, which also contained 10 mM* $Na_2S_2O_4$ *and 100 mM tris chloride, pH 8.5, and then examined by low temperature EPR spectroscopy as in Figure 4B.*

F. *Summary of structural ideas*

In Table III, we give the current evidence on the types of centres in the MoFe protein. There appears to be two clusters involving 8 Fe atoms, which give rise to the g=3.65 EPR signal and the M_{EPR} Mössbauer species. These do not seem to be the typical kind of clusters found in ferredoxins. There are two to three additional clusters which appear to be of the 4Fe-4S variety though they do not normally exhibit themselves in the EPR of reduced states of nitrogenase. This leaves four to eight iron atoms unaccounted for. The H^+- and N_2-reducing sites and the two kinds of CO binding sites may also be Fe centres involving some of these iron atoms. The two M_{EPR} centres probably do not serve any of those functions and M_{EPR} appears to be at or near the site on the MoFe protein which accepts electrons from the Fe protein. Studies of the metal centres in the MoFe protein are clearly at a primitive stage, but the foregoing discussion should indicate that several experimental approaches show promise of unravelling the structures of Mo and Fe centres and that new and interesting metal chemistry is anticipated in addition to the expectation that part of the FeS centres, at least, are variations on familiar themes.

Table IIIA

Evidence about Mo in the MoFe protein

Centre	Atoms Molecule	Clusters Molecule	Comments	References
Mo	1-2	?	heavy atom nearby (EXAFS spectroscopy)	Hodgson, Klein and coworkers (1976)
Mo-cofactor	?	?	guanidine extraction of Cp1, high MW fractions reactivate aponitrate reductase from *nit-1* mutant of *N. crassa*	Zumft (1975)
Mo-cofactor	?	?	solvent extraction of Av1, Cp1, other MoFe proteins, reactivates apoAv1 from UW-45 mutant of Av	Shah and Brill (1976)

V. THE FUNCTIONING ENZYME

Though it is generally agreed that the overall "mechanism" of nitrogenase is as shown in Figure 1, this simple scheme does

Table IIIB
Evidence about Fe in the MoFe protein

Centre	Atoms Molecule	Clusters Molecule	Comments	References
M_{EPR}	8	2	Centre that gives magnetic Mössbauer and EPR of Av1, reduced in fixing state, oxidized by thionine ($E^{\circ\prime}$ 0?)	Münck, *et al.* (1975) Smith and Lang (1974) Zumft, *et al.* (1974)
D and Fe^{2+}	~9 ~3	?	Mössbauer doublet and wide Mössbauer doublet; unchanged in fixing state, oxidized to $S \geqslant 3/2$ state by thionine	Münck, *et al.* (1975) Smith and Lang (1974)
S	1	?	Mössbauer species seen in Av1, Kp1	
M_D	~12	?	Produced from D and Fe^{2+} by thionine oxidation - 4 species of Fe, at least	
fd type 1	~8	~2	Cluster transfer and solvent perturbation; probably not M_{EPR}	
fd type 2	<4	<1	CO-induced signals; both Fd_{red} and $hipip_{ox}$ types observed; subset of type 1?	Davis, *et al.* (1976)
Co-binding sites	?	2?	Inferred from existence of fd type 2, and co-binding to FeS cluster in hydrogenase	Davis, *et al.* (1976)

not contain much information about the catalysis of nitrogen reduction. In general, one would expect that with six reactants, the overall reaction rate with a given substrate would show a complicated dependence on the concentration of the various components of the reaction, and this appears to be true. As will be appreciated, our knowledge of the molecular events in dinitrogen reduction is imperfect, but the studies described in the following sections allow some insight into various aspects of the process. In particular, it is becoming possible to discuss questions such as the composition of the active complex(s), the role of various components in regulating the rates of reduction and possible models of the redox structure of the complex which help explain the non-reciprocal behaviour of the two protein components, variable ratios of MgATP hydrolyzed per electron transferred and the question of how many and what kinds of substrate reducing sites are present.

A. *Kinetics of nitrogenase*

One view of nitrogenase (Hardy, *et al.*, 1971) might be that it has a number of partially overlapping sites for substrate reduction, including sites for H_2 evolution, N_2 reduction, and N_3^-, C_2H_2, CN^-, and $CNCH_3$ reduction. Hwang, *et al.* (1973) presented a large body of evidence interpreted in this way. An alternative view is that a series of states of nitrogenase exist that are capable of catalyzing the various reactions observed, perhaps at the same sites. Silverstein and Bulen (1970) observed that the level of ATP present influenced the competition between H^+ and N_2 for elections, low [ATP] favouring the former. The idea emerges that the level of ATP affects the distribution of nitrogenase molecules between more- and less-reduced forms. Davis, *et al.* (1975) have discussed this matter and pointed out that one ought to be able to decide between these models given that the function of the Fe protein is to reduce the MoFe protein, which then reduces substrates. The former model predicts that the competition between substrates should be a function of substrate competition only while the latter model predicts that the ratio of the protein components as well as ATP levels should affect these competitors. Using Av nitrogenase, they verified the results of Silverstein and Bulen and also demonstrated that component ratios affect the competition of C_2H_2 and N_2. They conclude that the effect of raising the Fe protein concentration (in the presence of an efficient ATP source) is to raise the proportion of highly reduced nitrogenase molecules, thus favouring the reduction of the 6-e^- oxidant, N_2. Lower Fe protein levels enhance the proportion of less reduced enzyme molecules and favour 2-e^- reductions (H^+ and C_2H_2). Similar conclusions were reached by Rivera-Ortiz and Burris (1975).

Thorneley (1975) studied the MgATP-induced electron flow between the components of nitrogenase, using the stopped-flow technique. Using a ΔA_{425} of 2000 M^{-1} cm^{-1} (*cf.* discussion in II-D), he showed that up to three equivalents of Kp2 are rapidly oxidized by Kp1 with a *unimolecular* rate constant of 2 x 10^2 s^{-1}. The important implication is that preformation of complexes between the reactants is rapid. Thorneley also reasons from the conclusion of Walker and Mortenson (1973) (that Cp2 is a $2e^-$ reductant) that up to $6e^-$ (the number required for N_2 reduction are rapidly transferred. The subsequent findings of Thorneley, *et al.* (1976) that the initial ΔA_{425} corresponds to a $1e^-$ transfer, as well as the findings of Ljones (1973) referred to earlier, make this number very doubtful. Whatever the true number, Thorneley's finding that electron transfer appears to be about 10^2 faster than N_2 reduction, as well as his conclusion that MgATP binding (to Kp2 or to the enzyme complex) is also very rapid ($k > 2.5 \times 10^6 M^{-1}s^{-1}$) compared to turnover, are important points.

Using a somewhat different form of the enzyme, the complex of Av1 and Av2 (*ca.*, 1:1) as isolated, Watt, *et al.* (1975) have studied the stoichiometry and ATP/$2e^-$ values over a range of temperatures. They found the latter parameter to go through a minimum (*ca.* 4.5 mole MgATP hydrolyzed per $2e^-$ transferred) at 20°C. This work clearly points to different rate constants, including binding constants, contributing to the limiting elementary processes, at different temperatures. This phenomenon might be useful, for example with freeze-quenched Mössbauer spectroscopy, in studying electron transfers between the metal centres as well as possibly distinguishing various classes of centres. The thermal sensitivity of complex formation (Thorneley, *et al.*, 1975) is discussed below.

Watt and Burns (unpublished) and Watt (this volume) have also examined the kinetics of dithionite utilization. They conclude that the reaction rate law depends on $[S_2O_4^{2-}]^{1/2}$, *i.e.*, that the actual reductant is $SO_2^{\cdot -}$, as independently deduced by Yates, *et al* (1975) from presteady-state measurements. This reinforces the conclusion that the Fe protein acts as a $1e^-$ *acceptor*. The rate law also contains a $[MgATP]^2$ term, consistent with the concept that the Fe protein binds *two* MgATP and transfers *one* electron to the MoFe protein. Hageman and Burris (personal communication) have verified the $[S_2O_4^{=}]^{1/2}$ dependence using *purified* Av components and have found that K/V for varying $[S_2O_4^{=}]$ depends on the concentration of Av2 and tends to a finite value as [Av2] is raised.

B. *The assay of nitrogenase*[3]

Since the acetylene reduction assay is often used as a rapid method for the estimation of nitrogenase both in the laboratory and in the field, it seems important to know what the rate of N_2 reduction by a given system would be given an observed rate of C_2H_2 reduction. Shah, *et al.* (1975) have examined this question with respect to Av nitrogenase. Using purified Av1 and Av2, they find that at Av1:Av2 = 0.13:1, the ratio of rate of C_2H_2 reduction to rate of N_2 reduction is 3.9, declining to 3.4 near a 1:1 ratio of components, and rising to 5.3 when Av1:Av2 = 30:1·Preferential inactivation of the more labile Fe protein in cells or extracts can skew the results is a constant ratio of acetylene:N_2 reduction is assumed. Acetylene is an inhibitor of nitrogenase as well as a substrate, and both the apparent Ki and Km depend on ratio Av1:Av2. At high ratios, inhibition sets in at 0.05 atm, well below the levels normally employed in such assays.

C. *Complex formation between the MoFe protein and the Fe protein*[4]

Because the MoFe component probably contains the substrate reduction site, one might assign the Fe protein two quite different alternative roles in the overall enzyme reaction. As one extreme, it may be treated as a substrate in a coupled reaction consistent with some kinetic studies (Bergersen and Turner, 1973). At the other extreme, one may assume that there is a tight complex formed between the protein components with dissociation occurring very rarely relative to the frequency of enzyme turnovers. This type of model has been presented by the Sussex group (Eady, 1973; Thorneley, *et al.*, 1975). Certainly the correct answer lies somewhere in between these extremes. We have begun some systematic explorations of the properties of various models for the component interactions in order to generate some predictions of nitrogenase behaviour as a function of several variables which might be tested.

It has been recognized for ten years that the functioning of nitrogenase requires the interaction of two easily separable protein components. There have been numerous attempts to estimate the stoichiometry of the interaction, based on the assumption that the active enzyme results from formation of a complex with a reasonably defined stoichiometry (Silverstein and Bulen, 1971; Eady, 1973; Vandecastelle and Burris, 1970). Technical difficulties in attempts to determine the interaction stoichiometry directly by mass transport methods (Eady, 1973) have led to attempts to determine the properties of the complex by inference from kinetic experiments. By developing a suitable computer model and fitting output from the model to the observed experimental points, one might be able to infer the nature of the interaction and the composition of the complex.

Any such process of modelling must meet certain criteria before being accepted. Firstly, it must provide a reasonable fit to all of the available experimental data. Secondly, it must not contain assumptions, which cannot be tested. Thirdly, it ought to provide some predictions, which can be tested. If the inference from kinetic experiments to mass transport experiments is to be valid, there ought not to be inexplicable discrepancies between the results of different kinetic experiments. A model failing any of these criteria is inadequate. In practice, these criteria are simultaneously applied and it can readily be demonstrated that any *simple* model of nitrogenase complex formation fails to meet these criteria.

(1) MoFe protein + Fe protein $\underset{1s^{-1}}{\overset{10^7 s^{-1} M^{-1}}{\rightleftharpoons}}$ binary complex (active species)

Thorneley *et al.* (1975)

(2) MoFe protein + Fe protein $\rightleftharpoons$ binary complex (inactive species)

binary complex + Fe protein $\rightleftharpoons$ ternary complex (active species)

(3) MoFe protein + Fe protein $\rightleftharpoons$ binary complex (active species)

binary complex + MoFe protein $\rightleftharpoons$ ternary complex (inactive species)

Fig. 10 *Alternative Schemes for Complexation of Nitrogenase Components.*

The simplest model possible for association of the nitrogenase components is to assume that one mole of each forms the active enzyme (Thorneley, 1975; Thorneley, *et al.*, 1975) as in Figure 10(1). This leads to relatively simple behaviour in which addition of aliquots of one component to a fixed amount of the other yields a hyperbolic increase in the concentration of active complex. The variation of either component holding the other constant, should give a comparable effect. Tightness of binding (Keq) alters the relative position but not the shape of the titration curve. For any fixed ratio of components, a "dilution effect" is obtained which can be overcome by addition of either component. Thorneley, *et al.* (1975)[5] used this type of model for association of nitrogenase components and derived therefrom an estimate of the association constant of the components (*ca.* $10^{-7} m^{-1}$). The inadequacy of this model is apparent on simple inspection. Titration of nitrogenase components against one another never yields true hyperbolic plots, nor is there any clear reciprocity between the shapes of titration curves obtained holding constant either the MoFe protein or the Fe protein (Figure 11).

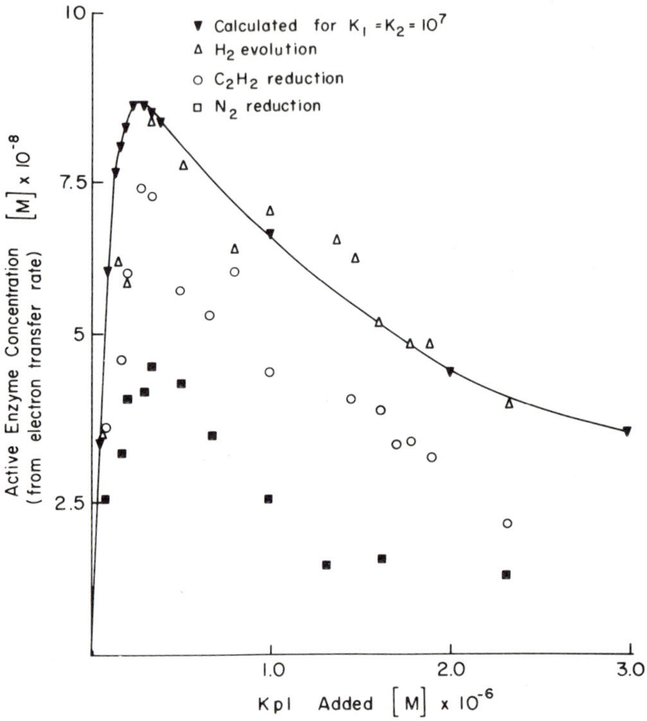

Fig. 11 *Titration of Kp2 by Kp1. Activities were derived as for Figure 11 with the assumption that the concentration of Kp2 is 0.36 μM rather than 0.42 μM as assumed by Eady,* et al. *(1972) from whose work the experimental data is taken. Solid inverted triangles are calculated results with K_1 and K_2 both equal to 10^7; open triangles, hydrogen evolution data; open circles, acetylene reduction data; solid squares, nitrogen reduction data.*

A model in much better agreement with observed results is one in which there is formation of a 1:1 complex which is inactive, followed by a second round of association to make a 1:2 complex which is the active enzyme (Figure 10(2)). Mortenson proposed this sort of model a number of years ago (lecture

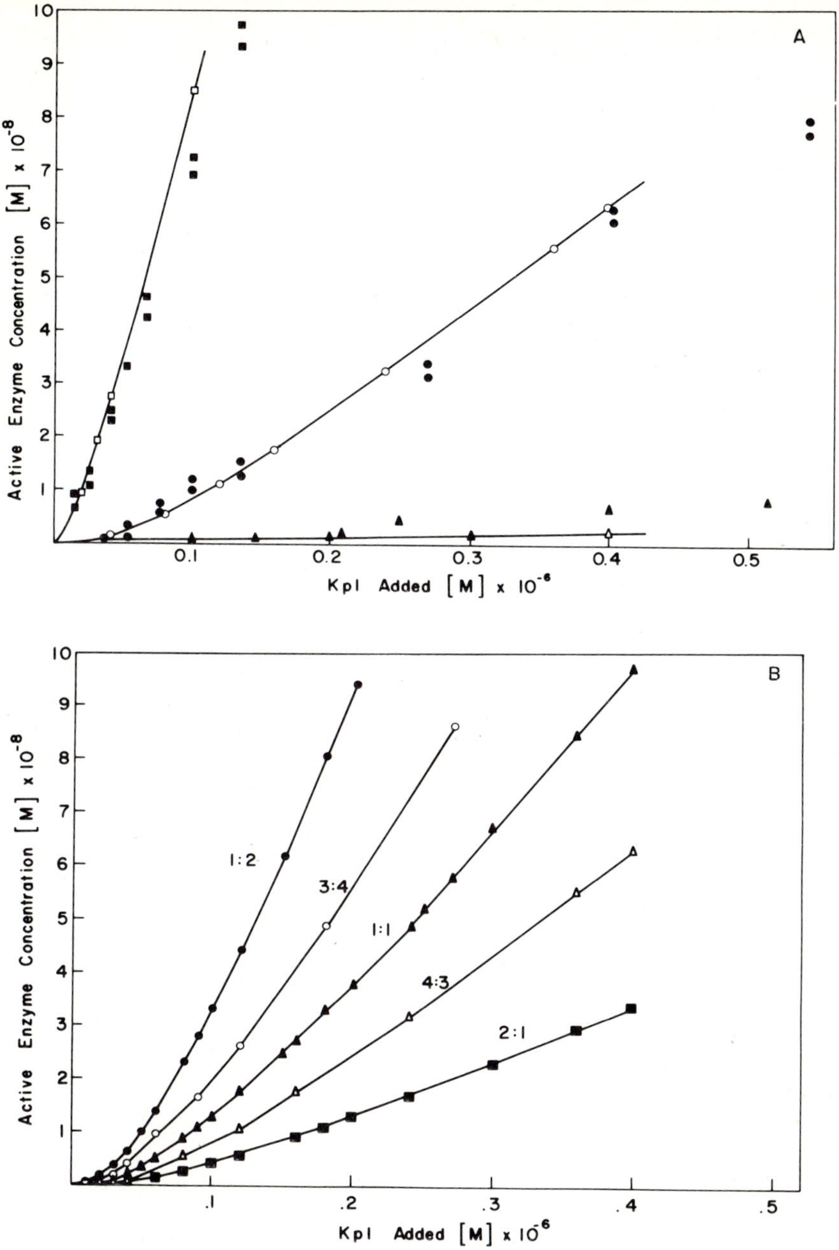
A
Active Enzyme Concentration [M] x 10⁻⁸
Kpl Added [M] x 10⁻⁶
B
1:2
3:4
1:1
4:3
2:1
Active Enzyme Concentration [M] x 10⁻⁸
Kpl Added [M] x 10⁻⁶

given at New York University, Oct. 1969). By appropriate choice of the association constants for the two successive reactions and of the apparent molar concentrations of the protein components, one can easily obtain a fit to a set of data for reduction of a single substrate by a particular enzyme preparation. The results of a fit to the data of Eady, *et al.* (1972) are seen in Figures 11 and 12.

The use of two equilibrium constants plus two molar concentration terms provides four adjustable parameters which makes fitting to a single simple substrate reduction curve rather easy. This can lead to an illusion of reliability in the fit. However, the adjustable parameters are not completely independent or unique so that several combinations can provide a rather similar result. The truly limiting factor for fitting becomes the analytical methods with a finite error (which may be random or systematic, constant magnitude or constant fraction error). One must be able to distinguish between several comparable fits to data. We have done fittings of nitrogenase activity curves for enzyme from Av and Cp as well as for Kp. Using rather similar association constants and assumptions for the activity and molar concentrations, we can get a reasonable fit in each case. The data for Kp is chosen for illustration as it is already published and is as complete as any available to us in our laboratory in that it includes both dilution

---

Fig. 12 (A) *Dilution effect in* Klebsiella pneumoniae *nitrogenase. Experimental data is taken from Thorneley,* et al. *(1975) and replotted with the assumption that the effective concentration of Kp2 is 3/4 of that proposed by those authors. Specific activities of pure components are assumed to be 3000 nmoles/min/mg protein and molecular weights are taken as 2000,000 and 60,000 for the MoFe and Fe protein components, respectively. Calculated molar concentrations of complex are for a molar ratio as indicated with association constants of* 10^7 *for both* K_1 *and* K_2 *(Figure 10B). Solid circles, experimental data for a nominal 1:1 mixture; open circles, calculated for a 4:3 mixture; solid squares, experimental data for nominal 1:12 mixture; open squares, calculated for a 1:9 mixture; solid triangles, for a nominal 8:1 mixture; open triangles, for a calculated 10.6:1 mixture.* (B) *Dilution effect as a function of small changes in protein component ratio. Calculated concentrations of active enzyme complex for reaction Kp1 + Kp2 Complex; Complex + Kp2 Enzyme (Figure 10(2)) with association constants of* 10^7 *for both reactions. Molar concentration of Kp1 as indicated. Solid circles, 1:2 ratio Kp1:Kp2; open circles, 3:4; solid triangles, 1:1; open triangles, 4:3; solid squares, 2:1.*

curves and titration patterns for three different substrates.

In order to increase confidence in the model, it is necessary to measure another enzyme variable, such as activity with a different substrate. In this case, if the proposed model is unique and correct, the enzyme activity profile should fit the previously developed set of parameters. This is the point at which simple nitrogenase models break down. In Figure 12, we show a fit of calculated complex concentration to experimental data (Eady, *et al.*, 1972) for several reactions of Kp nitrogenase. Assuming that hydrogen evolution is the most direct measure of the activity of nitrogenase as a two-electron transferring enzyme, we have made a "best fit" to that activity curve. Clearly, the fit to substrate reduction curves is only a qualitative one. Changing parameters to fit to the hydrogen evolution data does not improve the fit to the other data.

These results indicate some of the deficiences of simple models for nitrogenase. Acetylene is not a good substrate for kinetic studies of nitrogenase because in the case of *Azotobacter*, there is a substrate inhibition by high levels of acetylene at high ratios of MoFe:Fe proteins (Shah, *et al.*, 1975) as is found with *Klebsiella pneumoniae*. In *C. pasteurianum* by contrast, there is a "substrate activation" which also varies as a function of the component ratio, (Davis, *et al.*, unpublished). Thus, acetylene reduction results require different models for various types of nitrogenase. Nitrogen reduction clearly takes 3-times as many electrons as acetylene reduction so that whatever mechanism one chooses, the electron requirement per active centre per turnover is greater and nitrogen reduction has to compete with the alternate pathway of hydrogen evolution. Thus, one might expect that unless intramolecular electron transfers were rate-limiting, there would be conditions under which electron supply for nitrogen reduction would be more limited than that for hydrogen evolution or acetylene reduction. These turn out to be low concentration of reductant, low ATP and low relative level of Fe protein, in accord with the experimental observations (Silverstein and Bulen, 1970; Davis, *et al.*, 1975) but this cannot be accommodated in the *simple* model for component associations. By assuming that further rounds of association are required for nitrogen reduction or by changing the ratio of the component association constants for different substrates, one can obtain a better fit to the data for nitrogen reduction. The further approach makes a more complex model with more adjustable parameters and is harder to test. The latter introduces additional mechanistic complications in the form of "allosteric interactions."

We conclude that it is premature to propose a model for

nitrogenase component interactions on the basis of the available data for the functioning enzyme. In order to develop a model with any confidence, it will be necessary to measure many more variables than there are adjustable parameters in the model, or else to make use of mass transport methods to directly observe the association phenomenon. Given the lability of the enzyme and the complex kinetics which it exhibits, we favour the latter approach. An entirely different approach, employed by Emerich and Burris, is described below.

The fit of calculated to experimental data shown in the figures can be improved significantly by further refinements in the fitting procedure. Systematic variation of the molar ratio by increments of 5-10% and of the association constants by comparable amounts would allow one to find a set of numbers which provided a good least squares fit to all three of the curves in Figure 11. However the experimental unreliability of 5% or so (Thorneley, *et al.*, 1975) in determining protein specific activities sets a reliability limit on derived parameters. For the data of Figure 12, the variability is considerably larger so there would be no advantage to further refinement in the calculations.

D. *Heterologous component crosses as a kinetic tool*

The interaction of nitrogenase components from heterologous organisms was examined by Detroy, *et al.* (1968; see also Burris, 1971). They used nitrogen reduction as a criterion and obtained the results shown in Table IV. Since the work of Detroy, *et al.*, the relationship of the two components in the functioning of nitrogenase has become somewhat clearer and various workers have begun to study the properties of heterologous crosses for further insight into the mechanism of nitrogenase. Smith, *et al.* (1976) using purified components, have examined the activity of Kp1 + Cp2 and Cp1 + Kp2, both of which exhibit little or no nitrogen reduction capacity (Table IV). Smith, *et al.* find that the Cp1 plus Kp2 combination also does not reduce H^+ or C_2H_2 and has an ATPase activity only about 1/200*th* that of Kp nitrogenase. However, the Kp1 plus Cp2 combination has interesting and informative properties. The steady state rate of H^+ and C_2H_2 reduction rate was linear down to the resolution of the method (*ca.* 10 sec) while C_2H_2 reduction showed a 10 min lag at 30° C before the rate of reduction became linear, implying that an activation step precedes the appearance of the C_2H_2-reducing activity, a step not needed for H^+ reduction. Stopped flow measurements indicated that electron flow from Cp2 to Kp1 at the same rate ($2.5 \times 10^2 s^{-1}$) and the same dependence on [MgATP] (Km=0.4 mM) as from Kp2 to Kp1

(Thorneley, 1975). However, the ATP hydrolyzed/$2e^-$ ratio was *50* for H^+ reduction. Smith, *et al.*, suggested that there are at least two roles for ATP in nitrogenase activity. They propose that the Fe protein-MgATP complex can reduce the MoFe protein rapidly to the H^+-reducing form and that further events, perhaps requiring MgATP, occur before nitrogenase is competent to reduce C_2H_2 and presumably, N_2. The use of this technique in EPR and Mössbauer experiments is of obvious interest for further studies.

Table IV

Nitrogen reduction ability of mixtures of nitrogenase components from different organisms[a]

Fe protein	*MoFe Protein* Av1	Kp1	Bp1	Cp1
Av2	+	+	+	-
Kp2	+	+	+	-
Bp2	-	+	+	+
Cp2	-	-	-	+

[a]From Detroy, *et al.* (1968); partially purified extracts were used and the combinations were tested for N_2 reduction activity. + = complementary function.

Emmerich and Burris (1976) have undertaken a comprehensive study of heterologous mixtures, using homologous and heterologous proteins as competitors and analyzing steady state data by classical techniques (Cleland, 1970). When Av1 plus Av2 mixtures are assayed in the presence of various levels of Cp2, for example, analysis of the data indicates that two Cp2 molecules are bound tightly to each Av1 in the inhibited complex. All activities of nitrogenase (reduction of N_2, C_2H_2 and H^+, and hydrolysis of MgATP) are inhibited. Complex formation requires active components and oxygen-denatured components were ineffective as inhibitors. Some of the experiments suggested that a large number of Cp2 molecules can bind Av1. Overall, the results suggest that the Fe protein from each organism competes for the same site(s) on the MoFe protein, the implication being that the *active complex* contains two or more Fe proteins per MoFe proteins. This is of obvious interest given the current lack of agreement on the composition of active nitrogenase. It is critical in such studies, which are essentially titrations, that the proportion of active molecules in each preparation be known rather exactly. Emerich and Burris used Cp2 and Av1 with activities of around 2100 and Av2 with an activity of *ca.* 1000. They assume that the specific activities of fully active nitrogenase is 3000, for the purposes of their calculations. This figure is not unreasonable in light of recent reports (see Zumft and Mortenson, 1975; Orme-Johnson and Davis, 1976 for reviews), and it should be noted that for their results to be changed by a factor of 2, *i.e.*, to a 1:1 ratio of components in the active complex, the "ultimate" specific activity of nitrogenase components would have to approach 6000. Not all of the possible combinations of components are equally useful, since, for example, Cp1 and Av2 do not readily form a complex, but the technique introduced by Emerich and Burris should prove useful for a variety of chemical and physical studies.

E. *Some physiological consequences of the properties of nitrogenase*

It is apparent that an organism that carries out N_2 reduction must supply an anaerobic environment, a supply of electrons at $E°' < -400$mv, and an efficient MgATP generator. The *in vitro* studies suggest that the efficiency of the system is remarkably affected by the Fe protein:MoFe protein ratio. If we can accept for the moment that the Fe protein binds 2 MgATP and transfers 1e^- on a kinetically significant timescale, then it is apparent that the Fe protein has at least part of the apparatus (not expressed unless the MoFe protein is present) for energy transduction, *i.e.*, conversion of the hydrolysis energy of MgATP into a low potential for the electrons. This conversion makes H^+ reduction irreversible and makes possible the initial unfavourable steps in N_2 reduction. Since all substrates require two or more e^- and since H^+ reduction is thermodynamically favoured for $E°' < -400$mv (not counting the effective $\Delta E°'$ furnished by MgATP hydrolysis), it is not surprising that this energy transduction machine leaks electrons onto the solvent and that N_2 is not an efficient competitor for the electrons. How then does one get high MgATP hydrolyzed/e^- transferred ratios? It seems a reasonable hypothesis that, in addition to the above properties, there is also a finite probability of the occurrence of a futile cycle. Therefore, only 1 e^- is injected into the MoFe protein (suppose that only one Fe protein is bound or only one Fe protein is complexed at the moment to MgATP) then, if a further e^- is not forthcoming within a short (but unknown) time, there is some probability that the electron will simply return to the Fe protein, leaving 2 MgATP hydrolyzed with no resultant substrate reduction. This, and the evidence that 2 or more Fe proteins are required for the active complex, explains the variable hydrolysis/e^- transfer ratios and emphasizes that in "synthetic organisms" one should keep in mind that an excess of Fe protein, of the proper kind (*i.e.*, effective at reducing the MoFe protein), should be provided.

A second feature of interest is the remarkable effect of MgADP/MgATP concentration ratios on the functioning of nitrogenase. MgADP blocks the reduction of the Fe protein (Yates, *et al.*, 1975) which explains its potent inhibitory power. Haaker, *et al.*, (1974) and Veeger (this volume) proposed on the basis of *in vivo* experiments with Av, in which the ADP/ATP ratio and redox potential were manipulated more or less independently, that it is the nucleotide ratio that controls the rate of nitrogenase activity. Davis and Orme-Johnson (1976) examined Cp nitrogenase as a function of the MgATP generator capacity and showed (Figure 13) that, as proposed by Haaker, *et al.* and by Bergersen and Turner (1973), nitrogenase is appropriately sensitive over a physiologically reasonable range of ADP/ATP ratios. The reason is that the nitrogenase locale in the cell not only has to be anaerobic and be furnished with electrons and MgATP, but that the regeneratory apparatus for MgATP should be quite close by, if nitrogenase is to function at its maximum efficiency. The couple between nitrogenase and the ATP generator is highly self-regulating, a property that may be of considerable survival value to anaerobes but perhaps somewhat inconvenient in contrived situations of the sort currently aimed at by geneticists and agronomists.

FOOTNOTES

[1] Symbols: Kp, *Klebsiella pneumoniae;* Cp, *Clostridium pasteurianum;* Cv, *Chromatium vinosum;* Av, *Azotobacter vinelandii;* Ac, *Azotobacter chroococcum;* Rj, *Rhizobium japonicum;* Rr, *Rhodospirillum rubrum;* Sl, *Spirillum lipoferum.*

[2] Hexamethylphosphoramide is a carcinogen. All of the polar solvents useful for unfolding iron-sulfur proteins, such as dimethylsulfoxide, formamide, N-methylformamide, N,N,dimethylformamide, N-methyl-pyrrolidone, dimethyl acetamide, etc., should be handled with some care.

[3] Specific activities are conventionally given as nmoles substrate (often acetylene) reduced per min per mg protein. A given component of nitrogenase is assayed in the presence of an excess of the other; for the Fe protein, MoFe protein is titrated in until a plateau of activity is reached beyond which

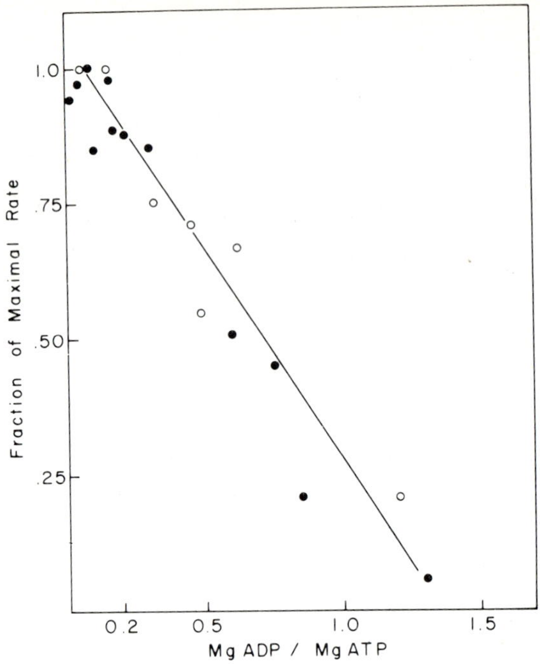

Fig. 13 *Control of the rate of Cp nitrogenase (dithionite oxidation) by the ADP/ATP ratio, which was set by adjusting the level of creatine kinase/phosphocreatine ATP-regenerating system. The actual nucleotide levels were determined for each condition. Solid circles, excess Cp2; open circles, limiting Cp2. From Davis and Orme-Johnson (1976).*

---

the activity declines. Assays of the MoFe protein (excess Fe protein) can be less carefully done as excess Fe protein does not inhibit. For a recent paper dealing with these and other complexities, see Shah, *et al.* (1975).

[4]The computer simulations of alternative models for complex formation were performed by L.C. Davis while at Kansas State University. Professor D.J. Cox is thanked for helpful discussions.

[5]These authors find that the association constant for Kp nitrogenase showed a sharp break at 17°C with $\Delta H=0$ above and $\Delta H=418$ kJ $mole^{-1}$ below this temperature. E‡ for acetylene reduction was 80 kJ $mole^{-1}$ between 12 and 40°C. These results are

probably somewhat less sensitive to the model employed than the actual composition of the complex, the main concern of this section.

ACKNOWLEDGEMENT

The NIH is thanked for support of original work reported herein under GM 17170, and the NSF is thanked for postdoctoral fellowship support to L.C.D. and B.A.A. P. Ludden, D. Emerich, R. Hageman, R.H. Burris, G. Watt, V.K. Shah, and W.J. Brill are thanked for permission to quote from work in advance of publication as well as for stimulating discussions of the ideas put forth here.

REFERENCES

Aasa, R. and Vanngard, T. (1975). *J.Mag.Res.* 19, 308

Albrecht, S.L. and Evans, M.C.W. (1973). *Biochem.Biophys.Res. Comm.* 55, 1009.

Bray, R.C. (1961). *Biochem.J.* 81, 189.

Burris, R.H. (1971). In "Chemistry and Biochemistry of Nitrogen Fixation" (J.R. Postgate, ed.), pp. 105-160. Plenum Press, London.

Burris, R.H. and Orme-Johnson, W.H. (1974). In "Microbial Iron Metabolism" (J. Neilands, ed.), p. 187. Academic Press, New York.

Cammack, R. and Evans, M.C.W. (1975). *Biochem.Biophys.Res. Comm.* 67, 544.

Cleland, W.W. (1970). In "The Enzymes" (P.D. Boyer, ed.), 3rd ed., Vol. 2, p. 1. Academic Press, New York.

Creutz, C. and Sutin, N. (1973). *Proc.Nat.Acad.Sci.U.S.A.* 70, 1701.

Davis, L.C., Shah, V.K. and Brill, W.J. (1975). *Biochim.Biophys.Acta* 384, 353.

Detroy, R.W., Witz, D.F., Parejko, R.A. and Wilson, P.W. (1968). *Proc.Nat.Acad.Sci.U.S.A.* 61, 537.

Dunham, W.R. and Bearden, A.J. (1971). *Biochim.Biophys.Acta* 253, 134.

Eady, R.R. (1973). *Biochem.J.* 135, 531.

Eady, R.R. and Postgate, J.R. (1974). *Nature* 249, 805.

Eady, R.R., Smith, B.E., Cook, K.A. and Postgate, J.R. (1972). *Biochemistry* 128, 655.

Emerich, D. and Burris, R.H. (1976). *Proc.Nat.Acad.Sci.U.S.A.* in press.

Erbes, D.L., Burris, R.H. and Orme-Johnson, W.H. (1975). *Proc. Nat.Acad.Sci.U.S.A.* 72, 4795.

Evans, M.C.W. and Albrecht, S.L. (1974). *Biochem.Biophys.Res. Comm.* 61, 1187.

Evans, M.C.W., Telfer, A. and Smith, R.V. (1973). *Biochim.Biophys.Acta* 310, 344.
Haaker, H., deKok, A. and Veeger, C. (1974). *Biochim.Biophys. Acta* 357, 344.
Hardy, R.W.F., Burns, R.C. and Parshall, G.W. (1971). In "Bioinorganic Chemistry" (R.F. Gould, ed.) 100, p. 219. Advances in Chem. Series, American Chemical Society Publications, Washington, D.C.
Holm, R. and Ibers, J.A. (1976). In "Iron Sulfur Proteins" (W. Lovenberg, ed.), Vol. 3, in press. Academic Press, New York.
Huang, T.C., Zumft, W.G. and Mortenson, L.E. (1973). *J.Bacteriol.* 113, 884.
Hwang, J.C., Chen, C.H. and Burris, R.H. (1973). *Biochim.Biophys.Acta* 292, 256
Kajiyama, S., Matsuki, T. and Nosoh, Y. (1969). *Biochem.Biophys.Res.Comm.* 37, 711.
Kennedy, C., Eady, R.R. and Kondorosi, E. (1976). *Biochem.J.* 155, 383.
Kleiner, D. and Chen, C.H. (1974). *Arch.Microbiol.* 98, 93.
Lambeth, D.O. and Palmer, G. (1973). *J.Biol.Chem.* 248, 6095.
Ljones, T. (1973). *Biochim.Biophys.Acta* 321, 103.
Ljones, T. and Burris, R.H. (1972). *Biochim.Biophys.Acta* 275, 93.
Ludden, P.W. and Burris, R.H. (1976). *Science*, in press.
Mathews, R., Carlton, S., Sands, R.H. and Palmer, G. (1974). *J.Biol.Chem.* 249, 4326.
Mortenson, L.E., Walker, M.N. and Walker, G.A. (1976). In "Proceedings of the First International Symposium on Nitrogen Fixation" (W.E. Newton and C.J. Nyman, eds.) Vol. 1, p. 117. Washington State University Press, Pullman, Washington.
Mortenson, L.E., Zumft, W.G. and Palmer, G. (1973). *Biochim. Biophys.Acta* 292, 422.
Munck, E., Rhodes, H., Orme-Johnson, W.H., Davis, L.C., Brill, W.J. and Shah, V.K. (1975). *Biochim.Biophys.Acta* 400, 32
Nagatani, H.H., Shah, V.K. and Brill, W.J. (1974). *J.Bacteriol.* 120, 697.
Newton, W.E. and Nyman, C.J. (eds.) (1976). "Proceedings of the First International Symposium on Nitrogen Fixation," Washington State Press, Pullman, Washington.
Orme-Johnson, W.H. (1973). *Ann.Rev.Biochem.* 42, 159.
Orme-Johnson, W.H. and Davis, L.C. (1976). In "Iron Sulfur Proteins" (W. Lovenberg, ed.), Vol. 3, in press. Academic Press, New York.
Orme-Johnson, W.H., Hamilton, W.D., Jones, T.L., Tso, M.-Y.T., Burris, R.H., Shah, V.K. and Brill, W.J. (1972). *Proc.*

Nat.Acad.Sci.U.S.A. 69, 3142.
Orme-Johnson, W.H., Hansen, R.E., Beinert, H., Tsibris, J.C.M., Bartholomaus, R.C. and Gunsalus, I.C. (1968). *Proc.Nat. Acad.Sci.U.S.A.* 60, 368.
Palmer, G., Multani, J.S., Cretney, W.C., Zumft, W.G. and Mortenson, L.E. (1972). *Arch.Biochem.Biophys.* 153, 325.
Pienkos, P., Shah, V.K., Gordon, J. and Brill, W.J. (1976). In "Recent Advances in Bioinorganic Chemistry" (K.N. Raymond, ed.) in press. American Chemical Society Publications, Washington, D.C.
Rivera-Ortiz, J.M. and Burris, R.H. (1975). *J.Bacteriol.* 123, 537.
Shah, V.K., Davis, L.C. and Brill, W.J. (1975). *Biochim.Biophys.Acta* 384, 353.
Silverstein, R. and Bulen, W.A. (1970). *Biochemistry* 9, 3809.
Smith, B.E., Lowe, D.J. and Bray, R.C. (1973). *Biochem.J.* 135, 331.
Smith, B.E., Thorneley, R.N.R., Eady, R.R. and Mortenson, L.E. (1976). *Biochem.J.* 157, 439.
Smith, B.E., Thorneley, R.N.F., Yates, M.G., Eady, R.R. and Postgate, J.R. (1976). In "Proceedings of the First International Symposium on Nitrogen Fixation" (W.E. Newton and C.J. Nyman, eds.). Vol. 1, p. 150. Washington State University Press, Pullman, Washington, D.C.
Smith, B.E. and Lang, G. (1974). *Biochem.J.* 137, 169.
Smith, B.E., Lowe, D.J. and Bray, R.C. (1972). *Biochem.J.* 135, 331.
Stombaugh, N.A., Burris, R.H. and Orme-Johnson, W.H. (1973). *J.Biol.Chem.* 248, 7951.
Stombaugh, N.A., Sundquist, J.E., Burris, R.H. and Orme-Johnson, W.H. (1976). *Biochemistry* 15, 2633.
Taylor, K.B. (1969). *J.Biol.Chem.* 244, 171.
Thorneley, R.N.F. (1974). *Biochim.Biophys.Acta* 333, 487.
Thorneley, R.N.F. (1975). *Biochem.J.* 145, 391.
Thorneley, R.N.F., Eady, R.R. and Yates, M.G. (1975). *Biochim. Biophys.Acta* 403, 269.
Thorneley, R.N.F. and Willison, K.R. (1974). *Biochem.J.* 139, 211.
Thorneley, R.N.F., Yates, M.G. and Lowe, D.J. (1976). *Biochem. J.* 155, 137.
Tso, M.-Y.W. (1974). *Arch.Microbiol.* 99, 71.
Tso, M.-Y.W. and Burris, R.H. (1973). *Biochim.Biophys.Acta* 309, 263.
Tso, M.-Y.W., Ljones, T. and Burris, R.H. (1972). *Biochim. Biophys.Acta* 267, 600.
Walker, M. and Mortenson, L.E. (1973). *Biochem.Biophys.Res. Comm.* 54, 669.

Walker, M. and Mortenson, L.E. (1974). *J.Biol.Chem.* 249, 6356.
Watt, G.D., Bulen, W.A., Burns, A. and Hadfield, K.L. (1975). *Biochemistry* 14, 4266.
Watt, G.D. and Bulen, W.A. (1976). In "Proceedings of the First International Symposium on Nitrogen Fixation" (W.E. Newton and C.J. Nymans, eds.), Vol. 1, p. 248. Washington State University Press, Pullman, Washington, D.C.
Winter, H.C. and Burris, R.H. (1976). *Ann.Rev.Biochem.* 45, 409.
Yates, M.G., Thorneley, R.N.F. and Lowe, D.J. (1975). *FEBS Lett.* 60, 89.
Yates, M.G. and Planque, K. (1975). *Eur.J.Biochem.* 60, 467.
Zumft, W.G., Cretney, W.C., Huang, T.C., Mortenson, L.E. and Palmer, G. (1972). *Biochem.Biophys.Res.Comm.* 48, 1525.
Zumft, W.G. and Mortenson, L.E. (1975). *Biochim.Biophys.Acta* 416, 1.
Zumft, W.G. (1974). *Ber.Dtsch.Bot.Ges.* 87, 135.
Zumft, W.G., Mortenson, L.E. and Palmer, G. (1974). *Eur.J. Biochem.* 46, 525.
Zumft, W.G., Palmer, G. and Mortenson, L.E. (1973). *Biochim. Biophys.Acta* 292, 413.

KINETIC STUDIES OF $MgATP^{2-}$ AND $S_2O_4{}^{2-}$ UTILIZATION BY *AZOTOBACTER VINELANDII* NITROGENASE

G. D. WATT

The Charles F. Kettering Research Laboratory, Yellow Springs, OH 45387, U.S.A.

INTRODUCTION

A systematic and detailed analysis of steady-state kinetic data for enzyme-catalyzed reactions provides a means for obtaining important information concerning the catalytic mechanism of the enzyme being examined (Cleland, 1963; Cleland, 1970). Such a thorough analysis of nitrogenase catalysis would be desirable but, because of the complexity of this enzyme system, only partial kinetic studies dealing with selected aspects of nitrogenase catalysis have so far been undertaken (Zumft and Mortenson, 1975).

The kinetics of competition of the various reducible substrates of nitrogenase for electrons provided by nitrogenase has been studied (Hwang *et al.*, 1973; Rivera-Ortiz and Burris, 1975; Davis *et al.*, 1975) with the aim of determining the nature and the number of active sites present on the enzyme. At present, three models offer plausible explanations for nitrogenase-catalyzed substrate reduction reactions. The first (Hardy *et al.*, 1971; Hwang *et al.*, 1973) indicates that perhaps as many as five separate reduction sites are located on nitrogenase. According to this view, the following five sites (or modified sites) are proposed: (1) N_2 reduction and/or H_2 inhibition site; (2) acetylene reduction site; (3) azide, cyanide and isonitrile reduction site; (4) CO inhibition site; and (5) H^+ (H_2O) reduction site. A second alternative interpretation (Newton *et al.*, 1976) explains substrate reduction at a single site by making use of the varying electronic structures of the substrates and their requirements for binding and reduction. A third model (Silverstein and Bulen, 1970; Davis *et al.*, 1975) suggests that fewer than five sites exist and that substrate reduction specificity results from the number of electrons available at the reducing site. According to this view, when low levels of electrons are present, two-electron reductions are favoured, whereas six-electron reductions occur only when higher levels of electrons are reached. The details of how high reduction capacity is attained by the enzyme are not at all clear but such factors as temperature, the Fe protein ratio, and the $MgATP^{2-}$ and reductant concentrations are all important.

The kinetic studies of substrate reduction discussed above are producing useful information concerning enzyme/reducible substrate interaction and electron allocation under competitive conditions. Our interest in nitrogenase kinetics, centres on the fundamental process of nitrogenase catalysis in which reducing equivalents are produced by nitrogenase from $MgATP^{2-}$ and $S_2O_4^{2-}$ rather than their allocation after having been produced. In this regard, we have begun to investigate the steady-state kinetics of $MgATP^{2-}$ and $S_2O_4^{2-}$ utilization by nitrogenase under various conditions using newly developed electrochemical and calorimetric techniques (Watt *et al.*, 1975). These sensitive instrumental methods allow continuous monitoring of the progress of the enzyme-catalyzed reaction by following $S_2O_4^{2-}$ disappearance or heat production, respectively. We report here the results so far obtained from these studies.

EXPERIMENTAL

Nitrogenase

Preparations of nitrogenase *complex* from *Azotobacter vinelandii* were obtained as previously described (Bulen and LeComte, 1966; Bulen and LeComte, 1972). The enzyme preparations possessed activities of 300-400 nmoles H_2 evolved per min per mg of protein and gave ATP/2e values of 4-5 at 25°C.

Methods

All solutions were prepared in airless glassware under argon which had been previously passed through hot copper turnings (600°C) to remove oxygen. A creatine phosphate-creatine kinase ATP-generating system was used in all of the kinetic studies reported here to maintain a constant ATP concentration and to prevent ADP from reaching inhibitory levels. No variation in the rate of $S_2O_4^{2-}$ utilization at any of the temperatures studied was observed upon increasing the components of the generating system by a factor of 4. The Pi released was measured as previously described (Bulen and LeComte, 1966) using controls lacking $S_2O_4^{2-}$.

The rates of $S_2O_4^{2-}$ utilization by the nitrogenase catalyzed H_2 evolution, acetylene reduction and N_2 reduction reactions were measured polarographically. The polarographic apparatus and procedure for measuring $S_2O_4^{2-}$ concentration have been described (Watt *et al.*, 1975). A standard assay solution of 5.0 ml was used which consisted of Tes or Tris buffer (250 moles) pH 8, creatine phosphate (260 μmoles), creatine kinase (5 nmole) and variable amounts of both $S_2O_4^{2-}$ and ATP. The Mg^{2+} concentration always exceeded 3.0 mM and had no effect on the rate of $S_2O_4^{2-}$ utilization in the range of 1.0 to 5.0 mM at a given ATP concentration. Under the above experimental conditions, the $MgATP^{2-}$ complex was the only important ATP-containing species present during the reactions (see Thorneley and Willison, 1974).

RESULTS

Fig. 1 shows a polarographic trace of $S_2O_4^{2-}$ utilization during the hydrogen evolution reaction catalyzed by nitrogenase. The anodic current before enzyme is added (arrow) gives the initial $S_2O_4^{2-}$ concentration in the assay mixture. The constancy of the anodic current with time demonstrates that no $S_2O_4^{2-}$-consuming reactions are occurring in the absence of enzyme catalysis. The insert to Fig. 1 is a plot of the $[S_2O_4^{2-}]^{1/2}$ against time and its linearity indicates that the nitrogenase-catalyzed

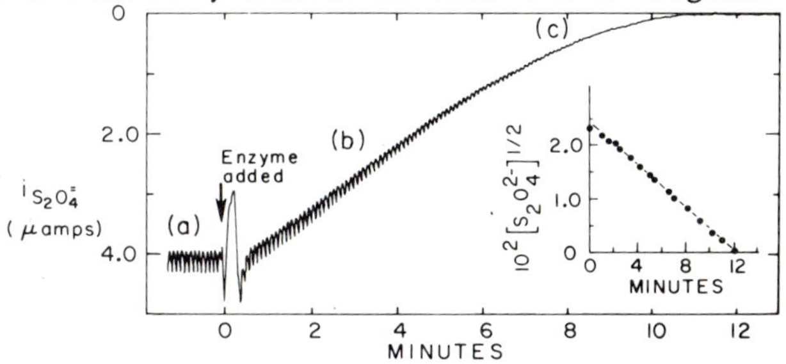

Fig. 1. *A polarographic trace resulting from $S_2O_4^{2-}$ utilization during the nitrogenase-catalyzed H_2 evolution reaction. The ordinate is the anodic diffusion current which is proportional to the $S_2O_4^{2-}$ concentration. Nitrogenase is added at the arrow beginning the reaction.*

H_2 evolution reaction, under conditions where [ATP] is maintained at a constant level by the generating system, is one-half order in $S_2O_4^{2-}$ throughout the entire reaction interval. A rate constant of 2.1×10^{-3} $M^{1/2}$ ml min^{-1} mg^{-1} at 25° is obtained from the slope of the straight line.

The one-half order rate constants at various temperatures for $S_2O_4^{2-}$ utilization by the nitrogenase-catalyzed H_2 evolution reaction, maintained at 5 mM ATP by the ATP-generating system, are given in Fig. 2 in the form of an Arrhenius plot. The rate constants for phosphate (P_i) production resulting from ATP hydrolysis at the same temperatures are also included in this figure. The rate constants for ATP hydrolysis at the temperatures indicated were calculated from equation (1) which relates the rate of $S_2O_4^{2-}$ utilization (measured in this study) and the ATP/2e values previously reported (Watt *et al.*, 1975) at these same temperatures. The one-half order rate constants

$$\frac{-d[ATP]}{dt} \quad - \frac{ATP}{2e}\,\frac{d[S_2O_4^{2-}]}{dt} \qquad (1)$$

for $S_2O_4^{2-}$ utilization produce a biphasic plot while the rate constants for P_i formation form a single straight line. Activation energies of +35 and +12.3 kcal/mole for $S_2O_4^{2-}$ utilization are obtained from the 10-20°C and the 20-35°C temperature regions, respectively.

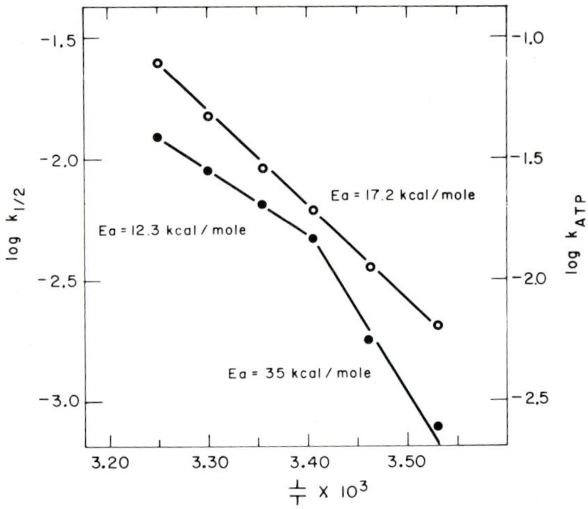

Fig. 2. *Arrhenius plots for the one-half order rate constants* ($k_{l/2}$) *for* $S_2O_4^{2-}$ *utilization ( ) and the rate constants* (k_{ATP}) *for ATP hydrolysis (O).*

The Arrhenius plot for the rate of P_i production (ATP hydrolysis) catalyzed by nitrogenase yields an activation energy of +17.2 kcal/mole. This result taken together with those for $S_2O_4^{2-}$ utilization discussed above clearly indicate that the rate of ATP hydrolysis and electron transfer from $S_2O_4^{2-}$ to products, catalyzed by nitrogenase, are affected differently by variations in temperature. In fact at 10°C, the rate of $S_2O_4^{2-}$ utilization is barely detectable while ATP hydrolysis occurs at an easily measurable rate. *At this temperature or below, nitrogenase is essentially an ATPase incapable of transferring electrons to substrate.*

The rate of $S_2O_4^{2-}$ utilization at 25°C for acetylene and nitrogen reduction produces identical curves to that for H_2 evolution shown in Fig. 1. Identical rate constants with the same temperature variations (same activation parameters) were observed for acetylene and nitrogen reduction as were found for H_2 evolution.

Fig. 3 is a plot of the rate at which $S_2O_4^{2-}$ is utilized by the nitrogenase-catalyzed H_2 evolution reaction *versus* the $S_2O_4^{2-}$ concentration at several fixed ATP concentrations. The curves are truly parabolic as the $[S_2O_4^{2-}]^{1/2}$ dependence requires and, within the $S_2O_4^{2-}$ concentration range studied (1×10^{-5} to 1×10^{-3} M), show no signs of saturation with respect to $S_2O_4^{2-}$. These effects are clearly demonstrated by the double reciprocal plot in the insert to Fig. 3, where reciprocal rate is plotted against the reciprocal of the square root of $[S_2O_4^{2-}]$. The resulting well defined straight lines all intersect at the origin indicating that, even at infinite $S_2O_4^{2-}$ concentration, the enzyme is not saturated with $S_2O_4^{2-}$.

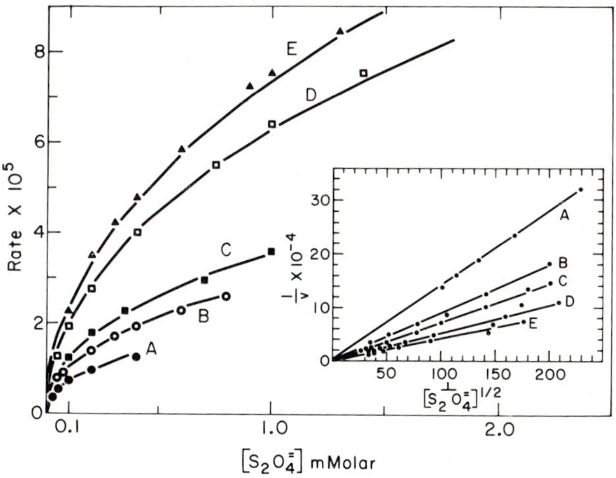

Fig. 3. *The rate of $S_2O_4^{2-}$ utilization at 25°C for the H_2 evolution reaction as a function of both ATP and $S_2O_4^{2-}$ concentration. The solid line is calculated from equation (5) and the points obtained experimentally. The ATP concentrations are: A = 5×10^{-5} M, B = 7.5×10^{-5} M, C = 1.3×10^{-4} M, D = 5×10^{-4} M and E = 1×10^{-3} M.*

Saturation of nitrogenase by ATP is evident in Fig. 4, where the rate of $S_2O_4^{2-}$ utilization is plotted against the ATP concentration. The curves at both 25° and 30°C are definitely sigmoidal.

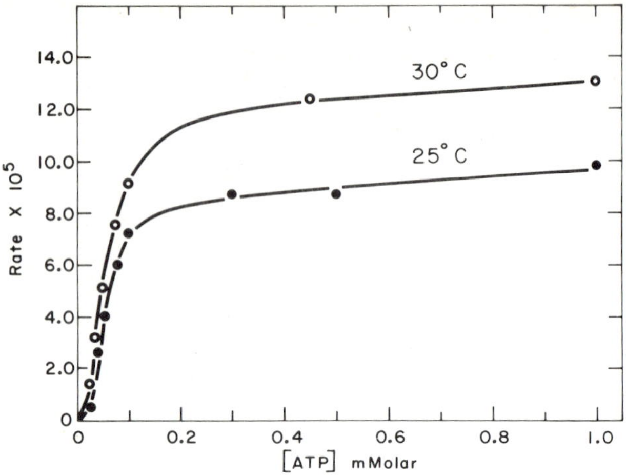

Fig. 4. *The initial rate of the H_2 evolution reaction as a function of ATP concentration. The $S_2O_4^{2-}$ was 10 mM and the ATP concentration was maintained constant by the ATP-generating system.*

DISCUSSION

General

The procedure for conducting a steady-state kinetic analysis of multi-substrate enzymes is well established (Cleland, 1963) and requires basically that the kinetic effects of varying each substrate be evaluated separately while the concentration of all remaining substrates are held constant. In principle, this procedure should be readily applied to nitrogenase which can be considered to require three substrates: (1) a reducible small molecule; (2) a reductant; and (3) $MgATP^{2-}$. However, this enzyme possesses an unusual feature which greatly complicates this type of steady-state analysis. Nitrogenase is not a simple single-protein enzyme but is composed of two dissociable, dissimilar proteins, the Fe protein (M.W. = 65,000) and the MoFe protein (M.W. = 220,000), which are composed of two and four subunits, respectively. The preferred stoichiometry of the "nitrogenase complex" presumably formed during the catalytic cycle by these two proteins has not been unequivocally established although values of 1 - 2 are reported (Eady *et al.*, 1972; Eady, 1973; Mortenson *et al.*, 1973; Vandecasteele and Burris, 1970; Shah *et al.*, 1975) most often for the Fe protein/MoFe protein ratio giving maximum activity. The role of these two proteins remains to be fully established although the emerging view is that the reduced, ATP-containing Fe protein is the "activated" electron transfer agent to the reducible substrate-binding MoFe protein. If this view is substantiated then perhaps this reduced ATP-activated Fe protein should be considered as a "substrate" itself and the kinetic effects of varying its concentration be evaluated according to the procedures for steady-state analysis.

Our steady-state kinetic studies have not yet progressed this far due, in part, to the uncertainty of the roles of the two protein components and to the difficulties of preparing the Fe protein $(ATP)_2$ in sufficient quantities and maintaining it at a high uniform activity for a sufficient time to do the kinetic experiments required for a thorough analysis. The approach reported here involves the use of a reproducibly prepared "nitrogenase complex" from *Azotobacter vinelandii* (Bulen and LeComte, 1972), consisting of a fixed Fe protein/MoFe protein ration near 1. With this enzyme preparation and protein ratio, the rate laws and, perhaps, insights into the mechanism for the utilization of the three substrates of nitrogenase can be obtained. The kinetic effects of varying the Fe protein concentration will be evaluated later. With this qualification concerning the ratio of the Fe and MoFe proteins, evaluation of the rate law for the utilization of $MgATP^{2-}$, $S_2O_4^{2-}$ and the various reducible substrates of nitrogenase and incorporation of this information into a partial mechanism of catalysis will be attempted.

Reducible substrate reduction

The polarographic technique developed in this laboratory (Watt *et al.*, 1975) is a sensitive, continuous method for following the rate of $S_2O_4^{2-}$ utilization during the nitrogenase-catalyzed substrate reduction reactions. Using this method, we have found that the rate of $S_2O_4^{2-}$ utilization, under otherwise identical conditions, is constant for the reduction of acetylene and molecular nitrogen and for the hydrogen evolution reaction run under argon, hydrogen and CO atmospheres. Similar results have been reported (Burns, 1968; Silverstein and Bulen, 1969; Hadfield and Bulen, 1969; Davis *et al.*, 1975) from studies of the rate of product formation instead of the rate of $S_2O_4^{2-}$ disappearance. Those studies have shown that as the partial pressure (or concentration) of a reducible substrate is reduced, the amount of reduced product formed decreases accordingly. However, a corresponding increase in H_2 evolution occurs keeping the total number of electrons transferred by nitrogenase constant. These results indicate that the reducible substrate concentration does not appear in the rate law for either $S_2O_4^{2-}$ utilization or reduced product formation. A reasonable interpretation suggests that the enzyme is first "activated", presumably by $S_2O_4^{2-}$ and $MgATP^{2-}$, and then becomes irreversibly committed to react with any available reducible substrate in a non-rate-limiting step. If one reducible substrate becomes limiting, an alternative substrate (or solvent) is reduced to such an extent that the total electron output by nitrogenase remains invariant. This interpretation, however, presents a problem in explaining the CO inhibition of all reactions except the H_2 evolution reaction because the formation of the two- (CH_2O), four- (CH_3OH) and six- (CH_4+H_2O)-electron reduction products of CO are much more favourable thermodynamically than, for example, the corresponding reduction products of N_2. Therefore, CO should be reduced if it binds at the N_2 reduction site but CO reduction has never been reported, suggesting that either CO inhibition occurs at a site remote from the reduction site or the formation of its reduction intermediates is not favoured for some reason. This latter point is discussed to some extent by Stiefel elsewhere in these Proceedings.

$S_2O_4^{2-}$ utilization

Fig. 1 clearly indicates that, with a constant ATP concentration maintained by the generating system (zero order with respect to ATP), the nitrogenase-catalyzed reactions have a one-half order dependence on the $S_2O_4^{2-}$ concentration over a wide concentration range. This one-half order dependence on $S_2O_4^{2-}$ has been previously observed with other reactions and has been discussed by Lambeth and Palmer (1973) and Creatz and Sutin (1974) as arising from SO_2^- (produced according to (2)) being the actual reducing species. Thorneley *et al.* 1976) have recently

$$S_2O_4^{2-} = 2SO_2^-, \ K = 1.4 \times 10^{-9} \text{ M} \qquad (2)$$

reported that the kinetics for reduction of the dye-oxidized Fe protein consists of several sequential reactions but that the fastest reaction, which is believed to be the physiological one, displays kinetic behaviour consistent with the SO_2^- reduction of oxidized Fe protein. A rate constant of $>10^8$ $M^{-1}sec^{-1}$ for reaction (3) was observed, where Fe_o and Fe_r are the oxididized

$$Fe_o + SO_2^- = Fe_r + SO_2 \qquad (3)$$

and reduced Fe protein, respectively.

The one-half order rate constant for $S_2O_4^{2-}$ utilization during the nitrogenase-catalyzed reactions observed in this study is consistent with reduction of the Fe protein component by SO_2^- in a single-electron reduction step. However, the rate constant of 3×10^5 $M^{-1}sec^{-1}$ for this process in terms of SO_2^-

being the reductant differs from the rate constant of $>10^8 M^{-1}$ sec^{-1} reported by Thorneley *et al.* (1976) for SO_2^- reacting with isolated, oxidized Fe protein. Though the conditions of the two experiments differ somewhat, it seems reasonable to conclude that the reduction of the Fe protein under turnover conditions is slower by a factor of $>10^3$ than the reduction of isolated Fe protein. Thus, we conclude, as did Thorneley *et al.* (1976) that some other reaction must limit the rate of reduction of the protein during enzyme catalysis.

According to Fig. 3, nitrogenase does not become saturated with $S_2O_4^{2-}(SO_2^-)$ within the concentration range studied. The reciprocal plot insert shows that, even at infinite $S_2O_4^{2-}$ concentration, enzyme saturation is not reached and that an infinite rate of enzyme turnover is to be expected under these conditions. The nitrogenase-SO_2^- interaction appears to be a chemical one in the sense that no enzyme-SO_2^- complex as such ever exists.

$MgATP^{2-}$ substrate

Fig. 4 shows that, at a fixed initial $S_2O_4^{2-}$ concentration, the rate of enzyme turnover at 25°C becomes saturated with ATP and that this saturation effect is definitely of a sigmoidal form. Similar behaviour is observed at 30°C but a less pronounced sigmoidal curve is obtained.

The sigmoidicity of these plots can be understood by analyzing more closely the results of the reciprocal plot insert of Fig. 3. This plot shows that the slopes of the straight lines are functions of the ATP concentration. This function was evaluated and is given by (4) where $K_1 = 5.05 \times 10^{-5} M^{-1}$ and $K_2 = 2.8 \times 10^{-9} M^{-1}$. The reciprocal of this function is the actual

$$\text{Slope} = \frac{[\text{ATP}]^2 + K_1[\text{ATP}] + K_2}{[\text{ATP}]^2} \qquad (4)$$

term in the rate law for enzyme catalysis involving ATP, with K_1 and K_2 being the equilibrium dissociation constants for the sequential binding of ATP to the enzyme. The values for K_1 and K_2 reported here for *Azotobacter vinelandii* nitrogenase determined under the "dynamic conditions" of enzyme turnover are in reasonable agreement with the values of 1.6×10^{-5} and 2.6×10^{-10} for the dissociation of ATP from the isolated Fe protein of *Clostridium pasteuranium* (Tso and Burris, 1973) determined under equilibrium conditions. Using the values of K_1 and K_2, the concentrations of the various nitrogenase species were calculated as a function of the ATP concentration and plotted in Fig. 5. This plot indicates: (1) that enzyme activity and the

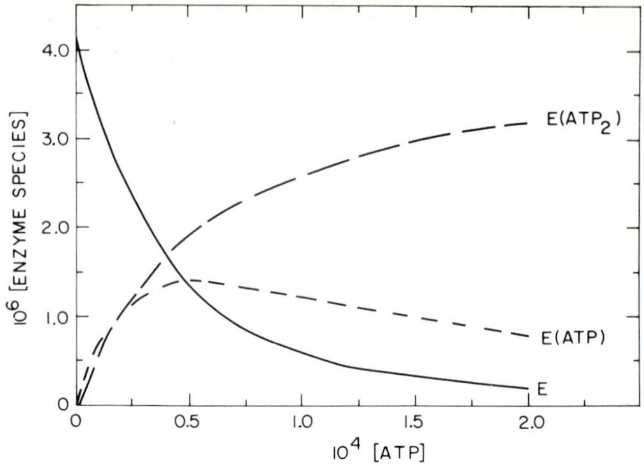

Fig. 5. *A species distribution plot for the various forms of nitrogenase as a function of ATP concentration.*

E $(ATP)_2$ concentration parallel each other; and (2) that the sigmoidal character of Fig. 4 results from the fact that, at low ATP concentrations, $E(ATP)_2$ is negligible but that it increases rapidly to prominence at the expense of both E and E $(ATP)_1$ in response to increasing ATP. The enzyme activity is consequently insignificant at low ATP levels but increases rapidly with increasing ATP concentrations as both measured by Fig. 4 and indicated by Fig. 5.

Rate law for nitrogenase catalysis

Combining the results of the three previous sections, the rate law (5) for nitrogenase catalysis results. Using this equation, the solid lines in Fig. 3 were calculated and agreement between them and the experimental points is very good over

$$\frac{-d[S_2O_4^{2-}]}{dt} \quad \frac{2.35 \times 10^{-3}\ E_T\ [S_2O_4^{2-}]^{1/2}\ [ATP]^2}{[ATP]^2 + K_1\ [ATP] + K_2} \tag{5}$$

a $S_2O_4^{2-}$ concentration range of 1000 and an ATP concentration range of 100. This rate law contains only the three parameters K_1, K_2 (which are the ATP-binding constants) and the numerical constant $2.35 \times 10^{-3}\ M^{1/2} min^{-1}\ ml\ mg^{-1}$ at 25°C (which is the one-half order rate constant for $S_2O_4^{2-}$ utilization).

The rate law (5) is consistent with the simplified and incomplete mechanism given by (6) - (8) which is also consistent with many other lines of evidence (Zumft and Mortenson, 1975).

We have left E_o unspecified, although there are reasons to believe it is the Fe protein. The point at which the MoFe protein enters the scheme and two protein components interact and what reactions occur following such an event are still obscure.

$$E_o + 2ATP = E_o(ATP)_2 \qquad (6)$$

$$E_o(ATP)_2 + SO_2^- \longrightarrow E_r(ATP)_2 + SO_2 \qquad (7)$$

$$E_r(ATP)_2 \longrightarrow \quad \longrightarrow E_o + Products + 2ADP + 2P_i \qquad (8)$$

The sequence of events given in (6) - (8) (the binding of ATP to the enzyme followed by reduction with SO_2^-) was deduced from an analysis of the kinetic patterns in Fig. 3 and from equation (5) according to the rules given by Cleland (1963) for steady-state analysis. The insert of Fig. 3 shows that all lines intersect at a common point, which in this case, is the origin of the plot. The ATP concentration was varied by a factor of 100 in this plot but this variation had no effect on the maximal rate (as evidenced by all lines intersecting at a common point on the reciprocal rate axis) when the $S_2O_4^{2-}$(SO_2^-) concentration is "saturating". Consequently, we conclude that the kinetic effects of ATP occur prior to SO_2^- reduction.

ACKNOWLEDGEMENTS

The helpful discussions with W. W. Cleland at the Biochemistry Department of the University of Wisconsin-Madison are greatly appreciated. Many of the polarographic measurements were made by Mrs. Andrea Burns.

REFERENCES

Bulen, W.A. and LeComte, J.R. (1972). *In:* Methods in Enzymology (A. San Pietro, ed.), Vol. XXIV Part B, pp. 456-70. Academic Press, New York and London.
Bulen, W.A. and LeComte, J.R. (1966). *Proc.Nat.Acad.Sci.U.S.A.*, 56: 979-986.
Burns, R.C. (1969). *Biochim.Biophys.Acta*, 171: 253-259.
Cleland, W.W. (1963). *Biochim.Biophys.Acta*, 67: 104-137.
Cleland, W.W. (1970). *In:* The Enzymes (P.D. Boyer, ed.), Vol.II, pp. 1-65. Academic Press, New York and London.
Creatz, C. and Sutin, N. (1974). *Inorg.Chem.*, 13: 2041-2043.
Davis, L.C., Shah, V.K. and Brill, W.J. (1975). *Biochim.Biophys.Acta*, 403: 67-78.
Eady, R.R. (1973). *Biochem.J.*, 135: 531-535.
Eady, R.R., Smith, B.E., Cook, K.A. and Postgate, J.R. (1972).

Biochem.J., 128: 655-675.

Hadfield, K.L. and Bulen, W.A. (1969). *Biochemistry*, 8: 5103-5108.

Hardy, R.W.F., Burns, R.C. and Parshall, G.W. (1971). *In:* Bioinorganic Chemistry, Advances in Chemistry series No. 100, pp. 219-247. American Chemical Society Publications, Washington D.C.

Hwang, J.C., Chen, C.H. and Burris, R.H. (1973). *Biochim.Biophys.Acta*, 292: 256-270.

Lambeth, D.O. and Palmer, G. (1973). *J.Biol.Chem.*, 248: 6095-6103.

Mortenson, L.E., Zumft, W.G., Huang, T.C. and Palmer, G. (1973). *Biochem.Soc.Trans.*, 4: 35-37.

Rivera-Ortiz, J.M. and Burris, R.H. (1975). *J.Bacteriol.*, 123: 537-545.

Shah, V.K., Davis, L.C. and Brill, W.J. (1975). *Biochim.Biophys.Acta*, 384: 353-359.

Silverstein, R. and Bulen, W.A. (1970). *Biochemistry*, 9: 3809-3815.

Thorneley, R.N.F. and Willison, K.R. (1974). *Biochem.J.*, 139: 211-214.

Thorneley, R.N.F., Yates, M.G. and Lowe, D.J. (1976). *Biochem. J.*, 155: 137-144.

Tso, M-Y.W. and Burris, R.H. (1973). *Biochim.Biophys.Acta*, 309: 263-270.

Vandecasteele, J.P. and Burris, R.H. (1970). *J.Bacteriol.*, 101: 794-801.

Watt, G.D., Bulen, W.A., Burns, A. and Hadfield, K.L. (1975). *Biochemistry*, 14: 4266-4272.

Zumft, W.G. and Mortenson, L.E. (1975). *Biochim.Biophys.Acta*, 416: 1-52.

SOME ASPECTS OF THE MECHANISM OF NITROGENASE

B.E. SMITH, R.R. EADY, R.N.F. THORNELEY,
M.G. YATES and J.R. POSTGATE

ARC Unit of Nitrogen Fixation, University of Sussex, Brighton BN1 9QJ, U.K.

INTRODUCTION

Earlier, electron paramagnetic resonance (e.p.r.)(Smith *et al.*, 1972,1973) and Mossbauer studies (Smith and Lang, 1974) enabled us to describe a preliminary scheme for the electron transfer processes within nitrogenase. In this scheme, which is now generally accepted, MgATP binds to the Fe protein and induces very rapid electron transfer to the MoFe protein, which binds and transfers electrons to reducible substrates. Stopped-flow spectrophotometric studies (Thorneley, 1975) have allowed a more detailed kinetic study of the protein-protein electron transfer but within this simple framework, many aspects of nitrogenase function require investigation. In this communication, three such aspects will be discussed in some detail: (1) The reduction of oxidised Fe protein and the effects of MgATP and MgADP on this process; (2) The complex between the MoFe and Fe proteins; and (3) Substrate reduction and the roles of ATP in nitrogenase function.

We have used nitrogenase components from three sources, *viz, Klebsiella pneumonia* (Kp), *Azotobacter chroococcum* (Ac), and *Clostridium pasteurianum* (Cp).

The MoFe protein will be designated protein 1 and the Fe protein 2, thus the Fe protein from *C. pasteurianum* will be called Cp2, etc.

REDUCTION OF THE OXIDISED Fe PROTEIN

In the steady-state, the MoFe protein is almost completely in the e.p.r.-silent reduced state. Thus, electron transfer from the MoFe protein to reducible substrates must be much slower than the protein-protein electron transfer and it has been suggested that substrate reduction is the rate-limiting step in enzyme turnover. However, this is unlikely because the rate of electron transfer through the enzyme to H^+, N_2 plus H^+ or

C_2H_2 plus H^+ is constant (R.R. Eady and B.E. Smith, unpublished results) although some step associated with electron transfer seems to be rate-limiting. E.p.r. studies (Smith *et al.*, 1973; Thorneley *et al.*, 1976) showed that, in the steady-state, the Fe protein is partially in the oxidised form and thus, re-reduction of the Fe protein is one possible candidate for the rate-limiting step in turnover.

Reduction of isolated Ac2ox

Ac2 protein can be oxidised by phenazine methosulphate to an e.p.r.-silent form, Ac2ox, with only about 20% loss in specific activity

The reduction of Ac2ox by $Na_2S_2O_4$ can be monitored by the decrease in absorbance of either the protein at 425 nm or the $S_2O_4^{2-}$ at 315 nm. Stopped-flow spectrophotometry at either wavelength showed the reaction to occur in four phases, Fig. 1c (Yates *et al.*, 1975; Thorneley *et al.*, 1976). The first phase of the reaction was very rapid ($\tau < 1$ms when $[Na_2S_2O_4]$ = 10 mM) and corresponded to the transfer of one electron from dithionite to Ac2ox. Rapid-freeze e.p.r. studies demonstrated that this first phase could be correlated with the appearance of the e.p.r. signal of reduced Ac2 protein (Fig. 1d), although integration of the e.p.r. signal corresponded to only 0.24 electrons per molecule. The second, third and fourth phases of the reaction together corresponded approximately to the transfer of one more electron but no e.p.r. changes were observed. A detailed kinetic analysis of the above reduction phases indicated that the reductant was $SO_2^{\cdot -}$, derived from the dissociation of $S_2O_4^{2-}$ (Thorneley *et al.*, 1976).

When the reduction of oxygen-inactivated Ac2ox by 10 mM dithionite was monitored at 425 nm, the time-constant of the first phase of the reduction was 80 msec compared with less than 1 msec for the active protein. This difference could indicate that the slower reduction phases observed with Ac2ox are associated with inactive protein. However, no correlation between the amplitude of the rapid, first phase of the reduction and activity of the protein was found.

Mossbauer spectroscopy (Smith and Lang, 1974) and cluster extrusion experiments (W.H. Orme-Johnson, unpublished; G. Gillum, L.E. Mortenson, and R.H. Holm, unpublished) indicated that the Fe protein of nitrogenase contained one Fe_4S_4 cluster. Such clusters are well known in the ferredoxins and high-potential iron proteins (Adman *et al.*, 1973; Carter *et al.*, 1972) and generally exist in the 1^- or 3^- e.p.r.-active or 2^- e.p.r.-inactive forms. The cluster in reduced Fe protein appears from e.p.r. and Mossbauer data to be in the 3^- form. Presumably, Ac2ox is in the 2^- form because it is e.p.r.-inactive and thus,

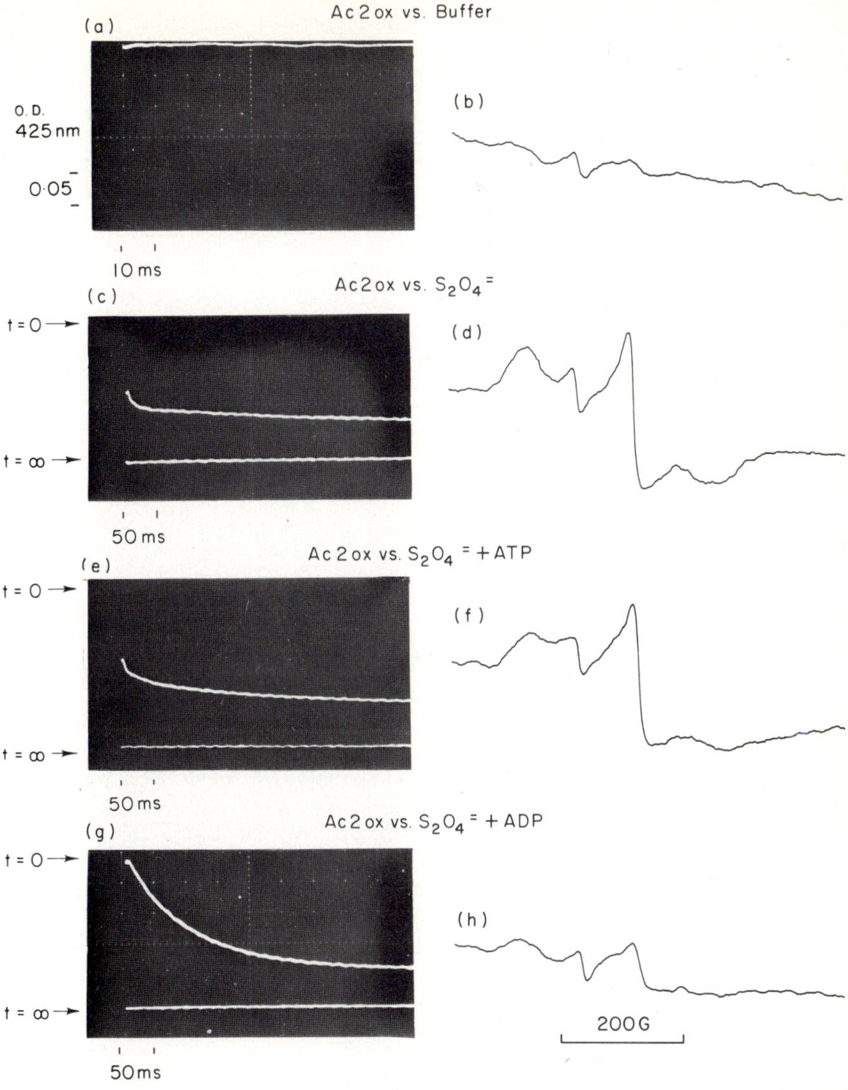

Fig. 1. *Stopped-flow spectrophotometric oscillographs of the optical changes at 425nm and e.p.r. spectra from rapid-freezing experiments on the reduction of Ac2ox by* $SO_2 \cdot^-$ *at 23°C. In each experiment, one syringe contained Ac2ox (78μM) in 25mM Tris-HCl buffer, pH7.4, containing 10mM* $MgCl_2$ *and 70μM dithiothreitol. The contents of this syringe were mixed with equal volumes from the other syringe of either Tris buffer (a) and (b); buffer plus 20mM* $Na_2S_2O_4$ *(c) and (d); buffer plus* $Na_2S_2O_4$ *plus 18mM ATP, (e) and (f); or buffer plus* $Na_2S_2O_4$ *plus 18 mM ADP, (g) and (h). Reaction time in the rapid-freezing experiments was 20msec. E.p.r. spectra were recorded at 9GH*$_z$*, 15°K and 15mW microwave power, using a modulating field of 12.5G.*

it should accept one electron per molecule on reduction. Since in the first phase of reduction the total number of electrons transferred was equivalent to the number of Ac2 protein molecules present, it is probable that each molecule of Ac2ox accepts one electron. However, integration of the e.p.r. spectrum of this and other nitrogenase Fe proteins is generally much less than one electron/mole (Smith *et al.*, 1973; Zumft *et al.*, 1973) and does not correlate with activities of the protein. In addition, each molecule of Ac2 protein appeared to accept up to one more electron in the three slower phases, which was accompanied by a further decrease in the absorbance at 420 nm (presumably associated with the Fe_4S_4 clusters) but no further e.p.r. changes. These observations indicate that a direct comparison of the Fe protein with the ferredoxins may not be valid.

Reduction of Ac2ox in the presence of MgATP or MgADP

The kinetics of the reduction of Ac2ox by $Na_2S_2O_4$ were essentially unaffected in the presence of a 9mM MgATP (Fig. 1e and f). However, 9mM MgADP slowed the first phase of the reduction to τ = 92 msec as monitored by stopped-flow spectrophotometry at 23°C or rapid-freeze e.p.r. (Fig. 1g and h) (Yates *et al.*, 1975). Thus, MgADP binds to Ac2ox in less than 1 msec and prevents the rapid reduction by $SO_2\cdot^-$. In the steady-state at 23°C, the turnover time of Ac2 protein is about 500 msec/electron, and the protein is about 50% oxidised. The rate of the ADP-inhibited reduction is too fast to accommodate the concentration of Ac2ox observed in the steady-state and presumably other reactions, such as complex formation with Ac1 protein, deadenylation or dephosphorylation, further inhibit the rate of Ac2ox reduction by $SO_2\cdot^-$ during nitrogenase turnover.

Inhibition of the rate of reduction of Ac2ox by MgADP (1mM) was not relieved by addition of MgATP (7mM), suggesting either that MgATP binds very weakly to Ac2ox or that its binding does not affect the binding of MgADP.

THE COMPLEX BETWEEN THE MoFe AND Fe PROTEINS

Before direct electron transfer from the Fe protein to the MoFe protein can occur, a complex between the two proteins must be formed. Stopped-flow spectrophotometric kinetic data (Thorneley, 1975) showed that, with the nitrogenase from *K. pneumoniae* this complex is formed rapidly ($k > 10^7 M^{-1} sec^{-1}$) with an association constant, $K_{complex} > 2 \times 10^6 M^{-1}$ at 23°C. The object of this section is to consider the nature of this complex, its life time and role in substrate reduction.

Ultracentrifuge studies

Eady (1973) studied the interaction of Kp1 and Kp2 proteins in the ultracentrifuge and found that at Kp2: Kp1 molar protein ratios greater than 1:1, uncomplexed Kp2 protein could be observed, indicating that a 1:1 complex was formed. However, $Na_2S_2O_4$ apparently prevented complex formation. Similar results were obtained with Ac1 and Ac2 proteins and with mixtures of the complementary *K. pneumoniae* and *A. chroococcum* proteins (Thorneley *et al.*, 1975) but $Na_2S_2O_4$ (5mM) did not entirely prevent complex formation in these cases.

The Dilution Effect

Disproportionately low activity at high dilutions has been observed with nitrogenase from a number of organisms (see Eady and Smith, 1976, for a review), Thorneley *et al.* (1975) obtained data on the dilution effect for acetylene reduction by *K. pneumoniae* (Fig. 2) and *A. chroococcum* nitrogenase and the heterologous nitrogenase formed from the complementary proteins of

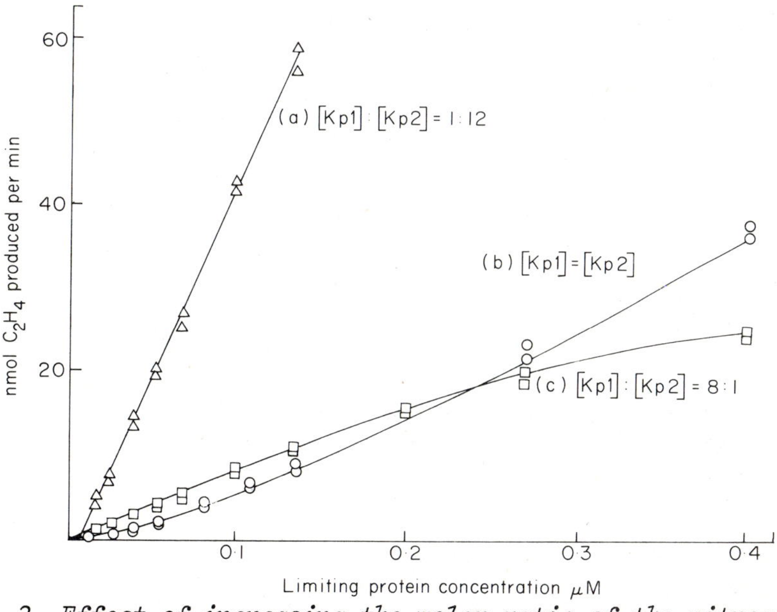

Fig. 2. *Effect of increasing the molar ratio of the nitrogenase component proteins from* K. pneumoniae *on the rate of acetylene reduction at 30°C. (a) Molar ratio of Kp1:Kp2 = 1:12, abscissa refers to concentration of Kp1 protein. (b) Molar ratio Kp1: Kp2 = 1:1. (c) Molar ratio Kp1: Kp2 = 8:1, abscissa refers to concentration of Kp2 protein. Curve b is a computer fit to the data.*

these two organisms. They found that the dilution effect observed with a 1:1 molar ratio of the MoFe and Fe proteins from *K. pneumoniae* could be overcome by the addition of excess of *either* protein (Fig. 2). This observation is difficult to reconcile with the suggestion that the dilution effect is due to a square dependence of the reaction rate on the concentration of Fe protein due to dissociation of a 2:1 Fe:MoFe protein complex (see Burns and Hardy, 1975, for a review).

Thorneley *et al.* (1975). analysed their data for acetylene reduction on the assumption that the dilution effect was caused by the dissociation of the 1:1 complex indicated by ultracentrifuge data. The analysis gave a good fit to the experimental data (Fig. 2) and yielded association constants of from 2 to 5.5 x $10^7 M^{-1}$ at 30°C for the nitrogenase studied. A plot of $\log_{10} K_{complex}$ against $\frac{1}{T}$°K for the Kp1:Kp2 complex (Fig. 3) showed

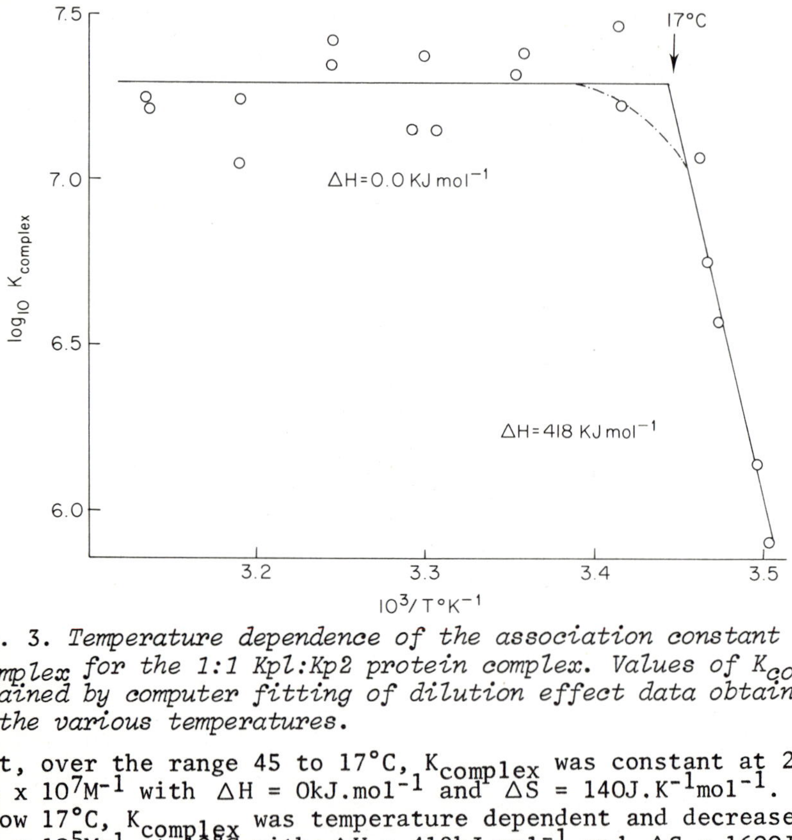

Fig. 3. *Temperature dependence of the association constant $K_{complex}$ for the 1:1 Kp1:Kp2 protein complex. Values of $K_{complex}$ obtained by computer fitting of dilution effect data obtained at the various temperatures.*

that, over the range 45 to 17°C, $K_{complex}$ was constant at 2.0 ± 0.6 x $10^7 M^{-1}$ with ΔH = 0kJ.mol^{-1} and ΔS = 140J.$K^{-1}mol^{-1}$. Below 17°C, $K_{complex}$ was temperature dependent and decreased to 8.3 x $10^5 M^{-1}$ at 12°C with ΔH = 418kJ.mol^{-1} and ΔS = 1600J.K^{-1} mol^{-1}. However, the specific activity of the complex gave a linear plot from 12 to 40°C with ΔH = 80kJ.mol^{-1}.

$K_{complex}$ can be equated to the ratio of the rate constants for the formation and dissociation of the complex :

$$\text{MoFe} + \text{Fe} \underset{k_D}{\overset{k_f}{\rightleftharpoons}} \text{MoFe.Fe}$$

$$K_{complex} = \frac{k_f}{k_D}$$

For the Kp1.Kp2 1:1 complex, $k_f > 10^7 M^{-1} sec^{-1}$ (Thorneley, 1975) and $K_{complex} = 2 x 10^7 M^{-1}$. Thus, $k_D > 0.5\ sec^{-1}$ at 23°C. Similarly, for the Ac1:Ac2 1:1 complex, assuming the same limit of k_f and with $K_{complex} = 5.5 \times 10^7 M^{-1}$, $k_D > 0.18\ sec^{-1}$ at 23°C.

The turnover numbers for these two enzymes at 23°C are: Kp = 0.56 moles sec^{-1} and Ac = 1.1 moles sec^{-1}. These calculations indicate that the complex is relatively long-lived and may have a role in nitrogenase function after protein-protein electron transfer. This suggestion is supported by the observation that, when the MoFe:Fe protein ratio is considerably greater than 1, the MoFe protein is largely reduced but apparently inactive (Smith *et al.*, 1973; Mortenson *et al.*, 1973). The similarity of k_D and the turnover number for the Kp1.Kp2 enzyme could indicate that complex dissociation is the rate-limiting step in turnover but this interpretation should be treated with extreme caution. Even with the assumption that k_f is very close to $10^7 M^{-1} sec^{-1}$, the measured $K_{complex}$ could be a combination of more than one equilibrium constant and thus, a reaction immediately preceding complex dissociation could be rate-limiting.

An upper limit for k_f can be calculated from diffusion theory, which gives the rate constant for collisions between Kp1 and Kp2, and is given by:

$$k = \frac{4\pi N.RDf}{1000}$$

where N is Avogadro's number, R is the distance of closest approach of the two reactants, D is the sum of the diffusion coefficients of the reactants and f is a factor which takes into account electrostatic interactions between reactants. Using the values of D and the partial specific volumes for Kp1 and Kp2 proteins from Eady *et al.* (1972), we obtain:

$$k = 5.25 \times 10^9 \times f\ M^{-1} S^{-1}$$

Current theories (Hammes and Alberty, 1959) are inadequate to calculate f but values of the order of 10^{-2} to 10^{-3} have been estimated for the dimerization of highly charged molecules. Both Kp1 and Kp2 protein have high negative charges at pH 7.4 since their pI values are close to 5 and 4 respectively (Eady *et al.*, 1972). In addition, a steric factor must be taken into account to convert the calculated collisional rate constant into one for effective collisions. Koren and Hammes (1976) estimated that only 2% of the total solid angle of 4π led to dimerization of insulin. Thus, the steric and electrostatic factors could reasonably be assumed to decrease the calculated collisional rate constant of $5.25 \times 10^9 M^{-1}s^{-1}$ to a value close to $10^7 M^{-1}s^{-1}$ as found by Thorneley (1975) and used in the above calculations.

SUBSTRATE REDUCTION

We have argued above that nitrogenase substrate reduction is probably carried out by the MoFe.Fe protein complex with the site(s) for substrate reduction on the MoFe protein (Smith *et al.*, 1973; Bui and Mortensen, 1968) but the nature and number of these sites has been the subject of much conjecture. CO inhibits the reduction of all substrates except the proton and thus, the proton-reducing site may well be distinct from the site(s) involved in reduction of other substrates. On the basis of the mutual inhibition of a number of substrate reductions, Hwang *et al.* (1973) concluded that there were five sites or modified sites for substrate reduction on nitrogenase. Below, we report our recent studies on substrate reduction by nitrogenase and present evidence that ATP may have more than one role in enzyme turnover.

Activity titration curves

We have examined in detail the activity titration curves obtained when the MoFe proteins from *K. pneumoniae* or *A. chroococcum* are titrated with the homologous or heterologous Fe proteins. The general shape of these curves (Fig. 4) is well-known (e.g. Eady *et al.*, 1972) and shows a sharp rise in activity with increasing Fe protein followed by a more gradual rise until a plateau is eventually reached with a large excess of Fe protein. However, the curves for H_2 evolution and C_2H_2 reduction are not identical (Fig. 4). C_2H_2 reducing activity rises more rapidly with increasing Fe protein concentration than that for H_2 evolution but the two curves eventually cross when the Fe:MoFe protein ratio is about 20:1. Similar behaviour with both the homologous and heterologous protein combinations was observed. At the plateaux of activity, the total electron flow to C_2H_2 plus H^+ was the same as that to H^+ alone. These data indicate that the C_2H_2 and H^+ reducing forms of the enzyme may not be identical. More definite evidence on this hypothesis is presented in the next section.

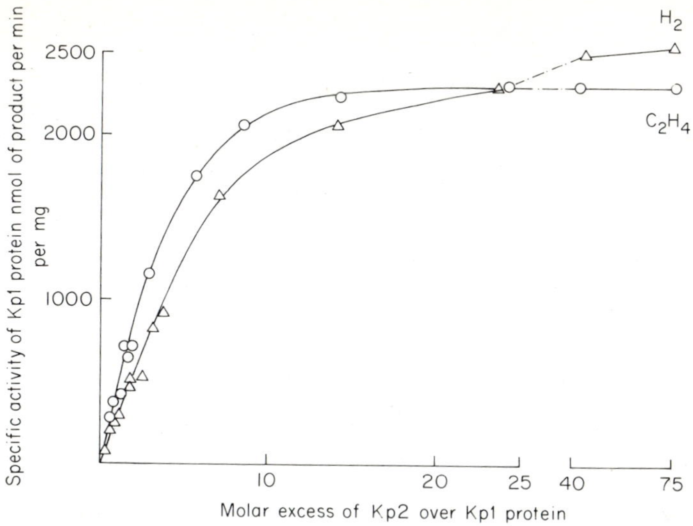

Fig. *Titration of Kp1 protein with Kp2 protein at 30°C as monitored by H_2-evolution under Argon, Δ, or C_2H_4 production under 20% C_2H_2/80% Argon, O. Each assay contained 0.49 nmoles Kp1 protein in 1 ml.*

The Kp1/Cp2 enzyme

A 1:1 molar mixture of Kp1 and Cp2 proteins had a specific activity for H_2 evolution of 53 nmole H_2 evolved/min/mg Cp2 protein, about 12% of the rate observed with the homologous *C. pasteurianum* proteins under similar conditions. The rate of H_2 evolution was linear with time. However, the rate of C_2H_2 reduction by the Kp1/Cp2 enzyme exhibited a lag of about 8 min at 30°C before becoming linear at 51 nmol ethylene produced/min/mg Cp2 protein, corresponding to a turnover time for C_2H_2 reduction of about 20 sec. Thus, the lag period cannot be accommodated within the turnover time and must correspond to an activation step which occurred only once during each experiment. This difference was explained by postulating different forms of the enzyme as being responsible for H_2 evolution and C_2H_2 reduction (Smith *et al.*, 1976) but only small amounts of N_2 reduction were reported. We have reinvestigated N_2 reduction by the Kp1/Cp2 enzyme and have established (Fig. 5) that N_2 is reduced but only after a 35 min lag. Thus, N_2 is apparently reduced by yet a third form of the enzyme presumably formed by a different and possibly subsequent activation step.

Stopped-flow spectrophotometric kinetic studies showed that the MgATP-induced electron transfer from Cp2 to Kp1 protein occurred with essentially the same rate and with the same MgATP

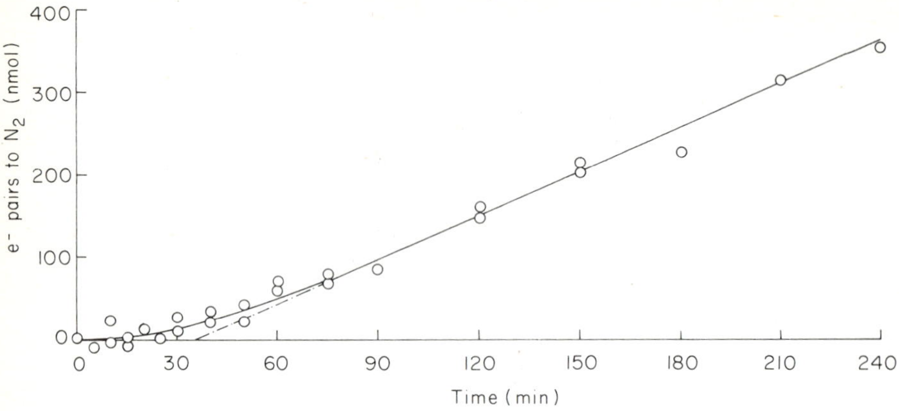

Fig. 5. *N_2 reduction by the Kp1/Cp2 enzyme under 100% N_2 at 30°C. Each assay contained 1.09 nmoles Cp2 protein and 0.58 nmoles Kp1 protein in 1 ml. NH_3 was determined by the indophenol method as described earlier (Eady* et al., *1972).*

dependence as was observed with the homologous *K. pneumoniae* (Thorneley, 1975) or *C. pasteurianum* enzymes.

ATP hydrolysis by the Kp1/Cp2 enzyme was linear with time (Fig. 6), occurred at about 40% of the rate observed with the homologous *C. pasteurianum* enzyme and gave an ATP/2e ratio of 50 for H_2 evolution. Thus, ATP hydrolysis and electron transfer are partially uncoupled in the hybrid enzyme. However, the ATP-induced protein-protein electron transfer was unaffected, indicating that ATP plays at least two roles in the enzyme turnover and that, in the Kp1/Cp2 enzyme, one site is unaffected whilst another is considerably perturbed. A similar conclusion can be drawn from the MgATP dependence of the protein-protein electron transfer since, even at 60 M MgATP, the rate of this reaction (τ = 40 msec) is much faster than enzyme turnover but, at this low concentration of MgATP, the enzyme activity is well below its maximum.

It is tempting to suggest that it is some process associated with this 'uncoupled' ATP site which has become rate-limiting in the Kp1/Cp2 enzyme, but we have no concrete evidence on this point.

CONCLUSIONS

Nitrogenase is an extremely complex enzyme and in order to understand its action, we have tried to identify and study partial reactions and interactions of various oxidation states of the proteins as well as the overall enzymic process. By gradually increasing the complexity of the system studied, we hope

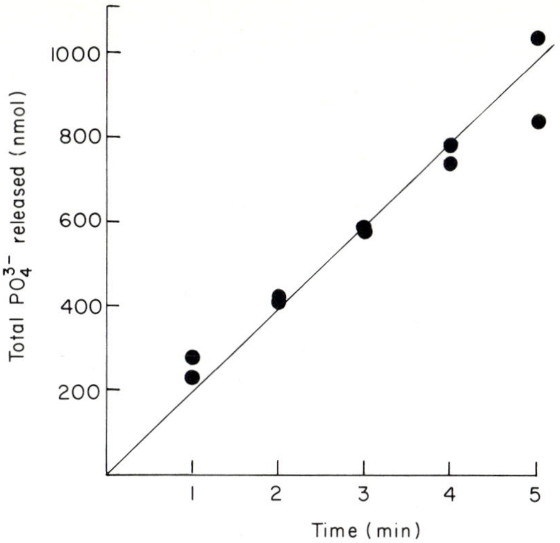

Fig. 6. Time course of ATP hydrolysis by the Cp2/Kp1 enzyme. *ATP hydrolysis was measured as* PO_4^{3-} *released in assays containing 10 mM ATP plus 10 mM* Mg^{2+}. *Individual assays were stopped at the times indicated. Each assay contained 1.73 nmoles Cp2 and 1.72 nmoles Kp1 in 1 ml.*

to identify critically important interactions.

We have found that the reduction of Ac2ox by $SO_2.^-$ is initially very rapid although there are slow subsequent phases. The rapid reduction is inhibited by MgADP but this inhibition alone is insufficient to account for either the relatively high level of Ac2ox in the steady-state or the inhibitory effect of MgADP on enzyme turnover. Thus, we must study other interactions in order to understand these phenomena.

Ultracentrifuge and dilution effect data indicate that a 1:1 MoFe:Fe protein complex is functionally active, although other complexes may also be active. Analysis of the activity data combined with stopped-flow spectrophotometric kinetic results indicate that this complex is relatively long-lived and thus, it may have a role in nitrogenase function after protein-protein electron transfer. The temperature dependence of the association constant for the complex showed a sharp break after 17°C, indicating a conformational change of one of the proteins at this temperature.

The cross-reaction of Cp2 with Kp1 protein enabled us to distinguish kinetically the formation of one form of the enzyme, which could only reduce protons, from others, formed after activation steps, which could also reduce C_2H_2 or N_2. Titration curves for H_2 evolution and C_2H_2 reduction with the homologous

Kp and Ac enzymes also indicate that different forms of the enzyme are responsible for these processes.

ATP hydrolysis was virtually uncoupled from electron transfer but the protein-protein electron transfer and its MgATP dependence were as effective as with the homologous enzymes. This observation indicated that ATP has at least two roles to play in enzyme turnover. Similar conclusions were drawn from work with ATP analogues (Eady *et al.*, 1975).

ACKNOWLEDGEMENTS

We thank Professor L.E. Mortenson for supplying the Fe protein from *C. pasteurianum* (Cp2 protein) and Dr. D.J. Lowe for running the e.p.r. spectra.

REFERENCES

Adman, E.T., Sieker, L.C. and Jensen, L.H. (1973). *J.Biol.Chem.*, 248: 3987-3996.

Bui, P.T. and Mortenson, L.E. (1968). *Proc.Nat.Acad.Sci.U.S.A.*, 61: 1021-1027.

Burns, R.C. and Hardy, R.W.F. (1975). *In:* Nitrogen Fixation in Bacteria and Higher Plants (A. Kleinzeller, G.F. Springer and H.G. Wittman, eds), pp. 138, Vol. 21. Springer-Verlag, New York, Heidelberg and Berlin.

Carter, C.W., Jr., Kraut, J., Freer, S.T., Alden, R.A., Sieker, L.C., Adman, E. and Jensen, L.H. (1972). *Proc.Nat.Acad. Sci.U.S.A.*, 69: 3526-3529.

Eady, R.R. (1973). *Biochem.J.*, 135: 531-535.

Eady, R.R. and Smith, B.E. (1976). *In:* Dinitrogen Fixation (R. W.F. Hardy, ed.), in press. Wiley Interscience, New York.

Eady, R.R., Smith, B.E., Cook, K.A. and Postgate, J.R. (1972). *Biochem.J.*, 128: 655-675.

Eady, R.R., Kennedy, C., Smith, B.E., Thorneley, R.N.F., Yates, M.G. and Postgate, J.R. (1975). *Biochem.Soc.Trans.*, 3: 488-492.

Hammes, G.G. and Alberty, R.A. (1959). *J.Phys.Chem.*, 63: 274-278.

Hwang, J.C., Chen, C.H. and Burris, R.H. (1973). *Biochim.Biophys.Acta*, 292: 256-270.

Koren, R. and Hammes, G.G. (1976). *Biochemistry*, 15: 1165-1171.

Mortenson, L.E., Zumft, W.G. and Palmer, G. (1973). *Biochim. Biophys.Acta*, 292: 422-435.

Smith, B.E. and Lang, G. (1974). *Biochem.J.*, 137: 169-180.

Smith, B.E., Lowe, D.J. and Bray, R.C. (1972). *Biochem.J.*, 130: 641-643.

Smith, B.E., Lowe, D.J. and Bray, R.C. (1973). *Biochem.J.*, 135: 331-341.

Smith, B.E., Thorneley, R.N.F., Eady, R.R. and Mortenson, L.E. (1976). *Biochem.J.*, 157: 439-447.
Thorneley, R.N.F. (1975). *Biochem.J.*, 145: 391-396.
Thorneley, R.N.F., Eady, R.R. and Yates, M.G. (1975). *Biochim. Biophys.Acta*, 403: 269-284.
Thorneley, R.N.F., Yates, M.G. and Lowe, D.J. (1976). *Biochem. J.*, 155: 137-144.
Yates, M.G., Thorneley, R.N.F. and Lowe, D.J. (1975). *FEBs Letts.*, 60: 89-93.
Zumft, W.G., Palmer, G. and Mortenson, L.E. (1973). *Biochim. Biophys.Acta*, 292: 413-421.

MOLYBDENUM STORAGE COMPONENT FROM *CLOSTRIDIUM PASTEURIANUM*

BARBARA B. ELLIOTT and L. E. MORTENSON

Department of Biological Sciences, Purdue University, West Lafayette, Indiana 47907, U.S.A.

INTRODUCTION

The importance of MoO_4^{2-} in biological dinitrogen fixation has been demonstrated (Bortels, 1930). It is now known that molybdenum (Mo) is a constituent of the MoFe protein of the nitrogenase complex (Bulen and LeComte, 1966; Eady *et al.*, 1972; Mortenson *et al.*, 1967). Recent attempts have been made to characterize the role of Mo in the catalytic reactions of the nitrogenase components and to establish its role in the regulation of the nitrogenase system (Benneman *et al.*, 1973; Brill *et al.*, 1974; Cardenas and Mortenson, 1975).

Approaching the problem from the level of permeation followed by incorporation into nitrogenase apoprotein, Elliott and Mortenson (1975) characterized an active molybdate transport system in the N_2-fixing anaerobe, *Clostridium pasteurianum*. Transport of MoO_4^{2-} in the organism is energy-dependent because sucrose is required in the minimal media and the system is inhibited by the glycolysis inhibitors, NaF, iodoacetic acid and arsenate. The cells accumulate MoO_4^{2-} against a gradient and the uptake shows a marked dependence on temperature (optimum 37°C) and pH (optimum 6.0). The rate of moO_4^{2-} uptake with increasing MoO_4^{2-} concentrations shows saturation kinetics with an apparent Km and Vmax of 4.8×10^{-5}M and 55 nmol/g dry cells per min, respectively. Inhibition studies with the anions SO_4^{2-}, $S_2O_3^{2-}$, WO_4^{2-} and VO_3^{2-} show that SO_4^{2-} and WO_4^{2-} competitively inhibit MoO_4^{2-} uptake (apparent Ki [SO_4^{2-}] is 3.0×10^{-5}M; apparent Ki [WO_4^{2-}] is 2.4×10^{-5}M), whereas $S_2O_3^{2-}$ and VO_3^{2-} have no inhibitory effect.

The absence of Mo has been shown to prevent the synthesis of nitrogenase (Cardenas and Mortenson, 1975) but little detail is available on how Mo regulates or on how MoO_4^{2-} accumulation is regulated. The present communication describes the regulation of MoO_4^{2-} accumulation by *C. pasteurianum* and shows that its regulation is closely associated with the regulation of nitrogenase activity in this organism.

Exchange experiments demonstrate that only a small percentage of MoO_4^{2-} taken up by the cells remains unbound. As a result, we have postulated that cells grown with low concentrations of molybdate synthesize a Mo-binding and -storage component. When Mo is made available to such cells, it binds to this component and is not easily exchanged by external MoO_4^{2-} (Elliott and Mortenson, 1975). The data presented here describes an isolation procedure for this Mo-binding and -storage component from *C. pasteurianum* and gives some of its properties. Further study of this small molecular weight Mo-containing protein should prove useful in elucidating the form of Mo in nitrogenase, how Mo is inserted into nitrogenase and its role in dinitrogen fixation.

RESULTS AND DISCUSSION

Coordinate control of nitrogenase and the molybdate transport system

Regulation of the molybdate transport system by NH_3. The N_2-fixing system of *Clostridium pasteurianum* is regulated by NH_3 (Daesch and Mortenson, 1972). When this organism is grown in the presence of excess NH_3, there is no detectable nitrogenase activity as measured by acetylene reduction, N_2 reduction or cross reaction with nitrogenase antisera. Since *C. pasteurianum* cells growing on NH_3 do not synthesize the MoFe protein and since no other major essential molybdoproteins are required, one would expect the need to transport MoO_4^{2-} to be small. Measurement of nitrogenase (acetylene reduction activity), MoO_4^{2-} transport and the ammonia concentration of the medium during various times of the diauxic growth of *C. pasteurianum* are recorded in Fig. 1. Compared with cells fixing N_2, cells grown in the presence of 2mM NH_3 are greater than 95% repressed for MoO_4^{2-} accumulation. Ability to accumulate Mo begins to increase just prior to NH_3 exhaustion (under Ar or N_2) and continues to increase throughout the lag period as cells shift from NH_3-growing to N_2-fixing conditions. Since *C. pasteurianum* is unable to synthesize MoFe protein in the absence of MoO_4^{2-} (Cardenas and Mortenson, 1975) it is more important to note that MoO_4^{2-} accumulation activity appears just prior to the onset of the diauxic lag and before the synthesis of the apoproteins of the MoFe and Fe components of nitrogenase.

When cells are shifted from N_2-fixing to NH_3-growing conditions, the Mo accumulation activity per fixed number of cells decreases because of the increase of cells in the absence of the synthesis of the Mo storage proteins. This dilution of

MoO_4^{2-} storage activity parallels the dilution of nitrogenase activity when cells are shifted from N_2-fixing to NH_3-growing conditions (Elliott and Mortenson, 1976).

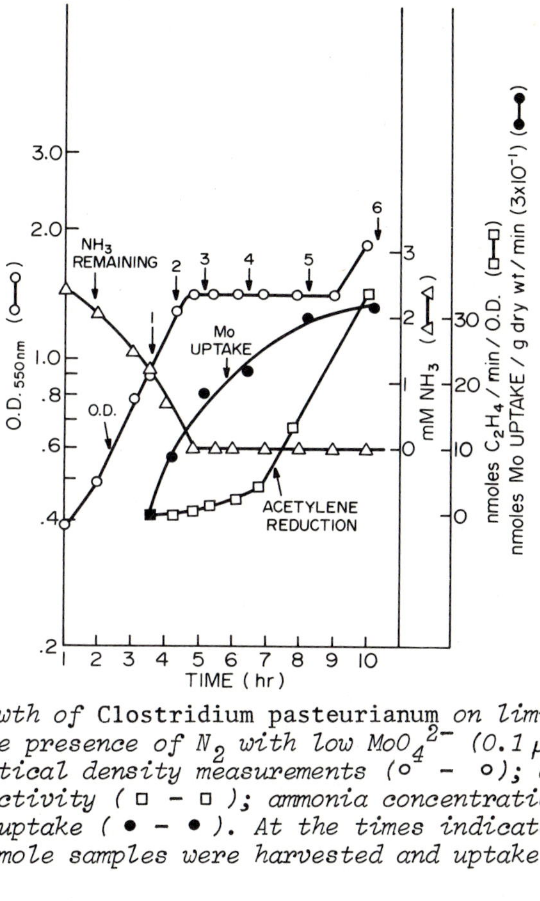

Fig. 1. *Growth of* Clostridium pasteurianum *on limiting ammonia (3mM) in the presence of* N_2 *with low* MoO_4^{2-} *(0.1 μM) and* SO_4^{2-} *(22 μM). Optical density measurements (○ - ○); acetylene reduction activity (□ - □); ammonia concentration (Δ - Δ) and* MoO_4^{2-} *uptake (● - ●). At the times indicated by the arrows, 8 nmole samples were harvested and uptake activity was measured.*

Protein synthesis as a definitive requirement for molybdate accumulation activity To demonstrate that new protein synthesis is required for the appearance of the MoO_4^{2-} storage protein, cells first were grown under NH_3-limiting conditions. Such cells, just prior to the exhaustion of NH_3 from the medium, are 95% repressed for molybdate accumulation activity 25 nmoles/g dry wt per 60 min). If these cells are assayed two hours after the onset of the diauxic lag, they show greatly increased levels of molybdate accumulation activity (223 nmoles/g dry wt per 60 min). If just prior to the exhaustion of NH_3 from the medium, chloramphenicol is added to a culture to make the final concentration 300 g/ml and the cells harvested two hours later assays for MoO_4^{2-} accumulation show no increase. This observation strongly suggested that new protein synthesis is required for the accumulation of MoO_4^{2-}. Similar experiments adding chloramphenicol at various times during diauxic growth and assaying for Mo accumulation two hours later, also showed no increase (Elliott and Mortenson, 1976).

Effects of carbamyl phosphate on molybdate accumulation Seto and Mortenson (1973, 1974) have demonstrated that the regulation of nitrogenase activity by NH_3 may occur partially *via* carbamyl phosphate, which acts as an end product inhibitor as well as a repressor of nitrogenase synthesis. Similarly, carbamyl phosphate (> 15mM), but not NH_3, inhibits 58% of the MoO_4^{2-} uptake activity *in vitro*. If similar experiments (II.A.2) with 1mM carbamyl phosphate added to the culture just prior to the exhaustion of NH_3 are carried out, the accumulation activity, measured 2 hr later, is 100% repressed. This repression of MoO_4^{2-} transport *via* carbamyl phosphate closely parallels the repression of nitrogenase activity in this organism (Elliott and Mortenson, 1976).

Effects of methionine sulfoximine on molybdate accumulation It has been demonstrated recently that methionine sulfoximine, an inhibitor of enzymes involved in NH_3 assimilation, causes the nitrogenases of *Azotobacter vinelandii* and *Klebsiella pneumoniae* to be synthesized even in the presence of excess NH_3 (Gordon and Brill, 1974). The results presented in Table I demonstrate that the addition of 2.5 mg/ml methionine sulfoximine to a culture of *C. pasteurianum* growing in the presence of 8mM NH_3 causes coordinate derepression of nitrogenase and the MoO_4^{2-} accumulation system.

Table I

Effects of MSX[a] on Nitrogenase and Molybdate Transport Derepression in *Clostridium pasteurianum*

N-source	Inhibition[a]	Acetylene[b] reduction	MoO_4^{2-}[c] uptake	O.D.[d]
NH_3	None	0.0	23	0.67
NH_3	MSX	21.4	110	0.52
N_2	None	56.4	223	0.62
N_2	MSX	56.3	234	0.57

[a]MSX = Methionine sulfoximine 2.5 mg/ml; [b]nmoles C_2H_4 formed/min per O.D.; [c]nmoles $^{99}MoO_4^{2-}$/g dry wt per 60 min; [d]$O.D._{550}$ of culture at time of assay.

Evidence for the existence of a molybdenum storage protein

Uptake and exchange studies with MoO_4^{2-} The ability of externally added unlabelled MoO_4^{2-} to exchange with $^{99}MoO_4^{2-}$ accumulated in clostridial cells is shown in Fig. 2. After 20 min, only 9% of the radioactivity taken up by the cells was

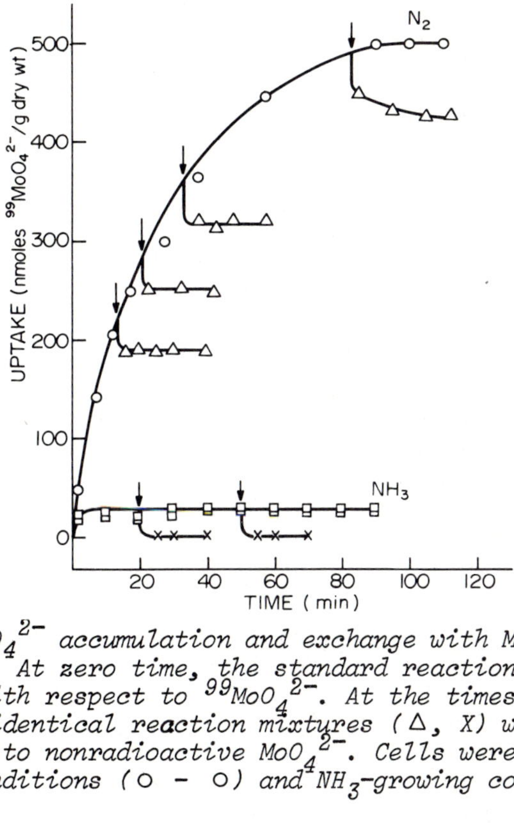

Fig. 2. *$^{99}MoO_4^{2-}$ accumulation and exchange with MoO_4^{2-} in* C. pasteurianum. *At zero time, the standard reaction mixture was made 10 μM with respect to $^{99}MoO_4^{2-}$. At the times indicated by the arrows, identical reaction mixtures (Δ, X) were made 1 mM with respect to nonradioactive MoO_4^{2-}. Cells were grown under N_2-fixing conditions (○ - ○) and NH_3-growing conditions (8 mM)(□ - □).*

"chased" by a 100-fold excess of externally added unlabelled MoO_4^{2-}. As the time of exposure to $^{99}MoO_4^{2-}$ increased, the amount that could not be "chased" increased. The amount of $^{99}MoO_4^{2-}$ that remained in the cells after the initial small chase remained the same throughout the secondary incubation period. The small amount of MoO_4^{2-} taken up by NH_3-grown cells is completely exchangeable. These exchange experiments with MoO_4^{2-} demonstrate that only a small percentage of the MoO_4^{2-} taken up by cells grown on low concentration of MoO_4^{2-} (0.7 μM) remains free and that all Mo in NH_3-grown cells is free. This result suggests that cells grown on low MoO_4^{2-} synthesize a

Mo-binding and -storage component to scavenge Mo and when excess MoO_4^{2-} is later made available to the cell, it binds to this component and cannot be exchanged easily.

Uptake and exchange studies with WO_4^{2-} The ability of WO_4^{2-} to chase radioactive MoO_4^{2-} from cells is shown in Fig. 3. The initial exchange of WO_4^{2-} for $^{99}MoO_4^{2-}$ was similar to that seen

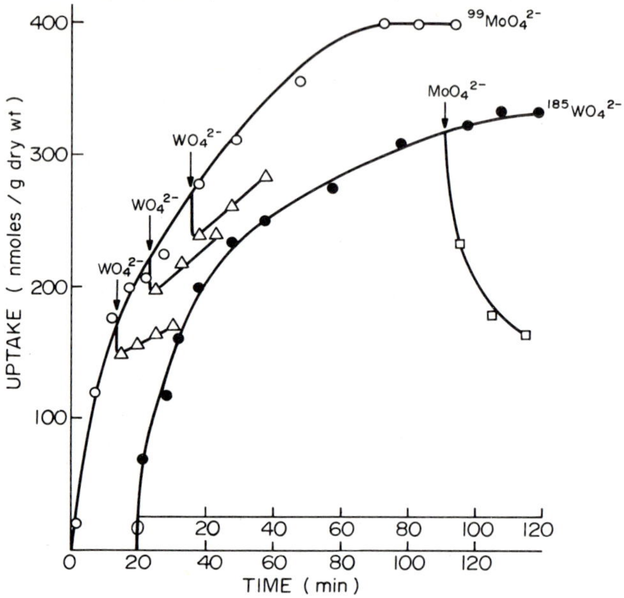

Fig. 3. $^{99}MoO_4^{2-}$ *accumulation by* C. pasteurianum *and its exchange with* WO_4^{2-} *and* $^{185}WO_4^{2-}$ *accumulation by* C. pasterianum *and its exchange with* MoO_4^{2-}. *At zero time, the standard reaction mixture was made 10 M with respect to* $^{99}MoO_4^{2-}$ (○ - ○) *or 10 μM with respect to* $^{185}WO_4^{2-}$ (● - ●). *At the times indicated by the arrows, identical reaction mixtures were made 1 mM with respect to nonradioactive* WO_4^{2-} (△ - △) *or nonradioactive* MoO_4^{2-} (□ - □).

in the MoO_4^{2-} chase experiments. With WO_4^{2-}, however, after the initial exchange, $^{99}MoO_4^{2-}$ accumulation resumed but at a slower rate than the initial uptake in the absence of WO_4^{2-}. If the affinity of the molybdenum-binding and -storage component for WO_4^{2-} were less than its affinity for MoO_4^{2-} or if WO_4^{2-} could not bind to it at all, the resumption of MoO_4^{2-} uptake after exchange of the free pool by WO_4^{2-} would be explained. Evidence that this is true comes from the reciprocal experiment (Fig. 3) in which cells were allowed to accumulate $^{185}WO_4^{2-}$ and the free pool was chased by adding MoO_4^{2-}. These experiments demonstrate that a much greater percentage of the $^{185}WO_4^{2-}$ accumulated by the cells remains free and that the cells can accumulate a higher concentration of free WO_4^{2-} than free MoO_4^{2-}

Distribution of Mo in clostridial cells

Repression of synthesis of the MoO_4^{2-} accumulation system by MoO_4^{2-} Cells grown in the presence of 1mM MoO_4^{2-} are 80% repressed for the maximum accumulation activity seen with 0.1 μM MoO_4^{2-} and at 0.1mM accumulation is 50% lower. Concentrations of MoO_4^{2-} lower than 0.1 μM in the growth medium did not increase accumulation whereas MoO_4^{2-} concentrations higher than 10mM completely inhibited growth of this organism (Elliott and Mortenson, 1976).

Distribution of Mo in clostridial cells grown in the presence of 10 M MoO_4^{2-} The total amount of Mo associated with 60g of cells grown in the presence of 10μM MoO_4^{2-} was 33.5 moles; the cell wall-membrane fraction contained 17.6 μmoles and the soluble cell components 16.3μ moles. After repeated washing of the cell wall-membrane fraction with 0.05M Tris pH 7.5, 9.5 moles of its associated Mo was released. This Mo probably was occluded with the cell wall-membrane fraction during the centrifugation process. The remaining 46% (8.1 moles) of the Mo in the cell wall-membrane fraction (or 24% of the total Mo of the cell) is tightly bound, can be released only by treatment with detergents (SDS, Triton X-100), and may be associated with a membrane bound Mo transport system (Table II).

Distribution of Mo in clostridial cells grown in the presence of 0.7 μM MoO_4^{2-} When cells were grown in the presence of 0.7 μM MoO_4^{2-}, the total amount of Mo associated with 60g of cells was 30 μmoles. In such cells, the wall-membrane fraction initially contained 60% of the Mo with the remainder in the crude extract. After repeated washing of the cell wall-membrane fraction, 13 moles or 44% of the total cell Mo remained tightly associated (Table III) which indicates that more MoO_4^{2-} transport system is synthesized when cells are grown under low Mo conditions or that more Mo storage protein is bound to the membrane.

Distribution of Mo in crude extracts

Molybdenum binding and storage protein. Crude extracts from cells grown in the presence of various concentrations of MoO_4^{2-} were fractionated anaerobically on a DEAE cellulose column previously equilibrated with 0.05M Tris-HCl pH 7.5 plus 1mM dithionite. Fig. 4 shows a typical elution profile of the protein, the Mo in the fractions and the position of the MoFe protein of nitrogenase.

Table II

Distribution of molybdenum in clostridial cells grown in the presence of 10 μM MoO_4^{2-}

Fraction	Volume	Protein mg/ml	Total Protein mg	Mo nmoles /ml	Total Mo nmoles
CW + CM[a]	300	N.D.	N.D.	58.7	17,596
1st wash of CW + CM	300	16.8	5,040	25.0	7,500
2nd wash of CW + CM	300	4.36	1,308	6.5	1,950
3rd wash of CW + CM	300	2.86	858	0	0
4th wash of CW + CM	300	0	0	0	0
5th wash + 2% of CW + CM	200	N.D.	N.D.	45.2	9,040
Crude Extract	290	32.3	9,367	56	16,240
Crude Extract + 2% PSO_4	280	32	9,960	57.1	15,988

[a]CW = cell wall; CM = cell membrane; cell walls were washed with 0.05M Tris pH 7.5. 25% of the total molybdenum in the cell is tightly associated with the cell wall-membrane fraction.

Table III

Distribution of molybdenum in clostridial cells grown in the presence of 0.7 μM MoO_4^{2-}

Fraction	Volume	Protein mg/ml	Total Protein mg	Mo nmoles /ml	Total Mo nmoles
CW + CM[a]	300	N.D.	N.D.	63.0	18,900
1st wash of CW + CM	250	23.0	5,750	20.0	5,000
2nd wash of CW + CM	270	7.0	1,890	7.0	1,890
3rd wash of CW + CM	340	2.86	972	0	0
4th wash of CW + CM	300	0	0	0	0
5th wash + 2% of CW + CM	300	N.D.	N.D.	43.2	12,970
Crude Extract	270	29	7,830	40.6	10,950

[a]CW = cell wall; CM = cell membrane; cell walls were washed with 0.05 M Tris pH 7.5. 44% of the total molybdenum in the cell is tightly associated with the cell wall-membrane fraction.

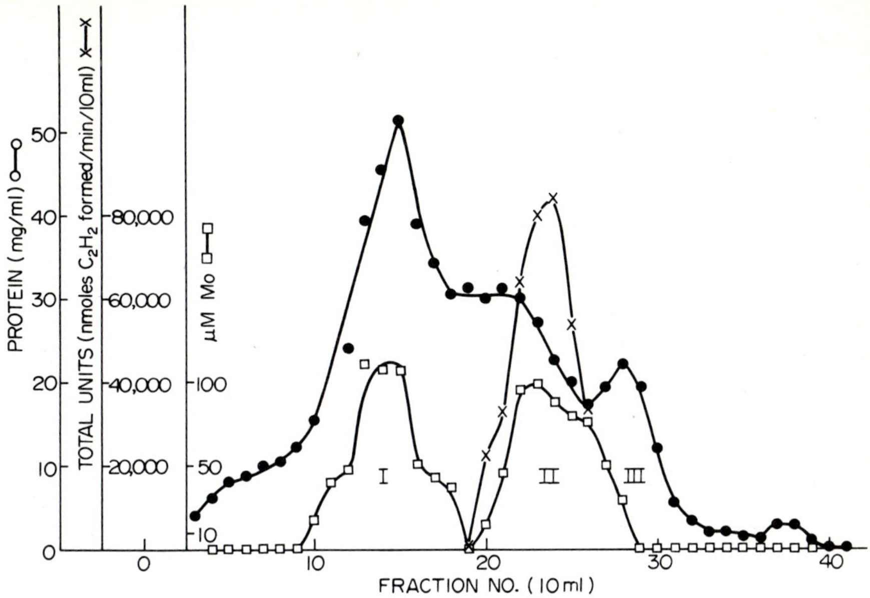

Fig. 4. *Fractionation of Mo-containing compounds in crude extracts of* C. pasteurianum. *Cells were grown on 0.7 μM MoO_4^{2-} until the O.D. reached 2.0 at which time the culture was adjusted to 100 μM MoO_4^{2-} and the cells were harvested 2 hr later. Extracts prepared from 50 g cells were treated with 200 mg protamine sulphate per g protein and after centrifuging, the supernatant solution was loaded on a 2.5 x 32 cm DEAE cellulose column previously equilibrated with 0.05 Tris-HCl pH 7.5 plus 1 mM sodium dithionite. The protein was eluted with a 0.1 to 1.0 M NaCl gradient (300 ml each). Ten ml fractions were collected anaerobically and assayed for protein (o - o), molybdenum (□ - □), and MoFe protein activity (x - x).*

Fig. 5 shows the Mo elution profiles from DEAE cellulose columns of extracts of cells grown with different concentrations of Mo. When cells are grown in the presence of 0.7 μM MoO_4^{2-} (Fig. 5A), about 28% (2.0 μmoles) of the total Mo in the crude extract loaded on the column appears in fraction I. The remainder is in fraction II, the MoFe protein of nitrogenase. If the crude extract is made 500 μM with respect to MoO_4^{2-} prior to the DEAE column chromatography (Fig. 5B) the amount of Mo in fraction I does not increase, demonstrating that fraction I is not free molybdate and that a component in it cannot bind free molybdate. The remaining Mo in the crude extracts elutes in two additional fractions, the MoFe protein of nitrogenase (fraction II) and fraction III which is dialyzable and could be free

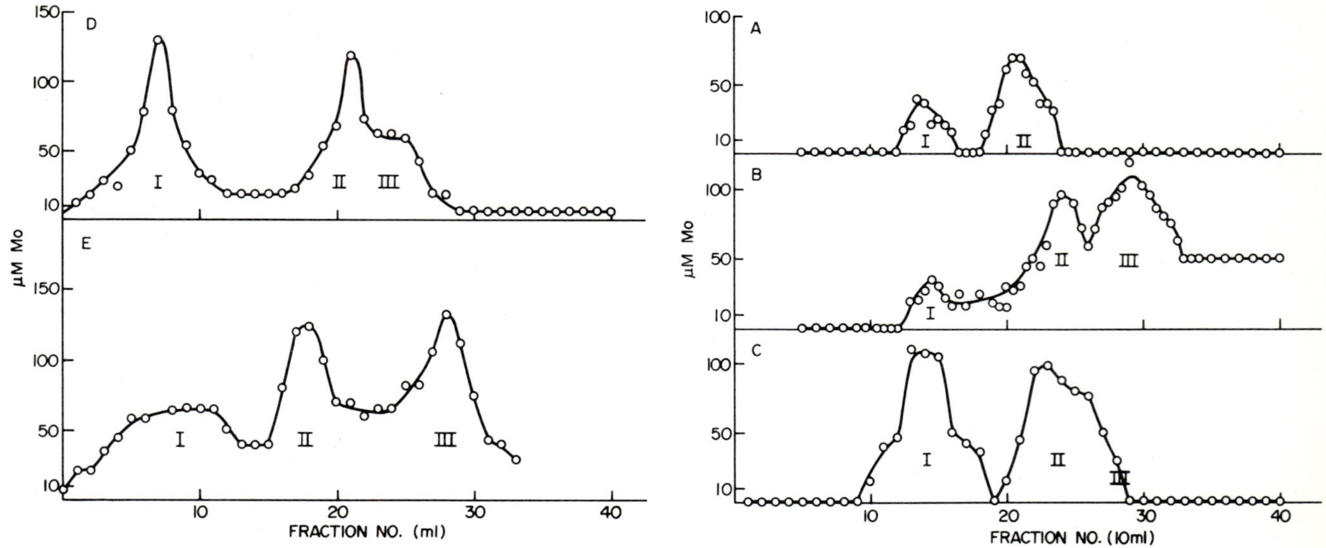

Fig. 5. *Fractionation of Mo-containing compounds of crude extracts of* C. pasterianum *grown in the presence of varying* MoO_4^{2-} *concentrations and with* MoO_4^{2-} *added before chromatography. Crude extracts from 50 g of each batch of cells were prepared and loaded on DEAE cellulose columns (see legend of Fig. 4). The elution profiles show the μM concentration of molybdenum. (A) represents cells grown on 0.7 μM* MoO_4^{2-}*, (B) cells grown in 0.7 μM* MoO_4^{2-}*, with 500 M* MoO_4^{2-} *added to the crude extract, (C) cells grown in 0.7 μM* MoO_4^{2-} *until the O.D. reached 2.0 at which time the culture was made 100 μM with respect to* MoO_4^{2-}*, 300 μg/ml of chloramphenicol was added, and the culture incubated, (D) cells grown in 10 μM* MoO_4^{2-} *and (E) cells grown in 500 μM* MoO_4^{2-}*.*

MoO_4^{2-} or a low molecular weight protein that weakly binds MoO_4^{2-}.

It can be postulated that, when cells are grown on low MoO_4^{2-}, they synthesize a large amount of a Mo-binding and -storage protein to allow them to make use of all Mo. If additional Mo were then added to the medium, it would bind strongly to this component and would not be easily exchanged. Thus, when cells were grown in the presence of 0.7 μM MoO_4^{2-} and two hours before harvesting, the culture was adjusted to 100 μM MoO_4^{2-} (Fig. 5C), the amount of Mo now associated with fraction I increased dramatically from 2.0 to 6.24 μmoles (50% of the total Mo in the crude extract). This increase occurs in the presence or absence of added chloramphenicol (300 μg/ml), indicating that the increase is the result of the binding of an inorganic species of Mo to preformed protein.

There are several possibilities to explain why this preformed binding and storage protein will not bind MoO_4^{2-} when excess MoO_4^{2-} is added to the crude extract. Firstly, an energy source may be needed for MoO_4^{2-} binding which would not be available in the crude extract. Secondly, a physical transfer of MoO_4^{2-} from the permease to the molybdenum storage protein may be required and thirdly, the MoO_4^{2-} may be converted to a new Mo species by the permease or a reductase and the storage protein may only bind this altered form of Mo.

When cells are grown with concentrations of MoO_4^{2-} that give optimum N_2 fixation activity (10 μM; Fig. 5D), about 6.0 moles or 50% of the total molybdenum in the crude extract is associated with fraction I. Further evidence that this may be a molybdenum storage protein can be deduced from the elution profile on DEAE cellulose (Fig. 5E) of extracts from cells grown on high levels of Mo (500 μM). Under these conditions, a smaller amount of the Mo is present in fraction I, 4.0 μM Mo or 22% of the total Mo of the crude extract. The fact that this "storage" protein is present even at these high concentrations of MoO_4^{2-} suggests that it also functions in the transfer of Mo to the MoFe protein. Hence, the term storage protein may be a misnomer. If fraction I is concentrated using an Amicon pM 10 filter, none of the Mo passes through the membrane, indicating that the Mo is associated with a protein >10,000 daltons. Fig. 6 shows an elution profile of this fraction from a Sephadex G-100 column. About 95% of the Mo put on the column was recovered in a single peak that eluted after a major protein peak of 60,000 daltons (hydrogenase). Sodium dodecyl sulfate electrophoresis of this fraction shows no contaminating Mo Fe protein. Recent study of the purification and properties of this protein, which is believed to be a Mo-binding, -storage and -transfer protein, has shown it to have a molecular weight of about 67,000 and to contain 6 or more Mo atoms. Further characterisation is in progress.

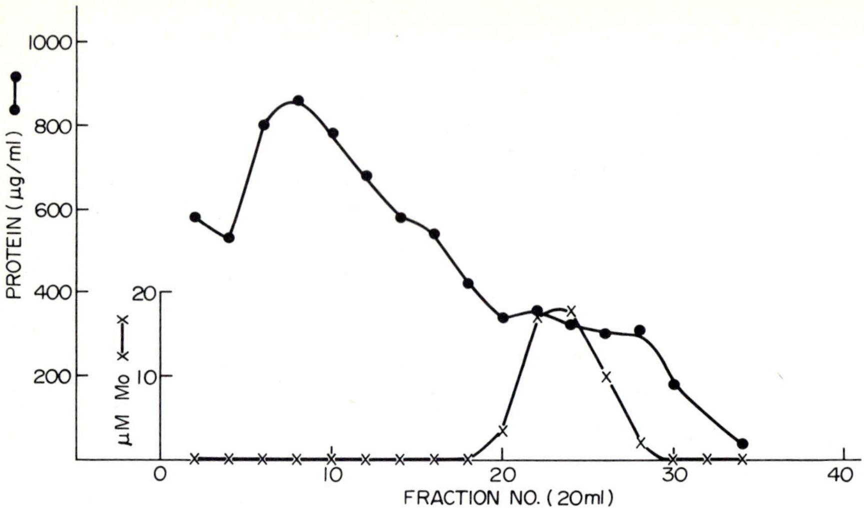

Fig. 6. *Elution profile of fraction 1 (Fig. 5C) after Sephadex G-100 filtration. Fraction I from Fig. 5C was concentrated on an Amicon pM 10 filter and the concentrate was loaded on a 5 x 70 cm G-100 Sephadex column previously equilibrated with 0.05M Tris-HCl pH 7.5 plus 1.0 µM dithionite. Twenty ml fractions were collected anerobically and assayed for protein (• - •) and molybdenum (x - x). The recovery of the Mo was 95%.*

The MoFe protein of nitrogenase For a typical crude extract preparation, the Mo associated with fraction II is 5.5μmoles. The Mo content of this fraction, which contains all the MoFe protein, is slightly lower when cells are grown on low concentrations of MoO_4^{2-}. Based on an acetylene reduction activity of 100 units/mg for the crude extract and a maximum activity of 2,200 units/mg for purified MoFe protein, 83% of the Mo of fraction II can be accounted for as the MoFe nitrogenase component.

Excess MoO_4^{2-} When cells are grown in low concentrations of MoO_4^{2-}, all of the Mo in the crude extract is associated with the Mo storage protein (fraction I) and the MoFe protein (fraction II). As the MoO_4^{2-} concentration of the medium increases, a third Mo peak appears (fraction III) that elutes after the MoFe protein from a DEAE cellulose column. If free MoO_4^{2-} is added to a crude extract from cells grown on low MoO_4^{2-} (Fig. 5B), this third fraction increases dramatically. The molybdenum associated with this fraction is dialyzable and may be: (i) free MoO_4^{2-}; (ii) a small molecular weight protein of $<10,000$ MW that is capable of binding free molybdate; or (iii) a larger molecular weight protein that is very unstable with respect to Mo binding.

ACKNOWLEDGEMENTS

This investigation was supported by Public Health Service grant A1-04865 from the National Institute of Allergy and Infectious Diseases. B. B. E. was a trainee of Public Health Service training grant T01 GM01392 from the National Institute of General Medical Sciences. Purified MoFe protein and Fe protein were kindly provided by Richard E. Bare. We would like to thank Richard E. Bare for technical assistance with the acetylene reduction assay and Dorothy Sheets for assistance with cell growth.

REFERENCES

Benneman, J.R., Smith, G.M., Kostel, P.J. and McKenna, C.E. (1973). *FEBs Letters*, 29: 219-221.

Bortels, H. (1930). *Arch.Mikrobiol.*, 1: 333-342.

Brill, W.J., Steiner, A.L. and Shah, V.K. (1974). *J.Bacteriol.*, 118: 986-989.

Bulen, W.A. and LeComte, J.R. (1966). *Proc.Nat.Acad.Sci.U.S.A.*, 56: 979-986.

Cardenas, J. and Mortenson, L.E. (1975). *J.Bacteriol.*, 123: 978 984.

Daesch, G. and Mortenson, L.E. (1972). *J.Bacteriol.*, 110: 103-109.

Eady, R.R., Smith, B.E., Cook, K.A. and Postgate, J.R. (1972). *Biochem.J.*, 128: 655-675.

Elliott, B.B. and Mortenson, L.E. (1975). *J.Bacteriol.*, 124: 1295-1301.

Elliott, B.B. and Mortenson, L.E. (1976). *J.Bacteriol.*, in press.

Gordon, J.K. and Brill, W.J. (1974). *Biochem.Biophys.Res.Commun.*, 59: 967-971.

Mortenson, L.E., Morris, J.A. and Jeng, D.Y. (1967). *Biochim. Biophys.Acta*, 141: 516-522.

Seto, B. and Mortenson, L.E. (1973). *J.Bacteriol.*, 117: 805-812.

Seto, B. and Mortenson, L.E. (1974). *J.Bacteriol.*, 120: 822-830.

PHYSIOLOGICAL ASPECTS OF NITROGEN FIXATION

M.G. YATES

A.R.C. Unit of Nitrogen Fixation,
University of Sussex, Brighton BN1 9QJ, U.K.

INTRODUCTION

Physiological aspects of nitrogen fixation cover a wide and increasing variety of topics: electron transport to nitrogenase; the means of protecting nitrogenase against oxygen damage *in vivo;* the energy requirement for nitrogen fixation; regulation of nitrogenase activity; location of nitrogenase within the cell; pathways of nitrogen assimilation, regulation of nitrogenase synthesis; enzymes related to and affecting nitrogenase activity (e.g. hydrogenase) and the effects of substrates for nitrogenase (e.g. acetylene) on bacterial metabolism. These topics need to be considered for each of the wide range of types of micro-organisms which contain nitrogenase: both free-living and symbiotic. Aspects of this wide-ranging topic have been reviewed by Dalton (1974), Mulder and Brotonegoro (1974); Postgate (1974), Evans and Phillips (1975) and discussed in a nitrogen fixation workshop (Evans, 1975). Since that time new knowledge has come to light in several aspects of the physiology of nitrogen fixation, including newly discovered nitrogen-fixing bacteria. Table I gives a reasonably up-to-date list of nitrogen-fixing organisms. I shall attempt to review briefly the advances in the aspects mentioned above over the last few years while knowing that many of these topics will be discussed more rigorously elsewhere in this volume. In particular, aspects of the enzyme nitrogenase which might seem pertinent to this review, such as the possible role of the activating factor for *Rhodospirillum rubrum* Fe protein (Ludden and Burris, 1976) and the possible existence of a common Mo-containing subunit for molybdo-enzymes (Nason *et al.*, 1970), will be left to others to discuss.

Table I

Biological systems that fix N_2

Symbiotic	lichens	*Nostoc, Anabaena, Tolypothrix* (P)
	angiosperms (gunnera)	*Nostoc*
	azolla	*Anabaena*
	tropical grasses	*Azotobacter paspalum, Spirillum lipoferum*
	root nodules	legumes *Rhizobium* non-legumes *'Frankia' actinomycetes*
Free-living	aerobic	*Nostoc, Anabaena, Plectonema,* etc. (P) *Azotobacter, Beijerinckia, Derxia* *Mycobacterium,* Methane-oxidizing bacteria *Corynebacterium autotrophicum* *Spirillum lipoferum* *Thiobacillus ferrooxidans*
	facultative	*Klebsiella, B. polymyxa, B. macerans* *E. coli** *Rhodospirillum, Rhodopseudomonas* (P)
	anaerobic	*C. pasteurianum, C. butyricum* *Desulfotomaculum, Desulfovibrio* *Chromatium, Chlorobium, Chloropseudomonas* (P)

*By genetic transfer of *nif* from *K. pneumoniae*.
(P) Photosynthetic.

ELECTRON TRANSPORT TO NITROGENASE

Electron carriers

Ferredoxins and flavodoxins are the only known biological electron carriers which are capable of transferring electrons to nitrogenase. Low potential cytochromes such as cytochrome c_3 from *Desulfovibrio* can replace ferredoxin in some reactions but they have not been tested exhaustively as possible carriers to nitrogenase. Ferredoxin was first shown to be an absolute requirement for nitrogen fixation in *C. pasteurianum* by Mortenson (1964) and that it linked directly between hydrogenase and nitrogenase by Walker and Mortenson (1974). Flavodoxin substituted for ferredoxin in iron-starved *C. pasteurianum* (Knight and Hardy, 1966). The ubiquitous distribution of these carriers has been revealed in the last ten years; they occur in most types of bacteria examined and in plants. Some flavodoxins are constitutive while others are induced or derepressed by iron starvation. Both classes of proteins have been reviewed recently: ferredoxins by Orme-Johnson (1974), Jensen (1974) and Palmer (1975) and flavodoxins by Mayhew and Ludwig (1975). Table II describes the properties of the carriers which are present in nitrogen-fixing bacteria including those known to mediate electron flow to nitrogenase either by coupling pyruvate or hydrogen metabolism (Mortenson, 1964) or heat-treated spinach chloroplasts to nitrogenase (Benemann *et al.*, 1969; Yoch *et al.*, 1969). There are no distinguishing properties within either of these two groups of carriers which can be associated with the ability to transfer electrons to nitrogenase: $[Fe_4S_4]_2$ eight-iron, and $[Fe_4S_4]$ four-iron ferredoxins are effective and even the $[Fe_2S_2]$ two-iron spinach ferredoxin will apparently act, though weakly in this role (Bothe and Yates, 1976). Molecular weights range from 6,000 to 24,000. The flavodoxins also vary in molecular weight and in stability towards oxygen of their semi-quinone radicals. Some ferredoxins are apparently more active than others as electron carriers to nitrogenase but a quantitative comparison of relative abilities of electron transfer is of little value since purified proteins may have lost their full *in vivo* potential during isolation. The observation that assays with physiological electron carriers usually do not realise the potential enzymic activity of the nitrogenase with sodium dithionite as the reductant supports this view. Perhaps one general comment that is safe to make is that ferredoxins are usually more active than flavodoxins.

Two features of the ferredoxins are worth mentioning in more detail. Several nitrogen-fixing organisms contain two ferredoxins capable of electron transfer to nitrogenase. Table II gives an indication of the relative amounts formed and the ability to transfer electrons to nitrogenase. Unfortunately, for any possible understanding of their roles, there is no apparent pattern within these parameters that can be related to the organisms' ability to fix nitrogen. Whether they all occur when the cells are grown with fixed nitrogen is not known. Only in *R. rubrum* is there a rationale as to their biological roles: ferredoxin I is produced only in the light; ferredoxin II is produced in the light and when the cells are grown heterotrophically in the dark (Shanmugam *et al.*, 1972). The two ferredoxins from *Azotobacter* are apparently similar (Yoch and Arnon, 1972) but those from *B. polymyxa* (Yoch, 1973) and *M. flavum* (Bothe and Yates, 1976) are sufficiently different to be regarded as separate proteins. No clear evidence has yet emerged of separate biological roles for these proteins.

Table II

Ferredoxins and flavodoxins in nitrogen-fixing bacteria

Organism	M.W.	$[Fe_nS_n]$**	E_m (mv)	Ferredoxins: Relative amounts*	Ferredoxins: Relative activity in nitrogenase assay*	Ferredoxins: NADP reduction by chloroplasts	References
C. pasteurianum	6,000	$[Fe_4S_4]_2$				+	Benemann and Valentine (1972)
A. vinelandii Ferredoxin I	14,500	$[Fe_4S_4]_2$	-450	5	1	+	Yoch and Arnon (1972)
			+340				Sweeney *et al.* (1975)
Ferredoxin II			-460	1	1	+	
M. flavum Ferredoxin I	13,500	$[Fe_4S_4]_2$	ND	5	1	-	Bothe and Yates (1976)
							Bothe and Yates (unpublished)
Ferredoxin II	17,500	$[Fe_4S_4]$	ND	1	5	+	
B. polymyxa Ferredoxin I	8,800	$[Fe_4S_4]$	-380	1	2	+	Yoch (1973a)
Ferredoxin II	8,800	$[Fe_4S_4]$	-420	2	3	+	Stombaugh *et al.* (1973)
R. rubrum Ferredoxin I	9,000	$[Fe_4S_4]_2$		formed in light	3		Shanmugam *et al.* (1972)
Ferredoxin II	15,000	$[Fe_4S_4]$		formed in light and dark	1		Yoch and Arnon (1975)

D. gigas						Bruschi *et al.*, (1976)
Ferredoxin I	18,000 (trimer)	$[Fe_4S_4]$ in monomer				
Ferredoxin II	24,000 (tetramer)	$[Fe_4S_4]$				
D. desulfuricans	6,000	$[Fe_4S_4]$	-330			Zubieta *et al.* (1973)
R. japonicum bacteroid	9,400	$[Fe_4S_4]$		+	-	Koch *et al.* (1970) Wong *et al.* (1971)
Chromatium	5,600	$[Fe_4S_4]_2$		+	+	Benemann and Valentine (1972)
A. cylindrica	10,000	$[Fe_2S_2]$		+	+	Benemann and Valentine (1972)
K. pneumoniae				+	-	Yoch (1974)
Gloeocapsa				+		Gallon *et al.* (1973)

Flavodoxins

Organism	M.W.	Redox Potentials (mv) E_1	E_2	Dependence upon iron-deficiency	Supports nitrogenase activity	Supports NADP reduction by chloroplasts	References
C. pasteurianum	14,600	-419	-132	+	+	+	Benemann and Valentine (1972)
A. vinelandii	23,000	-495	+ 50	-	+	+	Edmondson and Tollin (1971) Barman and Tollin (1973) Van-Lin and Bothe (1972)

Table II (cont.) Flavodoxins

Organism	M.W.	Redox potentials (mv) E_1	E_2	Dependence upon iron-deficiency	Supports nitrogenase activity	Supports NADP reduction by chloroplasts	References
K. pneumoniae	21,000			-	+	-	Yoch (1974)
D. gigas	16,000			+		+	Dubourdieu and Legall (1970)
D. vulgaris	16,300	-438	-149			+	Dubourdieu *et al.* (1975)
R. rubrum	22,800						Cusanovich and Edmondson (1971)
R. japonicum bacteroid				-	+	-	Koch *et al.* (1970)

* The figures for relative amounts and relative activity are comparisons of values within the same organism and are not comparisons between different species.

** Fe and acid-labile S values are corrected to the nearest whole cluster.

The second interesting property concerns the electron paramagnetic resonance (e.p.r.) spectroscopy of the ferredoxins. Ferredoxin I from *A. vinelandii*, the ferredoxin from *A. chroococcum*, both ferredoxins from *M. flavum* and ferredoxin II from *D. gigas* are paramagnetic when oxidised and diamagnetic when reduced (Fig. 1). In this respect they resemble the high potential iron protein (HIPIP) from Chromatium rather than the more usual type of ferredoxin which is paramagnetic when reduced. Yet their redox potentials are sufficiently negative to transfer electrons to nitrogenase. Sweeney *et al.* (1975), basing their deductions upon the e.p.r. signals obtained by oxidation of *A. vinelandii* ferredoxin I with potassium ferricyanide, suggested that it contained two [Fe_4S_4] clusters acting between the C and C^+ (-2 and -1) oxidation states (Carter *et al.*, 1972) but with redox potentials some 700 mv apart. Sweeney *et al.* considered that different polypeptide chains around the clusters control their redox behaviour. Howard *et al.* (1976) have analysed the amino terminal sequence for *A. vinelandii* ferredoxin I. This includes 8 cysteine residues, 7 of which can be correlated to cysteine residues in other bacterial ferredoxins but the insertion of cys-24 indicates a major change in the environment of one of the [Fe_4S_4] clusters. Bothe, Yates and Lowe (unpublished) have observed identical e.p.r. spectra in *M. flavum* and *A. chroococcum* ferredoxins which differ only slightly from the e.p.r. spectrum of *A. vinelandii* ferredoxin I published by Sweeney *et al.*, and which respond in a similar manner to oxidation by potassium ferricyanide. Fig. 1 shows the difference between the e.p.r. signals of *M. flavum* ferredoxin II and *A. chroococcum* ferredoxin and the more conventional ferredoxins. The role of these ferredoxins and the reasons for their anomalous iron-sulfur clusters need further investigation.

Nitrogen-fixing bacteria from which ferredoxins have not been isolated include *Spirillum lipoferum*, *Corynebacterium autotrophicum*, methane-oxidising bacteria and, surprisingly, *K. pneumoniae*. Yoch (1974) did obtain a brown fraction from *K. pneumoniae* which supported nitrogenase activity with the chloroplast system but he did not purify or characterise it further. Attempts in our laboratory to isolate such a fraction have met with little success.

Flavodoxins can be subdivided into two main types based upon their molecular weights: one type is approximately 14,000 and the other 23,000, Table II. Both types will transfer electrons to nitrogenase. All flavodoxins contain 1 FMN per mol which is involved in the electron transfer, and recent X-ray crystallography indicates the presence of an aromatic residue parallel to the plane of the isoalloxazine ring (Mayhew and Ludwig, 1975). Chemical binding experiments suggest that this is

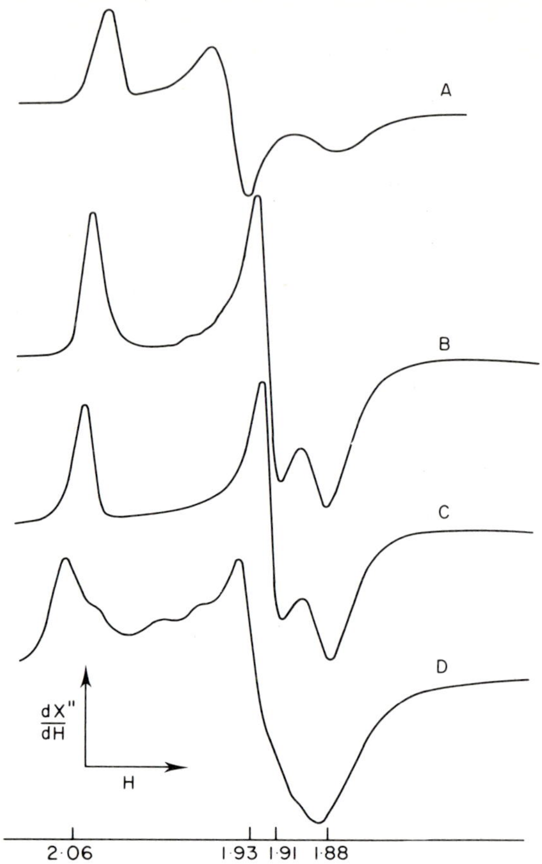

Fig. 1.a. *Electron paramagnetic resonance spectra of* reduced *samples of the iron-sulfur proteins. Conditions of EPR spectroscopy: microwave frequency, 9.2 Ghz; microwave power, 0.9 milliwatt; modulation frequency, 100 KHz; modulation amplitude, 10 G; magnetic field sweep rate, 260 G* min^{-1}*, time constant, 0.25 s; sample temperature, 13°K.* Abscissa, *linear in field (g values corresponding to several field positions are given);* ordinate, *an arbitrary function of the first derivative of the microwave absorption. A, spinach ferredoxin; B,* Bacillus polymyxa *fd II; C,* B. polymyxa *fd I, all reduced with excess sodium dithionite; D,* Clostridium acidi-urici *fd, in the presence of 0.06 atmosphere of* H_2*, hydrogenase, and 0.1 M Tris HCl buffer, pH 8.0. The last sample is approximately half-reduced. From Stombaugh* et al. *1974).*

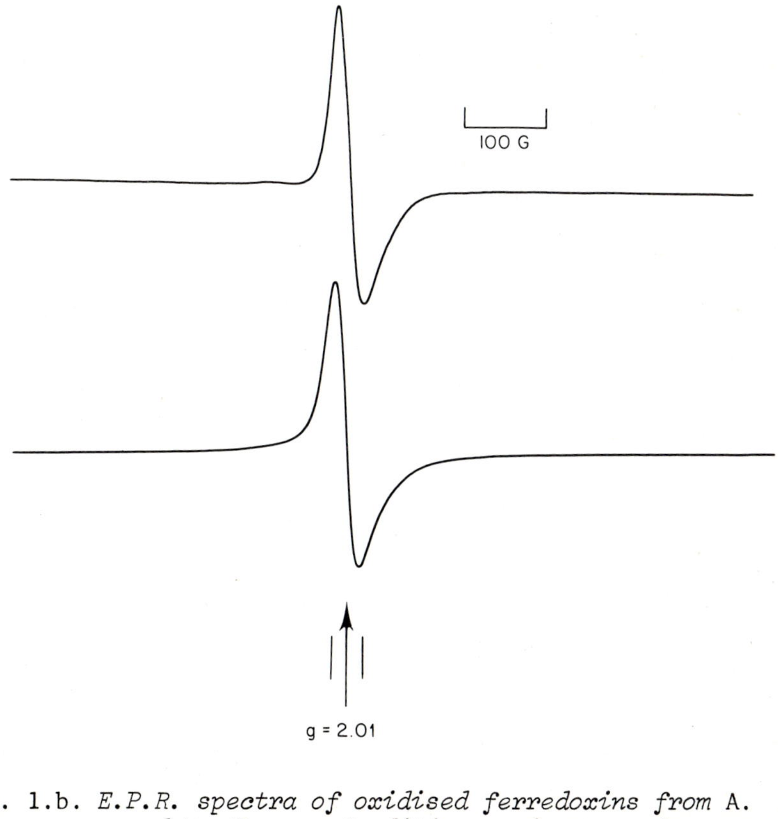

Fig. 1.b. *E.P.R. spectra of oxidised ferredoxins from* A. chroococcum *and* M. flavum. *Conditions: microwave frequency 9.3 GHz, modulation frequency 100 KHz, microwave power 10 mW, modulation amplitude 10 G. Time constant 0.3 sec., temperature 12°K. Top trace:* M. flavum *ferredoxin II; bottom trace:* A. chroococcum *ferredoxin.*

tryptophan (Ryan and Tollin, 1973). Sequence studies show identicalities between the proteins, these, presumably, being the peptide sequences involved in FMN binding. *Azotobacter vinelandii* flavodoxin has been sequenced recently (Tanaka *et al.*, 1975).

From the viewpoint of nitrogen fixation perhaps *Azotobacter* flavodoxin is most interesting. It has an extremely oxygen-stable semiquinone which prompted Benemann *et al.* (1969) to speculate that if the semiquinone form donated electrons to nitrogenase it would need to be resistant to oxygen in an obligate aerobe. This possibility became unlikely when Yates (1972b) observed direct electron transfer from fully reduced *A. chroococcum* flavodoxin to nitrogenase in an apparent 1 electron step resulting in the formation of the semiquinone. Until adequate kinetic studies have been performed the possibility of a two electron step followed by comproportionation to yield the semiquinone cannot be discounted. However, a one electron step is the most likely when one takes into consideration the redox potential of the hydroquinone-semiquinone couple and the knowledge that the active oxidised Fe protein takes up one electron when reduced (Thorneley *et al.*, 1976).

Electron donors

Knowledge of the reductants for nitrogen fixation *in vivo* is considerable but confusing and I refer the reader to reviews which covered advances in the field until 1973, (Benemann and Valentine, 1972; Yates and Jones, 1974, and, in particular, Evans and Phillips, 1975). Table III describes the electron donors known to serve nitrogenase, and the most active and therefore likely to be of most physiological significance, is underlined. Only new knowledge arising since the latest reviews will be discussed.

Pyruvate is the most likely source of electrons and ATP for nitrogen fixation in anaerobic bacteria although Jungermann *et al.* (1974) obtained kinetic data which suggested that as much as 25% of the electrons for nitrogenase were supplied through NADH in *C. pasteurianum* . Yoch (1974) showed that pyruvate was the major electron donor for nitrogenase in *K. pneumoniae* and that *E. coli* extracts also contained a pyruvate-metabolising system which donated electrons to *K. pneumoniae* nitrogenase.

The search for electron donors to nitrogenase in aerobes has centred around NADH and NADPH and two lines of evidence suggest that NADPH is the most likely source of electrons in *A. vinelandii* and blue-green algae. The first is direct evidence for NADPH-supported nitrogenase activity in cell-free extracts (Benemann *et al.*, 1971; Smith *et al.*, 1971; Gallon *et al.*, 1973). Indirect evidence involving NADPH is that NADP-dependent isocitrate dehydrogenase, would not fix nitrogen but grew adequately with fixed nitrogen. When grown on succinate the cells fixed nitrogen and contained isocitrate dehydrogenase.

The above evidence points to NADPH generated through isocitrate dehydrogenase as the electron donor for nitrogenase. An alternative possibility was suggested by Veeger and his colleagues (Bresters *et al.*, 1972; Haaker *et al.*, 1972) that pyruvate contributed both electrons and ATP for nitrogenase activity in *A. vinelandii* through a mechanism similar to that in *C. pasteurianum*. Pyruvate dehydrogenase, phosphotransacetylase and acetokinase are present in *A. vinelandii* (Bresters *et al.*, 1975). However, Campbell and Yates (1973) showed that acetokinase was absent from other *Azotobacter* strains and Benemann *et al.* (1971) obtained no nitrogenase activity with pyruvate in crude extracts of *A. vinelandii*.

Benemann and Valentine (1972) argued that the NAD(P)H/NAD(P) ratio must be high to poise the redox potential for nitrogen fixation. Haaker *et al.* (1974) invalidated this theory by showing an inverse relationship between the NAD(P)H/NAD(P) ratio and nitrogenase activity with increasing pO_2 in *A. vinelandii*. By using the uncoupler 4,-5,-6,7-tetrachloro-2-trifluoromethylbenzimidazole and the fluorescence probe 9-amino acridine they showed a direct correlation between nitrogenase activity and the energised state of the cytoplasmic membrane. Since the uncoupler inhibited nitrogenase activity but did not affect ATP levels, Haaker *et al.* argued that it affected electron transport. Such results imply that an intact chemiosmotic gradient across the cytoplasmic membrane is necessary for nitrogenase function. If this is so, then, bearing in mind that *Azotobacter* can produce ATP anaerobically (Yates and Daniel, 1970; Bresters *et al.*, 1972), if the pH is poised correctly, nitrogenase could be made to function anaerobically in whole cells. The need for an intact chemiosmotic gradient may account for the difficulty in obtaining cell-free extracts from *Azotobacter* that support nitrogenase with physiological substrates and the low nitrogenase activities of such extracts when obtained. Such a requirement may, in part, answer the criticism of Evans and Phillips (1975) who questioned the significance of experimental results with reconstituted electron transport sequences on the basis that activities were so low compared with the rates of nitrogen fixation *in vivo* and that all the components used were not obtainable from a single nitrogen-fixing organism. The need for a chemiosmotic gradient may also explain why such little activity as is obtainable in cell-free extracts is associated with particulate fractions (Yates and Daniels, 1970).

A second criticism of Benemann and Valentine's hypothesis is that high NAD(P)H/NADP ratios could inhibit several key enzymes in *Azotobacter*. Yates and Jones (1974) discussed these

Table III

Electron transport systems to nitrogenase in bacteria

Organism	Reductant	Enzyme	Electron carriers	Uncharacterised requirements	References
C. pasteurianum	*pyruvate*	phosphoroclastic	Ferredoxin	-	Mortenson (1964)
	H_2	hydrogenase	or		Knight and Hardy (1966)
	NADH	-fd.reductase	Flavodoxin		Jungermann *et al.* (1974)
	NADPH	-fd.reductase			
B. polymyxa	pyruvate	phosphoroclastic	Ferredoxin I	?	Benemann and Valentine (1972)
	formate	formate hydrogen lyase	Ferredoxin II		
	NADPH	-fd. reductase			Yoch (1973b)
K. pneumoniae	*pyruvate*	phosphoroclastic	Flavodoxin	?	Yoch (1974)
	formate	formate hydrogen lyase	Ferredoxin?		
	malate				
	NADPH	-fd. reductase?			
A. vinelandii	NADH	dehydrogenase	benzyl viologen		Klucas and Evans (1968)
	NADPH	-fd. reductase?	Ferredoxin I and Flavodoxin	Labile component	Benemann *et al.* (1971)
A. chroococcum	NADH	-dehydrogenase	endogenous	?	Yates and Daniel (1970)
	NADH	-dehydrogenase	benzyl viologen		Yates (1971)

	----	----	Flavodoxin hydroquinone	-	Yates (1972b)
M. flavum	NADH	dehydrogenase	benzyl viologen		Biggins and Postgate (1971)
	?	?	Ferredoxin I Ferredoxin II		Bothe and Yates (1976)
A. cylindrica	NADPH	-fd. reductase	Ferredoxin	?	Smith *et al.* (1971)
Gleoecapsa	NADPH	-fd. reductase	Ferredoxin	?	Gallon *et al.* (1973) Gallon *et al.* (1975)
R. japonicum bacteroids	NADH	dehydrogenase?	benzyl or methyl viologen		
	NADH	endogenous	FAD	labile component	Wong *et al.* (1971)
	NADPH	-fd. reductase?	FMN or FAD	?	Wong *et al.* (1971)

The most active substrates are in italics.

points in detail. It therefore seems both unlikely and undesirable that the NAD(P)H/NADP ratio should be high in nitrogen-fixing *Azotobacter;* perhaps association with active enzyme sites imposes constraints on the molecular structures and changes the redox potential of these carriers *in vivo* to allow electron donation to nitrogenase.

Sequence of electron transfer

Electron transfer to nitrogenase in the anaerobe *C. pasteurianum* has long been considered to operate in the sequence: pyruvate —— ferredoxin —— nitrogenase. H_2 could substitute for pyruvate if exogenous ATP was available. Recently, Walker, and Mortenson (1974) confirmed the direct link from hydrogenase —— ferredoxin —— oxidised nitrogenase Fe protein using purified proteins. Oxidised nitrogenase Fe protein was rapidly reduced in this system but when replaced by oxidised MoFe protein the reduction rate was considerably slower than the turnover time of nitrogenase. In this system a complete sequence of electron flow would be:

$$H_2 \text{ —— hydrogenase —— ferredoxin —— [Fe protein —— MoFe protein] —— } N_2$$

Other electron donors are listed in Table III. Other evidence that electron transfer occurs as indicated within nitrogenase has been reviewed elsewhere (see Zumft and Mortenson, 1975).

The scheme (NADPH —— ferredoxin —— x —— flavodoxin —— nitrogenase) proposed by Benemann *et al.* (1971) for electron transfer to nitrogenase in the aerobe *A. vinelandii* has not been seriously challenged. A similar scheme has been proposed for soybean bacteroids (see Evans and Phillips, 1975). Yates (1972b) obtained evidence in part support of this scheme by observing that purified nitrogenase from *A. chroococcum* functioned catalytically with substrate concentrations of *Azotobacter* flavodoxin hydroquinone. Reduced *Azotobacter* ferredoxin did not support nitrogenase activity except with catalytic amounts of methyl viologen as an intermediate electron carrier (Yates, unpublished). Unfortunately, there are several unsatisfactory aspects of this scheme. As mentioned earlier, nitrogenase activities with physiological substrates are very low. Secondly, no active NADPH-ferredoxin reductase has been observed in *Azotobacter* or bacteroids. Thirdly, when carrier free-illuminated chloroplasts are used as the reductant, either spinach ferredoxin (Bothe and Yates, 1976) or *Azotobacter* ferredoxin or flavodoxin (Yates, unpublished) will mediate electron transfer to purified nitrogenase. This implies that ferredoxin, with suitable reductant will transfer electrons directly to nitrogenase and that flavodoxin can be reduced to the hydroquinone (Bothe and Falkenberg, 1972) without the aid of ferredoxin. These criticisms must be answered if the scheme is to gain credibility. The fact that both ferredoxin and flavodoxin are constitutive suggests uses other than as electron transfer agents for nitrogenase. Kadam and Naik (1975) proposed ferredoxin as an electron carrier for nitrate reductase in *Azotobacter*.

How the ferredoxin and flavodoxin might associate with the energised membrane proposed by Haaker *et al.* (1974) is not clear. There is no doubt that both these proteins and nitrogenase associate with easily-sedimented material in *Azotobacter* although the nitrogenase is easily solubilised (Oppenheim and Marcus, 1970) whereas membranes have to be treated with organic solvents to solubilise ferredoxin and flavodoxin.

A second unresolved problem is the relative contribution of dark and light NADPH production to nitrogen fixation in blue-green algae. Smith *et al.* (1971) and Gallon *et al.* (1973, 1975) proposed that the major pathway of electrons to nitrogenase involved photosynthetically produced NADPH. Benemann and Weare (1973) showed that dark nitrogenase activity was substantial over a short period. Bothe and Loos (1972) offered a sensible solution when they observed substantial acetylene reduction in the presence of infra-red light or inhibitors of photosystem II. They proposed that ATP for nitrogenase activity was produced through photosynthesis but that reductant was produced by metabolism of carbon substrates: presumably either glucose-6-phosphate (Bothe, 1970), pyruvate (Bothe and Falkenberg, 1973) malate or isocitrate (Gallon *et al.*, 1975). These observations, together with the known ability of blue-green algae to fix nitrogen when grown heterotrophically in the dark (see Fay, 1976 and references therein) leave open the question of the intracellular location of nitrogenase in these organisms.

PROTECTION OF NITROGENASE AGAINST OXYGEN DAMAGE

There are several lines of evidence that nitrogenase is an anaerobic process from which it is mandatory to exclude oxygen: the susceptibility to oxygen damage of both nitrogenase proteins; the autooxidisability of ferredoxins and flavodoxins which act as electron carriers to nitrogenase; exclusion of oxygen is mandatory for optimum nitrogenase activity *in vitro;* growth and nitrogenase activity in nitrogen-fixing obligate aerobes is inhibited by excess oxygen. These and other lines of evidence which will emerge later, imply strongly that nitrogenase is susceptible to inhibition and damage by O_2 *in vivo*. The physiological processes whereby nitrogenase is protected from damage or inhibition by oxygen are of primary importance for the organisms' survival. This problem has been reviewed recently (Postgate, 1974; Yates and Jones, 1974) and will therefore be discussed only briefly here, detailing new evidence and unresolved hypotheses.

The mechanisms that have evolved in microbes to control access of oxygen and protect their nitrogenase vary in sophistication and effectiveness. The possibilities include: I, avoidance in anaerobes; II, respiratory protection in *Azotobacter*, III, conformational change in *Azotobacter;* IV, association with other biological macromolecules; V, microaerophily in *Mycobacterium flavum* and related species; VI, heterocyst formation in blue-green algae; VII, leghaemoglobin in legume nodules; VIII, slime production; IX, possession of superoxide dismutase, catalase and peroxidase.

Avoidance in anaerobes

There is no reported systematic study of the relative sensitivity to oxygen *in vivo* of nitrogenase in anaerobes although that of *Desulfovibrio desulfuricans*, strain Berre S, is more sensitive than that of *Clostridium pasteurianum* since, under similar conditions, the latter can be harvested without anaerobic precautions and retain activity while the former cannot (J. R. Postgate and T.H. Blackburn, unpublished observations). *Klebsiella pneumoniae* can grow and fix nitrogen with low partial pressures of O_2 in the atmosphere (Klucas, 1973; Hill, 1975, 1976a) but, in an N-free medium, only when the O_2 concentration in solution was zero (Hill, 1976a). Morris (1975) has expertly reviewed the physiology of obligate anaerobiosis and lists six possible mechanisms of oxygen toxicity to growth including two mechanisms which could actively damage proteins: oxidation of essential thiol (-SH) groups and damage by the products of oxygen utilisation, hydrogen peroxide, OH· radicals and $O_2^{\cdot -}$ radicals. Most anaerobes contain no catalase, peroxidase or superoxide dismutase, enzymes which detoxify these products. Significantly, perhaps, of eight aerotolerant anaerobes, seven contained superoxide dismutase. To my knowledge there is no report of a systematic study of the levels of superoxide dismutase in nitrogen-fixing bacteria. However, it is known that its activity is much greater at high than low oxygen tensions in N_2-fixing *A. chroococcum* (H. Lees, personal communication). Francis and Alexander (1972) found a correlation between ineffectiveness in rhizobia and low levels of catalase and peroxidase.

Respiratory protection in Azotobacter

Respiratory protection of nitrogenase simply means that the rate of O_2 uptake is sufficiently fast to maintain the O_2

concentration at the cell surface at zero and the environment of the nitrogenase anaerobic. There is considerable experimental support for this concept and it has been adequately discussed in recent publications (Postgate, 1971, 1974; Yates and Jones, 1974). It may be summarised as follows:

(i) Cultures of *A. chroococcum* are sensitive to excess oxygen when fixing nitrogen but not when grown on a fixed nitrogen source. This was shown to be an effect of excess oxygen rather than a lack of CO_2 (Dalton and Postgate, 1969a).

(ii) The growth and nitrogen-fixing efficiencies (yield per g of carbon substrate) of *A. chroococcum* in continuous culture are inversely proportional to the dilution rate and the pO_2. This implies that carbon source is 'wasted' during prolonged exposure to O_2, presumably to protect the nitrogenase from O_2 damage (Dalton and Postgate, 1969a,b).

(iii) Carbon- or phosphate-limited populations which cannot respire rapidly show specific damage to their nitrogenase when exposed to excess oxygen (Dalton and Postgate, 1969a; Lees and Postgate, 1973).

(iv) N_2-fixing *A. chroococcum* adapt to high pO_2 levels by markedly increasing their Q_{O_2} values and cytochrome a_2 content (Drozd and Postgate, 1970b). When *A. vinelandii*, grown at low aeration, was suddenly exposed to increased pO_2 their Q_{O_2} values, NADH- and NADPH dehydrogenase activities increased to the accompaniment of *de novo* synthesis of cytochrome a_2 and a sharp decrease in the phosphorylating efficiency of the respiratory chain, particularly at site I (Jones *et al.*, 1973).

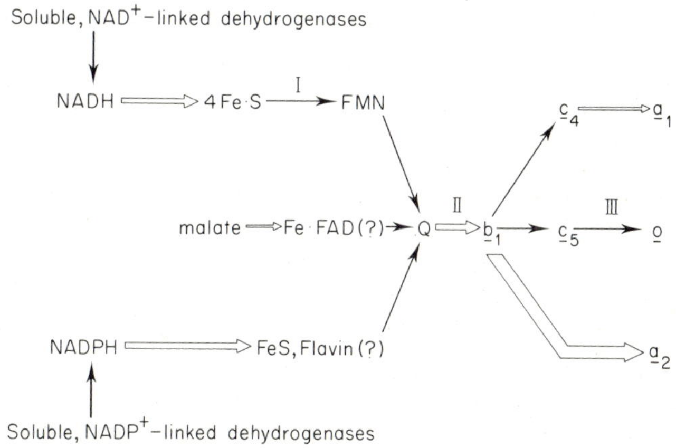

Fig. 2. *Branched respiratory chain of* A. vinelandii *from Jones and Redfearn (1967).*

(v) The branched respiratory chain of *A. vinelandii* (Fig. 2)(Jones and Redfearn, 1967) allows for considerable variation in electron transfer and phosphorylating efficiency. The increase in NAD(P)H dehydrogenases and cytochrome a_2 is evidence to suggest that this pathway operates at high pO_2. The high degree of uncoupling would ensure the highest Q_{O_2}.

(vi) The concept of respiratory protection for nitrogenase gives a rational basis for the very high Q_{O_2} values for which the Azotobacteraceae are famous (Williams and Wilson 1954).

The need for a branched respiratory chain to implement respiratory protection has been questioned by Haaker and Veeger (1976). Washed cells of *Azotobacter*, grown under O_2-limited conditions reduced considerably more acetylene with pyruvate than sucrose at high oxygen levels. They deduced that respiratory protection relied on the source of carbon substrate rather than a branched chain.

Conformational protection in Azotobacter

The ability of nitrogen-fixing aerobes to withstand oxygen stress depends on their nutritional status and the population density (Drozd and Postgate 1970b); in early stages of growth organisms are more susceptible to oxygen stress and under these conditions the nitrogenase has to be protected from O_2 damage to avoid excessive resynthesis of the enzyme. Dalton and Postgate (1969a) suggested that the enzyme was capable of adopting a conformation in which it was oxygen-tolerant and inactive. The subject has been discussed in detail (Postgate 1971, 1974; Yates and Jones 1974) and, once again, the evidence will only be summarised here:

(1) Nitrogenase preparations in crude extracts from *Azotobacter* are oxygen-tolerant whereas those in crude extracts from anaerobes are oxygen sensitive (Kelly 1969). All purified nitrogenases, including those from aerobes, are oxygen sensitive (Bulen and LeCompte 1966; Kelly 1969). Therefore, in *Azotobacter*, the nitrogenase is in a conformation which is possibly brought about by association with unknown components such as proteins, lipids, phospholipids, in which the O_2 sensitive sites are somehow protected from O_2 damage. Crude NADH-dehydrogenase protects Azotobacter nitrogenase from oxygen damage; purified NADH dehydrogenase does not (Yates 1970). In crude extracts from *Azotobacter* oxygen-tolerant nitrogenase is particulate; Oppenheim *et al.* (1970) and Oppenheim and Marcus (1970) obtained oxygen-sensitive, relatively soluble, nitrogenase, free from NADH-dehydrogenase, from *A. vinelandii* by osmotic shock. The implication of these results is that the oxygen-tolerant form of nitrogenase is brought about by association with other components and the enzyme becomes oxygen-sensitive when these unknown components are removed. Postgate (1971) also suggested

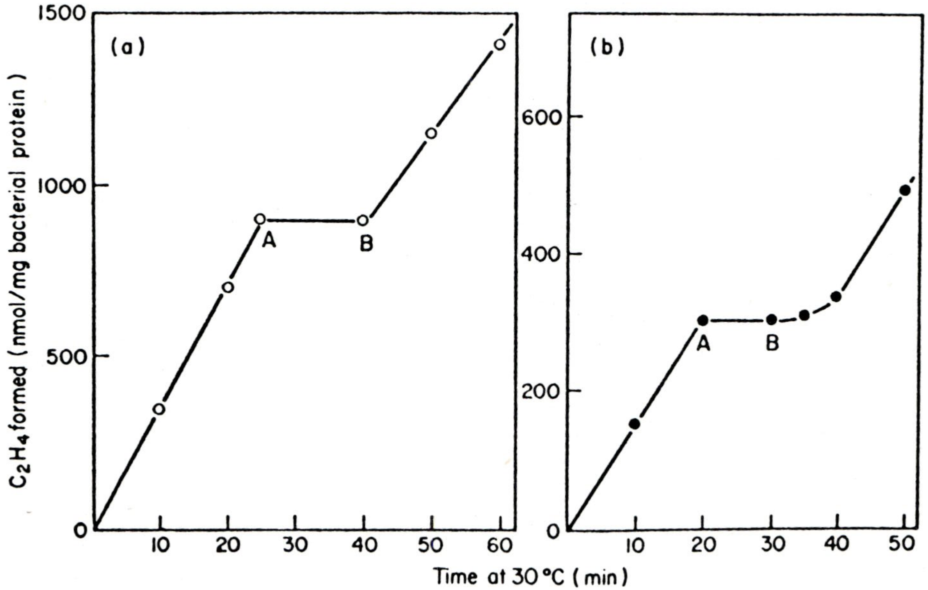

Fig. 3a and b *"Switch off" and "switch on" of nitrogenase activity by* Azotobacter chroococcum *in response to aeration. 2 ml samples of growing cultures shaken gently in argon + 0.1 atm* O_2 *containing* C_2H_2*. Shaking rate increased at A, returned to original value at B.* (a) *Shaking rates: 75 x 1.5 cm/min to 150 x 4 cm/ min and back.* (b) *Shaking rates: 75 x 1.5 cm/min to 250 x 1 cm/min and back. From Hill* et al. *(1972).*

that the enzyme is in the oxygen-sensitive state when active, a logical inference since anaerobic conditions are necessary for optimum nitrogenase activity *in vitro* (Yates and Daniel 1970).

(2) Nitrogenase activity can "switch on" immediately when reverting from a high oxygen solution rate where the enzyme is inactive to a low oxygen solution rate. Fig. 3 shows four types of "switch off-switch on" situations. In Fig. 3a instantaneous "switch on" was observed in the presence of chloramphenicol indicating that no synthesis *de novo* of nitrogenase was necessary (Drozd and Postgate 1970b). Presumably conformational protection

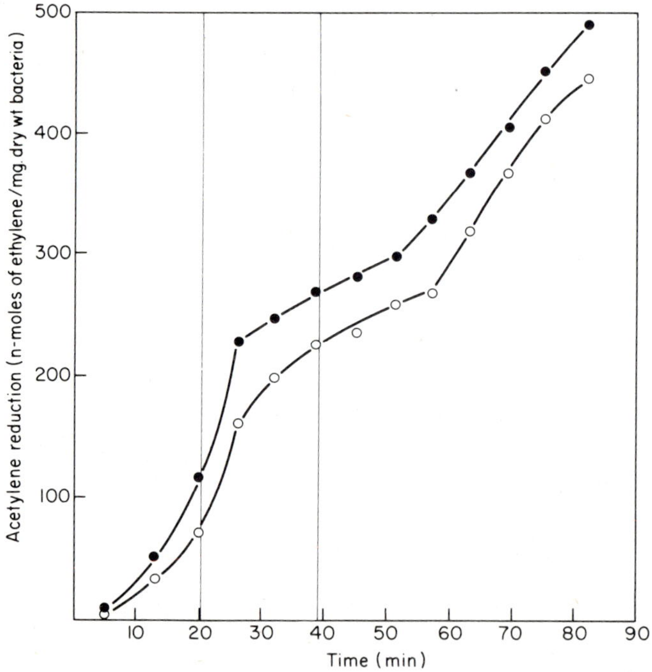

Fig. 3c *Inhibition of acetylene reduction in* Derxia gummosa *by an increase in aeration. Samples from O_2-limited N_2-fixing continuous culture grown in 0.1 atm O_2 at a dilution rate of $0.008h^{-1}$ were exposed to 0.05 atm (●) or 0.1 atm (O) with 0.03 atm C_2H_2. After 20.5 min incubation at 30° the shaking rate, initially at 100 strokes/min (amplitude 3.8 cm), was increased to 176 strokes/min. After a further 18.5 min incubation the shaking rate was returned to 100 strokes/min. From Hill (1971).*

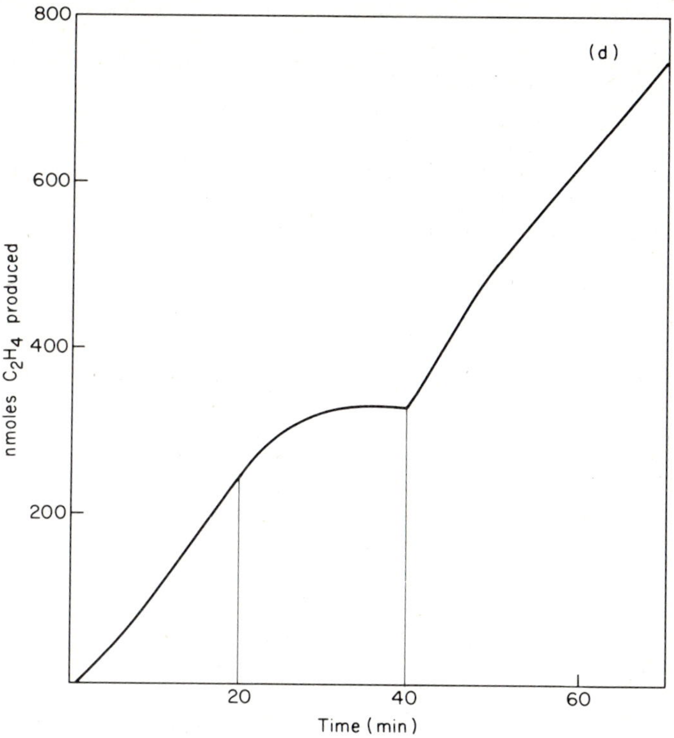

Fig. 3d *"Switch off" and "switch on" of nitrogenase activity by* Mycobacterium flavum *in response to aeration. Conditions as in Fig. 3a but using 10 ml of culture and 0.05 atm O_2 (L.A. Smith and M.G. Yates unpublished).*

occurs in *M. flavum* also (Fig. 3d) because "switch on" is apparently immediate when the oxygen solution rate reverts to the original low value. There are three possible explanations for the sluggish "switch on": I. that the nitrogenase has suffered oxygen damage and *de novo* synthesis does occur in these cases; II. that the respiratory chain is not sufficiently active to prevent the lag and III. that there is insufficient of whatever allosteric effector causes the enzyme to "switch on". Postgate (1971) suggested that this delay in achieving the original rate of nitrogenase activity argued against the likelihood that "switch off" of nitrogenase was due to a competition for electrons. Such an explanation was suggested to explain similar "switching off" phenomena with nitrate or tetrathionate reduction in coliform bacteria (Pichinoty and Bigliardi-Rouvier 1963).

The possibility of an allosteric effector which controls "switch off-switch on" has engaged interest. The iron protein is the most oxygen-sensitive in nitrogenase and it is known to bind both MgATP and MgADP (Bui and Mortenson 1968; Biggins and Kelly 1969; Tso and Burris 1973: Yates *et al.* 1975) which induce a comformational change (see reviews by Eady and Postgate 1974 and Zumft and Mortenson 1975). The iron proteins of *M. flavum*, *A. chroococcum* and *K. pneumoniae* are more oxygen-sensitive when associated with ATP (Biggins and Postgate 1971; Yates 1972a; Eady *et al.* 1975); ADP antagonised this effect. ATP also makes the sulfhydryl groups more reactive towards DTNB (Thorneley and Eady 1973) and the Fe atoms more reactive towards α.α-dipyridyl (Walker, G.A. and Mortenson 1974). This is convincing evidence for the association of MgATP and MgADP with the Fe protein of nitrogenase. Yates *et al.* (1975) showed that the oxidised and active Fe protein of nitrogenase reacted very rapidly with reductant [$SO_2^{\cdot -}$; (Thorneley *et al.* 1976)] (Fig. 4), and that MgADP but not MgATP inhibited this reduction. If the initial cause of "switch off" is due to competition for electrons between oxygen and nitrogenase or autoxidation of flavodoxin

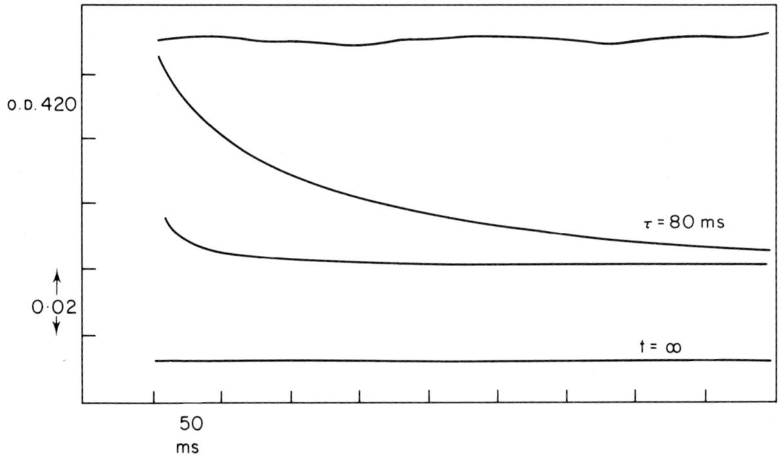

Fig. 4 *Reduction by sodium dithionite of enzymatically active or inactive oxidised Fe protein from* A. chroococcum *nitrogenase. The reaction was followed by stopped flow spectrophotometry. Top trace, reaction of active Ac2ox with degassed buffer. Second trace, reaction of oxygen-inactivated Ac2 with $Na_2S_2O_4$. Third trace, reaction of active Ac2ox with $Na_2S_2O_4$. Bottom trace, optical density at infinite time. From Yates* et al. *(1975).*

hydroquinone, the nitrogenase is likely to be oxidised in the "switch off" state and the possibility that oxidised protein or, in particular, oxidised protein in association with MgADP was oxygen-tolerant could not be ignored. However, neither oxidised Fe protein nor the ADP complexed protein was more oxygen-tolerant than the reduced Fe protein (Campbell and Yates unpublished). The reason for oxygen-tolerance of nitrogenase during "switch off" remains unclear. Finally it must be mentioned that Kurz *et al.* (1975) in studies with synchronised *A. vinelandii* observed "switch on-switch off" phenomena following the addition of various carbon compounds and suggested that it was merely a manifestation of nitrogenase sensitivity to changes in the environment. However they showed a drop in O_2 consumption coinciding with nitrogenase "switch off" upon adding acetate. A more likely explanation is that adding acetate temporarily inhibited the uptake of carbon substrate which lowered the O_2 uptake rate thereby preventing respiratory protection and inducing "switch off".

Location of nitrogenase in Azotobacter

As mentioned earlier, Azotobacter nitrogenase is particulate in crude extracts (Bulen *et al.* 1964) and this may play an important role in the protection of nitrogenase against oxygen damage. Yates and Daniel (1970) and Biggins and Postgate (1971) showed acetylene-reducing activity with physiological substrates was associated with easily sedimented particulate fractions in *A. chroococcum* and *M. flavum* respectively and although optimum acetylene reduction was obtained under anaerobic conditions some activity was observed in the presence of oxygen (Yates and Daniel 1970). This may mean that the nitrogenase was protected when "switched on" although it is more likely that respiratory protection was operating. Yakovlev and coworkers (Yakovlev and Levchenko 1964; Yakovlev *et al.* 1965; Levchenko *et al.* 1969) using the reduction of iodonitrotetrazolium to formazan in the presence of electron donors decided that nitrogenase was located in cell organelles which they called "mitochondria". Perhaps these mitochondria are the same cellular organelles that Reed *et al.* (1974) called azotophores. These vesicular membranes some 40-75 nm in diameter contained the bulk of cellular nitrogenase and were easily separated from respiratory membranes by sucrose gradient centrifugation. Oppenheim *et al.* (1970) and Oppenheim and Marcus (1970) showed that N_2-grown *A. vinelandii* contained an extensive network of internal membranes whereas NH_4^+-grown cells did not, an observation which was confirmed by Hill *et al.* (1972) for *A. chroococcum* and by Marcus and Kaneshiro (1972). However, Drozd *et al.* (1973) could find no difference in phospholipid content to match the appearance of membrane networks and Pate *et al.* (1974) claimed that membrane formation in both N_2-grown and NH_4^+-grown *A. vinelandii* was related to the age of the culture and oxygen content of the growth medium. Marcus and Kaneshiro (1972) found differences in phospholipids, anionic and neutral lipids and coenzyme Q content between N_2-grown and NH_4^+-grown *A. vinelandii* which suggested considerable differences in membrane structure. Despite the complicating effects of age and O_2 content of the medium on membrane structures it seems likely that nitrogenase is associated with membrane fractions in both *Azotobacter* and *M. flavum* which protect it from oxygen damage both by physical association and respiratory activity.

Microaerophily in M. flavum *and other species*

M. flavum 301 was first isolated and described by Federov and Kalininskaya (1961) and was thought only to fix nitrogen in symbiotrophic association with other organisms. Biggins and Postgate (1969) showed that it fixed in pure culture but only at low pO_2 (0.05 atm). Presumably its Q_{O_2} value was insufficient to afford respiratory protection at higher pO_2 levels. Nevertheless it has a degree of conformational protection judged by the rapid "switch on" shown in Figure 3d. *Corynebacterium autotrophicum* (Gogotov and Schlegel, 1974; Wiegel and Schlegel 1976; Berndt *et al.* 1976; DeBont and Leitjen 1976) is also a microaerophilic nitrogen fixer and so are nitrogen-fixing methane oxidisers (DeBont and Mulder 1972), nitrogen fixing spirilla and cowpea strains of rhizobia (see Dobereiner, Bergersen, this volume).

Slime production in aerobes

Jensen *et al.* (1960) first described *Derxia gummosa* and noticed that it had a tendency to form either massive or small colonies on nitrogen-free agar. This they attributed to competition for trace amounts of fixed nitrogen in the agar and they suggested that nitrogen fixation could not be initiated without fixed nitrogen in small amounts to stimulate growth. However, Hill (1971) showed clearly that massive colony type was related to the oxygen regime and suggested that massive colonies (Fig. 5) formed slime which protected nitrogenase against oxygen-damage by impeding diffusion of oxygen to the cell. The sluggish "switch off" exhibited by *D. gummosa* in Fig. 3c could also be attributed to slime retarding the expected increase in the oxygen solution rate upon increasing the shaking rate.

Fig. 3d demonstrates a similar sluggish "switch off" with

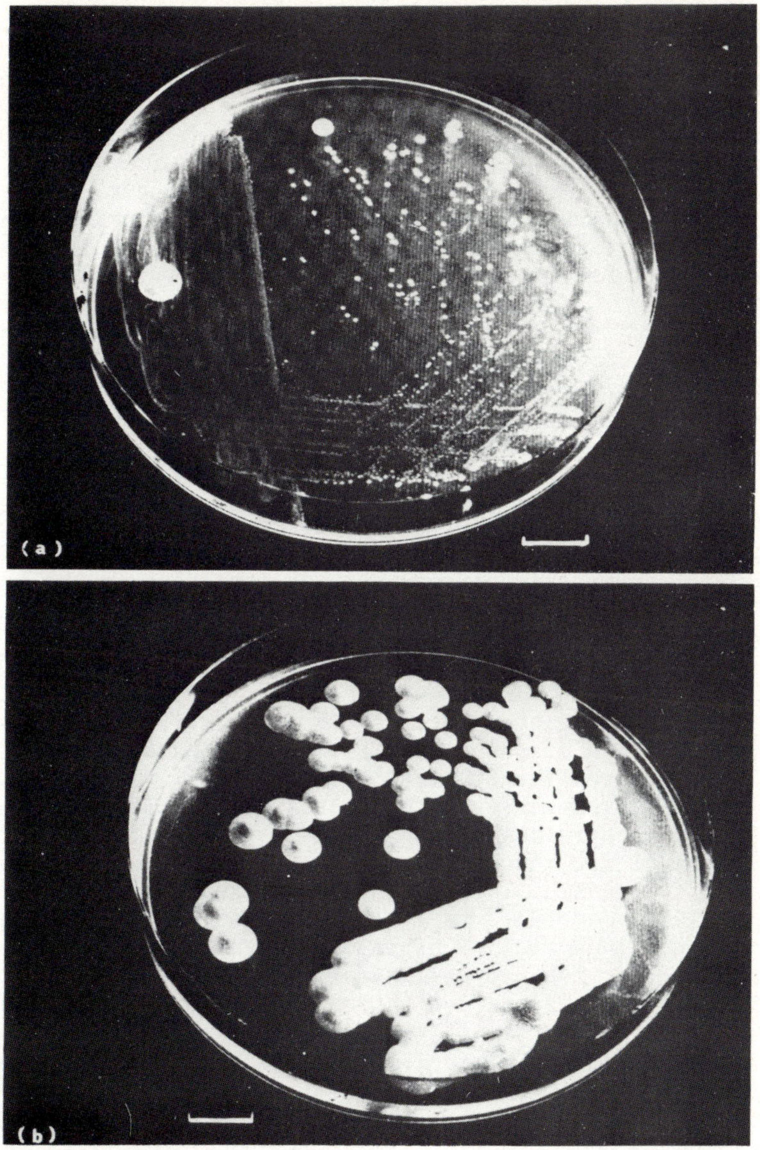

Fig. 5 *Colonial morphology of* Derxia gummosa *grown on N-free glucose agar for 13 days at 30°C.* (a) *In N_2+0.2 atm O_2, showing numerous small and one large N_2-fixing colony.* (b) *In N_2+ 0.05 atm O_2, showing uniformly large N_2-fixing colonies. Bar= 1 cm. From Hill* et al. *(1972).*

a gum-producing strain of *M. flavum*. When the gum was removed by centrifugation the "switch off" time was shorter but not instantaneous. When such cultures were tested for nitrogenase activity against increasing pO_2 levels, the optimum pO_2 for acetylene-reducing activity in the presence of extra-cellular polysaccharide was 0.025 atm; when the polysaccharide was removed the optimum pO_2 was 0.02 atm and in a gum-free strain with the same cell density the optimum pO_2 was 0.015 atm. One possible explanation for these observations is that extracellular polysaccharide retards the oxygen solution rate and that capsular polysaccharide impedes oxygen uptake into the cell. This phenomenon is not restricted to N_2-fixing organisms since NH_4^+-grown *M. flavum* responds to increased aeration by increasing polysaccharide formation. Wilkinson (1958) reviewed evidence of polysaccharide formation by micro-organisms in response to increased aeration.

Heterocysts in blue-green algae

Heterocysts form in heterocystous blue-green algae when the organisms have to fix N_2 as their nitrogen source. It thus was logical to assume that nitrogenase was localised in the heterocysts. The subject has been well reviewed by Stewart (1974) and will only be summarised here. Both Fay *et al.* (1968) and Stewart *et al.* (1968) suggested that heterocysts might be the site of nitrogen fixation and Stewart *et al.* (1969) confirmed this by showing that isolated heterocysts fixed nitrogen when supplied with ATP and a reductant. Fleming and Hazelkorn (1973, 1974) demonstrated the physical presence of nitrogenase in heterocysts. Heterocysts are ideally suited for fixing nitrogen: they photophosphorylate through photosystem I but they have no photosystem II and thus do not produce O_2 which could damage or inhibit nitrogenase. They also have thickened cell walls which Stewart (1974) describes as a massive synthesis of membranous material upon induction of nitrogenase synthesis and these membranes could provide protection for nitrogenase against oxygen. Heterocysts also respire and this serves a dual function of respiratory protection for nitrogenase and ATP synthesis for nitrogenase activity. Their one disadvantage is that they do not produce their own carbon substrate and are dependent upon the surrounding vegetative cells for electron donors. This could add meaning to the observations of Bothe and Loos (1972) and their suggestion that electron donors for nitrogenase are not derived from photosystem I. This influx of carbon substrate from neighbouring cells promotes the greatest reducing capacity near the poles of heterocysts (where they link with vegetative cells; Stewart *et al.* 1969) and here also the membranous system is most complex (Lang and Fay 1971). It is tempting to suggest that nitrogenase is also located here but the intracellular location of nitrogenase has not been determined.

There are views that heterocysts are the only sites of nitrogen fixation in heterocystous blue-green algae. These depend on the observations that heterocyst number and nitrogenase activity are closely related under aerobic and microaerophilic conditions and that nitrogenase activity dropped off rapidly when heterocysts were detached from vegetative cells (Stewart *et al.* 1969; Wolk and Wojcinck 1971a; Fay and Kulasooriya 1972; and Weare and Benemann 1973). Tel-or and Stewart (1976) isolated heterocysts from *Anabaena cylindrica* and observed that they produced ATP through photophosphorylation and oxidative phosphorylation and contained ferredoxin, ferredoxin-NADP reductase and 90% of the cell-free nitrogenase activity of aerobically-grown cultures.

However, evidence does exist that nitrogenase occurs in vegetative cells. Wolk and Wojcinck (1971b) calculated that only 30-40% of nitrogenase resided in heterocysts and Ohmori and Hattori (1971) reported that non-heterocystous Anabaena filaments reduced acetylene in air. Evidence that vegetative cells contain nitrogenase has come directly from Smith and Evans (1970) who isolated nitrogenase from vegetative cells and indirectly from Van Gorkum and Donze (1971) who found that phycocyanin, which acts as a rapidly mobilised nitrogen reserve, was produced only in vegetative cells near the heterocysts under aerobic conditions but in all vegetative cells uniformly under microaerophilic conditions. This and other evidence suggests strongly that nitrogenase is operative in heterocysts under aerobic conditions and in both heterocysts and vegetative cells under microaerophilic conditions. Whether nitrogenase is "switched off" or damaged or not synthesised in vegetative cells under aerobic conditions has not been reported.

Non-heterocystous blue-green algae do fix nitrogen but only under microaerophilic conditions and the nitrogenase of the unicellular blue-green alga *Gloeocapsa* is particulate (Rippka *et al.* 1971) and presumably protected against oxygen damage by association as in *Azotobacter*, although any protection offered in this manner must be slight since nitrogenase in crude extracts of *Gloeocapsa* is O_2-sensitive (Gallon *et al.* 1975). Perhaps any protection afforded by particles operates in the presence of a low oxygen regime since nitrogenase activity in *Gloeocapsa* occurs concomitantly with low photosystem II activity and a fall in nitrogenase activity in later growth stages corresponds with a marked increase in O_2 production (Gallon *et al.* 1975).

Carpenter and Price (1976) in an article entitled

'explanation for aerobic nitrogen fixation without heterocysts' state that the reason the marine organisms Trichodesmium can fix nitrogen under air is that its trichomes can aggregate in parallel clusters and the central cells lose photosystem II activity and gain the capacity to fix nitrogen. However, evidence exists that such clusters also harbour free-living nitrogen-fixing bacteria (R. Knowles, personal communication)

The role of leghaemoglobin in nodules

Kubo (1939) was the first to show that legume root nodules contained a soluble haemoglobin-type pigment which possessed the property of reversible oxygenation. This essential property was confirmed by Keilin and Wang (1945) and by Virtanen and his colleagues who also established the important relationship between leghaemoglobin content and nodule effectiveness (Virtanen *et al.* 1947). Nash and Schulman (1976) confirmed earlier evidence that leghaemoglobin is reduced throughout the life cycle of the nodule.

Conclusive evidence for the role of oxygenated leghaemoglobin ($Lb^{2+}O_2$) as an oxygen carrier to bacteroids to facilitate oxygen uptake by terminal oxidases and thereby to increase ATP production for nitrogenase function has been obtained by the experiments of Appleby, Bergersen and their colleagues (Bergersen *et al.* 1973; Bergersen and Goodchild 1973; Wittenberg *et al.* 1974; Appleby *et al.* 1975, 1976; Bergersen and Turner 1975a and b); Wittenberg 1977. Bergersen *et al.* (1973) added LbO_2 to respiring bacteroid suspensions and observed a doubling of O_2 uptake and a twenty-fold increase in nitrogenase activity, either by measuring N_2^{15} incorporation, acetylene reduction or H_2 evolution. This activity, including H_2 evolution, was abolished by CO which formed a stable LbCo complex and blocked the $LbO_2 \rightleftarrows Lb$ cycle. Wittenberg *et al.* (1974) investigated the effect on nitrogenase activity in bacteroid suspensions of a series of O_2-carrying proteins and concluded that the stimulation of nitrogenase activity was related to the kinetics of O_2-binding to the carriers. They suggested that the O_2 carriers facilitated oxygen diffusion across an unstirred layer which surrounded the bacteroids in suspension. This suggestion has been modified recently by Bergersen and Turner (1975a) who studied the effect of LbO_2 on O_2 uptake and nitrogenase activity by bacteroid suspensions in the absence of a gas phase, to eliminate the effect of a gas-liquid interface which complicated the interpretation of results when shaken assays under a gas phase were used. An experimental environment without a gas space was also more relevant to the *in vivo* system. They found that LbO_2 allowed maximum rates of O_2 consumption and nitrogenase activity at levels of O_2 as low as 1/25th of those which were limiting in the absence of the carrier. These results, according to the authors, showed that the role of LbO_2 was to maintain stabilised low O_2 concentrations in a range which was optimal for nitrogenase function. Later work (Bergersen & Turner 1975b) showed that this range of free O_2 concentrations was optimal for ATP production in the bacteroids and that the apparent nitrogenase optimum followed from this. The high level of ATP in the bacteroids at low free O_2 concentrations was due to the activities of a terminal oxidase system of high affinity for O_2, which produced ATP efficiently. This system was sensitive to inhibition by N-phenyl-imidazole and carbon monoxide. When oxyleghaemoglobin was present, the consumption of O_2 by this system was half maximal near 0.03 µM free O_2. In the absence of leghaemoglobin, bacteroid respiration was restricted by the diffusion resistance imposed below about 0.3 µM O_2. Leghaemoglobin apparently removed this restriction by macromolecular carrier-facilitated diffusion. At concentrations of free O_2 above 1-3 µM, in the presence or absence of oxyleghaemoglobin or oxymyoglobin, bacteroid respiration was mediated by a terminal oxidase system of lower affinity for O_2, which produced ATP less efficiently and which was insensitive to inhibition by N-phenylimidazole and carbon monoxide. These results are reminiscent of the split respiratory pathway in Azotobacter (Jones and Redfearn 1967). Appleby *et al.* (1976) suggested that the low affinity oxidase might act in a "protective manner" towards nitrogenase. Bergersen and Turner (1975b) suggest that the low nitrogenase activity in bacteroids at 1-10 µM free O_2 may be due to low ATP production rather than inhibition of nitrogenase *per se*. However, there is no clear evidence that nitrogenase is protected from oxygen damage in these systems. It is probable that such protective mechanisms as occur in *Azotobacter* are normally unnecessary in bacteroids since the consensus of opinion (Bergersen *et al.* 1973) is that bacteroids are O_2-limited *in vivo* and that leghaemoglobin serves to facilitate O_2 availability at the bacteroid surface at low O_2 tensions. Appleby *et al.* (1975; 1976) give evidence that cytochrome P_{450} acts as an intracellular O_2 carrier from leghaemoglobin to the high oxygen-affinity terminal oxidase system.

Nitrogenase activity in free-living Rhizobia

Rhizobia normally fix nitrogen in symbiotic association with legume root nodules and the lack of evidence for nitrogen fixation in free-living rhizobia, led Dilworth and Parker (1969) to suggest that genetic information from the plant was necessary for rhizobia to fix N_2. However, recently, several groups have obtained fixation by free-living rhizobia in the absence of plant cells, mainly with the slow growing types such as cowpea, but lately with quicker growing types (Pagan *et al.* 1975; Kurz and LaRue 1975a; McComb *et al.* 1975; Meister 1975; Tjepkema and Evans, 1976). Originally an important factor seemed to be a requirement for two carbon sources, a pentose and a dicarboxylic acid (e.g. arabinose + succinate). A small amount of fixed nitrogen as a starter is also helpful. These requirements together with the observation that nitrogen-fixing rhizobia produce slime (Kurz and LaRue 1975a) suggest that the real requirement is a very low oxygen regime and that the small amount of fixed nitrogen helps the organism to get a start and to respire sufficiently rapidly to protect nitrogenase, the gum retards oxygen diffusion and the dual carbon source aids increased respiration. Consistent with this probability, Keister, 1975, and Tjepkema and Evans (1975) obtained N_2 fixation by free-living cowpea strain 32H1 under microaerobic conditions using only one carbon substrate. Bergersen *et al.* (1976) showed that removal of succinate from a continuous culture of cowpea strain 32H1 caused a rise in dissolved O_2 following a diminished O_2 uptake and nitrogenase activity was lost: it was restored following restoration of the succinate supply and a return to low dissolved O_2. It seems, therefore, that free-living rhizobia are microaerophilic, slime-producing nitrogen fixers similar to *Derxia gummosa* (Hill 1971) and *M. flavum*; the apparent need for two carbon sources is unreal, the 4-carbon source (e.g. succinate) being an easily mobilisable substrate for respiratory protection. Finally, it may be pertinent to note that Phillips *et al.* (1973) showed that anaerobically-grown rhizobia produce proteins which can transfer electrons to nitrogenase whereas aerobically-grown cells did not. This implies that O_2 represses the synthesis of these proteins.

The ATP requirement for nitrogenase in vivo

The need for ATP for functioning nitrogenase *in vitro* is well established: evidence of its ability to induce a conformational change in the Fe protein, the ATP-dependent electron transfer from the Fe protein to the MoFe protein of nitrogenase (see the review by Zumft and Mortenson 1975, also Thorneley, 1975) and a possible role in reduction of substrate (Eady *et al.* 1975) and complex formation between nitrogenase proteins (Burns and Hardy 1975) is well documented and will not be discussed further in this review.

The requirement for ATP in nitrogenase function *in vivo* is based on the premise that during growth the yeild of microorganisms per mol of ATP is lowered if much ATP is required to synthesise a monomer such as NH_4^+ from N_2 (Stouthamer 1973) and the decrease in molar growth yield on replacing NH_4^+ with N_2

has been used to calculate the ATP utilised per N_2 reduced in nitrogen-fixing bacteria. If the metabolic pathway of substrate fermentation is known the yield of ATP per mol of substrate metabolised can be predicted and the yield of organism per mol of ATP determined for a particular substrate limitation. This figure (Y_{ATP}) is usually 10.5 g dry wt of bacteria per mol of ATP (Stouthamer 1969). On this basis Daesch and Mortenson (1967) determined that 20 mols of ATP were utilised per mol of N_2 reduced in *Clostridium pasteurianum* which is in reasonable agreement with the figure of ATP/N_2 ratio of 15 observed *in vitro* if one considers that NH_4^+ is assimilated *via* glutamine synthetase, which consumes one ATP/NH_3, in nitrogen-fixing organisms (see Dixon 1975). Dalton and Postgate (1969b) studied continuous cultures of *A. chroococcum* with N_2 or NH_4^+ as the nitrogen source and observed that the Y_{sub} values in the N_2 and NH_4^+-grown cultures drew closer together at faster growth and the extrapolated ATP/N_2 ratio at a hypothetical zero maintenance energy was 5. Postgate (1971) doubted the validity of this calculation but emphasised that Y_{sub} value differences between NH_4^+ and N_2-grown cultures of *Azotobacter* were much smaller than those in anaerobes and suggested that *A. chroococcum* possesses a physiological mechanism for recovering some of the ATP expended in nitrogen fixation. Evidence for such a mechanism will be given in the next section.

Hill (1976b) was critical of the above evaluations and that of Shanmugan and Valentine (1975) on the basis that Y_{ATP} values could vary according to nutritional status of bacterial populations. She calculated Y_{ATP} values of 10.5±1.5 and 4.2±0.2 respectively for glucose-limited NH_4^+- and N_2-grown chemostate cultures of *K. pneumoniae* and calculated an ATP/N_2 of 29. Changing the temperature, pH or phosphate concentration of the medium had little effect on the ATP/N_2 ratio but on the basis of C_2H_2 versus N_2 reduction Hill calculated that at least 45% of nitrogenase activity could be utilised to evolve H_2 and that the true ATP/N_2 molar ratio was as low as 16.

The role of hydrogenase

One of the earliest observations made during studies on nitrogen fixation was that all nitrogen-fixing organisms contained hydrogenase though the converse was not always true. This relationship has puzzled scientists for many years. Broadly speaking, there are three main types of hydrogenase: a reversible hydrogenase common in obligate anaerobes and facultative anaerobes which usually catalyses the reduction of protons as a terminal electron acceptor in energy-producing metabolic pathways during growth and thus maintains a steady state of oxidised ferredoxin (Mortenson and Chen 1975). Secondly, there is the ATP-dependent, unidirectional, H_2-evolving capacity of nitrogenase which accounts, in part, for the correlation between hydrogenase and nitrogenase. To my knowledge no useful metabolic role has been put forward for this activity of nitrogenase and perhaps one has to accept that it is an inescapable part of the nitrogenase mechanism which leads to energy wastage. The third type of hydrogenase is the unidirectional, hydrogen-oxidising enzyme found in *Azotobacter*, some bacterial systems (Dixon 1975) and, more recently, *M. flavum*. For the sake of brevity the ATP-independent hydrogenases which are separate enzymes from the nitrogenase will be called conventional hydrogenase and the hydrogenase activity associated with nitrogenase, ATP-dependent H_2 evolution. Three possible roles have been put forward for the conventional hydrogenase from *Azotobacter* which action is increased in N_2-grown over NH_4-grown cells (Lee and Wilson 1943). Dixon (1972) suggested that: I. hydrogenase was used as an oxygen scavenger to aid respiratory protection of nitrogenase; II. as a means of removing inhibitory H_2 from the nitrogenase site and, III, as a means of recycling reducing power wasted by nitrogenase through ATP-dependent H_2 evolution. There was little supporting evidence for any of these possibilities although H_2-supported phosphorylation has been observed in Azotobacter and bacteroids (Hyndman *et al.*, 1953; Dixon, 1972). Failure to observe H_2 evolution by Azotobacter cultures under Ar and CO, conditions favourable for ATP-dependent H_2 evolution *in vitro*, led Postgate (1971) to speculate that the H_2-evolving site of nitrogenase might be aprotic *in vivo*. However, Brotonegoro (1974) demonstrated H_2 evolution by N_2-fixing *A. chroococcum* cultures under CO and acetylene. Smith, L.A. *et al.* (1976) confirmed this observation and, in addition found that acetylene inhibited conventional hydrogenase (Table IV). The explanation for Brotonegoro's observation can thus be stated simply: Co inhibits the substrate-reducing site of nitrogenase and promotes ATP-dependent H_2 formation. This H_2 would normally be recycled by conventional hydrogenase but acetylene prevents this and hence H_2 is evolved. These experiments give clear indication that Azotobacter nitrogenase evolved H_2 *in vivo* and that this H_2 is normally recycled by conventional hydrogenase to provide reducing power for ATP production (Hyndman *et al.* 1953) and possibly N_2 reduction. Thus the third proposal by Dixon (1972) for hydrogenase function appears to be correct and this may explain the apparent low ATP/N_2 ratio in *Azotobacter* (Dalton and Postgate 1969b) compared with nitrogen-fixing anaerobes (Daesch and Mortenson 1967; Shanmugam and Valentine 1975; Hill 1976b) where any ATP-dependent H_2 evolution is not recycled because the conventional

Table IVa

Inhibition by acetylene and carbon monoxide of hydrogenase activity

	percent inhibition				
Type of hydrogenase:	ATP-independent reversible		ATP-independent unidirectional (H_2 uptake)		ATP-dependent
source:	*K. pneumoniae*		*A. chroococcum*		Crude preparation of nitrogenase from *A. chroococcum**
	Whole Cells	Crude Supernatant	Whole Cells	Crude Supernatant	
Inhibitor C_2H_2 % in gas phase					
3.7	67				
6.5	82				
5.0		45	51(7)	51(36)	0(0)
10.0		85	75	83	0
20.0			96	95	0

From Smith *et al.* (1976).
*This assay was run in the presence of 20% CO in the atmosphere to prevent acetylene reduction by nitrogenase. The figures in parenthesis represent inhibition by CO.

Table IVb

H_2 evolution and acetylene reduction by N_2-fixing *A. chroococcum*

Additions to the atmosphere	n moles C_2H_2 reduced and H_2 evolved in 1 hour	
	C_2H_2	H_2
none	0	0
10% C_2H_2	780	20
5% CO	0	440
10% C_2H_2 + 5% CO	5	760

From Smith *et al.* (1976)

Table V

H_2 evolution from aerobic N_2-fixing systems

Crop	% of nitrogenase activity wasted as evolved H_2.*	
Alfalfa	46 ± 4	
White clover	50 ± 2	
Subterranean clover	44 to 60	
Soybean	43 ± 3	
Pea	54 ± 4	
Cowpea	1 ± 0.3	
Alder	1 ± 0.1	
Azotobacter chroococcum	0)	unless CO and C_2H_2
	)	are present in
Mycobacterium flavum	0)	the atmosphere
*Anabaena cylindrica***	15	

* Adapted in part from Schubert and Evans (1976)
** Benemann and Weare (1974)

hydrogenase evolves H_2 during growth. H_2 evolution by nitrogenase *in vivo* may be considerable: Schubert and Evans (1976; this volume) investigated H_2 evolution by nodulated nitrogen-fixing symbionts and concluded that between 40 and 60% of electron flow to nitrogenase was transferred to H_2 evolution. Table V gives an abbreviated survey of these authors' results.

Two possible reasons for the evolution of so much H_2 by symbiotic systems are: firstly, a low available concentration of ATP for nitrogen fixation (H_2 evolution by nitrogenase *in vitro* is favoured at low ATP; Silverstein and Bulen 1970) and secondly a weak conventional hydrogenase which cannot reoxidise the H_2 formed. As mentioned earlier Hill (1976b) estimated that 45% of electron flow to *K. pneumoniae* nitrogenase could be transferred to ATP-dependent H_2 evolution.

Table IV compares ATP-dependent H_2 evolution by *Azotobacter* in the presence and absence of CO and C_2H_2 and also the inhibitory effects of C_2H_2 and CO on the three types of hydrogenase activity. Similar observations have been made with *M. flavum* (Smith and Yates unpublished) and with *Anabaena cylindrica* (Bothe *et al.* unpublished).

This inhibitory effect of acetylene on conventional hydrogenase gives rise to important considerations regarding the use of acetylene as a measure of nitrogen-fixing capacity. Since H_2 formation by nitrogenase *in vivo* is considerable, prevention of recycling by acetylene may have a measurable effect on the nitrogen fixing efficiency in aerobes. Burris (1971) claimed that acetylene prevents any ATP-dependent H_2 evolution but this may not always be so: in Brotonegoro's and some of our experiments considerable H_2 was evolved in the presence of 0.1 atm. acetylene. A similar consideration is necessary in anaerobes. Inhibition of the terminal electron-accepting mechanism must affect the cell's metabolism and hence nitrogen-fixing capacity. BRouzes and Knowles (1971) showed that acetylene inhibited growth of N_2-fixing and NH_4^+-grown *C. pasteurianum*; the N_2-fixing cells recovered in time presumably through the conversion of C_2H_2 to C_2H_4 (the latter was not inhibitory). Our observations show that C_2H_4 does not inhibit conventional hydrogenase.

In conclusion we can say that our experiments have resolved the puzzle of hydrogenase in aerobic bacteria and since its activity most likely contributes to the efficiency of nitrogenase *in vivo* the relationship between nitrogenase and hydrogenase now has a rational basis.

Inhibition by acetylene of metabolic processes

Acetylene inhibits other metabolic processes besides nitrogen fixation (Schollhorn and Burris 1967) and hydrogenase (section V): N_2O reduction by denitrifying bacteria (Oshinari and Knowles 1976, Baldeston *et al.* 1976), methane oxidation (Colby *et al.* 1975; DeBont and Mulder 1974, 1976; Dalton and Whittenburg, 1976) and methanogenesis (Ellaway *et al.* 1971; Ormeland and Taylor, 1975) are all inhibited by levels of acetylene which are commonly used in tests of nitrogenase activity.

REGULATION OF NITROGENASE ACTIVITY *IN VIVO*

The ATP/ADP ratio

ADP has long been recognised as both a product and an inhibitor of nitrogenase function *in vitro* (Bulen *et al.* 1965, Moustafa and Mortenson 1967) and, by assumption, *in vivo*. Bui and Mortenson (1968) observed that both ADP and ATP bind to the reduced Fe protein; Tso and Burris (1973) using a Sephadex gel equilibration method calculated a binding constant of 5.2 μM for ADP. ADP also binds to the oxidised Fe protein and retards its rate of reduction by sodium dithionite (Yates *et al.* 1975). The apparent K_m for ATP, although varying slightly with protein concentration (Yates unpublished), is at least an order of magnitude higher than the apparent K_i for ADP (values of Burns 1969, Biggins and Kelly 1970, Gallon *et al.* 1972). This is in keeping with classical kinetics for regulation through product inhibition and would be particularly useful for a process using large amounts of ATP such as nitrogenase: assuming that the ATP+ADP pool in the vicinity of nitrogenase remained relatively constant then nitrogenase would not function without, at least, 10 times more ATP than ADP. Under such circumstances ATP would be released for essential maintenance functions rather than wasted on unnecessary NH_4^+ production. Evidence for control of nitrogenase by the ATP/ADP ratio in nitrogen-fixing organisms is slight, possibly due to the difficulty of measuring pool nucleotide levels accurately. But Haaker *et al.* (1974) showed that the internal ATP/ADP ratio increased in proportion to nitrogenase activity in cultures of *A. vinelandii* and both Bergerson and Turner (1975b) and Appleby *et al.* (1976) suggested, from indirect evidence using inhibitors, that the ATP/ADP ratio was important for nitrogen fixation in bacteroids. However, in a later communication, Haaker and Veeger (1976) failed to observe any clear relationship between the ATP/ADP ratio and nitrogenase activity in either O_2-limited or N_2-limited *A. vinelandii*.

Effect of carbamyl phosphate

Seto and Mortenson (1974) found that carbamyl phosphate both inhibited the activity and apparently repressed synthesis of nitrogenase in *C. pasteurianum*. Carbamyl phosphate bound to the MoFe protein and optimum binding occurred only with functioning nitrogenase. The binding constant was equal to the endogenous carbamyl phosphate concentration which suggested that carbamyl phosphate could regulate nitrogenase activity closely. Only N_2 or acetylene reduction was inhibited; H_2 evolution and ATP hydrolysis remained unaffected.

Inhibition of nitrogenase activity by NH_4^+

Although NH_4^+ does not inhibit purified nitrogenase preparations, regulation of nitrogenase activity by ammonium ions was suggested by L'vov *et al.* (1971) who showed that nitrogenase in freshly prepared particulate fractions from carefully disrupted *A. vinelandii* was inhibited by NH_4^+ions. Freezing and thawing abolished the effect. Drozd *et al.* (1972) observed that ammonium ions, at concentrations below that which repressed nitrogenase synthesis depressed nitrogenase activity of intact *A. chroococcum*; NO_3^- which also repressed nitrogenase synthesis was ineffective. The effect amounted to only some 30% inhibition in both cases and therefore cannot be regarded as convincing evidence for regulation of activity by production of NH_4^+ and would not account for "switch off" of nitrogenase activity. Tubb and Postgate (1973) found that NH_4^+ did not immediately inhibit nitrogenase activity in *K. penumoniae.*

Regulation by oxygen concentration

Oxygen inhibits *Azotobacter* nitrogenase *in vitro* (Wong and Burris 1972) and evidence has already been described showing that nitrogenase activity *in vivo* can be "switched off" for long periods by high oxygen tensions. Evidence that this phenomenon can be used to regulate nitrogenase activity was obtained by Dalton and Postgate (1969b) who observed that in "nitrogen limited" cultures (cultures in which the nitrogenase activity limits growth) the nitrogenase produces the same concentration of NH_4^+ at fast or slow dilution rates i.e. nitrogenase functions at the same rate however long the cells are in contact with the medium. This implies quite strongly that nitrogenase activity is influenced directly by the dissolved oxygen concentration

Regulation of nitrogenase synthesis by O_2

Since O_2 inhibits and inactivates nitrogenase the difficulty facing investigators on the effect of oxygen on nitrogenase synthesis is to distinguish between its effect on synthesis and synthesized enzyme. Pengra and Wilson (1958) and Goerz and Pengra (1961) showed that 0.05 atm O_2 inhibited growth of *K. penumoniae* and *Achromobacter* when fixing nitrogen. Pratt and Frenkel (1959) showed reversible inhibition of nitrogenase in *Rhodospirillum rubrum* by low levels of O_2 and Strandberg and Wilson (1968) showed an adaptive lag after *A. vinelandii* was depleted in repressor (NH_4^+ or NO_3) and this lag was increased by hyperbaric oxygen. In all these accounts it is not possible to distinguish between repression of synthesis or damage to the enzyme.

It is necessary to study the effect of oxygen on the appearance of nitrogenase proteins by enzymic activity, immunological techniques and gel electrophoresis. St. John, Shah and Brill (1974) shook NH_4^+-grown *K. pneumoniae* aerobically in N-free medium and found no antigenically detectable nitrogenase components. When shaken anaerobically nitrogenase was formed. Because the rate at which the Fe protein was inactivated by oxygen *in vivo* was slower than its synthesis during derepression the effect of O_2 was deduced to be on synthesis of nitrogenase rather than inactivation of preformed enzyme. Issack and Eady followed the synthesis of nitrogenase in *K. pneumoniae* by monitoring the nitrogenase proteins in crude extracts by gel electrophoresis. Their conclusions are that O_2 represses nitrogenase and that the kinetics of repression are different from the kinetics of NH_3^+ repression (see Fig. 5) which implies that these repressors act at different loci (Issack and Eady unpublished).

Regulation of nitrogenase synthesis by NH_4^+

Repression of nitrogenase synthesis by ammonium ions in all types of nitrogen-fixing organisms is well documented. Nitrate is also a co-repressor, but the probability is that it is first converted to ammonia (Drozd *et al.* 1972). That this is a repression phenomenon rather than that of prevention of induction by N_2 was substantiated by the observation by several workers that more nitrogenase is synthesised under an inert atmosphere (Ar) than under N_2 during NH_4^+ limitation (Daesch and Mortenson 1968; Munson and Burris 1969; Dalton and Postgate 1969a). Shah *et al.* (1972), Davis *et al.* (1972) and Drozd *et al.* (1972) all studied repression of nitrogenase synthesis by ammonia in Azotobacter species and the general conclusions were that nitrogenase activity was lost faster than the nitrogenase proteins disappeared, that both proteins were repressed at the same rate and that turnover of nitrogenase involving the removal of Fe might occur because the e.p.r. signal of the MoFe protein disappeared with activity but immunologically cross-reacting material remained. Tubb and Postgate (1973) in studies on sulphate-limited *Klebsiella penumoniae* found that nitrogenase activity was repressed and derepressed at rates faster than the multiplication time. They also studied the kinetics of repression using the protein synthesis inhibitors, rifampicin and chloramphenicol and found that NH_4^+ and rifampicin allowed a time lag before effecting repression whereas chloramphenicol acted immediately. This implied that NH_4^+, like rifampicin, acted at the transcriptional level and not, as chloramphenicol, at the translational level.

Ammonium ions repress other enzymes besides nitrogenase: enzymes affecting nitrate, urea, histidine and proline

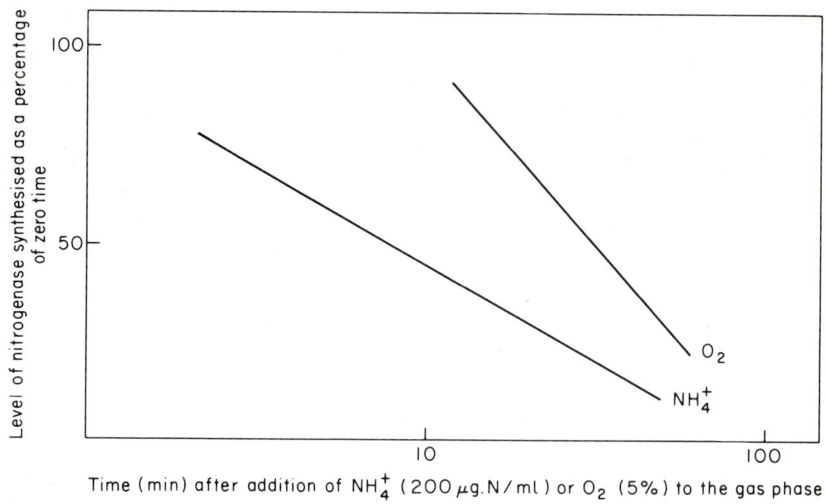

Fig. 6 *The development of nitrogenase repression in* Klebsiella pneumoniae *by NH_4^+ or O_2. Nitrogenase synthesis was determined by pulse-labelling of the culture with a mixture of ^{14}C amino acids and all cell proteins solubilised and subjected to SDS electrophoresis. The labelled proteins were visualised by autoradiography of the gel and quantitated by densitometry (R. Issack and R.R. Eady unpublished).*

utilisation are all repressed (Streicher and Valentine 1974) and glutamate dehydrogenase is induced (Tempest *et al.* 1970) by adding NH_4^+ to the growth medium. These may represent important control mechanisms in nitrogen-fixing as well as other bacteria. Both Tempest *et al.* (1970) and Nagatani *et al.* (1971) showed that NH_4^+ was assimilated in two pathways tightly regulated by NH_4^+ ion concentration and that NH_4^+ repressed the synthesis of glutamine synthetase, a key enzyme in one of the pathways. Prival and Magasanik (1971) and Prival *et al.* (1973) found a correlation between the levels of histidase and glutamine synthetase in *K. aerogenes* under NH_4^+ limitation and mutant strains deficient in glutamine synthetase were unable to derepress histidase activity. Since then other investigators (e.g.: Tubb 1974) have found a similar relationship between nitrogenase activity and glutamine synthetase in *K. pneumoniae*, and a popular current hypothesis is that the non-adenylylated form of glutamine synthetase induces nitrogenase synthesis. Repression of glutamine synthetase or its conversion into the adenylylated form causes repression of nitrogenase synthesis.

This hypothesis is not without its detractors. Shanmugam *et al.* (1975) obtained glutamine synthetase negative mutants of *K. pneumoniae* which fixed nitrogen; Drozd *et al* (1972) found no difference in the glutamine synthetase levels between N_2- and NH_4^+-grown *A. chroococcum.* Such complications should provide fascinating problems for investigation in the next few years.

The effect of ammonium ions can thus be described as similar to catabolite repression and the analogy may, apparently, extend to involving cyclic AMP: Lepo and Wyss (1974) claimed that derepression of nitrogenase in *A. vinelandii* was accelerated by cyclic AMP but their experiments took no account of the possible effects of O_2 on nitrogenase activity. On the other hand, Streicher and Valentine (1974) postulated that cyclic AMP may affect nitrogenase synthesis indirectly by repressing the level of glutamate synthetase or increasing that of glutamate dehydrogenase. If ammonia is assimilated by the glutamine synthetase-glutamate synthase pathway in nitrogen-fixing organisms (see section 7) then cyclic AMP could have an indirect effect on nitrogen-fixation. However, Drozd *et al.* (1972) failed to observe an effect of cyclic-AMP on intact nitrogen-fixing organisms and this possibility remains unresolved.

Regulation of nitrogenase synthesis by molybdenum

Molybdenum was recognised as a necessary element for nitrogen fixation in both symbiotic and free-living systems many years ago (see Wilson 1958). This appeared to be rationalised in the discovery that one of the nitrogenase proteins contains Mo which is apparently necessary for its function (the most recent analysis of proteins of high specific activity show 2 mols of Mo per mol MoFe protein: Zumft and Mortenson 1975; Smith, B.E. *et al.* 1976; Yates and Planqué 1975). Recently, however, evidence has been published indicating that Mo has a regulatory role in nitrogenase synthesis. In derepression studies with NH_4^+-grown nitrogen fixers lack of Mo when the cells were resuspended in N-free medium prevented synthesis of both nitrogenase proteins in *K. pneumoniae* (Brill *et al.* 1974) and *C. pasteurianum* (Cardenas and Mortenson 1975) but only of the MoFe protein in *A. vinelandii;* normal amounts of the Fe protein were produced (Nagatani and Brill, 1974). Addition of tungstate to Mo-free *A. vinelandii* induced formation of inactive apoprotein of the MoFe component and subsequent addition of molybdate with chloramphenicol (which inhibited further protein synthesis) yielded enzymically active protein. These experiments suggest that molybdenum, as well as being essential for nitrogenase activity also induces synthesis of the enzyme. The differences in responses to lack of Mo (presumably genetically controlled) between the anaerobic and aerobic nitrogen-fixers need to be explained.

Pathways of NH_4^+ assimilation

The pathways of ammonia assimilation in bacteria have been reviewed by Brown *et al.* (1974). They include aspartate ammonia lyase (aspartase), amino acid dehydrogenases such as alanine and glutamate dehydrogenase and the glutamine synthetase/glutamate synthase pathway. This latter pathway, first discovered in *Aerobacter aerogenes* (Tempest *et al.* 1970) is present in *K. pneumoniae* (Nagatani *et al.* 1971), *E. coli* (Berberich 1972), *B. subtilis*, *B. megaterium* and *B. licheniformis* (Meers *et al.* 1970a,b; Elmerich and Aubert 1971; Meers and Pedersen 1972), *A. chroococcum* (Drozd *et al.* 1972) and in the blue-green alga *Anabaena cylindrica* (Haystead *et al.* 1973; Lea and Miflin 1975; Lawrie *et al.* 1976). The current favoured hypothesis that this enzyme system is important for ammonia assimilation in nitrogen-fixing bacteria is based on six lines of evidence: I. the K_m values for ammonia with alanine dehydrogenase and glutamate dehydrogenase are high in bacteria (see Brown *et al.* 1974) and in *Anabaena cylindrica* (Rowell and Stewart 1976) while those for glutamine synthetase are low. This would greatly favour this enzyme's activity in the ammonia-limited conditions of nitrogen fixation; II. ammonia represses glutamine synthetase and nitrogenase synthesis and induces the formation of glutamate dehydrogenase (Tempest *et al.* 1970; Nagatani *et al.* 1971); III. tracer studies with ^{15}N (Stewart *et al.* 1975) and ^{14}C (Lawrie *et al.* 1976) show that glutamine is the most abundantly labelled of amino acids and amides subsequent to N_2-fixation in *A. cylindrica;* IV. mutants of *K. pneumoniae* are known (*asm*$^-$) with low glutamate synthase activity which can excrete NH_4^+ (Streicher *et al.* (1973). V. enzymically active glutamine synthetase regulates the synthesis of nitrogenase (see section 6.6); VI. the addition of L-methionine-DL-Sulfoximine (MSX) an inhibitor of glutamine synthetase activity to cultures of *A. vinelandii* or *A. cylindrica* allowed a high proportion of the newly fixed NH_4^+ to be excreted in the medium. This evidence clearly implies a role for glutamine synthetase as a major ammonia-assimilating pathway in these organisms (Gordon and Brill 1974; Stewart and Rowell 1975). In both investigations MSX prevented repression of nitrogenase synthesis by ammonium ions which implied that a product of ammonium assimilation rather than ammonium was the actual repressor; Gordon and Brill (1974) suggested glutamine or glutamate as possibilities.

To conclude on a note of uncertainty, Kurz *et al.* (1975) observed that although glutamine synthetase activity was high in bacteroids from pea (*Pisum sativum*) and bean (*Vicia faba*) glutamate synthase activity was low and so the authors concluded that NH_4^+ was not assimilated by this pathway. Glutamate dehydrogenase activity was also low which led to the suggestion that ammonia assimilation takes place outside the bacteroid.

CONCLUDING REMARKS

It is perhaps fitting in conclusion to draw together this wide spectrum of topics and pinpoint unresolved problems of general interest.

I. The nature of electron transport to nitrogenase in aerobes, the reasons for the low activity in cell-free preparation compared to anaerobic systems, the role of the unusual ferredoxins and the nature of the electron donors still requires elucidation.

II. What is the nature of conformational protection and how do we obtain conditions to prove or disprove that it occurs? Perhaps fluorimetric reagents will aid in the study of nitrogenase-containing membranes under different oxygen regimes.

III. If the relationship between hydrogenase and nitrogenase is as energetically important to nitrogen fixation as recent results imply (see section 5) development of strains of rhizobia which can maintain an active hydrogenase in symbiosis should be important in agriculture, particularly if the nitrogenase in bacteroids is ATP-limited (section 3.8). A widescale investigation of conventional hydrogenase activities in symbiotic and aerobic systems is indicated and genetic studies should be extended to include genes controlling hydrogenase as well as nitrogenase activities.

IV. Investigations into the regulation of nitrogenase synthesis should be extended. Why does nitrogenase require so many regulators? Is it because so much nitrogenase is synthesised normally that it becomes an energentically wasteful process of significance to the organism's survival, so much so, that those organisms with the most regulatory systems possess an advantage for survival. If glutamine or glutamate regulates nitrogenase synthesis, then the balance of activities between the NH_4^+-assimilating enzymes becomes critical.

REFERENCES

Appleby, C.A., Turner, G.L. and MacNicol, P.K. (1975). *Biochim. Biophys. Acta* 387, 461-474.

Appleby, C.A., Bergersen, F.J., MacNicol, P.K., Turner, G.L.,

Wittenberg, B.A. and Wittenberg, J.B. (1976). An International Symposium on Nitrogen Fixation (W.E. Newton and C.J. Nyman), Washington University Press. pp. 274-292.
Barman, B.G. and Tollin, G. (1972). *Biochemistry* 11, 4755-4759.
Benemann, J.R. and Valentine, R.C. (1972). *Adv.Microbial Physiol.* 8, 59-104.
Benemann, J.R. and Weare, N.M. (1974). *Science* 183, 174-175.
Benemann, J.R., Yoch, D.C., Valentine, R.C. and Arnon, D.I. (1969). *Proc.Nat.Acad.Sci.U.S.A.* 64, 1079-1086.
Benemann, J.R., Yoch, D.C., Valentine, R.C. and Arnon, D.I. (1971). *Biochim.Biophys.Acta* 226, 205-212.
Bennett, K.J., Silvester, W.B. and Brown, J.M.A. (1975). *Arch. Microbiol.* 105, 61-66.
Berberich, R. (1972). *Biochem.Biophys.Res.Commun.* 47, 1498-1503.
Bergersen, F.J. and Goodchild, D.J. (1973). *Aust.J.Biol.Sci.* 26, 729-740.
Bergersen, F.J., Turner, G.L. and Appleby, C.A. (1973). *Biochim.Biophys.Acta* 292, 271-282.
Bergersen, F.J. and Turner, G.L. (1975). *J.Gen.Microbiol.* 91, 345-354.
Bergersen, F.J. and Turner, G.L. (1975). *J.Gen.Microbiol.* 89, 31-47.
Bergersen, F.J., Turner, G.L., Gibson, A.H. and Dudman, W.F. (1976). *Biochim.Biophys.Acta* submitted.
Berndt, H., Ostwal, K.-P., Lalucat, J., Schumann, C., Mayer, F. and Schlegel, H.G. (1976). *Arch.Microbiol.* 108, 17-26.
Biggins, D.R. and Kelly, M. (1970). *Biochim.Biophys.Acta* 205, 288-299.
Biggins, D.R. and Postgate, J.R. (1969). *J.Gen.Microbiol.* 56, 181-189.
Biggins, D.R. and Postgate, J.R. (1971). *Eur.J.Biochem.* 19, 408-415.
Biggins, D.R., Kelly, M. and Postgate, J.R. (1971). *Eur.J.Biochem.* 20, 140-143.
Bothe, H. (1970). *Ber.Deut.Botan.Gen.* 83, 421-432.
Bothe, H. and Falkenberg, B. (1973). *Plant Sci.Lett.* 1, 151-156.
Bothe, H. and Loos, E. (1972). *Arch.Microbiol.* 86, 241-254.
Bothe, H. and Yates, M.G. (1976). *Arch.Microbiol.* 107, 25-31.
Bresters, T.W., DeAbieu, R.A., DeKok, A., Visser, J. and Veeger, C. (1975). *Eur.J.Biochem.* 59, 335-345.
Bresters, T.W., Krul, J., Scheepens, P.C. and Veeger, C. (1972). *Febs.Lett.* 22, 305-309.
Brill, W.J., Steiner, A.L. and Shah, V.K. (1974). *J.Bact.* 118, 986-989.

Brouzes, R. and Knowles, R. (1971). *Can.J.Microbiol.* 17, 1483-1489.
Brotonegoro, S. (1974). *Thesis Agric.Univ.Wageningen.*
Brown, C.M., MacDonald-Brown, D.S. and Meers, J.L. (1974). *Adv.Microbiol.Physiol.* 11, 1-52.
Bruschi, M., Hatchikian, E.C., Le Gall, J., Moura, J.J.G, and Xavier, A.V. (1976). *Biochim.Biophys.Acta,* 449 275-284.
Bui, P.T. and Mortenson, L.E. (1968). *Proc.Nat.Acad.Sci.U.S.A.* 61, 1021-1027.
Bulen, W.A., Burns, R.C. and LeComte, J.R. (1964). *Biochem. Biophys.Res.Commun.* 17, 265-271.
Bulen, W.A., Burns, R.C. and LeComte, J.R. (1965). *Proc.Nat. Acad.Sci.U.S.A.* 53, 523-539.
Bulen, W.A. and LeComte, J.R. (1966). *Proc.Nat.Acad.Sci.U.S.A.* 56, 979-986.
Burns, R.C. (1969). *Biochim.Biophys.Acta* 171, 253-259.
Burns, R.C. and Hardy, R.W.F. (1975). In "Nitrogen Fixation by Free-Living Micro-Organisms (W.D.P. Stewart, ed.), Cambridge University Press, pp. 421-436.
Burris, R.H. (1971). In "The Chemistry and Biochemistry of Nitrogen Fixation" (J.R. Postgate, ed.), Plenum Press, London, pp. 105-160.
Campbell, F.C. and Yates, M.G. (1973). *Febs.Lett.* 37, 203-206.
Cardenas, J. and Mortenson, L.E. (1975). *J.Bact.* 123, 978-984.
Carpenter, E.J. and Price, C.C. (1976). *Science* 191, 1278-1280.
Carter, C.W., Kraut, J., Freer, S.T., Alden, R.A., Sieker, L.C., Adman, E. and Jensen, L.H. (1972). *Proc.Nat.Acad. Sci.U.S.A.* 69, 3526-3529.
Chung, A.E. and Franzen, J.S. (1969). *Biochemistry* 8, 3175-3184.
Colby, J., Dalton, H. and Whittenbury, R., (1975). *Biochem.J.* 151, 459-462.
Cusanovich, M.A. and Edmondson, D.E. (1971). *Biochem.Biophys. Res.Commun.* 45, 327-336.
Daesch, G. and Mortenson, L.E. (1968). *J.Bact.* 96, 346-351.
Dalton, H. (1974). C.R.C. Critical Reviews in Microbiology, Feb. 1974, 183-220.
Dalton, H. and Postgate, J.R. (1969). *J.Gen.Microbiol.* 54, 463-473.
Dalton, H. and Postgate, J.R. (1969). *J.Gen.Microbiol.* 56, 307-319.
Davis, L.C., Shah, V.K., Brill, W.J. and Orme-Johnson, W.H. (1972). *Biochim.Biophys.Acta* 256, 512-523.
DeBont, J.A.M. and Leitken, M.W.M. (1976). *Arch.Microbiol.* 107, 235-240.
DeBont, J.A.M. and Mulder, E.G. (1974). *J.Gen.Microbiol.* 83, 113-121.

DeBont, J.A.M. and Mulder, E.G. (1970). *Appl.Environ.Microbiol.* 31, 640-647.
Dilworth, M.J. and Parker, C.A. (1969). *J.Theoret.Biol.* 25, 208-215.
Dixon, R.A. and Postgate, J.R. (1972). *Nature* 237, 102-103.
Dixon, R.O.D. (1972). *Arch.Microbiol.* 85, 193-201.
Dixon, R.O.D. (1975). In Nitrogen Fixation by Free-living Micro-organisms (W.D.P. Stewart, ed.), Cambridge University Press, pp. 421-435.
Drozd, J.W. and Postgate, J.R. (1970a). *J.Gen.Microbiol.* 60, 427-429.
Drozd, J.W. and Postgate, J.R. (1970b). *J.Gen.Microbiol.* 63, 63-73.
Drozd, J.W., Tubb, R.S. and Postgate, J.R. (1972). *J.Gen. Microbiol.* 73, 221-232.
Dubourdieu, M. and LeGall, J.R. (1970). *Biochem.Biophys.Res. Commun.* 38, 965-972.
Dubordieu, M., LeGall, J. and Favaudon, V. (1975). *Biochim. Biophys.Acta* 376, 519-533.
Eady, R.R. (1973). *Biochem.J.* 145, 391-396.
Eady, R.R., Kennedy, C., Smith, B.E., Thorneley, R.N.F., Yates, M.G. and Postgate, J.R. (1975). *Biochem.Soc.Trans.* 3, 488-492.
Eady, R.R. and Postgate, J.R. (1974). *Nature*, London 249, 805-810.
Eady, R.R., Smith, B.E., Thorneley, R.N.F., Yates, M.G. and Postgate, J.R. (1975). In "Nitrogen Fixation by Free-living Micro-organisms (W.D.P. Stewart, ed.), Cambridge University Press, pp. 377-388.
Edmondson, D.E. and Tollin, G. (1971). *Biochemistry* 10, 124-132.
Elleway, R.F., Sabine, J.R., Nicholas, D.J.D. (1971). *Arch. Microbiol.* 76, 277-291.
Elmerich, C. and Aubert, J.P. (1971). *Biochim.Biophys.Res. Commun.* 42, 371-376.
Evans, H.J. and Phillips, D. (1975). In "Nitrogen Fixation by Free-living Micro-organisms" (W.D.P. Stewart, ed.), Cambridge University Press, pp. 389-420.
Fay, P. (1970). *Biochim.Biophys.Acta* 216, 353-356.
Fay, P. (1976). *Appl.Environmental Microbiol.* 31, 376-379.
Fay, P. and Koolasuriya, S.A. (1972). *Arch.Microbiol* 87, 341-352.
Fay, P. and Lang, R.M. (1971). *Proc.Roy.Soc.B.* 178, 185-192.
Fay, P. and Walsby, A.E. (1966). *Nature*, London 209, 93-95.
Federov, M.W. and Kalaniskaya, T.O. (1961). *Mikrobiologiya* 30, 9-14.
Fleming, H. and Haselkorn, R. (1973). *Proc.Nat.Acad.Sci.U.S.A.*

70, 2727-2731.
Fleming, H. and Haselkorn, R. (1974). *Cell* 3, 159-170.
Francis, A.J. and Alexander, M. (1972). *Arch.Microbiol.* 85, 294-303.
French, J.R.J., Turner, G.L. and Bradbury, J.F. (1976). *J.Gen. Microbiol.* 95, 202-206.
Gallon, J.R., Kurz, W.G.W. and LaRue, T.A. (1973). *Can.J.Microbiol.* 19, 461-465.
Gallon, J.R., LaRue, T.A. and Kurz, W.G.W. (1972). *Can.J.Microbiol.* 18, 327-332.
Goerz, R.D. and Pengra, R.M. (1961). *J.Bacteriol.* 81, 568-572.
Gogotov, J.M. and Schlegel, H.G. (1974). *Arch.Microbiol.* 97, 359-362.
Gordon, J.K. and Brill, W.J. (1974). *Biochem.Biophys.Res.Commun.* 59, 967-971.
Haaker, H., Bresters, T.W. and Veeger, C. (1972). *Febs.Lett.* 23, 160-162.
Haaker, H., DeKok, A. and Veeger, C. (1974). *Biochim.Biophys. Acta* 357, 344-357.
Haaker, H. and Veeger, C. (1976). *Eur.J.Biochem.* 63, 499-507.
Haystead, A., Dharmawardene, M.W.N. and Stewart, W.D.P. (1973). *Plant Sci.Lett.* 1, 439-445.
Hill, S. (1971). *J.Gen.Microbiol.* 67, 77-83
Hill, S. (1975). *J.Gen.Microbiol.* 91, 207-209.
Hill, S. (1976a). *J.Gen.Microbiol.* 93, 335-345.
Hill, S. (1976b). *J.Gen.Microbiol.* in the press.
Hill, S., Drozd, J.W. and Postgate, J.R. (1972). *J.Appl.Chem. Biotechnol.* 22, 541-558.
Howard, J.B., Lorsbach, T. and Que, L. (1976). *Biochem.Biophys.Res.Commun.* 70, 589-595.
Hyndman, L.A., Burris, R.H. and Wilson, P.W. (1953). *J.Bacteriol.* 65, 522-531.
Jeng, D.Y., Devanathan, T. and Mortenson, L.E. (1969). *Biochem. Biophys.Res.Commun.* 35, 625-633.
Jensen, L.H. (1974). *Ann.Rev.Biochem.* 43, 461-474.
Jensen, H.L., Petersen, E.J., De, P.K. and Bhattachary, R. (1960). *Archiv.Microbiol.* 36, 182-195.
Jones, C.W., Brice, J.M., Wright, V. and Ackrell, B.A.L. (1973). *Febs.Lett.* 29, 77-81.
Jones, C.W., Erickson, S.K. and Ackrell, B.A. (1971). *Febs. Lett.* 13, 33-35.
Jones, C.W. and Redfearn, E. (1967). *Biochim.Biophys.Acta* 143, 340-365.
Jungermann, K., Kirchnairuy, H., Katz, N. and Thauer, R.K. (1974). *Febs.Lett.* 43, 203-206.
Kadam, S.S. and Naik, M.S. (1975). *Ind.J.Biochem. and Biophys.* 12, 317-321.

Kajiyama, S. and Nosoh, Y. (1973). *J.Biochem.* 73, 889-891.
Keilin, D. and Wang, Y.L. (1945). *Nature* 155, 227-229.
Keister, D.L. (1975). *J.Bacteriol.* 123, 1265-1268.
Kelly, M. (1969). *Biochim.Biophys.Acta* 191, 527-540.
Kennedy, C.K., Eady, R.R., Kondorosi, E. and Rekosh, D.K. (1976). *Biochem.J.* 155, 383-389.
Knight, E., Jr. and Hardy, R.W.F. (1966). *J.Biol.Chem.* 241, 2752-2756.
Koch, B., Wang, P., Russell, S.A., Howard, R. and Evans, H.J. (1970). *Biochem.J.* 118, 773-781.
Kondorosi, A., Barabas, I., Svab, Z., Orosz, L., Sik, T. and Hotolkiss, R.D. (1973). *Nature New Biol.* 246, 153-154.
Krieg, N.R. (1976). *Bacteriol.Rev.* 40, 55-115.
Kubo, H. (1939). *Acta Phytochim.* 11, 195-200.
Kurz, W.G.W. and LaRue, T.A. (1973). *Can.J.Microbiol.* 19, 321-324.
Kurz, W.G.W. and LaRue, T.A. (1975a). *Nature* 256, 407-409.
Kurz, W.G.W. and LaRue, T.A. (1975b). *Can.J.Microbiol.* 21, 738-741.
Kurz, W.G.W., Rokosh, D.A. and LaRue, T.A. (1975). *Can.J. Microbiol.* 21, 1009-1012.
Lawrie, A.C., Codd, G.A. and Stewart, W.D.P. (1976). *Arch. Microbiol.* 107, 15-24.
Lea, P.J. and Miflin, B.J. (1975). *Biochem.Soc.Trans.* 3, 381-384.
Lee, S.B., Wilson, J.B. and Wilson, P.W. (1942). *J.Biol.Chem.* 144, 273-381.
Lees, H. and Postgate, J.R. (1973). *J.Gen.Microbiol.* 75, 161-166.
Lepo, J.E. and Wyss, O. (1974). *Biochem.Biophys.Res.Commun.* 60, 76-80.
Levchenko, L.A., Ivleva, I.N. and Yakovlev, V.A. (1969). *Izvest.Akad.Nauk.S.S.S.R.Ser.Biol.* No. 1, 165-169.
Ludden, P.A. and Burris, R.H. (1976). *Science*, 194, 424-426.
L'vov, N.O., Sergeev, N.S., Veinova, M.K., Shaposhnikov, G.L. and Kretovich, W.L. (1971). *C.R.Acad.Sci.U.S.S.R.* 206, 1493-1496.
McComb, J.A., Elliott, J. and Dilworth, M.J. (1975). *Nature* 256, 409-410.
Marcus, L. and Kaneshiro, T. (1972). *Biochim.Biophys.Acta* 288, 296-303.
Mayhew, S.G. (1971). *Biochim.Biophys.Acta* 235, 276-288.
Mayhew, S.G. and Ludwig, M.L. (1975). In "The Enzymes" (P.D. Boyer, ed.), Academic Press, New York, pp. 57-118.
Meers, J.L. and Pedersen, K.L. (1972). *J.Gen.Microbiol.* 70, 277-286.

Meers, J.L., Tempest, D.W. and Brown, C.M. (1970a). *J.Gen. Microbiol.* 64, 187-194.
Morris, J.G. (1975). *Adv.Microbial Physiol.* 12, 169-246.
Mortenson, L.E. (1964). *Proc.Nat.Acad.Sci.U.S.A.* 52, 272-279.
Mortenson, L.E. (1966). *Biochim.Biophys.Acta* 127, 18-25.
Mortenson, L.E. and Chen, J-S. (1974). In "Microbial Iron Metabolism"(J.B. Neilands, ed.), Academic Press, New York. pp. 232-282.
Moustafa, E. and Mortenson, L.E. (1967). *Nature* 216, 1241-1242.
Munson, T.O. and Burris, R.H. (1969). *J.Bacteriol.* 97, 1093-1098.
Nagatani, H.H. and Brill, W.J. (1974). *Biochim.Biophys.Acta* 362, 160-166.
Nagatani, H.H., Shah, V.K. and Brill, W.J. (1974). *J.Bact.* 120, 697-701.
Nagatani, H., Shimizu, M. and Valentine, R.C. (1971). *Arch. Microbiol.* 79, 164-175.
Nash, D.T. and Schulman, H.M. (1976). *Biochem.Biophys.Res.Commun.* 68, 781-785.
Nason, A., Antoine, A.D., Ketchum, P.A., Frazier, W.A. and Lee, D.K. (1970). *Proc.Nat.Acad.Sci.U.S.A.* 65, 137-144.
Ohmori, K. and Hattori, A. (1972). *Plant Cell Physiol.* 13, 589-599.
Oppenheim, J., Fisher, R.J., Wilson, P.W. and Marcus, L. (1970). *J.Bact.* 101, 292-296.
Oppenheim, J. and Marcus, L. (1970). *J.Bact.* 101, 286-291.
Orme-Johnson, W.H. (1973). *Ann.Rev.Biochem.* 42, 159-204.
Ormeland, R.S. and Taylor, B.F. (1975). *Appl.Microbiol.* 30, 707-713.
Pagan, J.D., Child, J.J., Scowcroft, W.R. and Gibson, A.H. (1975). *Nature* 256, 406-407.
Oshinari, T. and Knowles, R. (1976). *Biochem.Biophys.Res.Commun.* 69, 705-710.
Palmer, G. (1975). In "The Enzymes" (3rd Edition, P.D. Boyer, ed.), pp. 2-56.
Pate, S.L., Shah, V.K. and Brill, W.J. (1973). *J.Bact.* 114, 1346-1350.
Pengra, R.M. and Wilson, P.W. (1958). *J.Bacteriol.* 75, 21-25.
Phillips, D.A., Daniel, R.M., Appleby, C.A. and Evans, H.J. (1973). *Plant Physiol.* 51, 136-138.
Pichinoty, F. and Bigliardi-Rouvier, J. (1963). *Biochim.Biophys.Acta* 67, 366-378.
Postgate, J.R. (1971). In "The Chemistry and Biochemistry of Nitrogen Fixation" (J.R. Postgate, ed.), Plenum Press, London, pp. 161-190.
Pratt, D.C. and Frenkel, A.W. (1959). *Plant Physiol.* 34, 333-337.

Prival, M.J., Brenchley, J.E. and Magasanik, B. (1973). *J.Biol. Chem.*, 248: 4334-4344.
Prival, M.J. and Magasanik, B. (1971). *J.Biol.Chem.*, 246: 6288-6296.
Reed, D.W., Toia, R.E. and Raveed, D. (1974). *Biochem.Biophys. Res.Commun.*, 58: 20-26.
Rippka, R., Neilson, A., Kunisawa, R. and Cohen-Brazire, G. (1971). *Arch.Microbiol.*, 76: 341-348.
Rowell, P. and Stewart, W.D.P. (1976). *Arch.Microbiol.*, 107: 115-124.
Schöllhorn, R. and Burns, R.H. (1967). *Proc.Nat.Acad.Sci.U.S.A.*, 57: 1317-1321.
Schubert, K. and Evans, H.J. (1976). *Proc.Nat.Acad.Sci.U.S.A.*, 73: 1207-1211.
Seto, B. and Mortenson, L. E. (1974). *J.Bact.*, 157: 805-812.
Shah, V.K., Davis, L.C. and Brill, W.J. (1972). *Biochim.Biophys. Acta* 256: 498-511.
Shanmugam, K.T., Buchanan, K.B. and Arnon, D.I. (1972). *Biochim. Biophys.Acta* 256: 477-486.
Shanmugam, K.T., Chan, I. and Morandi, C. (1975). *Biochim.Biophys.Acta* 4898: 101-112.
Shanmugam, K.T. and Valentine, R.C. (1975). *Proc.Nat.Acad.Sci. U.S.A.*, 72: 136-139.
Silverstein, R. and Bulen, W.A. (1970). *Biochemistry* 3809-3815.
Smith, B.E., Thorneley, R.N.F., Yates, M.G., Eady, R.R. and Postgate, J.R. (1976). *In:* An International Symposium on Nitrogen Fixation (W.E. Newton and C.J. Nyman, eds), pp. 150-176. Washington University Press.
Smith, L.A., Hill, S. and Yates, M.G. (1976). *Nature* 262: 209-210.
Smith, R.V. and Evans. M.C.W. (1970). *Nature (Lond.)*, 225: 1253-1256.
Smith, R.V., Noy, R.J. and Evans, M.C.W. (1971). *Biochim.Biophys. Acta* 253: 104-109.
Stewart, W.D.P. (1974). *In:* The Biology of Nitrogen Fixation (A. Quispel, ed.), pp. 696-718. North Holland Publishing Company, Amsterdam.
Stewart, W.D.P., Fitzgerald, G.P. and Burris, R.H. (1968). *Arch. Microbiol.*, 73: 250-260.
Stewart, W.D.P., Haystead, A. and Pearson, H.W. (1969). *Nature* 226: 226-228.
Stewart, W.D.P. and Rowell, P. (1975). *Biochem.Biophys.Res. Commun.*, 65: 846-855.
Stouthamer, A.H. (1969). *In:* Methods in Microbiology (J.R. Norris and D.W. Ribbons, eds), p. 629. Academic Press, London.
Stouthamer, A.H. (1973). *Antonie van Leeuwenhoek* 39: 545-565.

St. John, R.T., Shah, V.K. and Brill, W.J. (1974). *J.Bacteriol.* 119, 266-269.

Streicher, S.L. and Valentine, R.C. (1974). In "Microbial Iron Metabolism" (J.B. Neilands, ed.), Academic Press, New York. pp. 211-229.

Stombaugh, N.A., Burris, R.H. and Orme-Johnson, W.H. (1973). *J.Biol.Chem.* 248, 7951-7956.

Strandberg, G.W. and Wilson, P.W. (1968). *Can.J.Microbiol.* 14, 25.

Sweeney, W.V., Rabinowitz, J.C. and Yoch, D.C. (1975). *J.Biol. Chem.* 250, 7842-7848.

Tanaka, M., Haniu, M., Yasanabu, K.T. and Yoch, D.C. (1975). *Biochem.Biophys.Res.Commun.* 66, 639-645.

Telor, E. and Stewart, W.D.P. (1976). *Biochim.Biophys.Acta* 423, 189-195.

Tempest, O.W., Meers, J.L. and Brown, C.M. (1970). *Biochem.J.* 117, 405-407.

Thomas, J., Wolk, C.P., Shaffer, P.W., Austin, S.M. and Galonsky, A. (1975). *Biochem.Biophys.Res.Commun.* 67, 501-507.

Thorneley, R.N.F. and Eady, R.R. (1973). *Biochem.J.* 133, 405-408.

Thorneley, R.N.F. (1975). *Biochem.J.* 145, 391-396.

Thorneley, R.N.F., Yates, M.G. and Lowe, D.J. (1976). *Biochem. J.* 155, 137-144.

Tjepkema, J. and Evans, H.J. (1975). *Biochem.Biophys.Res.Commun.* 65, 625-627.

Tso, M-Y.W. and Burris, R.H. (1973). *Biochim.Biophys.Acta* 309, 263-270.

Tubb, R.S. (1974). *Nature*, London 251, 481-485.

Tubb, R.S. and Postgate, J.R. (1973). *J.Gen.Microbiol.* 79, 103-117.

Van Lin, B. and Bothe, H. (1972). *Arch.Microbiol.* 82, 155-172.

Van Gorkum, A.J. and Donze, M. (1971). *Nature* 234, 231-232.

Virtanen, A.I., Erkama, J. and Linkola, H. (1947). *Acta Chem. Scand.* 1, 861-870.

Walker, G.A. and Mortenson, L.E. (1974). *Biochemistry* 13, 2382-2388.

Walker, M.N. and Mortenson, L.E. (1974). *J.Biol.Chem.* 249, 6356-6358.

Weare, N.M. and Benemann, J.R. (1973). *Arch.Microbiol.* 93, 101-112.

Werner, D., Wilcockson, J. and Kalkowski, B. (1975). *Z.Naturforch* 30c, 687-688.

Wiegel, J. and Schlegel, H.G. (1976). *Arch.Microbiol.* 107, 139-142.

Williams, A.M. and Wilson, P.W. (1954). *J.Bact.*, 67: 353-360.
Wilson, P.W. (1958). Encyclopedia of Plant Physiology (W. Ruhland, ed.), pp. 9-47.
Wittenberg, J.B. (1977). The Professional Information Library, Dallas, Texas. In press.
Wittenberg, J.B., Bergersen, F.J., Appleby, C.A. and Turner, G.L. (1974). *J.Biol.Chem.*, 249: 4056-4066.
Wolk, C.P. and Wojcinck, E. (1971a). *Planta* 97: 124-134.
Wolk, C.P. and Wojcinck, E. (1971b). *J.Phycol.*, 7: 339-344.
Wong, P.P. and Burris, R.H. (1972). *Proc.Nat.Acad.Sci., U.S.A.*, 69: 672-675.
Wong, P., Evans, H.J., Klucas, R. and Russell, S. (1971). *In:* Biological Nitrogen Fixation (T.A. Lie and E.G. Mulder, eds), p. 525. Plant and Soil, special edition.
Yakovlev,V.A. and Levchenko, L.A. (1964). *Dokl.Akad.Nauk.U.S.S.R.*, 159: 1173-1174.
Yakovlev,V.A., Vorobev, L.V., Levchenko, L.A., Linde, V.R., Slepko, G.I. and Syrtsova, L.A. (1965). *Biokhimiya* 30: 1167-1177.
Yates, M.G. (1970a). *J.Gen.Microbiol.*, 60: 393-401.
Yates, M.G. (1970b). *Febs.Lett.*, 8: 281-285.
Yates, M.G. (1971). *Eur.J.Biochem.*, 24: 347-357.
Yates, M.G. (1972a). *Eur.J.Biochem.*, 29: 386-392.
Yates, M.G. (1972b). *Febs.Lett.*, 27: 63-67.
Yates, M.G. and Daniel, R.M. (1970). *Biochim.Biophys.Acta* 197: 161-169.
Yates, M.G. and Jones, C.W. (1974). *Adv.Microbial Physiol.*, 11: 97-136.
Yates, M.G. and Planque, K. (1975). *Eur.J.Biochem.*, 60: 467-476.
Yates, M.G., Thorneley, R.N.F. and Lowe, D.J. (1975). *Febs.Lett.*, 60: 89-93.
Yoch, D.C. (1972). *Biochem.Biophys.Res.Commun.*, 49: 335-342.
Yoch, D.C. (1973a). *Arch.Biochem.Biophys.*, 158: 633-640.
Yoch, D.C. (1973b). *J.Bacteriol.*, 116: 384-391.
Yoch, D.C. (1974). *J.Gen.Microbiol.*, 83: 153-164.
Yoch, D.C. and Arnon, D.I. (1972). *J.Biol.Chem.*, 247: 4514-4520.
Yoch, D.C. and Arnon, D.I. (1975). *J.Bact.*, 121: 743-754.
Yoch, D.C., Arnon, D.I. and Sweeney, W.V. (1975). *J.Biol.Chem.*, 250: 8330-8337.
Yoch, D.C., Benemann, J.R. Valentine, R.C. and Arnon, D.I. (1969). *Proc.Nat.Acad.Sci., U.S.A.*, 64: 1404-1410.
Zubieta, J.A., Mason, R. and Postgate, J.R. (1973). *Biochem.J.*, 133: 851-854.
Zumft, W.G. and Mortenson, L.E. (1975). *Biochim.Biophys.Acta* 416: 1-52.

ADDENDUM

References added in proof

Gallon, J. R., Kurz, W.G.W. and LaRue, T.A. (1975). *In:* Nitrogen fixation by free-living microorganisms. (W.D.P. Stewart, ed.), pp. 159-173. Cambridge University Press.

Klucas, R.V. and Evans, H.J. (1968). *Plant Physiol.*, 43: 1458-1460.

Wilkinson, J.F. (1958). *Bact.Revs.*, 22: 46-73.

AEROBIC NITROGEN FIXATION IN *A. VINELANDII*

H. HAAKER, G. SCHERINGS and C. VEEGER

Department of Biochemistry, Agricultural University, De Dreijen 11, Wageningen, The Netherlands

INTRODUCTION

Organisms which fix dinitrogen in the presence of oxygen have to overcome two difficulties: 1) Protect the oxygen-sensitive nitrogenase against irreversible inactivation. 2) Synthesize electron carriers with a low redoxpotential necessary for nitrogenase activity with oxygen present in the medium. With regard to the first problem Dalton and Postgate (1970) proposed a respiration and conformation protection. About the mechanisms of both protection systems some information is available (Drozd and Postgate, 1970; Jones *et al.*, 1973; Haaker and Veeger, 1976; Yates, 1970). Electron carriers such as ferredoxin and flavodoxin, capable of donating reducing equivalents to the nitrogenase (Benemann *et al.*, 1969; Yoch, 1969, 1972a,b; Yates, 1972), have been isolated from *Azotobacter*. Benemann *et al.* (1971) proposed an electron transport chain between NADPH via ferredoxin and flavodoxin to the nitrogenase. In our opinion some reservations must be made. 1) In whole cells nitrogen fixation takes place at different ratios of $NAD(P)H/NAD(P)^+$ (Haaker *et al.*, 1974; Haaker and Veeger, 1976), while it is doubtful that the intracellular redoxpotential of pyridine nucleotides is low enough to generate reducing equivalents for the nitrogenase. 2) Ferredoxin-$NADP^+$ oxidoreductase has never been found in *Azotobacter*. In this report we present some recent results in relation to the oxygen protection of, and the electron donation to, the nitrogenase of *Azotobacter*.

MATERIALS AND METHODS

Growth conditions and enzyme preparations

Azotobacter vinelandii was either grown in a batch culture, harvested during the logarithmic phase and stored as described earlier (Bresters *et al.*, 1970) or grown in a chemostat N_2-

limited with a dilution rate of 0.1 h^{-1}, harvested, washed and used directly (Haaker and Veeger, 1976).

Crude extracts were prepared in two ways. Mechanically, using a Manton Gaulin homogenizer (CEM) and by an osmotic (CEO) treatment as described by Shah *et al.* (1972).

Nitrogenase complex was purified as described by Bulen and LeComte (1972); depending on the extract they were derived from, the terminology M and O was added to the terminology of Bulen and LeComte. The following preparations were tested; crude extract (CEM or CEO) and C-42-I(M and O). The crude components I (crude I) and II (crude II) were prepared from C42-I and separated on a DEAE cellulose column as described by Shah *et al.* (1972). C 42-I was also eluted from a DEAE cellulose column as a single fraction with 90 mM $MgCl_2$ (C-I-90). Chemostat-grown cells were used to prepare a Tris-EDTA-toluene preparation. 100-200 mg bacterial protein was suspended in 50 mM Tris-HCl, 10 mM EDTA, 0.2 ml toluene, final pH 8.0, final volume 20 ml. The suspension was held anaerobically at 0°C for 15 min. After centrifugation (10,000 x g, 5 min.) at 4°C the pellet was washed with 40 ml 25 mM Tris-HCl pH 7.4 and 5 mM $MgCl_2$ and centrifuged again. The pellet was resuspended in the same buffer and stored at 0°C. (TET preparation). A control preparation was prepared without adding toluene (TE preparation).

Analytical methods

Acetylene reduction assays were run at 30°, shaking in 6.5 ml bottles sealed with Subaseal, which contained a mixture of 25 mM Tricine buffer, 2 mM ATP, 4 $MgCl_2$, 10 mM creatine phosphate and 5 U. creatine kinase, final pH 7.5. This mixture was thoroughly flushed for at least 30 min. with purified argon which passed a heated BASF catalyst. The acetylene (10%) and 20 mM dithionite were added. The mixture was equilibrated for 10 min. with the gas phase to remove the last traces of oxygen, after which the nitrogenase preparation was added and the ethylene produced was measured as described earlier (Haaker *et al.*, 1974). When other reducing agents instead of dithionite were used, the last traces of oxygen were removed by 10 mM glucose, 10 U glucose oxidase and 40 g catalase/ml, which was added to the assay mixture.

Pyruvate dehydrogenase was measured in the assay mixture described by Bresters *et al.* (1975), but the activity was measured by following the oxygen consumption caused by the presence of excess NADH oxidase activity using 5 mM pyruvate plus 5 mM oxaloacetate. The NADH, NADPH and malate oxidase activities were measured according to Ackrell and Jones (1971), glucose-6-phosphate dehydrogenase according to Senior and Dawes (1971). Acridine fluorescence was measured as described by Haaker *et al.*,

(1974) using 5 μM atebrin instead of 9-amino-acridine. For a TET preparation the fluorescence assay mixture contained a pyruvate dehydrogenase assay mixture plus 0.2 mg catalase, final volume 2 ml. The fluorescence of a TE preparation was performed in a mixture containing 20 mM Tris-HCl, 1 mM $MgCl_2$, 5 mM pyruvate and 0.2 mg catalase, final pH 7.4, final volume 2 ml.

Photoreactions were performed in the nitrogenase assay mixture under the following additions: 10 or 40 μM deazaflavin (3, 10 dimethyl-5-deazaisoalloxazine), flavodoxin purified according to Yoch (1972a). The light intensity was 15 mW/cm^2 measured between 400-500 nm. Anaerobic spectrophotometry was performed with "Subaseal" stoppered cuvettes on a Cary-16 spectrophotometer, final volume 1 ml. Oxygen inactivation was followed by exposing nitrogenase to air as described by Kelly (1969) in a medium containing 25 mM Tris-HCl pH 7.4 and 5 mM $MgCl_2$, final volume 2 ml. 0.3 ml samples were taken and examined for nitrogenase activity. Anaerobic gel electrophoresis was performed according to Kedinger *et al.* (1974) except that thioglycolic acid was omitted and 2 mM $Na_2S_2O_4$ was added to the upper buffer, after flushing with argon for 30 min. Samples were prior to application diluted in a medium containing: 25 mM Tris-HCl, 2 mM $Na_2S_2O_4$, 5 mM $MgCl_2$, 20% sucrose and 0.04 mM indigusulfonic acid, final pH 7.6. About 50 μg of protein was applied gel.

Biochemicals

All enzymes and cofactors were purchased from Boehringer. Flavodoxin from *P. elsdenii* and *D. vulgaris* were gifts from S. G. Mayhew; deazaflavin was a gift from V. Massey and S. Ghisla.

RESULTS AND DISCUSSION

Oxygen protection

Oppenheim *et al.* (1970) and Drozd and Postgate (1970) suggested that a nitrogenase-cytoplasmic membrane complex or the presence of cytoplasmic membranes is responsible for an oxygen-tolerant nitrogenase. When *Azotobacter* cells are ruptured by osmotic shock, the cell free extract is relatively free of cytoplasmic membranes and the nitrogenase was oxygen sensitive (Oppenheim *et al.*, 1970). This contrasts with a cell extract prepared after disruption with a French pressure cell. Fig. 1 gives a comparison of the oxygen sensitivity of nitrogenase preparations prepared by different cell-lysis methods. Addition of 0.1 M KCl to the cell-free extract of osmotically lysed cells increases the oxygen-sensitivity considerably. Since in the experiments presented by Oppenheim *et al.* (1970) in the osmotically lysed cell extract 0.1 M KCl was present, we think that the observed oxygen-sensitivity is due to the ionic strength

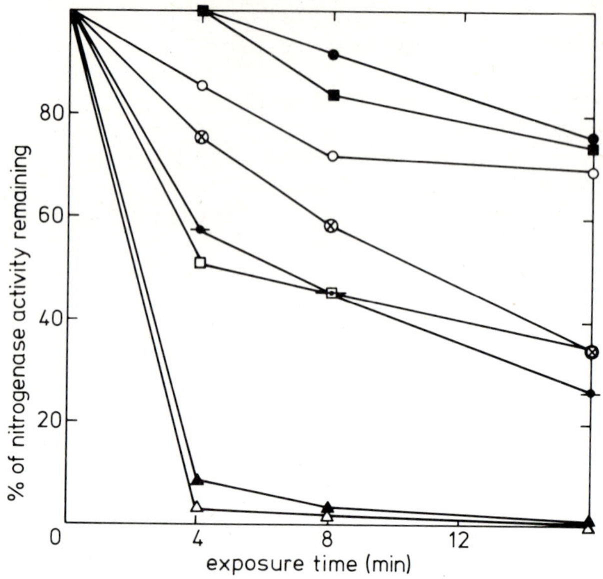

Fig. 1. *Oxygen stability of different nitrogenase preparations of* A. vinelandii. *Conditions as described in Materials and Methods. For every tested preparation a nitrogenase activity of about 100 nmoles ethylene produced per min. was used.* ●—● , *CEM;* ○—○, *CEO;* ⊗—⊗,*CEO+ 0.1 M KCL;* ■—■, *C 42-I-M;* □—□ , *C 42-I-0;* ✦—✦, *C 42-I-90;* △—△, *crude I + crude II (1:6);* △—△ *crude I + crude II (1:6) plus cytoplasmic membranes 3.6 mg.*

and not to the method of cell-rupture. Fig. 1 also shows that during the purification of the nitrogenase complex according to Bulen and LeComte (1972) the oxygen stability is relatively constant. After eluting the nitrogenase as a single fraction with 90 mM $MgCl_2$ from a DEAE-cellulose column, the nitrogenase is more oxygen-sensitive than the original complex. The nitrogenase complex prepared by mixing the crude components is extremely oxygen-sensitive and only a slight stabilization against oxygen inactivation can be obtained by adding washed cytoplasmic membranes. This is in contrast with the effect of a crude or purified NADH dehydrogenase reported by Yates (1970). The first phase shows a rapid inactivation; the extent depends on the type of preparation. The second phase is slower and occurs in all preparations.

Fig. 2 shows anaerobic polyacrylamide gel electrophoresis of different nitrogenase preparations. The nitrogenase components are easily detectable; these main bands were shown to be

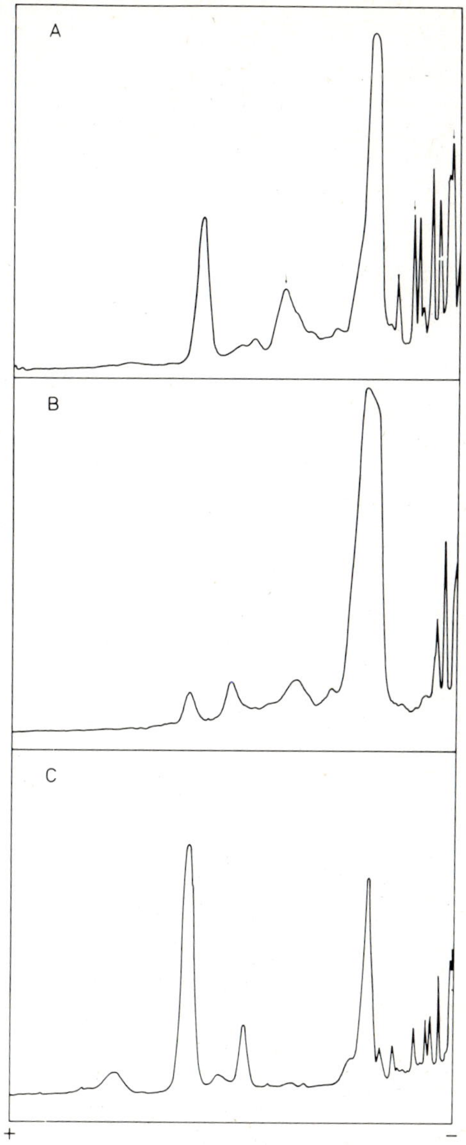

Fig. 2. *Anaerobic non-denaturating polyacrylamide gel electrophoresis of "C 42-I" nitrogenase preparations. Gels were prepared and run as described in Materials and Methods. Patterns are from the following preparations:*

A, C 42-I; B, crude component I; C, crude component II.

Fe-containing by reaction with dipyridyl reagent according to Brill *et al.* (1974). In no case was protein retained on the stacking gel, therefore the oxygen-protecting proteins must be detectable on the gels. The first pattern is that of a nitrogenase complex purified according to Bulen and LeComte (1972) up to the second Mg precipitation, which has high NADH dehydrogenase activity (NADH oxidase activity also was measurable). The activity of the complex is stable for at least 3 months at 0°C under Ar. When this preparation was applied to a DEAE-column and eluted successively with 0.25 M and 0.35 M NaCl, the eluates showed patterns 2 and 3. These eluates or combinations have no NADH oxidase or NADH dehydrogenase activity and upon storing the components 2-3 weeks separately, no nitrogenase activity could be induced by combining them. Arrows in Fig. 2A indicate the proteins which are only present in the oxygen-stable complex and these proteins may be involved in stabilization and oxygen protection of the nitrogenase. The results presented in Fig. 1 and 2 allow us to conclude that oxygen-stability of *Azotobacter* nitrogenase is not explained by the presence of cytoplasmic membranes but is caused by a high molecular weight protein factor that forms more or less a complex with the nitrogenase. The method of cell-rupture determines only partly the formation of such a complex.

Properties of toluene-treated Azotobacter *cells*

Jackson and Demoss (1965) showed that low concentrations of toluene make bacterial membranes permeable to small molecules. Since the possibility exists that toluene attacks the permeation barrier and does not influence the original structure, such a preparation may be useful for studying the generation of reducing equivalents for the nitrogenase *Azotobacter*.

Table I gives some information about the influence of toluene on the permeation of important metabolites and cofactors. All cofactors tested like ATP, TPP, CoASH, NAD^+ permeate into a TET preparation. Probably the diffusion limit for molecules with a mol. wt. less than 1000 is relatively small; however, since cytoplasmic enzymes do not leak into the medium, enzymes which are added externally probably do not pass the membranes. Since the diffusion limit for small molecules also occurs with the products, one must be careful with the interpretation of the observed velocities in a TET preparation. Nitrogenase activity produces ADP, a strong inhibitor (Ljones, 1973). Since ADP has to be regenerated on the other side of the membrane, the diffusion of ADP out of the TET vesicle probably determines the maximum nitrogenase activity in a TET preparation. The increase in oxidase activity is due to a membrane destruction by sonication. Freezing and thawing of *Azotobacter* membranes

Table I

Some enzymatic properties of *A. vinelandii* cells after a toluene treatment. TET preparation was prepared and enzyme activities were measured as described in the text. The activity is expressed as nmoles $min^{-1}.\ mg^{-1}$.

	TET	sonicated TET
Nitrogenase		
complete	1.6	3.0
no-ATP regenerating system	0.3	0
no-dithionite	0	0
Pyruvate dehydrogenase complex		
complete	105	66
no-pyruvate	4.0	7.3
no-TPP	6.0	7.8
no-CoASH	5.3	6.9
no-NAD^+	0	0
Glucose-6-phosphate dehydrogenase		
complete	410	440
no-NAD^+, $NADP^+$	0	0
no-glucose-6-P, NAD^+	13	12
no-glucose-6-P, $NADP^+$	17	4
no-glucose-6-P	30	100
Oxidases		
NADH-oxidase	347	659
NADPH-oxidase	95	280
malate-oxidase	2000	3500

also increases the oxidase activities (Jones and Redfearn, 1966). As shown in Table I glucose-6-phosphate and other substrates like isocitrate and $NAD(P)^+$ have no diffusion limit into a TET preparation in spite of the very high enzyme activities. Another aspect of a TET preparation is the enzyme organization. In a TET preparation one can expect a higher degree of organization than in ruptured cell preparations. The pyruvate dehydrogenase overall reaction, measured by the oxygen uptake *via* NADH oxidase, is an example of this higher degree of enzyme organization in a TET preparation. Because NADH and especially AcCoA strongly inhibit the overall reaction (Bresters *et al.*, 1975b), it is necessary to generate CoASH and NAD^+ efficiently to obtain a maximum steady-state activity. Because it is unlikely that enzymes such as the pyruvate dehydrogenase complex or citrate synthase are inactivated by sonication, a higher degree of enzyme organization must be the reason for the higher activity of the combined action of pyruvate dehydrogenase complex, citrate synthase and NADH oxidase in a TET preparation as compared with a sonicated TET preparation. However, one cannot exclude the possibility that, in the case of nitrogenase, the higher degree of organization within the membrane leads to lower activity, as discussed on theoretical grounds by Goldman and Katchalski (1971).

Reaction of flavodoxins with nitrogenase

Complete reduction of flavodoxins, especially the flavodoxin of *Azotobacter*, has long been a problem. Photoreduction of flavodoxin in the presence of EDTA, acetate and Tricine was not possible but with catalytic amounts of deazaflavin it is possible to photoreduce flavodoxins fast and completely (V. Massey, personal communication). With this system, we were able to reduce different types of flavodoxins completely and studied their reactivities as electron donor for the nitrogenase.

Reduced flavodoxins from *A. vinelandii*, *P. elsdenii* and *D. vulgaris* were treated with different amounts of partly purified nitrogenase and rates of electron transfer were measured spectrophotometrically as shown in Fig. 3. These rates differ widely and one can observe that of the three types of flavodoxin the *Azotobacter* flavodoxin is by far the best electron donor. Continuous illumination with a fixed limiting amount of carrier, depending on the mol. wt. 21-32 μg/ml a non-rate-limiting amount of nitrogenase complex (0.24 mg/ml) and 20% C_2H_2 present gave rates of 2, 0.3 and 0.2 nmoles ethylene produced min^{-1}. $nmol^{-1}$ of flavodoxin, respectively. On the other hand we have observed that when *P. elsdenii* flavodoxin (0.42 mg/ml) was continuously photoreduced by chloroplasts and a cell-free extract of *A. vinelandii* was added (with endogenous electron carriers present) rates of C_2H_2-reduction approached that of the dithionite-driven reaction. *A. vinelandii* flavodoxin 0.55 mg/ml is hardly active (10%) in this system because it is not reduced beyond the semiquinone state (c.f. Bennemann *et al.*, 1969). So presumably the redox potential as well as the "fit" with the nitrogenase are important in determining the rate of electron transfer mediated by a flavodoxin. Lastly, the importance of redox potential is especially clear when the extent of oxidation of reduced methylviologen by nitrogenase was compared with that of reduced benzylviologen. Benzylviologen, photo-

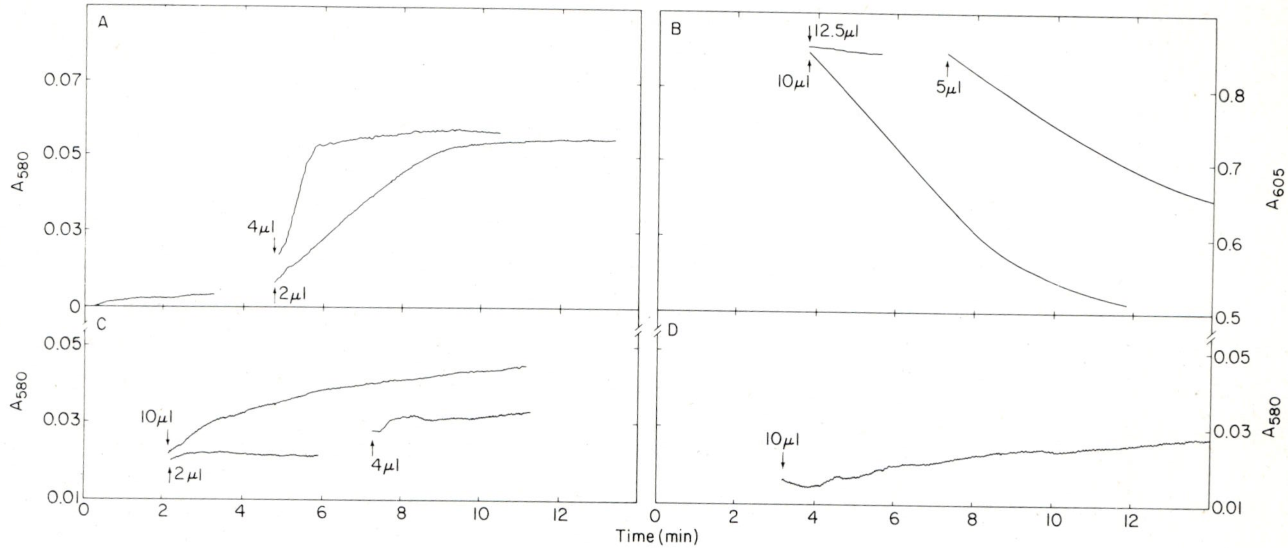

Fig. 3. *Flavodoxins are from the following organisms: A,* A. vinelandii *(0.3 mg/ml); C,* P. elsdenii *(0.2 mg/ml); D,* D. vulgaris *(0.2 mg/ml); B, of methylviologen (0.2 mM). Reactions were performed with 10 μM deazaflavin as described in the text. Time course of enzymatic and non-enzymatic oxidation of photo-reduced electron carriers. Different amounts of enzyme C 42-I, 12.5 mg/ml were added at the times indicated.*

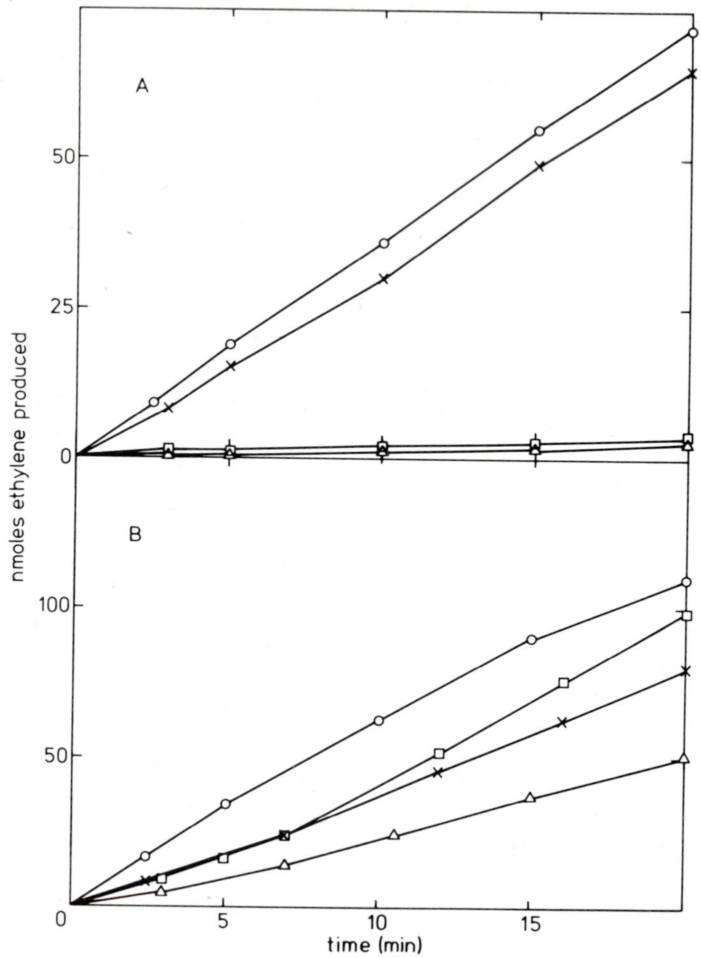

Fig. 4. *Effect of different electron donors on the nitrogenase activity in toluene-treated* A. vinelandii *cells or sonicated toluene-treated* A. vinelandii *cells. Preparations were made and activities measured as described in the text. A, 0.93 mg TET preparation was suspended in the acetylene assay mixture. Electron donors: ○—○, dithionite; ×—×, light, V, 40 μM deazaflavin benzylviologen; △—△ , light, 40 μM deazaflavin; □—□ , light, 40 μM deazaflavin, 3.8 nmoles* A. vinelandii *flavodoxin. B, 1.1 mg sonicated TET preparation was suspended in the acetylene assay mixture, same symbols as A.*

reduced as far as possible with our system, can be oxidised by nitrogenase only slightly whereas Fig. 3B shows that reduced methylviologen can be oxidized to a high extent. By continuous illumination, however, no difference in reaction rates with either methylviologen or benzylviologen as carriers could be observed.

Localisation of flavodoxin in a TET preparation

As shown in Fig. 3A fully-reduced flavodoxin from *Asotobacter* acts as a very good electron donor for purified nitrogenase. Under constant illumination photo-reduced flavodoxin has for at least 20 min. the same activity as dithionite as electron donor for the nitrogenase; also in a crude extract (Fig. 4B).

Fig. 4A, B shows the low nitrogenase activity with photochemically reduced deazaflavin in a TET preparation in contrast with the activity of a sonicated TET preparation. With a purified nitrogenase complex, photochemically reduced deazaflavin has an activity of about 10% of the dithionite activity. This means that the observed nitrogenase activity with deazaflavin in a sonicated TET preparation, about 50% of that of the dithionite activity, is due to the presence of endogenous flavodoxin, ferredoxin or another unknown low-potential electron donor.

As already shown in Table I dithionite can act as electron donor for nitrogenase in a TET or sonicated TET preparation (Fig. 4). The same is true for photochemically reduced benzylviologen. In a TET preparation the rate-limiting step in the overall nitrogenase activity is probably not the donation of electrons by the reducing agent but the diffusion of the strong inhibitor ADP out of the vesicle, because the rate of ethylene production is nearly the same with dithionite as with reduced benzylviologen. With a purified nitrogenase complex the rate of ethylene production with reduced benzyl- or methylviologen (see also Fig. 3) is about 1/3 of that of the dithionite- or reduced flavodoxin-driven nitrogenase activity. The observed activity with benzylviologen in a sonicated TET preparation is of course the sum of the activity with photochemically reduced benzylviologen plus that with the photochemically reduced endogenous electron carriers.

The results indicate that benzylviologen and dithionite react in the TET preparation directly with the nitrogenase or a site close to it. Furthermore, the lack of activity of the TET-preparation with deazaflavin indicates that flavodoxin and ferredoxin are buried inside the membrane and thus are not reduceable. The lack of activity of a TET-preparation in the presence of deazaflavin plus flavodoxin is explained by the fact that the flavodoxin does not penetrate through the membrane.

Electron transport system to nitrogenase in Azotobacter

Benemann *et al.* (1971) suggested that reduced pyridine nucleotides donate electrons *via* ferredoxin and flavodoxin to the nitrogenase. We tried to repeat their experiments with extracts of mechanically-ruptured cells, but were unsuccessful. The reason for the failure of these experiments in our hands is not known, but it might be connected with the fact that the redox properties of flavodoxin from the strain of *Azotobacter* used by Benemann *et al.* (1971) differ from those of flavodoxin from the strain used by us and others (Yoch, 1975).

Because a TET preparation certainly has organized structures which could provide an electron-donating system for nitrogenase, we tested such preparations for acetylene reduction with organic substrates as the ultimate source of reducing equivalents. However, in the presence of the substrates and cofactors used by Benemann *et al.* (1971) the rate of ethylene production was very low (0.1 nmoles. min^{-1}. mg^{-1}). In earlier publications (Haaker *et al.*, 1974; Haaker and Veeger, 1976) we proposed that the so called energized state of the cytoplasmic membrane is important in nitrogen fixation by whole cells and possibly coupled to the generation of reducing equivalents. We therefore investigated whether it is possible to energize the cytoplasmic membrane in a TET preparation by using the change of acridine fluorescence as a probe (Haaker *et al.*, 1974). Intact cells (TE preparation) show a high nitrogenase activity with pyruvate as substrate (Haaker and Veeger, 1976) and under aerobic conditions they cause a large increase in atebrin fluorescence indicating that the cytoplasmic membrane is highly energized (Fig. 5).

Table I and Fig. 5 show that TET preparations also oxidize pyruvate, but that there is no concomitant energization of the membrane. This lack of energization is not surprising since TET preparations are not selectively permeable for substrates and are thus permeable for protons. This could be the reason why no nitrogen fixation can be observed aerobically or anaerobically in a TET preparation while the conditions for nitrogen fixation are optimal: organic substrates which introduce high ratios of reduced to oxidized pyridine nucleotides, together with a high ratio of ATP to ADP. Such conditions are not possible even in whole cells.

A further attempt was made to obtain a preparation in which the oxidation of organic substrates could be coupled to nitrogenase. Since it was observed that crude nitrogenase preparations become very sensitive to oxygen in the presence of reducing agents (dithionite or oxidizable substrates), we worked

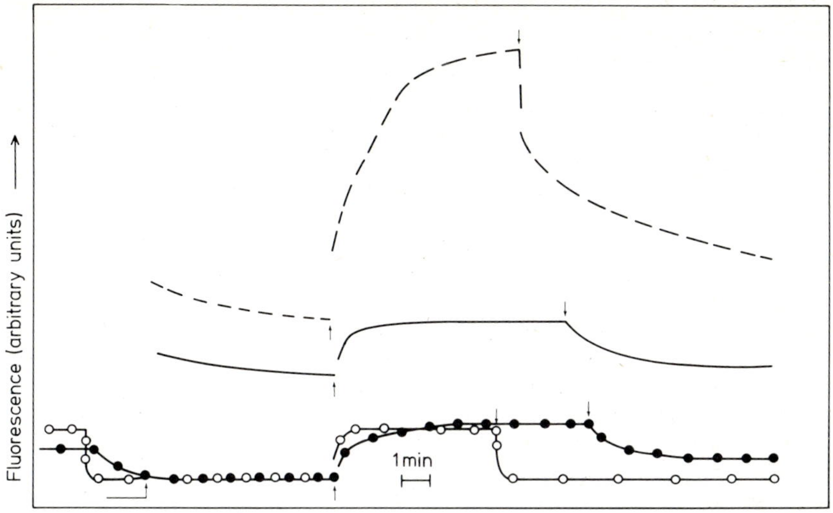

Fig. 5. *Effect of toluene on the enhancement of the atebrin fluorescence upon energization of* A. vinelandii *cells.* A. vinelandii *cells were treated with toluene and fluorescence was measured as described in the text.* •—•, *8 mg TET preparation in 2 ml medium;* o—o, *8 mg TE preparation in 2 ml medium. At the arrow* ⌐ *5 μM atebrin was added, and fluorescence followed* ——, *TET preparation;* ------, *TE preparation. At arrow* ↑ *0.34 μmole H_2O_2 was added to a TET preparation and 3.4 μmole H_2O_2 to a TE preparation. At* ↓ *anaerobiosis was reached. The fluorescence of both preparations in the absence of atebrin are given in the corresponding* •—• *and* o—o *curves. The corresponding arrows indicate the time of H_2O_2 addition (corresponding amounts) and anaerobiosis.*

under anaerobic conditions. Pyruvate was chosen as substrate because it was found that under anaerobic conditions pyruvate oxidation can be coupled to nitrogenase, although the activity is low (1% of the activity with pyruvate under optimal aerobic conditions). Cells were broken gently by treating them with lysozyme in the presence of Tris-EDTA. During cell-lysis the activity with pyruvate decreased to zero indicating that even these mild conditions finally caused complete loss of coupling between pyruvate oxidation and nitrogenase activity. Coupling was not restored by adding of a variety of cofactors, but full nitrogenase activity was observed with dithionite and ATP. However, there was some evidence for intermediate structures which responded to substrates that are thought to influence the ratio Acetyl CoA/CoASH. During the course of the lysis reaction oxaloacetate caused a 2-fold stimulation of activity and -hydroxy-butyrate and acetylphosphate caused an inhibition. None of these three components had any effect on the activity of whole cells. It appears therefore that, at an intermediate state of lysis, the cell becomes permeable to low mol. wt. compounds. These observations emphasise the strong dependence of nitrogenase activity with physiological substrates on an intact cytoplasmic membrane and suggest that the ratio Acetyl CoA/CoASH could be important in the coupling with pyruvate oxidation.

ACKNOWLEDGEMENTS

We thank Dr. S. G. Mayhew for his very useful comments during this investigation and Miss L. A. M. Wolbers for the help in some experiments. We also thank V. Massey for permission to apply his method of flavodoxin induction prior to publication. The present investigation was supported by the Netherlands Foundation for Chemical Research (S.O.N.) with financial aid from the Netherlands Organization for the Advancement of Pure Research (Z.W.O.).

REFERENCES

Ackrell, B.A.C. and Jones, C.W. (1970). *Eur.J.Biochem.*, 20: 29-35.

Benemann, J.R., Yoch, D.C., Valentine, R.C. and Arnon, D.I. (1969). *Proc.Nat.Acad.Sci.*, 64: 1079-1086.

Benemann, J.R., Yoch, D.C., Valentine, R.C. and Arnon, D.I. (1971). *Biochim.Biophys.Acta*, 226: 205-212.

Bresters, T.W., De Abreu, R.A., De Kok, A., Visser, J. and Veeger, C. (1975). *Eur.J.Biochem.*, 59: 335-345.

Brill, W.J., Westphal, J., Stieghast, M., Davis, L.C. and Shah, V.K. (1974). *Anal.Biochem.*, 60: 237-241.

Bulen, W.A. and LeComte, J.R. (1972). *Meth.Enzym.*, XXIV: 256-470.

Dalton, H. and Postgate, J.R. (1970). *J.Gen.Microbiol.*, 54: 463-479.

Drozd, J. and Postgate, J.R. (1970). *J.Gen.Microbiol.*, 63: 63-73.

Goldman, R. and Katchalski, E. (1971). *J.Theor.Biol.*, 32, 243-257.

Haaker, H., De Kok, A. and Veeger, C. (1974). *Biochim.Biophys. Acta*, 357: 344-357.

Haaker, H. and Veeger, C. (1976). *Eur.J.Biochem.*, 63: 499-507.

Jackson, R.W. and Demoss, J.A. (1965). *J.Bacteriol.*, 90: 1420-1425.

Jones, C.W., Brice, J.M., Wight, V. and Ackrell, B.A.C. (1973). *FEBS Lett.*, 29: 77-81.

Jones, C.W. and Redfearn, E.R. (1966). *Biochim.Biophys.Acta,* 113: 467-481.
Kedinger, C., Gissinger, F. and Chamban, P. (1974). *Anal.Biochem.,* 60: 237-241.
Kelly, M. (1969). *Biochim.Biophys.Acta,* 191: 527-540.
Ljones, T. (1973). *Biochim.Biophys.Acta,* 321: 103-113.
Oppenheim, J., Fisher, R.J., Wilson, P.W. and Marcus, L. (1970). *J.Bacteriol.,* 101: 292-296.
Senior, P.J. and Dawes, E.A. (1971). *Biochem.J.,* 125: 55-66.
Shah, V.K., Davis, L.C. and Brill, W.J. (1972). *Biochim.Biophys. Acta,* 256: 498-511.
Yates, M.G. (1970). *FEBS Lett.,* 8: 281-285.
Yates, M.G. (1971). *Eur.J.Biochem.,* 24: 347-357.
Yates, M.G. (1972). *FEBS Lett.,* 27: 63-67.
Yoch, D.C., Benemann, J.R., Valentine, R.C. and Arnon, D.I. (1969). *Proc.Nat.Acad.Sci.,* 64: 1404-1410.
Yoch, D.C. and Arnon, D.I. (1972a). *J.Biol.Chem.,* 247: 4514-4520.
Yoch, D.C. (1972b). *Biochem.Biophys.Res.Commun.,* 49: 335-342.
Yoch, D.C. (1975). *Arch.Biochem.Biophys.,* 170: 326-333.

CELLULAR PHYSIOLOGY AND THE ECOLOGY OF N_2-FIXING BLUE-GREEN ALGAE

W.D.P. STEWART, P. ROWELL and S.K. APTE

Department of Biological Sciences, University of Dundee, Dundee, Scotland, U.K.

INTRODUCTION

Blue-green algae are photosynthetic prokaryotes which can grow on inorganic medium, obtaining their necessary carbon and nitrogen from photosynthetic CO_2 fixation and N_2 fixation respectively. It is now clear that not only do they have sophisticated biochemical mechanisms which allow them to carry out both processes simultaneously but that they are important agents in a variety of habitats ranging from arctic tundra to tropical rice paddy soils (Stewart, 1975a). In this paper we consider some recent findings on the cellular interactions of N_2-fixing algae, and point to some outstanding gaps in our knowledge of the ecology of the group.

CELLULAR INTERACTIONS

Blue-green algae are unique amongst N_2-fixing prokaryotes in evolving O_2 during photosynthesis. Thus, for nitrogenase activity to occur there must be little, or no O_2 available to inactivate the O_2-sensitive enzyme. In heterocystous algae, which invariably fix N_2, active nitrogenase occurs in air in the non-O_2-evolving heterocysts; in non-heterocystous filamentous forms and in the vegetative cells of heterocystous forms the nitrogenase is active only under anaerobic or microaerobic conditions, and in the few unicellular forms which fix N_2 in air there is a temporal separation of O_2-evolving photosynthesis and O_2-sensitive nitrogenase activity. In all algae, however, the electrons for dinitrogen fixation are provided by photosynthate and the major source of ATP is photophosphorylation (see Stewart, 1976; Stewart *et al.*, 1975). We have investigated the detailed route of their provision in *Anabaena cylindrica* and the data obtained are presented here. Experimental methods, if not detailed in referenced papers, are given in the legends to the Tables and Figures.

The photosynthesis electron transport chain of heterocysts of Anabaena cylindrica

Anabaena cylindrica shows a typical higher plant type of photosynthesis with two photosystems linked in series. The results obtained in a study of the components of the photosynthetic electron transport chain which occur in heterocysts and some of the reactions in which these components may be involved are summarised in Fig. 1. Several points of interest emerge.

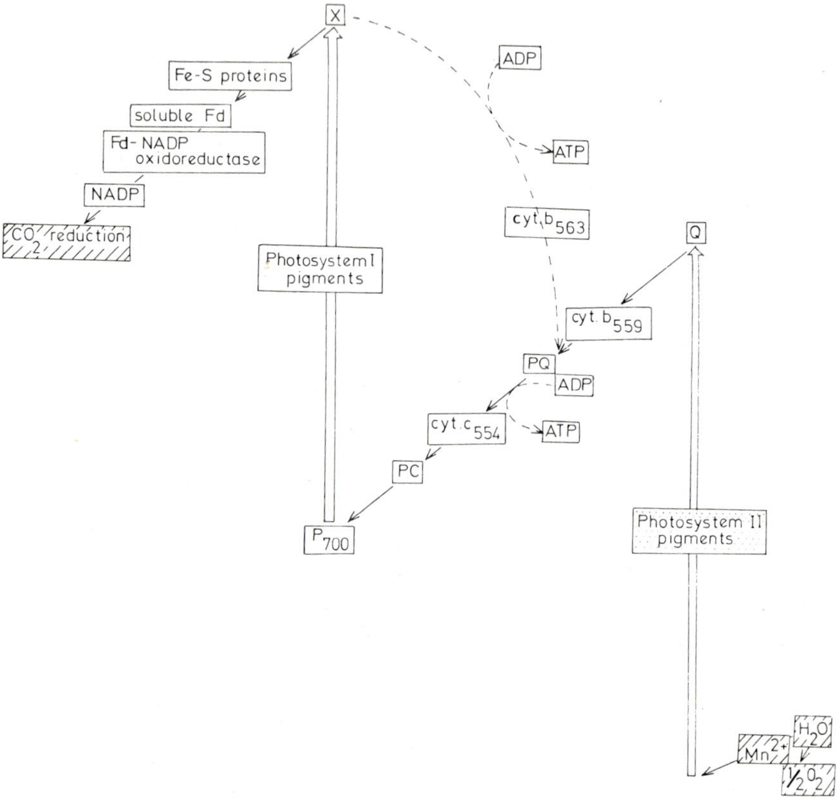

Fig. 1. *Components of the photosynthetic electron transport chain which are present (▭), undetectable (▨) or modified compared with those of vegetative cells (▤) in heterocysts of* Anabaena cylindrica.

First, the photosystem II of heterocysts is modified and, as others noted earlier (Carr and Bradley, 1973; Tel-Or and Stewart, 1975), there is no photosynthetic O_2 evolution. We have obtained evidence that the lack of O_2 evolution may be attributable not only to a deficiency in the photosystem II accessory pigments

(Thomas, 1970) but also to a deficiency in the Mn-containing O_2-evolving photosystem II reaction centre. On exposing Mn-deficient *Anabaena cylindrica* to ^{54}Mn over 8 days there was a marked incorporation of the radioisotope into the vegetative cells with incorporation into the heterocysts being only 9% of that in the vegetative cells, on a chlorophyll *a* basis. Since the ratio of Mn atoms : chlorophyll molecules in O_2-evolving organisms and preparations is 1:50:100 (Cheniae, 1970) and is only 1:877 in heterocysts (compared with 1:77 in vegetative cells) the results suggest that an O_2-evolving photosystem II reaction centre is inactive in heterocysts and this correlates with our failure to demonstrate O_2-evolution in such heterocyst preparations.

Other components of the photosynthetic electron transport chain of heterocysts are present however including cytochromes b_{559}, c_{554}, b_{563} and the photosystem I reaction centre P700, all of which were observed by difference spectroscopy. Electron paramagnetic resonance spectroscopy also provided evidence for the presence of bound ferredoxins corresponding to the A and B ferredoxins of higher plant chloroplasts, and biochemical evidence that plastoquinone, plastocyanin, soluble ferredoxin and ferredoxin-NADPH oxidoreductase were present or active in heterocysts was also obtained (see Tel-Or and Stewart, 1977).

The inability of heterocysts to fix CO_2 photosynthetically

Since photosynthetic CO_2 fixation might be expected in the presence of reduced ferredoxin and ATP, as in photosynthetic bacteria and chloroplast systems, heterocysts were examined for the presence of the key CO_2-fixing enzyme of the Calvin cycle, ribulose biphosphate carboxylase. The enzyme was depleted in heterocysts as evidenced by the lack of enzymic activity (see also Lex and Carr, 1974) and by the absence from heterocysts of the carboxysomes which appear to be packets containing ribulose biphosphate carboxylase in *Anabaena cylindrica* (Stewart and Codd, 1975; Codd and Stewart, 1976). Thus, as for O_2-evolution, there is a sophisticated biochemical lesion which in this case ensures that nitrogenase does not have to compete with photosynthetic CO_2 fixation for the available ATP and reduced ferredoxin.

Phosphorylation in heterocysts

The ability of heterocysts to photophosphorylate has been demonstrated directly in a preliminary report by Scott and Fay (1972) and by Tel-Or and Stewart (1975), using isolated heterocysts. This is in keeping with our observation, mentioned above, that several of the key components involved in photophosphorylation, e.g. P700, cytochrome b_{563} and plastocyanin occur in heterocysts. In our studies using ^{32}P as tracer, the observed rates of ATP formation were 55 to 159 μmoles ATP mg chl $a^{-1}h^{-1}$ in N_2-grown cultures, this rate of phosphorylation being several times higher than the observed rates of oxidative phosphorylation.

The reduction of ferredoxin in heterocysts

Since carbon is not fixed in heterocysts, it has to be transported there from the photosynthesising vegetative cells, and such a transport is well established (Wolk, 1968; Stewart *et al.*, 1969). There is now evidence (Juttner and Carr, pers. comm.) that much of the fixed carbon is transferred to the heterocysts as maltose and electron donation from such a carbon compound to ferredoxin is thus of key importance (see Tel-Or and Stewart, 1977). Since the Embden Meyerhof pathway is largely inactive in heterocysts due to a deficiency there of glyceraldehyde-3-phosphate dehydrogenase (Winkenbach and Wolk, 1973; Lex and Carr, 1974) ferredoxin reduction *via* pyruvate ferredoxin oxidoreductase is unlikely to be of major physiological importance in such N_2-fixing cells although this enzyme supports nitrogenase activity in extracts of whole filaments of *Anabaena cylindrica* (Codd *et al.*, 1974). Since the photosystem I electron transport chain of heterocysts appears to be intact this could possibly be a source of reductant to ferredoxin (Tel-Or and Stewart, 1977) but there is no evidence so far that the photoreduction of ferredoxin and nitrogenase occurs *in vivo* in heterocysts.

We have investigated the flow of electrons from glucose-6-phosphate to ferredoxin in heterocysts of *Anabaena cylindrica*. This compound can be formed readily from maltose via maltase and hexokinase. As Table I shows, the activity of glucose-6-phosphate dehydrogenase was enriched in heterocysts of Winkenbach and Wolk (1973) and Lex and Carr (1975). Extracts of NH_4^+-grown cultures without heterocysts showed activity comparable to that obtained in the vegetative cells of N_2-grown material. On incubating the extracts in the light, there was a marked inhibition of vegetative cell activity but heterocyst activity was less affected, suggesting a specific light inhibition of the enzyme in the vegetative cells only.

The donation of electrons from NADPH to ferredoxin *via* ferredoxin-NADP oxidoreductase was then investigated. As Table II shows this enzyme was present in the heterocysts of *Anabaena cylindrica* since the reductions of dichlorophenol-indophenol, and cytochrome *c* (a ferredoxin-dependent reaction) could both be readily observed in the light and dark. Vegetative cell extracts on the other hand, while showing cytochrome *c* reduction in the dark, showed no such reduction in the light.

Table I

The distribution of Glucose-6-Phosphate Dehydrogenase activity in different fractions of *Anabaena cylindrica* and the effect of light on enzymic activity

Fraction	n moles $NADP^+$ reduced min^{-1} mg $protein^{-1}$ 0 h	2 h in dark	2 h in light
Whole filaments (N_2-grown culture)	42.00	36.23	22.07
Vegetative cells (N_2-grown culture)	16.90	12.56	8.69
Heterocysts (N_2-grown culture)	120.80	109.51	105.55
Whole filaments (NH_4^+-grown culture)	18.40	15.94	10.14

The algae were grown in N_2 or NH_4^+ (3.0 mM) at 3000 lux and 25°C in steady state culture. The heterocyst and vegetative cell material was prepared, when required, according to Tel-Or and Stewart (1976), then broken by French pressure cell treatment at 16,000 p.s.i. and activity assayed in the 35,000 x g x 30 min supernatant fractions. Extracts were prepared in dim light (< 100 lux) then incubated at 3000 lux or in the dark at 25°C for 120 min prior to assay. The complete reaction mixture contained in a total volume of 3 ml: 0.1 M Tris-maleate buffer, pH 6.8; $MgCl_2$, 10 mM; $NADP^+$, 0.2 mM; glucose-6-phosphate, 30 mM and 0.5 - 1.0 mg protein. Activity was measured as increase in absorbance at 340 nm.

However, if the thylakoid membranes of the vegetative cells were separated from the enzyme-containing supernatant, cytochrome *c* reduction occurred both in the light and in the dark (Table III) but on re-adding the membranes to the soluble extract, activity was once again inhibited in the light. These findings suggest that in the vegetative cells where CO_2 is fixed photosynthetically, a light-dependent mechanism exists which ensures that the reduction of ferredoxin from NADPH does not occur, but that in the heterocysts where the prime function of NADPH may be to generate reduced ferredoxin for dinitrogen reduction, this mechanism is lacking.

Thus there is evidence of an active pathway in heterocysts by which ferredoxin can be reduced at equal rates in the light

Table II

The effect of light on ferredoxin-NADPH-oxidoreductase mediated reactions in vegetative cells and heterocysts of *Anabaena cylindrica*

Reaction	Vegetative cells		Heterocysts	
	Dark	Light	Dark	Light
NADPH → DCPIP reduction	0.42	6.42	12.95	12.99
NADPH → Fd → Cytochrome *c* reduction	5.54	0.0	8.70	8.83

The algae were grown and fractions prepared as in legend to Table I. The vegetative cells (NH_4^+-grown culture) and heterocysts (from N_2-grown culture) were broken at 16,000 p.s.i. and the crude extracts were measured for activity. Extracts were prepared in dim light (< 100 lux) then incubated at 3000 lux, or in the dark, at 25°C for 120 min prior to assay. The complete reaction mixture contained in a total volume of 3 ml: 30 mM HEPES buffer, pH 7.5; NADPH, 50 μM; dichlorophenol indophenol, 50 M or cytochrome c, 22 μM. Reactions were started by the addition of DCPIP or cytochrome *c* and NADPH oxidation was measured as decrease in absorbance at 340 nm. Activities are expressed as nmoles NADPH oxidised min^{-1} mg $protein^{-1}$.

Table III

The effect of thylakoid membrane preparations from the vegetative cells of *Anabaena cylindrica* on the activity of soluble ferredoxin-NADP oxidoreductase

Fractions	Oxidant	NADPH oxidation dark	NADPH oxidation light
soluble enzyme (100,000 x g x 60 min supernatant)	DCPIP	18.84	18.84
thylakoid membranes (100, 000 x g x 60 min pellet)	DCPIP	0.89	0.89
enzyme + membrane	DCPIP	6.03	6.10
soluble enzyme (100,000 x g x 60 min supernatant)	cytochrome *c*	7.24	7.24
thylakoid membranes (100, 000 x g x 60 min pellet)	cytochrome *c*	0.63	0.00
enzyme + membranes	cytochrome *c*	3.38	0.00

The alga was grown on NH_4^+ (3 mM NH_4Cl) and extracts prepared by French pressure cell treatment, as outlined in the legend to Table I. Further fractionation was carried out as indicated in the table, and when enzyme and thylakoids were mixed this was done in the same proportions as were present in the original extract. Activities are expressed as n moles NADPH oxidised by DCPIP or cytochrome *c* min^{-1} mg $protein^{-1}$ (see legend to Table II for further experimental details).

and dark by electrons transferred there from glucose-6-phosphate via glucose-6-phosphate dehydrogenase and ferredoxin-NADP oxidoreductase. Such results are in general agreement with the early studies *in vitro* of Bothe (1970) who, using extracts of whole filaments of *Anabaena cylindrica,* obtained nitrogenase activity when glucose-6-phosphate dehydrogenase, ferredoxin-NADP oxidoreductase and ferredoxin were available. While the route of electron donation to ferredoxin in the light is not fully worked out it is clear that the glucose-6-phosphate dehydrogenase ferredoxin-NADP oxidoreductase pathway can provide all the necessary reduced ferredoxin in the dark and that it could also do so in the light. Thus, the generation of ATP by cyclic photophosphorylation may be the main role of the photosynthetic electron transport chain of heterocysts (Fig. 2).

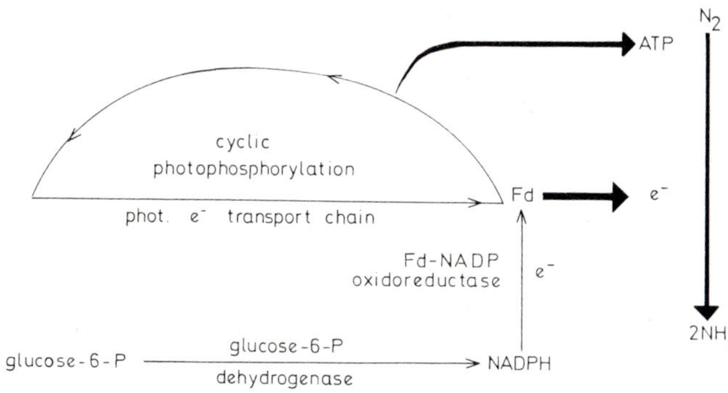

Fig. 2. *Possible inter-relations between glucose-6-phosphate dehydrogenase, ferredoxin-NADP oxidoreductase, ferredoxin, cyclic photophosphorylation and nitrogenase activity in heterocysts of* Anabaena cylindrica.

Glutamine synthetase (GS) and its regulation in Anabaena cylindrica

GS acts as the key NH_3 assimilating enzyme in N_2-fixing prokaryotes and there is evidence that in addition it may play an important role in the regulation of the *nif* operon (Tubb, 1974; Shanmugam *et al.*, 1975). In *Anabaena cylindrica* it is present both in heterocysts and vegetative cells, has a low K_m for NH_4^+ (1 mM) (Dharmawardene *et al.*, 1973) and is the primary route of NH_3 assimilation (Stewart and Rowell, 1975) with glutamine being one of the first ^{15}N-labelled compounds formed on exposure of the alga to $^{15}N_2$ (Stewart *et al.*, 1975). Thomas *et al.* (1975) have confirmed, using ^{13}N as tracer, that newly fixed NH_3 is incorporated into glutamine. We have speculated (Stewart *et al.*, 1975) that glutamine may be the compound via which newly fixed nitrogen is transferred from heterocysts to vegetative cells of *Anabaena cylindrica*.

The role of GS as a regulator of nitrogenase in *Anabaena cylindrica* by the mechanism proposed for *Klebsiella* by Shanmugam *et al.* (1975) is less clear since in an earlier study we failed to obtain any evidence for an adenylylation/deadenylylation of the *Anabaena* GS (Dharmawardene *et al.*, 1973). We have since carried out an additional study in search of an adenylylation/deadenylylation mechanism. First, tests were carried out on the effect of snake venom phosphodiesterase on the activity of GS using enzyme extracted from N_2 and NH_4^+-grown cultures. As Table IV shows there was no marked difference in activity with or without diesterase treatment. An

Table IV

The effect of snake venom phosphodiesterase on the activity of glutamine synthetase of *Anabaena cylindrica*

	Mg^{2+}-biosynthetic activity		Mn^{2+}-transferase activity
	N_2-grown	NH_3-grown	NH_3-grown
Untreated	47	26	196
Phosphodiesterase treated	41	23	174

The algae were grown in steady state culture on N_2 or NH_4^+ (3.0 mM NH_4Cl), ruptured by French pressure cell treatment, and 0.4 μmoles of snake venom phosphodiesterase (Sigma) added to 3 ml of a 35,000 x g x 20 min supernatant fraction, in the presence of 0.1 M Tris/HCl buffer, pH 7.5; 5 mM $MgCl_2$ and 2 mM mercaptoethanol. Activity was monitored over a 3 h period at 37°C using both biosynthetic (Rowell and Stewart, 1976) and transferase (Dharmawardene *et al.*, 1973) assays. Activities are expressed as nmoles product formed min^{-1} mg $protein^{-1}$.

increase in activity would have been expected if the enzyme had been adenylylated. Second, as Fig. 3 shows, on adding NH_4^+ which should adenylylate GS rapidly, and thus cause an inhibition of

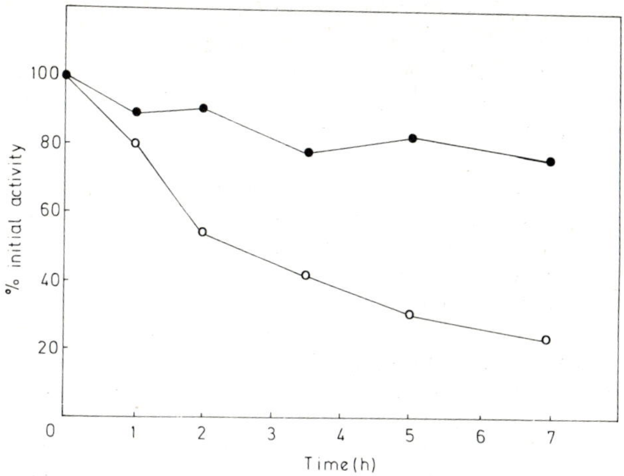

Fig. 3. *The effect of* NH_4^+ *(3.0 mM* NH_4Cl*) on the activity of glutamine synthetase (biosynthetic activity, Rowell and Stewart, 1976)(● - ●) and acetylene reduction (○ - ○) in steady state* N_2*-grown cultures of* Anabaena cylindrica *incubated at 3000 lux and 25°C, and to which the* NH_4^+ *was added at 0 time.*

GS activity in N_2-fixing cultures, no marked change in activity occurred over several hours, indicative of a lack of an adenylylation mechanism. Third, on cross-reacting extracts of *Anabaena cylindrica* with antiserum to *E. coli* GS no evidence of immunoprecipitation was obtained although the antiserum reacted with its homologous antigen (Fig. 4) and with GS obtained from various other N_2-fixing micro-organisms which possess an adenylylation mechanism (Tronick *et al.*, 1975). Furthermore, there was no inhibition of GS activity on adding the *E. coli* antiserum to extracts of *Anabaena cylindrica* possessing GS activity, although crude extracts of *E. coli* GS were inhibited (Rowell *et al.*, 1977). Such results provide further evidence that an adenylylation/deadenylylation mechanism is unlikely to be important in *Anabaena cylindrica*, although the possibility that it may occur, for example, in the purified enzyme, or in heterocysts but not in extracts of intact filaments, cannot be entirely ruled out.

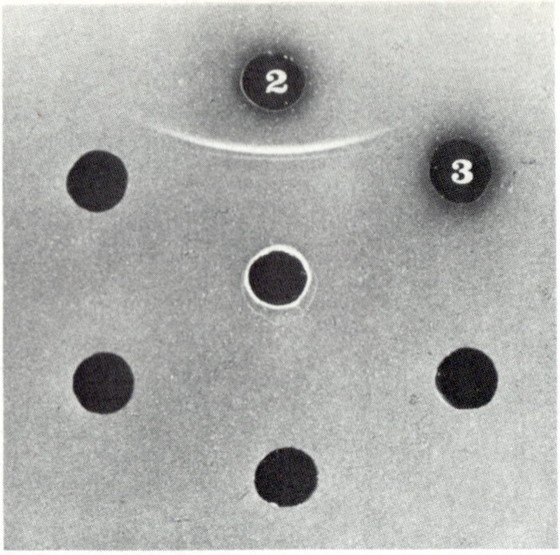

Fig. 4. *Ouchterlony double immunodiffusion plate showing the lack of immunoprecipitation on cross-reacting extracts of* Anabaena cylindrica *containing glutamine synthetase (well 3) with antiserum to* E. coli *glutamine synthetase (centre well). Immunoprecipitation between extracts of* E. coli *containing glutamine synthetase (well 2) and the homologous antiserum is seen. The antiserum to* E. coli *glutamine synthetase was kindly provided by Dr. S. Francis in the laboratory of Dr. E.R. Stadtman.*

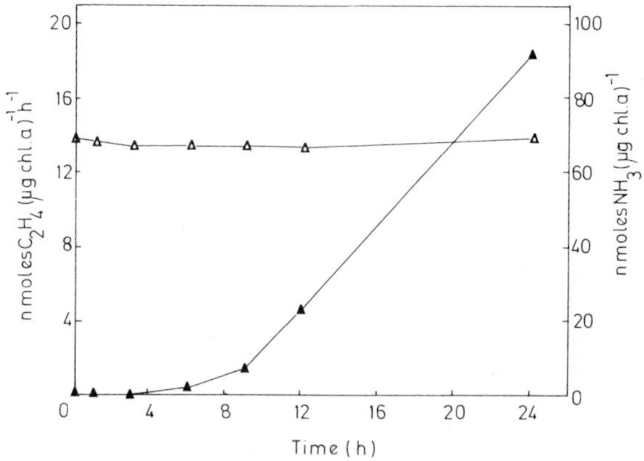

Fig. 5. *The effect of methionine sulphoximine (MSO)(1.0 M) on acetylene reduction (Δ - Δ) and NH_3 production (▲ - ▲) by log phase batch cultures of* Anabaena cylindrica *grown at 25°C and 3000 lux and to which the analogue was added at 0 time. For further details see Stewart and Rowell (1975).*

Glutamine synthetase and nitrogenase in Anabaena cylindrica

The possible relationship between nitrogenase activity and glutamine synthetase was then investigated. First, as Fig. 5 shows, on adding the analogue methionine sulphoximine (1.0 M) which inhibits GS/glutamate synthase activity, to N_2-fixing *Anabaena cylindrica,* the newly fixed NH_3 was excreted into the medium, but nitrogenase activity was sustained, indicating that NH_3 *per se* is not the regulator of nitrogenase synthesis. Second, as Fig. 3 shows, on adding NH_4^+ (3 mM) to a N_2-fixing culture of *Anabaena cylindrica,* the inhibition of C_2H_2 reduction which occurred preceded the reduction in GS activity, which suggests that GS *per se* may not be responsible for the regulation of nitrogenase. Third, on measuring the relationship between various ammonia assimilating enzymes, their products and nitrogenase activity in steady state NH_4^+-grown, N_2-grown and nitrogen-starved cultures of *Anabaena cylindrica,* the only marked correlations with nitrogenase activity increased, and inverse correlations between the rates of C_2H_2 reduction and the size of the glutamine and aspartate pools (Table V). Since we previously obtained evidence (Fig. 3) that change in GS activity did not necessarily precede change in nitrogenase activity it is possible that it is a product(s) of GS activity which

Table V

The effect of nitrogen source on the steady state amino acid pools and enzyme activities of *Anabaena cylindrica*

Reaction or amino acid	NH_4^+-grown	N_2-fixing	N-starved
Acetylene reduction	0	17.5	56.0
Glutamine synthetase	21.3	50.5	62.3
Glutamate-aspartate-aminotransferase	44.8	61.6	84.7
Glutamine	0.32	0.20	0.01
Aspartate	1.03	0.78	0.17
Glutamate	0.87	1.38	0.68
Alanine	0.20	0.16	0.25
Glycine	0.20	0.17	0.21

The algae were grown on N_2 or NH_4^+ (3.0mM) at 3000 lux at 25°C in steady state culture or in the case of N-starved algae N_2-fixing steady state cultures were batched and incubated under $A/O_2/CO_2$ (77.96/22.00/0.04, v/v) for up to 24 h. Glutamine synthetase was measured as biosynthetic activity; enzymic activities are expressed as nmoles product formed min^{-1} mg $protein^{-1}$; amino acid pools are expressed as nmoles μg chl a^{-1}. Full details of enzymic assays and the extraction and measurement of the amino acid pools are given in Rowell *et al.* (1977).

is important, and of the amino acids which we measured, glutamine and aspartate are possible candidates, although other molecules are not ruled out. It is of interest that Shanmugam *et al.* (1976) also noted an inhibition of nitrogenase activity by amino acids.

Nitrogenase activity and glutamine synthetase in symbiotic N_2-fixing algae

We have recently extended our study area from *Anabaena cylindrica* to algae growing symbiotically in association with liverworts (Rodgers and Stewart, 1977; Stewart and Rodgers, 1977) and lichens. In such symbioses the algae fix N_2 efficiently but scarcely grow, and transfer most of the nitrogen fixed to the eukaryotic partner (Fig. 6). We have further shown that in

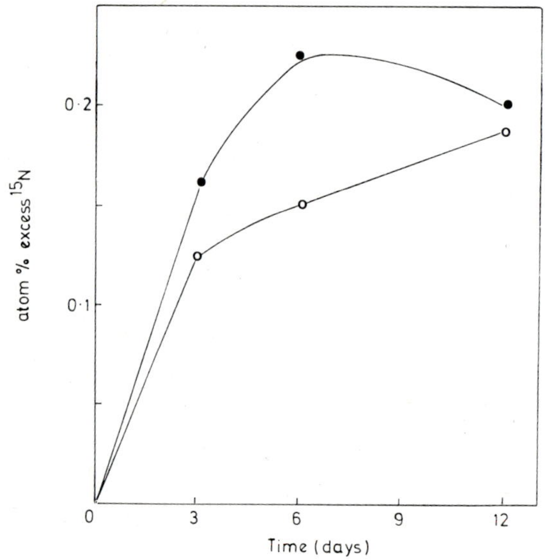

Fig. 6. *The incorporation of ^{15}N-label into gametophytes (● - ●) and sporophytes (○ - ○) of the liverwort* Anthocerus punctatus *after exposure of the thalli to $^{15}N_2$ (after Stewart and Rodgers, 1977).*

Blasia pusilla most of the nitrogen fixed by the alga is excreted as NH_3, but that on isolating the alga free from the host, NH_3 release ceases within 72 h and the alga begins to grow normally as a free-living organism. That is, the control mechanism which causes a virtual cessation of algal growth, and which causes NH_3 excretion by the alga in the symbiotic state is lost when the alga is separated from its eukaryotic host.

In studies on the lichen *Peltigera canina* we have used another approach to investigate the extracellular products liberated by the symbiotic algae. First, experiments were carried out in which lichen discs were incubated under argon or N_2 in the presence and absence of digitonin (0.01%, w/v) which affects the ability of the fungal partner to assimilate various compounds (Chambers *et al.*, 1976). As Fig. 7 shows, NH_3 excretion occurred under N_2 but to a much lesser extent under argon and only when digitonin was present, indicating that NH_3 excretion is largely dependent on N_2 reduction. Second, experiments were carried out in which the lichen discs were incubated under $^{15}N_2$

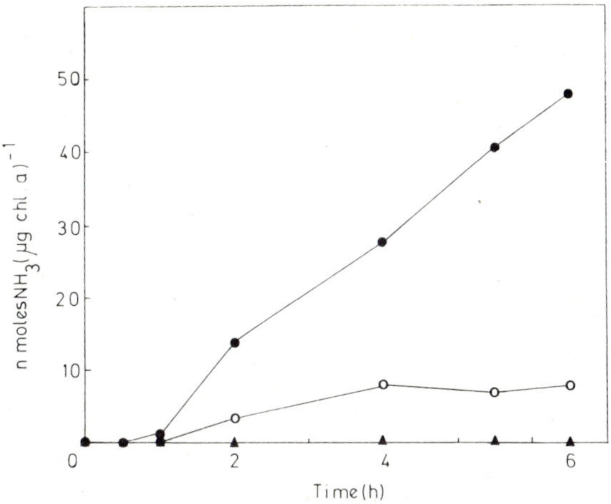

Fig. 7. *The production of extracellular NH_3 by discs of the lichen* Peltigera canina *on treatment of the discs with 0.01% (w/v) digitonin at 0 time and incubation under air or $A/O_2/CO_2$ (-7.96/22.00/0.04, v/v) at 3000 lux and 25°C.* •–•, *air + digitonin;* ○–○ , *$A/O_2/CO_2$ + digitonin;* ▲–▲, *air without digitonin.*

in the presence and absence of digitonin, and the labelling of the extracellular products with ^{15}N noted. As Table VI shows, in such experiments 50% of the total extracellular ^{15}N produced was present as NH_3, the remaining 50% being present as organic nitrogen. Thus the *Peltigera canina* data support the finding with *Blasia pusilla* in showing that NH_3 is a major form in which newly fixed nitrogen is transferred from the N_2-fixing algae. Although it seems likely that the extracellular organic nitrogen noted in the medium is due to leakage from the fungus on the addition of digitonin, the possibility cannot be entirely ruled out that in *Peltigera*, but not in *Blasia*, an excretion of organic compounds may also occur. The pathway of transfer of

Table VI

The distribution of ^{15}N label upon exposure of *Peltigera canina* discs to $^{15}N_2$ in the presence of digitonin

Sample	mg N	Atom % excess ^{15}N	% of total ^{15}N fixed
Discs	70	0.022	50
Extracellular organic N	3.9	0.205	25
Extracellular NH_3	1.1	0.727	25

Digitonin was used at a concentration of 0.01% (w/v); the exposure period to $^{15}N_2$ was 21 h; the ^{15}N label of the gas phase was 44 atom % ^{15}N.

fixed nitrogen from the alga to the fungus is uncertain, but it may be noted that under the electron microscope the heterocyst walls of the symbiotic algae appear irregular in density, perhaps indicating that such walls are "leaky" and these may thus facilitate the excretion of nitrogen.

Table VII shows the comparative results obtained on examining the symbiotic and free-living *Nostoc* and other components

Table VII

A comparison of the activities of various NH_3-assimilating enzymes in intact thalli, symbiotic *Nostoc* and free-living *Nostoc* from *Peltigera canina*

Enzyme	Intact thallus	Symbiotic alga	Free-living *Nostoc*
Glutamine synthetase	1	4	72
Glutamic dehydrogenase	202	2	n.d.
Alanine dehydrogenase	4	6	23
Glutamate-aspartate aminotransferase	41	17	40

The symbiotic alga was separated from the fungus using the method of Millbank. The free-living *Nostoc* was obtained from Dr. J.W. Millbank. Enzymic activities are expressed as n moles product formed mg protein^{-1} min^{-1}. GS (biosynthetic assay) was measured according to Kingdon *et al.* 1968); glutamate-aspartate aminotransferase and alanine dehydrogenase according to Rowell and Stewart (1976) and glutamate dehydrogenase was measured by following the oxidation of NADPH at 340 nm in a reaction mixture of 3.0 ml which contained: 300 μmoles Tris/HCl buffer, pH 7.5; 0.3 μmoles NADPH; 10 μmoles 2-oxoglutarate and 400 μmoles NH_4Cl; n.d., non-detectable.

of the *P. canina* symbiosis for GS, glutamic dehydrogenase, alanine dehydrogenase, and glutamate-aspartate aminotransferase activities. When the *Nostoc* phycobiont was grown free from the lichen it had high GS activity, undetectable glutamic dehydrogenase activity, and good glutamate-aspartate aminotransferase and alanine dehydrogenase activities. In the symbiotic state, on the other hand, GS activity decreased by 94%, while glutamate-aspartate aminotransferase and alanine dehydrogenase activities decreased by 50 to 60% and glutamate dehydrogenase activity actually increased slightly. In the intact thallus, on the other hand, which is mainly fungal hyphae, the major NH_3-assimilating enzyme is glutamic dehydrogenase, and there was negligible GS activity. Thus in the symbiotic state, but not in the free-living state, there is a very substantial decrease in the activity of the key algal NH_3-assimilating enzyme GS, and in the absence of effective levels of glutamic dehydrogenase, this appears to result in little growth of the alga in the symbiotic state, and excretion into the surroundings of much of the newly fixed NH_3. On carrying out histological studies in which sections of *Peltigera canina* were provided with the glutamic dehydrogenase substrates, NADPH (1mg ml^{-1}) and glutamate (20mg ml^{-1}), as well as blue tetrazolium (British Drug Houses, 0.25 mg ml^{-1}) the fungal hyphae adjacent to the algal layer showed an intense reduction of the tetrazolium salts, which was dependent on the addition of these substrates indicative of high glutamate dehydrogenase activity (Fig. 8). Thus the extracellularly produced NH_3 could be rapidly assimilated by the fungal hyphae adjacent to the algae.

These results thus show clearly that in the symbiotic state a major modification of the alga is its decreased GS activity, NH_3 excretion into the medium and sustained, or in the case of *Peltigera aphthosa* and *Blasia pusilla* increased, nitrogenase activity. The effect obtained is similar to that obtained on adding the analogue methionine sulphoximine to the free-living N_2-fixing *Anabaena cylindrica* but whether this is due to the provision by the fungus of an analogue type compound which modifies GS activity or whether it is due to internal regulation within the symbiotic algae as a result of the presence of the fungus is equivocal at present.

The ecological role of N_2-fixing algae

We showed above that in N_2-fixing blue-green algae, photosystem II was unnecessary for nitrogenase activity and that its

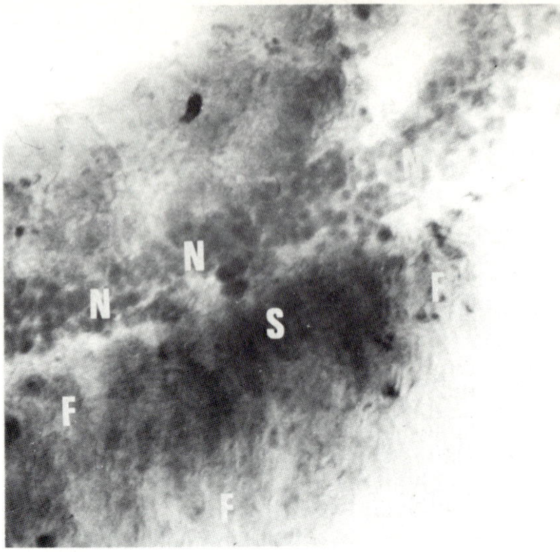

Fig. 8. *Tetrazolium blue reduction observed on treating a section of* Peltigera canina *with tetrazolium blue, and the glutamic dehydrogenase substrates NADPH and glutamate. Note that the intensity of staining (S) is greatest in the fungal hyphae (F) close to the* Nostoc *layer (N).*

role was simply to provide fixed carbon, that ATP could be provided by photophosphorylation, that ferredoxins associated with N_2 fixation were similar to those of higher plant chloroplasts, that a protective mechanism against O_2 inactivation was required, that newly fixed NH_3 was assimilated via GS and, if not so assimilated, was excreted into the medium. Thus the blue-green algae offer an interesting challenge as sources of fixed nitrogen from inorganic salts and water in the presence of sunlight and we have evidence that NH_3 liberated extracellularly by algae sustains a rapid germination and development of rice (Fig. 9). While the production of mutants capable of growth and NH_3 excretion is thus of considerable interest, there is increasing evidence that in nature the blue-green algae already provide an important source of fixed nitrogen in a variety of habitats. First, since the discovery by Stewart and Lex (1970) that the non-heterocystous filamentous *Plectonema boryanum* fixed N_2 anaerobically, it has been realised that this group of organisms may be an important hitherto unrecognised source of fixed nitrogen. The data in Table VIII provide information on some of the species which we have shown to reduce C_2H_2 under anaerobic but not aerobic conditions and it is likely that this list will be expanded considerably in the near future. Thus in habitats

Fig. 9. *The germination of rice (*Oryzae sativa *var* Ratna*) seedlings on incubating the seeds in: 1) a filtrate of* Anabaena cylindrica *which had been treated with methionine sulphoximine (1.0 μM) to cause the release of extracellular NH_3. The NH_3-containing filtrate was removed from the alga once the latter had removed all the methionine sulphoximine from the solution; 2) algal culture medium to which methionine sulphoximine (1.0 μM) but no algae had been added; 3) filtrate from a culture of* Anabaena cylindrica *which had not been treated with methionine sulphoximine. All photographs were taken 3 weeks after the addition of the different solutions. The total nitrogen contents of the algal filtrates with and without added methionine sulphoximine were similar. (Photograph by courtesy of K. Dadhich).*

such as salt marshes, rice paddy soils, mangrove swamps, etc. where de-oxygenated conditions occur and where non-heterocystous algae often predominate, N_2-fixation by these algae may be particularly important. Indeed we have already obtained evidence

Table VIII

Nitrogenase activity by various non-heterocystous filamentous blue-green algae

Alga	n moles C_2H_4/ sample/30 min
Plectonema boryanum 1462/2	375
Plectonema boryanum 1463/2	159
Plectonema boryanum 1463/1	3750
Plectonema boryanum 1462/4	1140
Plectonema boryanum 1446/3	570
Plectonema boryanum 581	192
Plectonema boryanum 790	27
Plectonema boryanum 597	279
Plectonema boryanum 594	246
Lyngbya sp. 1459/2	450
Phormidium foveolarium 1462/1	720
Oscillatoria tenuis 1459/4	4350
Anacystis nidulans	0

The algae had been grown aerobically on combined nitrogen (0.5 g l^{-1} of $NaNO_3$) in batch culture and during log-phase growth were washed free of combined nitrogen, and resuspended in medium free of combined nitrogen and gassed with A/CO_2 (99.96/0.04, v/v) for 3 days at 2000 lux and 25°C prior to assay for C_2H_2 reduction under anaerobic conditions.

of their possible significance in Scottish salt marshes (Stewart, 1975b). Second, it is becoming appreciated in the Western World, as it has long been appreciated in China, for example, that N_2-fixation by symbiotic *Anabaena* in association with *Azolla* (Fig. 10) may be particularly important (see Becking, 1975; Stewart, 1977). Third, free-living and symbiotic blue-green algae appear to be particularly important sources of fixed nitrogen in tundra ecosystems (see Stewart, 1975). Fourth, N_2-fixation by blue-green algae, particularly *Trichodesmium* (Dugdale *et al.*, 1961; Carpenter and Price, 1976; Gundersen *et al.*, 1976) and by *Richelia* endophytic in the diatom *Rhizosolenia* (Mague *et al.*, 1974) may be the major N_2-fixing agents in oceanic waters, but as yet we have little quantitative data to assess their importance on a global basis.

ACKNOWLEDGEMENTS

This work was made possible through S.R.C. and N.E.R.C. support to W.D.P.S. S.K.A. thanks the International Atomic

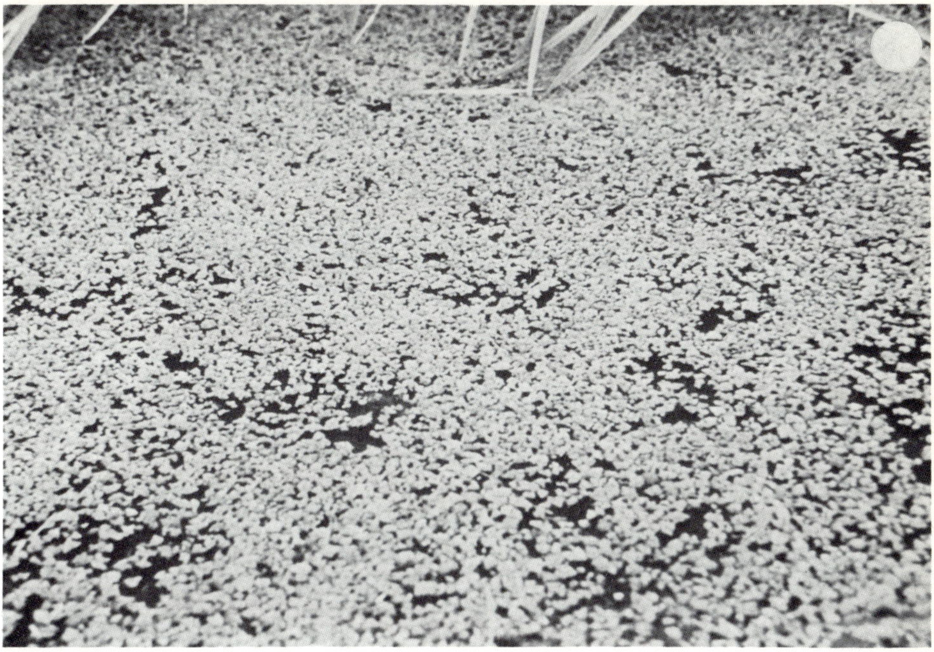

Fig. 10. Azolla *containing a symbiotic* Nostoc *growing on the water surface in an African rice paddy.*

Energy Agency for support. We thank Miss Gail Alexander for expert technical assistance, and Dr. M. Davies for the data in Table VIII.

REFERENCES

Becking, J.H. (1975). *In:* Symposium on Dinitrogen Fixation (W.E. Newton and C.J. Nyman, eds), pp. 556-561. Washington University Press, Pullman.

Bothe, H. (1970). *Ber.deut.bot.Ges.*, 83: 421-432.

Carpenter, E.J. and Price, C.C. (1976). *Science*, 191: 1278-1280.

Carr, N.G. and Bradley, S. (1973). *Symp.Soc.Gen.Microbiol.*, 23: 161-188.

Chambers, S., Morris, M. and Smith, D.C. (1976). *New Phytol.*, 76: 485-500.

Cheniae, G.M. (1970). *Ann.Rev.Plant Physiol.*, 21: 467-498.

Codd, G.A., Rowell, P. and Stewart, W.D.P. (1974). *Biochem. Biophys.Res.Commun.*, 61: 424-431.

Codd, G.A. and Stewart, W.D.P. (1976). *Planta (Berl.)*, 130: 323-326.

Dharmawardene, M.W.N., Haystead, A. and Stewart, W.D.P. (1973). *Arch.Microbiol.*, 90: 281-295.

Dugdale, R.C., Menzel, D.W. and Ryther, J.H. (1961). *Deep-Sea Research*, 7: 298-300.
Gundersen, K.R., Corbin, J.S., Hanson, C.L., Hanson, M.L., Hanson, R.B., Russel, D.J., Stollar, A. and Yamada, O. (1976). *Pacific Science*, 30: 45-68.
Kingdon, H.S., Hubbard, J.S. and Stadtman, E.R. (1968). *Biochemistry*, 7: 2136-2142.
Lex, M. and Carr, N.G. (1974). *Arch.Microbiol.*, 101: 161-167.
Rodgers, G.A. and Stewart, W.D.P. (1977). *New Phytol.*, in press.
Rowell, P., Enticott, S. and Stewart, W.D.P. (1977). *New Phytol.*, submitted.
Rowell, P. and Stewart, W.D.P. (1976). *Arch.Microbiol.*, 107: 115-124.
Scott, W.E. and Fay, P. (1972). *Brit.Phycol.J.*, 7: 283-284.
Shanmugam, K.T. and Morandi, C. (1976). *Biochem.Biophys.Acta*, 437: 322-332.
Shanmugam, K.T., Streicher, S.L., Morandi, C., Ausubel, F., Goldberg, R.B. and Valentine, R.C. (1975). *In:* Symposium on Dinitrogen Fixation (W.E. Newton and C.J. Nyman, eds), pp. 313-320. Washington University Press, Pullman.
Stewart, W.D.P. (1975a)(ed.). *Nitrogen fixation by free-living micro-organisms.* International Biological Programme, Vol. No. 6, Cambridge University Press, Cambridge.
Stewart, W.D.P. (1975b). *Proc.9th Europ.mar.biol.Symp.*, 667-660.
Stewart, W.D.P. (1976). *In:* A Treatise on Dinitrogen Fixation. Vol. 2, Wiley-Interscience, New York.
Stewart, W.D.P. and Codd, G.A. (1975). *Br.phycol.J.*, 10: 273-278.
Stewart, W.D.P., Haystead, A. and Dharmawardene, M.W.N. (1975). *In:* Nitrogen Fixation by Free-Living Micro-Organisms, IBP, Vol. 6, (W.D.P. Stewart, ed.), pp. 129-158. Cambridge University Press, Cambridge.
Stewart, W.D.P., Haystead, A. and Pearson, H.W. (1969). *Nature (Lond.)*, 224: 226-228.
Stewart, W.D.P. and Lex, M. (1970). *Arch.Microbiol.*, 73: 250-260.
Stewart, W.D.P. and Rodgers, G.A. (1977). *New Phytol.*, in press.
Stewart, W.D.P. and Rowell, P. (1975). *Biochem.Biophys.Res. Commun.*, 65: 846-856.
Stewart, W.D.P., Rowell, P. and Tel-Or, E. (1975). *Biochem.Soc. Trans.*, 3: 357-362.
Streicher, S.L., Shanmugam, K.T., Ausubel, F., Morandi, C. and Goldberg, R.B. (1974). *J.Bacteriol.*, 120: 815-821.
Tel-Or, E. and Stewart, W.D.P. (1975). *Nature(Lond.)*, 258: 715-716.
Tel-Or, E. and Stewart, W.D.P. (1975). *Biochem.Biophys.Acta*, 423: 189-195.

Tel-Or, E. and Stewart, W.D.P. (1977). *Proc.Roy.Soc.Lond.B*, submitted.

Thomas, J. (1970). *Nature(Lond.)*, 228: 181-183.

Thomas, J., Wolk, C.P. and Shaffer, P.W. (1975). *Biochem.Biophys.Res.Commun.*, 67: 501-507.

Tronick, S.R., Ciardi, J.E. and Stadtman, E.R. (1975). *J.Bacteriol.*, 115: 858-868.

Tubb, R.S. (1974). *Nature(Lond.)*, 251: 481-485.

Winkenbach, F. and Wolk, C.P. (1973). *Plant Physiol.*, 52: 480-483.

Wolk, C.P. (1968). *J.Bacteriol.*, 96: 2138-2143.

NITROGENASE IN CHEMOSTAT CULTURES OF RHIZOBIA

F.J. BERGERSEN

Division of Plant Industry, CSIRO, Canberra, Australia

INTRODUCTION

Recent work on nitrogenase in agar cultures (Kurz and La-Rue, 1975; McComb *et al.*, 1975; Pagan *et al.*, 1975) shows that the old belief that rhizobia fix nitrogen only within the tissues of the host plant, is no longer tenable. Nor is the symbiosis with rhizobia absolutely confined to the Leguminosae (Trinick, 1973). These observations open up new approaches to understanding the legume nodule system but they also pose important questions about the general applicability of the findings. For example, does the existence of nodules on *Trema cannabina* indicate that nodule systems involving rhizobia may exist more widely among non-legumes or be able to be developed artificially? Again, is the strain 32H1, which is so convenient for laboratory studies of nitrogenase in rhizobia, typical of rhizobia generally or is it an example of the constitutive mutants predicted by Dilworth and Parker (1969)? Perhaps rhizobia of this type were not found previously because inappropriate means were used to detect them. It is too early to do more than guess at the answers to these questions but the ability to grow cultures of rhizobia with active nitrogenase has provided the opportunity to study some aspects of their physiology.

In this work, we have concentrated initially on understanding the effects of conditions of oxygen supply. It was convenient to do this because we had just completed a study with soybean nodule bacteroids, relating to the role of leghaemoglobin in nodules. Some of the results of this study were reported at the first International Symposium on Nitrogen Fixation (Appleby *et al.*, 1976). It was shown that bacteroids must receive an adequate flux of oxygen but with the supply concentration in the range 0.01-0.1 μM. In this low range, high rates of ATP production were sustained efficiently by a high-affinity oxidase system, thus maintaining optimum nitrogenase activity. At higher concentrations of free dissolved oxygen,

ATP was produced only by a less efficient system. The presence of partially oxygenated leghaemoglobin in the surrounding solution ensured that optimum free oxygen concentrations were sustained (Appleby *et al.*, 1976; Bergersen and Turner, 1975a,b; Wittenberg *et al.*, 1974). We have utilized these results in developing convenient and reliable assays for nitrogenase in cultured rhizobia.

Whilst I was on study leave at the A.R.C. Unit of Nitrogen Fixation, University of Sussex, in 1973, Professor J.R. Postgate persuaded me to attempt to induce active nitrogenase in a chemostat culture of rhizobia. We failed to detect any nitrogenase in a glutamate-limited culture of *Rhizobium japonicum* CB1809, but on looking back, with our present knowledge, one steady state grown at low dissolved oxygen concentration undoubtedly had a little nitrogenase activity. We did not know how to run assays which would detect it.

The work on agar cultures with active nitrogenase cited above, opened the way for the experiments of Tjepkema and Evans (1975) and Keister (1975), who concluded that low oxygen concentrations were essential for the detection of nitrogenase in rhizobia. The studies with continuous cultures of rhizobia which are outlined in this paper confirm and amplify their results. Some of our results have already been reported (Bergersen *et al.*, 1976). Newer and current results are also described.

CULTURES

Chemostat

The apparatus was similar to that described by Baker (1968). The culture volume was 600 ml. It was equipped with pH control and a probe for measuring dissolved oxygen concentrations (Borkowski and Johnson, 1967). The culture was stirred by a teflon coated magnet bar at the highest rate which did not lead to gas entrappment. The head space was supplied with air or admixtures of air and nitrogen, at measurable rates. In addition, oxygen could be supplied from a submerged teflon tube (1 mm bore, 0.5 mm wall, 6 m long) wound as a coil on stainless steel baffles. When filled with an atmosphere of oxygen, the coil passed 600 nmoles per min of oxygen into sterile, deoxygenated medium. The maximum total flux of oxygen supply into 600 ml of sterile medium containing 1 μM dissolved oxygen was about 2 μmoles per min (dissolved from the head space + diffusion from the coil + oxygen dissolved in inflowing medium). It would be higher when culture was present and estimates from the respiration rates of samples from continuous cultures indicate about 6 μmoles of oxygen could be supplied per minute when the steady state dissolved oxygen was about 1 M. Samples of culture for analysis could be withdrawn through a hypodermic needle attached to a covered tube, into evacuated, rubber capped bottles. Culture effluent could be collected in ice-cooled receivers within a few seconds of overflow from the culture vessel. In most cases, the temperature of the culture was 28°C.

The Bacteria

The cowpea rhizobia (*Rhizobium* sp.) strains 32H1 and CB 756 and *R. japonicum* strain CB1809 have been used in this work, but most experiments have used 32H1. Stock cultures were maintained on yeast extract-mannitol agar. Continuous cultures were checked for purity and authenticated by serological and symbiotic tests at intervals throughout. Contamination was encountered only once at the end of a prolonged continuous culture. The work has involved nine separate cultures of 32H1, totalling almost a year of continuous culture of this strain.

Media

The basic medium was 30 mM KH_2PO_4 made to pH 6.25 with Na OH, containing 0.14 mM $MgCl_2$, 0.7 mM $CaCl_2$ and a mixture of trace elements. The nitrogen source was usually 1 to 2 mM glutamine; glycerol, inositol, arabinose or gluconate separately at 0.5% were used as major carbon sources in various experiments. Sodium succinate (5 to 25 mM) was necessary in addition. for the maintenance of low concentrations of dissolved oxygen. Preliminary batch cultures established that the best exponential growth rates were obtained at pH 6.2 to 6.8 and that the strains used had no growth factor requirements. The use of sodium gluconate, a cheap carbon source, was in many ways satisfactory in continuous cultures of rhizobia, considering the large volumes of medium used. However, the accumulating cations although presenting no pH problems when a controller was used, caused more rapid fouling of the apparatus due to the production of insoluble salts of the acidic 32H1 polysaccharide (Dudman, personal communication). In contrast, glycerol, supplemented with succinate, was a satisfactory carbon source and seemed to assist in dispersing the polysaccharide, thus reducing fouling problems. Arabinose, an excellent carbon source for 32H1 and CB756, was seldom used in continuous cultures because of its relatively high cost.

NITROGENASE ASSAYS

Preliminary experiments with suspensions prepared anaerobically from agar cultures of 32H1 and CB756 with active nitrogenase, established that, like soybean bacteroids, cultured rhizobia had greatest nitrogenase activity during deoxygenation of added oxyleghaemoglobin (*cf.* Bergersen and Turner, 1975a,b). We therefore developed a standard assay, based on that used by Wittenberg *et al.* (1974), in which samples of culture (1.75 ml) were shaken with 0.25 ml of 2 mM oxyleghaemoglobin, purified from soybean nodules. Rubber capped vials were used, containing a gas phase of 10% acetylene and 0.02 to 0.5% oxygen (according to the culture density, which affected oxygen demand); the balance was argon. The presence of leghaemoglobin stimulated nitrogenase activities about five-fold compared with assays from which it was omitted (Bergersen *et al.*, 1976). With 0.15 to 0.3 mg (dry wt) of bacteria per assay, rates were linear for at least 1 h. Throughout the work described in this report, activities in continuous cultures are given for such assays. The results thus indicate maximum or potential activities and do not necessarily indicate rates actually occurring in the culture. The preliminary results referred to above, with suspensions from agar cultures and added leghaemoglobin, have recently been confirmed with samples from continuous cultures. Increments of nitrogenase activity were measured in optical cuvettes in which consumption of oxygen from oxyleghaemoglobin was being monitored spectrophotometrically, as described in the experiments with no gas phase used for soybean bacteroids (Bergersen and Turner, 1975a). The results (Fig. 1) clearly show that accelerated nitrogenase activity occurs when the free oxygen concentration is between 0.1 and 0.01 μM, i.e. during deoxygenation of leghaemoglobin, and declines as the rate of oxygen consumption declines with exhaustion of oxygen. The agreement between measurements, in different cuvettes during the reactions validates the use of the nitrogenase increments to describe the total time course, and the relationship between free oxygen concentration and nitrogenase rate. The maximum nitrogenase rate agreed well with rates obtained in shaken standard assays done at the same time.

RESULTS FROM CONTINUOUS CULTURES

Glutamine-Limited Cultures

Nitrogenase activity was first detected in samples of a glutamine-limited continuous culture of 32H1, following reduction of the dissolved oxygen concentration to near 1 μM (Table I). The increase in yield, following reduction of the oxygen concentration (Table I) was often observed with 1 mM glutamine in the medium supply. It is possible that this was due to nitrogen fixation in the culture, but activities measured do not seem to be high enough for this and no analysis of nitrogen balance was made. It seems more likely that the increase was

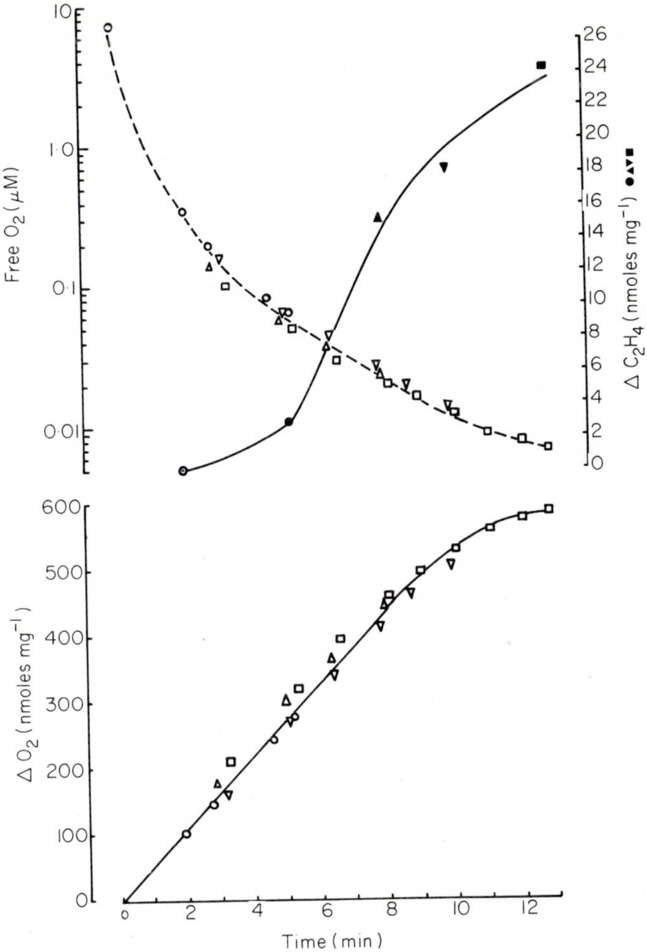

Fig. 1. *Respiration, nitrogenase activity and free oxygen concentrations in no gas phase reactions with 32H1 from oxygen-limited continuous culture. The reactions were as described in Bergersen and Turner (1975a) for reactions in optical cuvettes, using the oxygenation of leghaemoglobin (determined spectrophotometrically) to describe both free oxygen concentration and oxygen consumption. The reaction mixture was 4.25 ml of culture (0.16 mg/ml) and 0.25 ml of 2 mM oxyleghaemoglobin. Nitrogenase activity was measured as increments of acetylene reduced after a sample taken at 2 min. Symbols of like shape identify individual reactions.*

Table I

Characteristics of a glutamine-limited continuous culture of 32H1. The dilution rate (D) was 0.07 h^{-1}; the culture volume was 600 ml. Carbon sources were glycerol (0.5%), plus inositol (0.1%) and Na-succinate (25 mM)

Glutamine supplied	(mM) residual	Dissolved O_2 (μM)	Yield (mg/h)[a]	Nitrogenase (nmoles C_2H_4 /mg/h)[a]
2.00	0.04	39	6.3	0
1.00	0.05	34	3.6	0
1.00	0.03	0.9	4.8	28

due to the more efficient production of ATP at low concentrations of dissolved oxygen (Postgate, 1975; Bergersen and Turner, 1975b).

Nitrogenase activities up to about 100 nmoles ethylene per mg (dry wt) per h, were obtained with glutamine-limited continuous cultures.

When the supply of succinate was discontinued, leaving glycerol as the sole carbon source, good nitrogenase activity persisted until the residual succinate was exhausted. Dissolved oxygen then rose, over about 5 h, to 24 μM and all nitrogenase activity was lost. Upon restoration of the succinate supply, dissolved oxygen quickly fell to a low concentration (0.3 μM), but nitrogenase was restored to its former level only after about 50 h (Fig. 2). This culture was growing at a dilution rate of 0.07 h^{-1}, i.e. the generation time was about 14 h. Nitrogenase activity doubled in about 7.5 h. The relatively brief exposure to 24 μM oxygen (about 10% of air saturated medium) was apparently sufficient to destroy all nitrogenase. Recovery of activity was apparently the result of nitrogenase synthesis. At no time in our cultures of rhizobia has reversible "switch off" of nitrogenase been observed as reported for *Azotobacter chroococcum* (Postgate, 1975) in response to exposure of oxygen.

Oxygen-Limited Cultures

Inactivation by oxygen versus synthesis of nitrogenase Higher nitrogenase activities have been obtained with oxygen-limited continuous cultures of 32H1. In these, the medium contained glycerol (0.5%), succinate (5 mM) and glutamine (2 mM). The time course of one such culture is shown in Fig. 3. We conclude

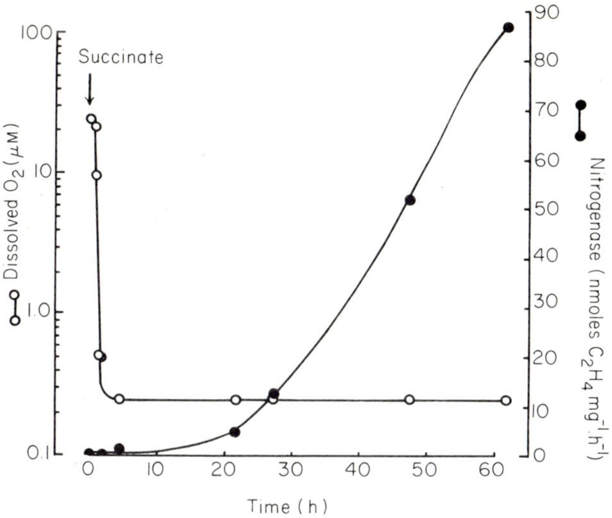

Fig. 2. *The time course of restoration of nitrogenase activity. A glutamine-limited culture of 32H1 was briefly exposed to 24 μM O_2, following exhaustion of succinate. Supply of succinate was restored at zero time. All culture parameters were kept constant throughout and the cell density was constant (0.1 mg/ml) after 21 h, having fallen slightly following exhaustion of succinate. Residual glutamine was less than 0.1 mM throughout.*

that these cultures were oxygen-limited because the changes recorded took place in response to changes in oxygen supply, all other conditions being held constant. There was no pH control in this experiment, but the pH was 6.6 to 6.7 until the onset of high oxygenation rates when it rose to 6.9 (day 28). The relationship between culture yield and oxygen concentration in the culture is not direct because of effects on efficiency of carbon utilization and because the rate of oxygen supply (oxygen solution rate as distinct from steady-state concentration in the culture) was not readily determinable. Residual glutamine in these cultures was higher (0.8 to 1.2 mM) than in glutamine-limited cultures (Table I).

The steady state concentration of dissolved oxygen again appeared to be the dominant factor in control of nitrogenase activity (Fig. 3). At low levels (1 μM O_2 or less) synthesis of nitrogenase exceeds inactivation and rates of activity near 300 nmoles C_2H_4/mg per h can be sustained almost indefinitely. However, at 10 μM O_2, the rate of inactivation apparently just exceeded the rate of synthesis and activity declined more slowly than expected from wash-out.

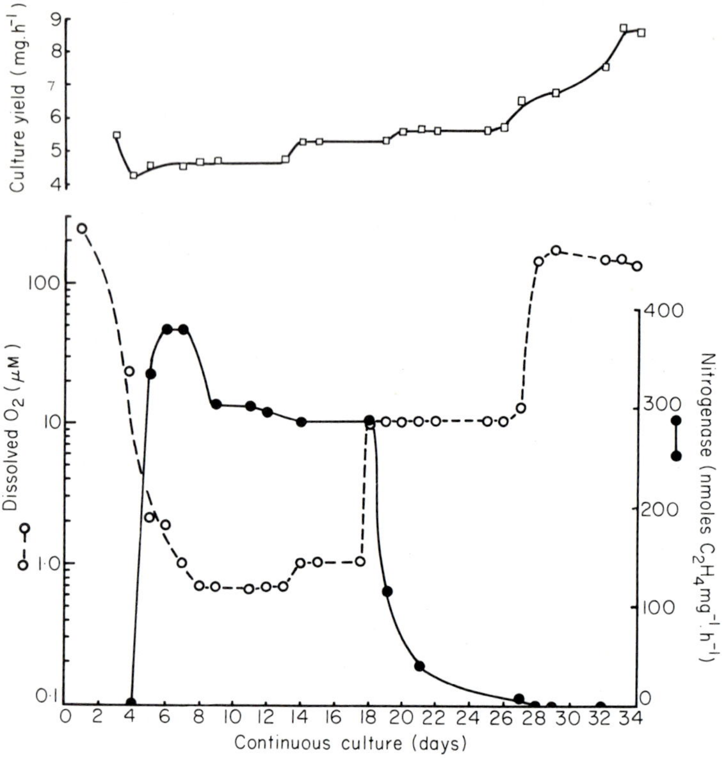

Fig. 3. *The time course of an oxygen-limited continuous culture of 32H1. Small changes in dissolved oxygen in the culture were the result of changes in the head gas; the final high concentration was the result of air bubbling and increased stirring. Residual glutamine was 0.8 - 1.2 mM. D = 0.05h*$^{-1}$.

One oxygen-limited culture ran for four weeks with a yield of 4.8 to 6.0 mg (dry wt) per h and nitrogenase activity of 200 to 296 nmoles C_2H_4 mg per h. The cells were used in a study of the high affinity oxidase systems of 32H1, which will be described elsewhere.

Can 32H1 fix sufficient N_2 for growth During one oxygen-limited continuous culture of 32H1, (D=0.05 h^{-1}) when a steady nitrogenase activity of 250 nmoles C_2H_4/mg per h was established, the glutamine supply was withdrawn to see if the nitrogenase activity would increase to the point of providing sufficient nitrogen to sustain growth. Sixteen hours later, the culture density was unchanged and nitrogenase had increased to 309 nmoles C_2H_4/mg per h. After 22 h, the culture density and activity were unchanged, but the next morning (39 h after withdrawal of glutamine supply) the culture was washing out, the oxygen demand had declined and dissolved oxygen was 102 μM; there was no nitrogenase activity remaining. Apparently, during the first 22 h, residual glutamine was assimilated from the medium and then the intracellular nitrogen pool was depleted to the point when growth rate declined, thus allowing the oxygen concentration to increase.

We have not been successful to date in sustaining actively growing cultures of rhizobia without added combined nitrogen, although Gibson (private communication) has been able to sustain periods of active nitrogenase in agar cultures grown on millipore discs and transferred to N-free agar when nitrogenase was fully developed. The reasons for failure with continuous cultures can be seen from the following considerations:

a) The cell-nitrogen yield of the above culture was 592 μg N per h (21.2 moles N_2 per h).

b) The maximum activity recorded in standard assays of culture samples was 300 nmoles C_2H_4 mg per h. Assuming a 3:1 relationship between this measurement and N_2 fixation, and that rates in the assays applied in the culture, which contained 600 ml at 0.18 mg per ml, this activity is equivalent to 11 moles N_2 per h. This is only about one half of the nitrogen yield of the culture.

An analysis of the medium supplied to and effluent from an oxygen limited culture is given in Table II. During the period

Table II

Analysis of total nitrogen in medium supply and effluent from an oxygen-limited continuous culture of 32H1[a]

Sample	Total N (μg/ml)[b]
medium	50.1 ± 0.5
effluent	56.9 ± 0.6
N_2 fixed	6.8 ± 1.1
washed cells	26.2 ± 0.3
supernatant (effluent)	30.8 ± 3.3

[a]The culture was supplied with medium containing 1.8 mM glutamine, 5 mM Na-succinate, 54 mM glycerol, at a rate of 32.7 ml/h (D= 0.054 h^{-1}). The effluent contained 0.23 mg (dry wt washed cells)/ml. Nitrogenase activities in standard assays with oxyleghaemoglobin were 201±15 nmoles C_2H_4/mg per h before and after effluent collection.

[b]Means and errors for triplicate samples of 10 ml.

in which the effluent was collected, the nitrogenase activity was steady at 202 nmoles C_2H_4 mg per h. Using the same assumptions as before, this would provide about 31% of the cell N. From Table II, apparent N_2 fixation was 25% of cell N. In this case then, the standard assay slightly over-estimated the fixation which was actually occurring in the culture.

Comparisons between Chemostat Cultures and Bacteroids

Several comparisons have been made between cells of 32H1 from continuous cultures with active nitrogenase and bacteroids prepared anaerobically (Bergersen and Turner, 1973) from nodules on Poona cowpeas inoculated with the same strain of rhizobia. Some results are listed in Table III. Although there

Table III

Comparison of 32H1 from a continuous culture with active nitrogenase and 32H1 bacteroids from cowpea nodules

Measurement		Chemostat	Nodules
Maximum nitrogenase (nmoles C_2H_4/mg/h)[a]		300	450
Respiration rate (nmoles O_2/mg/h)[b]	at 5 μM O_2	65	30
	at 1 μM O_2	65	18
	at 0.5 μM O_2	55	15
	at 0.1 μM O_2	30	12.5
Dry weight (g/cell)		2.9×10^{-13}	1.2×10^{-13}
Visible poly-β-hydroxy-butyrate[c]		none	+++
DNA[d]	(% dry wt)	4.4	2.3
	(g/cell)	1.3×10^{-14}	0.28×10^{-14}
RNA[e]	(% dry wt)	10.1	4.2
	(g/cell)	2.9×10^{-14}	0.5×10^{-14}

[a]Nitrogenase measured simultaneously in standard shaken assays with added leghaemoglobin.

[b]Measured in an oxygen electrode (Bergersen and Turner, 1975a); culture was used with no addition; washed bacteroids were supplied with 11 mM Na-succinate as substrate.

[c]Low contrast inclusions seen by phase contrast microscope and assumed by comparison with soybean bacteroids to be poly-β-hydroxy-butyrate.

[d]By diphenylamine method with thymus DNA standard.

[e]By orcinol method, yeast RNA standard.

are some points of resemblance between the two types of cells, most of the parameters which we have measured show distinct differences. Many of these are consistent with the differences to be expected from the contrast between exponentially growing bacteria and bacteria whose activities have been curtailed within the special environment of the host cells.

Nitrogenase Oxygen-limited cultures approached the activity of bacteroids on a cell dry weight basis, using the standard shaken assays with added oxyleghaemoglobin. On a cellular basis, chemostat cells were more active (Table III).

Respiration On a cell dry weight basis, rates of oxygen consumption were greater for samples of continuous cultures than for bacteroids supplied with succinate as substrate. Both types of cells possessed high-affinity oxidase systems which were active in the range of free dissolved oxygen which prevails during deoxygenation of oxyleghaemoglobin (*cf.* Fig. 1). Cells from continuous culture were saturated with respect to oxygen at about 1 μM while bacteroid respiration increased up to about 5 to 7 μM oxygen. The details of these comparisons will be published elsewhere.

DNA and RNA concentrations The diminished DNA and RNA content of bacteroids relative to chemostat cells (Table III) is in accord with the results of Dilworth and Williams (1967) who described the decline in these components in bacteria during nodule development in lupins, but our values are generally lower. Sutton (1974) reported similar values for DNA content but detected a diminished concentration only in bacteroids of a fast-growing *R. lupini* strain, relative to the DNA content of still and shaken broth cultures.

CONCLUSIONS

The results outlined in this paper indicate that the concentration of free dissolved oxygen is of great importance in the production and functioning of nitrogenase in the cultures of rhizobia which have been studied so far. However, rates of nitrogen fixation attained do not seem to be sufficient to sustain exponential growth on nitrogen-free media. If further work confirms this limitation, it introduces a possible new class of N_2-fixing micro-organisms, *viz.* those only able to supplement combined nitrogen for growth. Such organisms would not be detected by presently-used methods of isolating N_2-fixing micro-organisms from natural habitats.

ACKNOWLEDGEMENTS

It is a pleasure to record the participation of Mr. G.L. Turner in many of the experiments outlined in this report. Dr. E. A. Schwinghamer analyzed the DNA and RNA content of rhizobia recorded in Table III. Technical Assistance was provided by Harry Tantala, Judy Flanigan and latterly, Pat Atkinson.

REFERENCES

Appleby, C.A., Bergersen, F.J., Macnicol, P.K., Turner, G.L., Wittenberg, B.A. Wittenberg, J.B. (1976). *In:* Proceedings of the 1st International Symposium on Nitrogen Fixation (W.E. Newton and C.J. Nyman, eds), pp. 247-292. Washington State University Press.

Baker, K. (1968). *Lab. Practice,* 17: 817-824.

Bergersen, F.J. and Turner, G.L. (1973). *Biochem.J.,* 131: 61-75.

Bergersen, F.J. and Turner, G.L. (1975a). *J.Gen.Microbiol.,* 89: 31-47.

Bergersen, F.J. and Turner, G.L. (1975b). *J.Gen.Microbiol.,* 91: 345-354.

Bergersen, F.J., Turner, G.L., Gibson, A.H. and Dudman, W.F. (1976). *Biochim.Biophys.Acta,* 44: 164-174.

Borkowski, J.D. and Johnson, M.J. (1967). *Biotech.Bioeng.,* 9: 635-639.

Dilworth, M.J. and Williams, D.C. (1967). *J.Gen.Microbiol.,* 48: 31-36.

Dilworth, M.J. and Parker, C.A. (1969). *J.Theoret.Biol.,* 25: 208-218.

Keister, D.L. (1975). *J.Bacteriol.,* 123: 1265-1268.

Kurz, W.G.W. and LaRue, T.A. (1975). *Nature,* 256: 407-408.

McComb, J.A., Elliott, J. and Dilworth, M.J. (1975). *Nature,* 256: 409-410.

Pagan, J.D., Child, J.J., Scowcroft, W.R. and Gibson, A.H. (1975). *Nature,* 256: 406-407.

Postgate, J.R. (1975). *In:* The Biology of Nitrogen Fixation (A. Quispel, ed.), pp. 663-686. North Holland, Amsterdam.

Sutton, W.D. (1974). *Biochim.Biophys.Acta,* 366: 1-10.

Tjepkema, J.D. and Evans, H.J. (1975). *Biochem.Biophys. Res. Comm.,* 65: 625-628.

Trinick, M.J. (1973). *Nature,* 244: 459-460.

Wittenberg, J.B., Bergersen, F.J., Appleby, C.A. and Turner, G.L. *J.Biol.Chem.,* 249: 4057-4066.

GENETIC CONTROL OF NITROGEN FIXATION (Nif)

K. T. SHANMUGAM, F. O'GARA, K. ANDERSEN,
C. MORANDI and R. C. VALENTINE

Plant Growth Laboratory, Department of Agronomy and Range Science, University of California, Davis, California 95616, U.S.A.

INTRODUCTION

Nitrogen-fixing bacteria, such as *Klebsiella pneumoniae* have evolved an elaborate system for monitoring available nitrogen in their environment. In the presence of an available nitrogen source (e.g. NH_4^+, NO_3^-, amino acids etc.), nitrogenase biosynthesis is repressed; nitrogenase is synthesized when nitrogen is not available.

Much effort has been devoted to examining the possible role of glutamine synthetase, linked to NH_4^+ via an adenylylation enzyme cascade system (Shanmugam *et al.*, 1976a), as regulator of nitrogen fixation (Streicher *et al.*, 1974; Shanmugam, Chan and Morandi, 1975; Shanmugam and Morandi, 1976; Shanmugam, Morandi and Valentine, 1976; Tubb, 1974). The best evidence implicating glutamine synthetase in nitrogenase regulation is the finding that nitrogenase derepressed mutants (in the presence of excess NH_4^+) often occur as the result of structural alterations of glutamine synthetase.

Recent evidence (Shanmugam and Morandi, 1976) suggests that glutamine synthetase may represent only one element in a sophisticated control network involving amino acid in nitrogenase biosynthesis:

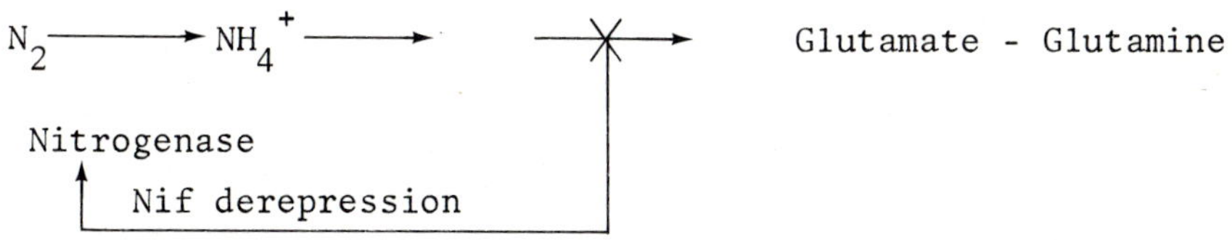

In this communication, some recent findings regarding the mechanism of Nif control in *Klebsiella* as well as possible applications of this work towards the regulation of nitrogen fixation and NH_4^+ assimilation in root nodule bacteria (*Rhizobia*) will be discussed.

Klebsiella pneumoniae

Ammonium ion, the product of the nitrogenase reaction is well known to be a powerful repressor of nitrogenase biosynthesis. Mutants of *K. pneumoniae*, which do not respond to NH_4^+, have been isolated as either glutamate or glutamine auxotrophs (see Shanmugam, Chan and Morandi, 1975,for a detailed description of these strains). These strains produce nitrogenase in the presence of NH_4^+ (15 mM) at levels as high as 100% compared to the control. However, Shanmugam and Morandi (1976) found that a mixture of amino acids (casamino acids) as low as 100 μg/ml in the culture medium repressed nitrogenase biosynthesis.

Table I

Repression of nitrogenase biosynthesis in *Klebsiella pneumoniae* by amino acids

Strain	Nitrogenase activity (μmoles/hr mg cell protein)					
	[Glutamine] (100 μg/ml)			[Glutamine] (1 mg/ml)		
	Glutamine alone	$+NH_4^+$	+L-Asp	Glutamine alone	$+NH_4^+$	+L-Asp
M5A1	1.92	0	0.71	0.27	0	0
Asm-1	1.75	0	0.03	1.77	0	0
SK-24	3.18	2.62	0.00	3.70	1.88	0
SK-26	3.53	2.97	0.73	1.99	1.61	0.21
SK-27	2.93	2.69	0.30	2.17	0.16	0.006
SK-55	3.56	2.71	0.28	2.04	0.67	0.01
SK-56	2.35	2.03	0.31	2.01	0.19	0.003

Concn. of NH_4^+ was 15 mM and L-Asp was 1 mg/ml. Nitrogenase activity was determined as described by Shanmugam, Loo and Valentine (1974).

In further experiments, they found that certain combinations of amino acids, such as L-glutamine and L-aspartate and in some cases NH_4^+ in combination with L-glutamine completely repressed nitrogenase biosynthesis (see Table I). In strains SK-24, SK-26, SK-27, SK-55 and SK-56, nitrogenase activity in the presence of NH_4^+, ranged from 76 to 92% of the control value when the glutamine concentration in the assay medium was 100 μg/ml. At this glutamine concentration, addition of 1 mg/ml of L-aspartate to the medium, completely repressed nitrogenase biosynthesis in SK-24 and in the parent Asm-1. A significant decrease in the levels of nitrogenase activity was also observed with other strains. In the presence of higher levels of glutamine (1 mg/ml), NH_4^+ acted as a strong repressor of nitrogenase biosynthesis in strains SK-27 and SK-56. Aspartate, in combinations with L-glutamine (1 mg/ml) completely repressed nitrogenase induction in strains SK-24, SK-27, SK-55 and SK-56. Glutamate (1 mg/ml) at both concentrations of glutamine did not have significant effect.

Under conditions in which nitrogenase biosynthesis was repressed by NH_4^+, NH_4^+ was assimilated to the level of glutamate either through glutamate dehydrogenase or glutamine synthetase and glutamate synthase. Furthermore, nitrogenase biosynthesis was always derepressed when the utilization of NH_4^+ by the cell was blocked.

The simplest interpretation of these results is that the overall mechanism of nitrogenase regulation involves the conversion of NH_4^+ to the level of amino acids and this conversion is a prerequisite for repression of nitrogenase biosynthesis by NH_4^+.

"Artificial Nodule"

The major function of bacteroids in root nodules of leguminous plants is believed to be to supply fixed nitrogen to the host plant. In confirmation of this, NH_4^+ produced from N_2 was shown by Bergersen and Turner (1967) to be excreted by isolated bacteroids. Shanmugam and Valentine (1975) described nitrogenase deprepressed mutants of *K. pneumoniae* which excrete fixed N_2 as NH_4^+. Using these mutants, we have developed a convenient model system to study the parameters affecting NH_4^+ excretion. Studies with the artificial nodule may therefore provide an insight into the mechanisms governing NH_4^+ excretion by the root nodule bacteria.

The Nif derepressed mutants are grown inside a dialysis bag which is suspended in a reservoir of medium, permitting fresh medium to diffuse in and products such as NH_4^+ to escape (Shanmugam *et al.*, 1976b). Using these growth conditions, the production of NH_4^+ by a variety of nitrogenase derepressed strains of *K. pneumoniae* have been tested (Table II). NH_4^+ was produced at a high rate (strain 37), and a final NH_4^+ level as high as 320 moles per mg cell protein was observed with strain SK-57. This higher value was accomplished primarily through selection of improved Nif derepressed strains and use of the dialysis culture technique (maximum NH_4^+ produced by strain SK-24 was almost doubled, from 50 to 92 μmoles/mg protein, by the use of the dialysis culture technique, Table II). We have no explanation for the considerable variation in NH_4^+ yields observed with the different strains. However, it is clear that strain selection is an important parameter in improving NH_4^+ yields. Since the strains used are either glutamate or glutamine auxotrophs, growth-limiting quantities of amino acids are added to the bacterial growth chamber at the beginning of the experiment to support growth. Usually, the cells exhaust their supply of exogenous amino acids during the first 1 to 1½ days of incubation after which time the cells are in an essentially non-growing state.

These studies are referred to in the title of this section as 'Artificial Nodule' since they might duplicate some of the regulatory events that take place during symbiotic nitrogen fixation. Some of the key requirements (considerations) for microbial production of ammonium ion from N_2 as determined from these studies are: i) derepression of the Nif genes; ii) repression of ammonia assimilating enzymes; iii) availability of large quantities of energy (not discussed here, see Streicher and Valentine, 1973; Shanmugam and Valentine, 1975 for more details); iv) considerable genetic variability has been observed among NH_4^+-exporting strains; the best strains in terms of longevity of export of fixed nitrogen appear to be those capable of maintaining a slow rate of protein synthesis.

Root Nodule Bacteria (Rhizobia)

The mechanism of genetic regulation of symbiotic nitrogen fixation has been difficult to study previously because of the complex interactions of the symbiont with its host plant. However, the discovery of nitrogenase biosynthesis by free-living *Rhizobia* has provided a fresh approach to this problem (Pagan *et al.*, 1975; Kurz and LaRue, 1975; McComb, Elliott and Dilworth, 1975; Tjepkema and Evans, 1975; Keister, 1975; see LaRue, Child and Kurz, 1976, for a historical perspective of this important discovery). O'Gara and Shanmugam (1976) found using ^{15}N analysis that free-living *Rhizobia* excrete a majority of their fixed nitrogen, presumably as NH_4^+. For example, cultures of free-living *Rhizobium japonicum* induced for nitrogenase exposed to gaseous $^{15}N_2$ showed relatively higher ^{15}N enrichment in the culture supernatants indicates that as much as 94% of the fixed nitrogen was exported by these cultures.

To test whether the excreted NH_4^+ derived intracellularly from an organic nitrogen source such as L-histidine, O'Gara and Shanmugam (1976) determined NH_4^+ produced from histidine (as a by-product of histidine catabolism) in portions of the culture medium of several different rhizobial strains comprising different nodulation groups. Values ranged from 1.3 μmoles NH_4^+/mg

Table II

Production of NH_4^+ from N_2 using Nif derepressed strains of *Klebsiella pneumoniae*

Nif derepressed strain	Phenotype	NH_4^+ production: Maximum rate moles NH_4^+ / mg protein hr	NH_4^+ production: Period of production (days) with rate larger than 50% of maximum rate	NH_4^+ production: Total (net) production moles NH_4^+ / mg protein
SK-24	Glu^-, R^+	1.1	6	92
SK-24*	Glu^-, R	0.85	2.5	50
SK-28	Glu^-, R	2.2	5	260
SK-37	Gln^-, R	2.8	5	280
SK-56	Gln^-, S	2.1	4	190
SK-57	Glu^-, R	2.0	7.5	320

The cells were grown at 25° in the artificial nodule in glucose-salts medium as described in the text. 50 μg/ml glutamate or glutamine was added inside the dialysis bag to allow growth.

+R = Often reverts during the course of an experiment; 0.3 mg/ml penicillin G was added after 1.5 day of incubation to prevent growth of revertants on NH_4^+.

S = Genetically stable - not observed to revert.

* = Cells incubated in stationary flask, data from Shanmugam and Valentine (1975).

protein per hr for *R. japonicum* to 7.0 μmoles/mg protein per hr for *R. leguminosarum*. Therefore, it appears that export of NH_4^+ from histidine is a general phenomenon among *Rhizobia*.

Table III

The effect of different nitrogen sources on the growth and assimilation of NH_4^+ by various strains of *Rhizobium*

Strain	N-source	Generation time (h)	Rate of assimilation of NH_4^+ (nmole/ml h)
R. trifolii T1	Glutamate + NH_4^+	4.5	0
	Leucine + NH_4^+	6.0	30
R. leguminosarum g44	Glutamate + NH_4^+	5.0	0
	Leucine + NH_4^+	5.5	20
Rhizobium sp. 'cowpea' strain 32H2	Glutamate + NH_4^+	5.75	0
	Leucine + NH_4^+	6.0	21
R. japonicum 110	Glutamate + NH_4^+	11.0	0
	Leucine + NH_4^+	14.0	7

The inocula grown in mannitol-salts-glutamate medium were freed of organic nitrogen by washing at 4°C. Utilization of NH_4^+ by the cultures was determined as described by O'Gara and Shanmugam (1976).

There is indirect evidence that available glutamate may play a role in the regulation of NH_4^+ metabolism leading to export of fixed nitrogen as NH_4^+. For example, when free-living *Rhizobia* belonging to different nodulation groups were presented simultaenously with glutamate and NH_4^+, they showed a marked preference for glutamate as nitrogen source to the exclusion of NH_4^+ (Table III). These cultures failed to utilize any detectable NH_4^+ as long as glutamate was present in the medium. In contrast, when glutamate was replaced in the growth medium by L-leucine, all rhizobial strains tested were able to utilize NH_4^+ for cellular growth. These values ranged from 7 nmoles/ml hr for *R. japonicum* to up to 30 nmoles/ml hr for *R. trifolii* (Table III). The major conclusions from these experiments is that the fate of NH_4^+ may be determined by the organic nitrogen source present in the medium, such as L-glutamate.

Table IV

Levels of NH_4^+ assimilatory enzymes in *Rhizobium trifolii* T1 as a function of the nitrogen source

Nitrogen source	Glutamine synthetase	Glutamate synthase NADPH	Glutamate synthase NADH	Glutamate dehydrogenase NADPH	Glutamate dehydrogenase NADH
	Specific activity (nmoles/min/mg protein)				
Glutamate	1244	8	18	5	18
Proline	1684	7	9	5	30
Aspartate	2345	12	47	5	22
Leucine	790	20	6	n.d.	n.d.
Glutamate $+NH_4^+$	397	8	22	5	34

Details for the preparation of cell-free extracts and the enzyme assay procedures described by O'Gara and Shanmugam (1976).

n.d. = not determined.

Biochemical experiments substantiate this view. For example, the glutamate-forming enzyme, glutamate synthase, was partially repressed by L-glutamate or by L-proline, which gives rise to L-glutamate in the cell (Table IV). Also, ammonium ion, which would tend to accumulate under these circumstances, functioned as a repressor of glutamine synthetase, which catalyzes the production of glutamine, the immediate precursor of glutamate.

Based on these experimental findings, O'Gara and Shanmugam (1976) proposed a working model for rhizobial symbiosis in which the *nif* genes, contributed by *Rhizobia*, are somehow derepressed during symbiotic N_2 fixation. Aside from the fact that derepression does occur, there is little available evidence regarding the mechanism of Nif regulation in root nodule bacteria. Perhaps the most interesting finding is that nitrogenase is induced in free-living root nodule bacteria in the presence of NH_4^+ (Kurz and LaRue, 1975; Keister, 1975), conditions which result in repression of nitrogenase in all free-living organisms such as *Klebsiella*. Like Nif derepressed strains of *Klebsiella*, discussed earlier, nitrogenase synthesis by root nodule bacteria in the presence of NH_4^+ may be derepressed.

The second aspect of the model which requires discussion concerns the mechanism of inhibition of NH_4^+ assimilation in root nodule bacteria. In *Klebsiella*, amino acids themselves may repress the utilization of NH_4^+ for biosynthesis of glutamate (Shanmugam and Morandi, 1976). In *Rhizobium*, O'Gara and Shanmugam (1976) found that: i) glutamate supplied exogenously blocks the utilization of NH_4^+ (Table III); ii) in the presence of glutamate, a majority of fixed nitrogen is exported apparently as NH_4^+; iii) export of large quantities of NH_4^+ from L-histidine may occur as the result of repression of NH_4^+-assimilating enzymes by metabolically generated glutamate (from histidine catabolism); iv) biochemical studies support the role of glutamate as repressor. One can speculate that root nodule bacteroids behave as physiologically-induced, glutamate-requiring auxotrophs depending on enzyme systems of the host plant to supply this vital amino acid. For example, it has already been established that a variety of auxotrophs of root nodule bacteria may be supplied by the host plant with their specific nutritional requirements such as tryptophan, glycine or methionine (Denarie, Truchet and Bergeron, 1976; Schwinghmer, 1969). In these cases, fixed nitrogen donated by root nodule bacteria is apparently metabolized by the amino acid synthetic machinery of the host with a portion of the amino acids being cycled back to the bacteroids for maintenance of protein synthesis.

It is obvious that our understanding of the mechanism of Nif regulation in *Rhizobia* is still fragmentary. However, it seems probable that progress will be made in the near future, perhaps leading to practical applications such as derepressed strains of root nodule bacteria less sensitive to inhibition by soil nitrogen.

ACKNOWLEDGEMENTS

This work was supported by Grants BMS72-02243 and APR75-09577 from the National Science Foundation, and Contract AG No. 12-14-1001-397 from the U.S. Department of Agriculture to Dr. R. C. Valentine, who is also a recipient of a USPHS Career Development Award (451440). K. Andersen is a recipient of fellowships from the Royal Norwegian Council for Scientific and Industrial Research, and the American-Scandinavian Foundation. Carlo Morandi is a recipient of a European Molecular Biology Organization fellowship.

REFERENCES

Bergersen, F.J. and Turner, G.L. (1967). *Biochem.biophys.Acta,* 141: 507.

DeNarie, J., Truchet, G. and Bergeron, B. (1976). *In:* Symbiotic Nitrogen Fixation in Plants (P.S. Nutman, ed.), Vol. 7, Intl. Biol. Programme, Cambridge Univ. Press, Cambridge, England, p. 47.

Keister, D.L. (1975). *J.Bacteriol.,* 123: 1265.

Kurz, W.G.W. and LaRue, T.A. (1975). *Nature,* 256: 407.

LaRue, T.A., Child, J.J. and Kurz, W.G.W. (1976). *In:* Proc. World Soybean Res. Conf., Urbana, Illinois, in press.

McComb, J.A., Elliott, J. and Dilworth, M.J. (1975). *Nature,* 256: 409.

O'Gara, F. and Shanmugam, K.T. (1976). *Biochim.Biophys.Acta,* in press.

Pagan, J.D., Child, J.H., Scowcroft, W.R. and Gibson, A.H. (1975). *Nature,* 256: 406.

Schwinghamer, E.A. (1969). *Can.J.Microbiol.,* 15: 611.

Shanmugam, K.T. and Morandi, C. (1976). *Biochim.biophys.Acta,* in press.

Shanmugam, K.T. and Valentine, R.C. (1975). *Proc.Natl.Acad.Sci. U.S.A.,* 72: 136.

Shanmugam, K.T. and Valentine, R.C. (1975). *Science,* 187: 919.

Shanmugam, K.T., Loo, A. and Valentine, R.C. (1974). *Biochim. Biophys.Acta,* 338: 545.

Shanmugam, K.T., Chan, I. and Morandi, C. (1975). *Biochim.Biophys.Acta,* 408: 101.

Shanmugam, K.T., Morandi, C. and Valentine, R.C. (1976). *In:* Iron-Sulfur Proteins (W. Lovenberg, ed.), Vol. III, Acad. Press, New York, NY, in press.

Shanmugam, K.T., Streicher, S.L., Morandi, C., Ausubel, F. and

Goldberg, R. (1976a). *In:* Proc. Intl. Symp. on Nitrogen Fixation (W.E. Newton and C.J. Nyman, eds), Vol. II, p. 313, Washington State Univ. Press, Pullman, Wash.

Shanmugam, K.T., Morandi, C., Andersen, K. and Valentine, R.C. (1976b). *In:* Proc. III Intl. Congress on Enzyme Engineering (K. Pye, ed.), Academic Press, New York, NY, in press.

Streicher, S.L. and Valentine, R.C. (1973). *Ann. Rev. Biochem.*, 42: 279.

Streicher, S.L., Shanmugam, K.T., Ausubel, F., Morandi, C. and Goldberg, R. (1974). *J. Bacteriol.*, 120: 815.

Tjepkema, J. and Evans, H.J. (1975). *Biochem. Biophys. Res. Commun.* 65: 625.

Tubb, R.S. (1974). *Nature*, 251: 481.

NIF⁻ MUTANT STRAINS OF *KLEBSIELLA PNEUMONIAE*, *AZOTOBACTER VINELANDII* AND *RHIZOBIUM JAPONICUM*

TANYA MALAVICH, MARY BACHHUBER, MARTHA HOWE, E. BISHOP, J. MAIER and W.J. BRILL

Department of Bacteriology and Center for Studies of Nitrogen Fixation, University of Wisconsin, Madison, Wisconsin 53706, U.S.A.

KLEBSIELLA PNEUMONIAE

Nif$^-$ mutant strains of this organism have been isolated and their phenotypes described previously (St. John *et al.*, 1975). The order of several mutations was obtained from three-factor crosses by transduction with phage P1, a phage that can be used for generalized transduction in *Klebsiella* (Streicher *et al.*, 1971). Since the *his* genes are closely linked to *nif* genes (Streicher *et al.*, 1971; Dixon and Postgate, 1971), *his* was used as an outside marker in reciprocal crosses. Mapping mutations by this technique is difficult, especially with closely linked mutations. In order to confirm the published map order and more easily establish a fine-structure map, we decided to use the technique of deletion mapping.

Bacteriophage Mu has been used to isolate deletion mutations in *Escherichia coli* (Howe and Bade, 1975). Mu can integrate at many sites on the chromosome and often causes mutations due to integration of the prophage. When lysogens of a heat-inducible Mu prophage are grown at high temperature, most of the cells die. Some of the heat-resistant survivors contain deletions that have removed the immunity and killing functions of the Mu prophage and some neighbouring bacterial DNA.

We decided to use Mu to generate deletions of the *nif* genes of *K. pneumoniae*. Since *K. pneumoniae* is normally resistant to Mu, we first isolated mutant strains that were sensitive to Mu. Mutants of *Klebsiella* resistant to each of six *Klebsiella* phage were isolated and tested for sensitivity to Mu. Mu-sensitive strains were obtained at a frequency of about 25% with four of the six phage. One strain, UN729, was chosen for use since it still remained sensitive to P1 (Bachhuber *et al.*, 1976).

Strains in which Mu was inserted into *his* were isolated by infecting UN729 with a heat-inducible Mu and then enriching for His auxotrophs by penicillin treatment. As expected, transduction of these strains to His^+ resulted in the loss of the prophage. When cultures of these lysogens were incubated at 42°C, most of the cells died. Among the heat-resistant survivors were presumptive deletion strains which had simultaneously lost Mu immunity and *nif* functions. The extent of the deletions into *nif* were mapped by P1 transduction, using recipient strains containing nif^- point mutations (St. John *et al.*, 1975).

Table 1. Restoration of N_2-Fixing Capacity of Nif^- Mutant Strains by RP41 Plasmids

Plasmid	Mutation in Recipient Strain					
	nif 4106	*nif* 4066	*nif* 4026	*nif* 4113	*nif* 4083	*nif* 4116
RP41	+	+	+	+	+	+
del3	+	+	-	-	-	-
del4	-	-	-	-	-	-
del5	+	+	+	-	+	+
del7	+	-	-	-	-	-

Dixon *et al.* (1976) constructed a plasmid, RP41, containing the *nif* region of *K. pneumoniae*. Deletion mutations in this plasmid would be extremely useful for fine-structure mapping and complementation analyses of *nif* mutations. In order to generate deletions we have isolated strains of *E. coli* containing a heat-inducible Mu prophage integrated in the RP41 plasmid and subjected those strains to high temperature. Again, many heat-stable survivors were found to be Nif^- (Bachhuber *et al.*, 1976). The deletions were mapped by conjugation with a variety of strains of *K. pneumoniae* carrying *nif* point mutations. Table I shows several deletions obtained from a single Mu lysogen that was Nif^+. Further analyses of strains carrying *nif* deletions in the chromosome or plasmid should yield information concerning the number of *nif* genes, the role of regulatory genes, and the polarity of *nif* operons.

AZOTOBACTER VINELANDII

Page and Sadoff (1976) recently described a method for obtaining transformation in *A. vinelandii*. This technique was used to conduct a genetic analysis of the many Nif^- mutant strains that we had previously isolated and characterized with respect to EPR signals, component activities, antigenic cross-reactive material, and iron-staining ability on polyacrylamide gels (Shah *et al.*, 1973; Brill *et al.*, 1974).

Linkage relationships between various *nif* mutations were determined by the ratio test (Hayes, 1968). This method involved crossing each mutant strain with every other mutant strain in which the donor strain carried a mutation conferring resistance to rifampicin. This marker enabled us to standardize the results of each cross. The number of Nif^+ transformants resulting from crosses between two mutations is roughly proportional to the distance between them on the chromosome. Thus, a reduced number of Nif^+ transformants resulting from a cross between strains carrying different *nif* mutations, as compared to the control cross employing donor DNA from a Nif^+ strain, would indicate that the two mutations are linked.

The preliminary results (Bishop and Brill, 1976) of the ratio method indicate that mutations which simultaneously eliminate both nitrogenase components are tightly linked. This class of mutations, however, does not appear to be closely linked to mutations in genes specifying the structure of the components. The chromosomal region which codes for the molybdenum cofactor that is a part of active component I (Nagatani *et al.*, 1974), is not close to the structural gene for component I. This situation is in contrast to that found in *K. pneumoniae*, where the two loci are cotransducible (St John *et al.*, 1975).

Some Nif^+ revertants of the pleiotropic-negative strain, UW2, are derepressed for nitrogenase synthesis (Gordon and Brill, 1972). DNA from one such revertant, strain UW590, was used to transform UW2 to the Nif^+ phenotype. Of the Nif^+ transformants obtained, only 8% were derepressed for nitrogenase synthesis. This suggests that simultaneous reversion to Nif^+ and the derepressed phenotype may be rather complex since one might have expected that all of the Nif^+ transformants would be derepressed.

RHIZOBIUM JAPONICUM

In order to get an insight into the soybean - *R. japonicum* symbiosis, we have been isolating mutant strains of *R. japonicum* that are unable to form an effective symbiosis. Since many of the strains of interest were expected to be defective in the initial recognition steps between the bacterium and plant, we required that plants be used for the mutant screening. We have described a rapid assay for screening the effectiveness of the symbiosis by use of the acetylene-reduction technique (Wacek and Brill, 1976). A mutagenized culture was grown in a variety of minimal and rich media to select against auxotrophs and slow-

Table II

Mutant Strains of *Rhizobium japonicum* unable to form an effective Symbiosis with Soybean

Strain	Size of Nodules	Presence of leghaemoglobin	Asymbiotic Acetylene Reduction
Wild type	large	yes	yes
SM1	-	-	yes
SM4	small	no	no
SM5	large	yes	no

Table III

Complementation of Extract from Strain SM5 with Components from *Azotobacter vinelandii*

Source of Extract	Additions to Extract	nmoles acetylene reduced per min per mg protein
Wild type (0.9 mg protein	-	2.3
SM5 (0.3 mg protein)	-	0.0
SM5 (0.3 mg protein)	Component I	0.0
SM5 (0.3 mg protein)	Component II	2.1

growing mutants. The culture was then diluted and plated on rich media. Portions of individual colonies were inoculated onto surface-sterilized seedlings. After two weeks incubation in a plant-growing room, the plants were tested for acetylene reduction.

Strains that were not capable of forming acetylene-reducing nodules were isolated and characterized by a variety of techniques. Table II summarizes the results obtained with three of these strains (Maier and Brill, 1976). Strain SM1 does not produce any nodules at all but reduces acetylene at the same rate as the wild type when tested by the asymbiotic assay (Kurz and LaRue, 1975); therefore, this strain presumably has a lesion in a gene specifically responsible for a step involved in the initiation of infection. One mutant, strain SM4, forms small green nodules. This strain also is unable to reduce acetylene asymbiotically. Such a strain should be useful for determining the steps involved in establishing a root nodule capable of N_2 fixation.

Strain SM5 produces nodules that contain leghemoglobin but the nodules do not reduce acetylene. When the rhizobia inside the nodule were separated from plant tissue and cell extracts prepared, the extracts, which were supplemented with ATP and sodium dithionite, did not reduce acetylene. Components I and II from *A. vinelandii* or *K. pneumoniae* cross react with respect to activity and serological properties with the components from *R. japonicum*. When purified component II from *A. vinelandii* was added to the extract of strain SM5, acetylene-reducing activity was observed (see Table III). Also, antiserum prepared against component II produced a significant cross reaction with the extract (Maier and Brill, 1976). This suggests that strain SM5 has a lesion in the structural gene for component II such that a defective protein is produced. When strain SM5 was tested by the asymbiotic assay, no activity was observed; therefore, it may be that the gene that codes for asymbiotic component II activity is the same gene that codes for component II activity that is found in root nodules.

REFERENCES

Bachhuber, M., Malavich, T. and Howe, M.M. (1976). *Abst.Ann. Meet.Amer.Soc.Microbiol.*, p.108.

Bishop, P.E. and Brill, W.J. (1976). *Abst.Ann.Meet.Amer.Soc. Microbiol.*, p.163.

Brill, W.J., Westphal, J., Stieghorst, M., Davis, L.C. and Shah, V.K. (1974). *Anal.Biochem.*, 60: 237-241.

Dixon, R., Cannon, F. and Kondorosi, A. (1976). *Nature*, 260: 268-271.

Dixon, R.A. and Postgate, J.R. (1971). *Nature*, 234: 47-48.

Gordon, J.K. and Brill, W.J. (1972). *Proc.Nat.Acad.Sci.*, 69: 3501-3503.

Hayes, W. (1968). "The Genetics of Bacteria and Their Viruses." John Wiley & Sons, New York.

Howe, M.M. and Bade, E.G. (1975). *Science*, 190: 624-632.

Kurz, W.G.W. and LaRue, T.A.G. (1975). *Nature*, 256: 407-409.

Maier, R.J. and Brill, W.J. (1976). *J.Bacteriol.* In press.

Nagatani, H.H., Shah, V.K. and Brill, W.J. (1974). *J.Bacteriol.*, 120: 697-701.

Page, W.J. and Sadoff, H.L. (1976). *J.Bacteriol.*, 125: 1080-1087.

St John, R.T., Johnston, H.M., Seidman, C., Garfinkel, D., Gordon, J.K., Shah, V.K. and Brill, W.J. (1975). *J.Bacteriol.*, 121: 759-765.

Shah, V.K., Davis, L.C., Gordon, J.K., Orme-Johnson, W.H. and Brill, W.J. (1973). *Biochim.Biophys.Acta*, 292: 246-255.

Streicher, S., Gurney, E. and Valentine, R.C. (1971). *Proc.Nat. Acad.Sci.*, 68: 1174-1177.

Wacek, T.J. and Brill, W.J. (1976). *Crop Science*, In press.

GENETIC ANALYSIS OF NITROGEN FIXATION IN *KLEBSIELLA PNEUMONIAE*

R.A. DIXON, C.K. KENNEDY, A. KONDOROSI*, V. KRISHNAPILLA** and J.R. POSTGATE

A.R.C. Unit of Nitrogen Fixation, University of Sussex, Brighton BN1 9QJ, England.

INTRODUCTION

Future progress in our understanding of the molecular biology of nitrogen fixation is undoubtedly dependent on a more complete analysis of the organisation, function and regulation of the *nif* genes. It is now well established that a series of *nif* genes are clustered close to the histidine operon in *Klebsiella pneumoniae* (Streicher, Gurney and Valentine, 1972) and that the probably order of genes in this region is *rfb gnd his nif shiA* (Shanmugam, Loo and Valentine, 1974). Transductional analysis has been used to determine the order of some *nif* mutations (Streicher, Gurney and Valentine, 1972; St John *et al.*, 1975) although the data for closely linked markers are not particularly convincing. The number of genes involved in nitrogen fixation is unknown, although it has been suggested that at least five genes are necessary; in addition to genes coding for the nitrogenase component proteins, of which there are three, there is evidence for ancillary genes which may determine on electron transport protein and a molybdenum cofactor (St John *et al.*, 1975).

We have developed several procedures to facilitate study of the *nif* system. The construction of plasmids carrying *K. pneumoniae nif* genes (Cannon, Dixon and Postgate, 1976; Dixon, Cannon and Kondorosi, 1976) has enabled us to carry out a diploid analysis of the *nif* region and to examine the expression

*Present address. Institute of Genetics, Biological Research Center, Hungarian Academy of Sciences, H-6701 Szeged Pf521, Hungary.

**Present address. Department of Genetics, Monash University, Clayton, Victoria 3168, Australia.

of *nif* in various genetic backgrounds. In this paper we present our recent findings on the *nif* genes of *Klebsiella pneumoniae* and discuss data obtained from transduction, complementation and biochemical analysis.

ISOLATION OF *NIF* MUTANTS

Host Background

The choice of host genetic background for mutant isolation can be extremely important in terms of further genetic analysis of a given set of mutants. For example, St John *et al.* (1975) isolated *nif* mutants in a His^+ strain of *K. pneumoniae* M5a1. In order to provide a suitable nearby marker for linkage studies they found it then necessary to isolate *his*D derivatives of each *nif* mutant independently. Consequently it was not possible to map the *nif* mutations precisely because each mutant strain contained a different *his*D mutation. We have therefore isolated all our chromosomal *nif* mutants in the same *his*D background (*his*D2, see Streicher and Valentine, 1974).

We have constructed two plasmids which carry the *K. pneumoniae nif* genes; FN68, an F'prime factor (Cannon, Dixon and Postgate, 1976) and RP41, a P plasmid (Dixon, Cannon and Kondorosi, 1976). The choice of background for isolation of plasmid mutants presented some initial difficulty, in that we were uncertain as to whether a given deletion strain of *K. pneumoniae* would be lacking all the *nif* genes. Indeed, we have observed that a P2 eductant of M5a1, strain Kp52 which was presumed to be deleted for the entire *nif* region (Shanmugam, Loo and Valentine, 1974) does in fact contain some *nif* genes. We have isolated some nif^- mutants of FN68 in *Escherichia coli* K12, although there is the possibility that this background may influence the classes of mutant obtained, since it is difficult to distinguish the Nif^- colony phenotype in this strain. The majority of our nif^- plasmid derivatives have been isolated in a mutant of *K. pneumoniae*, Kp107 which is deleted for *nif his* and *gnd*.

The F'*nif* factor FN68 is not as useful for complementation as the P plasmid RP41 since it dissociates in intergeneric matings. Accordingly, some *nif* mutations on FN68 were transferred to RP4 by recombination in a manner similar to that used to construct RP41.

Localised Mutagenesis

Many of our chromosomal Nif^- mutants have been isolated by a conventional procedure which involves mutagenesis, antibiotic enrichment in nitrogen-free medium and screening for Nif^- colonies on low nitrogen plates. However, this method can often

Table I
Temperature sensitivity *nif*⁻ mutants of *Klebsiella pneumoniae*

Designation	Nitrogenase Specific Activity[1] 22°	34°	Activity 34°/22°	Linkage to *his*D	Complementation group
Kp5022 (parent)	0.170	1.760	10.4	-	-
VK1	0.063	0.216	3.4	28%	*nif*K
VK3	0.096	0.096	1.0	26%	*nif*K
VK4	0.052	0.061	1.2		n.t.
VK5	0.051	0.061	1.2	28%	u.a.
VK6	0.037	0.290	7.8	n.t.	n.t.
VK7	0.159	1.530	9.6	n.t.	n.t.
VK8	0.057	0.147	2.6		n.t.
VK10	0.048	0.083	1.7	36%	*nif*K or H
VK13	0.170	1.150	6.8	n.t.	n.t.

[1]Whole cell activity is expressed as nmol ethylene formed per O.D. of culture per 30min. The mutants which were derived from Kp5022 (*hsp*R1, *his*D2, Streicher *et al.*, 1974) were grown in nitrogen-free medium supplemented with histidine and aspartate for 21hr. at 22°C. Chloramphenicol (100 g/ml) was then added and after 10min, the cultures were shifted to 34° for 1hr; parallel cultures were maintained at 22°C. Acetylene was then added and ethylene measured 30 and 120min later.

Mutant derivatives sensitive to phage P1 were selected by the method of Goldberg *et al.*, (1974) using phage P1Km *cts* 100. , indicates that P1 sensitive derivatives could not be isolated. P1 transduction was carried out according to Goldberg *et al.* (1974).

u.a. indicates mutant gave positive complementation with every *nif* plasmid mutant so far tested and therefore could not be assigned to a complementation group.

n.t. indicates strain was not tested.

be extremely tedious particularly when many independent mutants are required.

Since nitrosoguanidine frequently induces closely-linked multiple mutations it is possible to obtain mutations in localised regions by initially selecting for a particular mutation and then scoring for a second mutation in its vicinity (Oeschger and Berlyn, 1974). If the initial mutation selected is dominant, this procedure has the advantage that many independent mutants can be selected directly after mutagenesis and avoids the "grow-out" period normally necessary for segregation of mutant phenotypes, which gives rise to many sibs in the mutated population. This procedure has been adapted to isolate *nif* mutants by initially selecting for *his* operator-constitutive mutations using the histidine analogues 1,2,4-triazole-3-alanine and 3-amino 1,2,4 triazole (Roth, Anton and Hartman, 1966) or by initial selection for mutants resistant to phage TW3 (which presumably selects for *rfb* mutations (see Shanmugam *et al.*, 1974). We find that approximately 5% of the *rfb* or *his*O^c mutants selected after NTG mutagenesis are *nif*$^-$.

Temperature Sensitive Mutants

Although there is a wealth of data indicating that *nif* genes map close to *his* in *K. pneumoniae*, evidence suggesting that the nitrogenase structural genes are located in this region is only inferential. It should be possible to isolate temperature-sensitive (t.s.) *nif* mutants and demonstrate that the mutant protein is thermosensitive *in vitro*. If these mutations are linked to *his*, one would have direct evidence that the nitrogenase structural genes map in this region. *Nif* t.s. mutants can be isolated by screening colonies for growth under N_2 at 22° and 34°. Using the co-mutagenesis procedure we find that about 1% of *rfb* mutants are t.s. *nif*$^-$; both heat sensitive and cold sensitive mutants have been obtained.

CHARACTERISATION OF MUTANTS

T.S. Mutants

It was necessary to distinguish among the t.s. mutants whether the synthesis of nitrogenase proteins or nitrogenase activity itself was affected. Therefore temperature shift experiments were performed to determine nitrogenase specific activity at the permissive and non-permissive temperature. Typical results for nine heat sensitive mutants are shown in Table 1. Presumptive structural mutants such as VK1,3,5,8 and 10 show either none or only a small increase in specific activity following the "shift-up" from 22° to 34°C whereas presumptive synthesis mutants such as VK6,7 and 13 show a large

Table II
Properties of selected *nif* chromosomal mutants

Designation	Proteins present on SDS polyacrylamide gels		% co-transduction with *his*D2	Nif^+ transductants with Kp52 as donor	Complementation group
	Mo-Fe	Fe			
CK263	-	-	73	-	A
CK282	-	-	66	-	A
UN150	+	+	61	-	B
UN328	+	+	61	-	B
UN587	+	+	57	-	?
CK296	-	-	42	+	E
CK279	+	-	37	+	H
UN142	+	+	21*	+	H
CK290	+	+	31	+	u.a.
UN318	-	+	25*	+	K
UN316	+	+	25*	+	D

Mutants with prefix UN were provided by Dr. Winston J. Brill; those prefixed CK are derivatives of Kp5022 (*his*D *hsp*R1). * indicates co-transduction frequencies reported by St John *et al.* (1975) for other *his*D alleles. u.a. indicates mutant was unassignable in complementation tests.

increase in specific activity as does the parent strain.

The possible structural mutants were further studied using transduction, complementation and biochemical studies. The mapping results given in Table 1 show that the mutations are 28-36% linked to *his*D2. Complementation data indicate that VK1 and VK3 are *nif*K, VK5 complements every mutant tested so far and is unassignable at present and VK10 does not complement *nif*K or *nif*H mutants (a point discussed later); these four mutants all revert to Nif^+

So far, two of the suspected t.s. structural gene mutants have been tested for restoration of activity *in vitro* using purified nitrogenase components. Both mutants VK3 and VK5 gave cell-free extract preparations in which ATP- and dithionite-dependent acetylene-reducing activity decreased about 10-fold after incubation at 34° for 90 min. In both VK3 and VK5 extracts, activity lost after incubation at 34° could be restored by the addition of purified Mo-Fe protein. Both these mutants therefore have lesions in genes determining this component. As yet there is no way of determining individual sub-unit activity but it is possible that VK3 and VK5 have mutations in two different cistrons since it was not possible to assign a complementation group to VK5.

Other Chromosomal Mutants

About 50 chromosomal *nif* mutants isolated in this laboratory have been screened for the presence of nitrogenase proteins by electrophoresis of cell extracts on SDS polyacrylamide gels. Mutants isolated by St John *et al.* (1975) were included for comparison. Results obtained with selected mutants are shown in Table II. Both component proteins are present in some mutant extracts, some mutants lack only one of the components, others do not synthesise either of the nitrogenase proteins. These results may be confused by polarity effects and the possibility that a mutant sub-unit may affect the stability or synthesis of other proteins in the nitrogenase complex, but these data together with transduction and complementation evidence, allow us to make certain firm conclusions

In agreement with other workers who have used transduction with phage P1 to map the *nif* mutations with *his*D varies from 25 to 75%. It is not possible to order *nif* mutations precisely on the basis of co-transduction frequency. However, we have found that the P2 eductant strain Kp52 (isolated by Streicher *et al.*, 1972), which is deleted for *rfb gnd his* and *nif*, provides a useful mapping tool for the *nif* region. P1 lysates prepared on Kp52 are unable to produce Nif^+ transductants with recipient *nif* markers which are proximal to *his*, these include

Table III
Selected complementation tests with Nif^-/Nif^- diploids

Plasmid Donor	*Nif* mutation	Recipient mutant								
		Kp52	UN150 *nif*B	UN328 *nif*B	CK263 *nif*A	UN142 *nif*H	UN318 *nif*K	CK296 *nif*E	UN316 *nif*D	UN582 *nif*G
RP41	wild-type	+	+	+	+	+	+	+	+	-
RD7	*nif*A7	-	+	+	-	+	+	+	+	-
VK108	*nif*B108	-	-	-	+	+	+	+	+	-
VK61	*nif*E61	+	+	+	+	+	+	-	+	-
VK109	*nif*D109	+	+	+	+	+	+	+	-	-
VK75	*nif*K75	+	+	+	+	+	-	+	+	-
VK60	*nif*H60	+	+	+	+	-	±	+	+	-
VK107	*nif*-107	-	+	±	-	+	+	+	-	-

Results were recorded as units of ethylene produced per O.D. unit of culture per hr. and expressed as a % of the wild-type activity (RP41) in each background (except for UN582 where results are expressed as a % of RP41 activity in Kp107, a *his nif* deletion mutant). +, indicates 20%; ±, indicates 10%; -, indicates haploid level. Mutants prefixed UN were provided by W. Brill. In order to avoid confusion we have wherever possible adopted the symbols for *nif* loci used by St John *et al.* (1975).

mutants UN150 and UN328 lacking the presumed molybdenum cofactor, UN587 lacking the presumed electron transport protein and *nif*A mutants which do not synthesise either of the nitrogenase component proteins. However, Nif^+ transductants are produced with *nif* mutations which are more distal to *his* (see Table II); these mutants most probably have lesions in nitrogenase structural genes. Hence, Kp52 is only deleted for some *nif* genes located proximal to *his*.

COMPLEMENTATION ANALYSIS

Nif^-/Nif^- heterogenotes were constructed by transferring Nif^- derivatives of the plasmid RP41 into *K. pneumoniae* recipients having a chromosomally-located *nif* mutation. It was found that plate tests for complementation on nitrogen-free agar incubated anaerobically were unreliable; therefore diploids were tested for acetylene reduction and results were expressed as the % of the activity given by the wild-type plasmid RP41 in each mutant background. A high incidence of recombination made some complementation tests difficult to evaluate and it was necessary to test several individual colonies of each merodiploid in order to avoid ambiguity.

Over 600 distinct heterogenotes have now been tested for complementation and some representative results are shown in Table III. We have so far been able to define 6 *nif* cistrons and have assigned 10 mutants to *nif*A, 3 mutants to *nif*B, 2 mutants to *nif*E, 4 mutants to *nif*D, 5 to *nif*K and 9 to *nif*H. Although *nif*K plasmid mutants readily complement *nif*H recipients, we observe that *nif*H mutations located on the plasmid complement some *nif*K recipients only poorly. These negative effects exhibited in some *nif*K backgrounds most probably reflect complex formation between mutant and wild-type component proteins. The ration of hybrid to normal complexes formed by complementation is most probably dependent on whether a given mutation is located on the plasmid or on the chromosome and will be affected by the number of plasmid copies present. Some mutants show a complex complementation pattern and are difficult to classify. A typical example is mutant VK107 which does not complement mutations in *nif*D, *nif*A and one *nif*B mutant UN328. This strain presumably carries a single *nif* mutation since it readily reverts to Nif^+. It is unlikely that this complex pattern can be the result of a polar mutation since preliminary mapping data indicate that *nif*D, *nif*B and *nif*A are not contiguous on the genetic map. We have isolated a further 7 mutants which show the same complex pattern. In agreement with transductional evidence we have observed that the *his nif* deletion strain Kp52 is complemented by mutants carrying *nif*E, *nif*D, *nif*K and *nif*H but not by *nif*B or *nif*A; thus this deletion extends through *nif*B and *nif*A.

It has not been possible to classify a number of our *nif* mutants since they give complementation in every combination so far tested. Although there are possibly other *nif* cistrons which remain to be defined, it is most likely that our results are complicated by intracistronic complementations since both nitrogenase components are multimeric proteins.

Most of the recipient mutations tested were recessive since they gave positive acetylene reduction when carrying the wild-type plasmid RP41. An exception was strain UN582 which was clearly trans-dominant negative. This mutation maps distal to *his* compared with other *nif* mutations and has been defined as *nif*G by St John *et al.* (1975). Our results indicate that the *nif*G product might be a *nif* specific regulatory protein, although there may be other interpretations since UN582 is purported to have a second mutation affecting the level of glutamine synthetase (Brill, personal communication).

CONCLUSIONS

Biochemical analysis and mapping data employing P1-mediated transduction indicate that the nitrogenase structural genes map distal to *his* relative to more proximal *nif* genes whose function is ancilliary or regulatory. We have defined six *nif* cistrons by complementation analysis, three of which, *nif*A, *nif*E and *nif*K were not described in the scheme of St John *et al.* (1975). *Nif*A mutants show a pleiotropic negative phenotype and do not produce either of the nitrogenase proteins. It seems likely that the *nif*A product can activate nitrogenase synthesis in *trans* since *nif*E, *nif*D, *nif*K and *nif*H mutants can complement both Kp52 (which is deleted for *nif*A and does not produce nitrogenase) and *nif*A point mutants. We therefore suspect that the *nif*A cistron has a regulatory role and possibly determines a specific *nif* activator protein. Three of our chromosomal mutants were apparently partially dominant to RP41; nevertheless they could still be assigned to *nif*A, thus providing additional evidence that this cistron determines a regulatory product. *Nif*E mutants have a phenotype similar to *nif*A but form a distinct complementation group. *Nif*E is not deleted in strain Kp 52, unlike *nif*A. Our data support previous evidence that *nif*H determines Fe protein, but as yet we have no additional evidence to suggest that *nif*B produces a molybdenum co-factor. We have been unable to assign any of our mutants to *nif*F which supposedly codes for an electron transport protein, however, Kp52 has a deletion which extends through *nif*B and the *nif* mutation in UN587 (see Table II). The latter strain is the only *nif*F mutant described so far and has about 6% of wild-type activity, so it may be difficult to distinguish a negative result in complementation tests employing this background.

We observe that mutations affecting the synthesis or activity of Mo-Fe protein can be assigned to two complementation groups, *nif*D and *nif*K. These cistrons most probably determine the two different sub-unit types of the Mo-Fe protein (Kennedy *et al.*, 1976). We have isolated *nif*K mutants which are t.s. for Mo-Fe protein activity, confirming that this is a structural gene probably coding for one of the sub-unit types.

We recognise that intracistronic complementation could complicate our results, giving rise to an over-estimate of the number of cistrons involved. We hope to resolve this problem by accurate fine-structure mapping of specific *nif* mutations.

REFERENCES

Cannon, F.C., Dixon, R.A. and Postgate, J.R. (1976) *J.gen.Microbiol.*, 93: 111-125.

Dixon, R.A., Cannon, F.C. and Kondorosi, A. (1976) *Nature*, 260: 268-271.

Goldberg, R.B., Bender, R.A. and Streicher, S.L. (1974). *J.Bacteriol.*, 118: 810-814.

Kennedy, C.K., Eady, R.F., Kondorosi, E. and Rekosh, D.K. (1976) *Biochem.J.*, 155: 383-389.

Oeschger, M.P. and Berlyn, M.K.B. (1974) *Molec.gen.Genet.*, 134: 77-83.

Roth, J.R., Anton, D.N. and Hartman, P.E. (1966). *J.Mol.Biol.*, 22: 305-323.

St John, R.T., Johnston, M.H., Seidman, C., Garfinkel, D., Gordon, J.K., Shah, V.K. and Brill, W.J. (1975). *J.Bacteriol.*, 121: 759-765.

Shanmugam, K.T., Loo, A.S. and Valentine, R.C. (1974). *Biochim. Biophys.Acta*, 338: 545-553.

Streicher, S.L., Gurney, E.G. and Valentine, R.C. (1971) *Proc. Natl.Acad.Sci.* (U.S.A.), 65: 74-80.

Streicher, S.L., Gurney, E.G. and Valentine, R.C. (1972) *Nature*, 239: 495-499.

Streicher, S.L. and Valentine, R.C. (1974) *In*: Microbiol Iron Metabolism" (J.B. Nielands, ed.) pp. 211-229, Academic Press, New York and London.

Streicher, S.L., Shanmugam, K.T., Ausubel, F., Morandi, C. and Goldberg, R.B. (1974) *J.Bacteriol.*, 120: 815-821.

MUTANTS OF *KLEBSIELLA PNEUMONIAE* IN WHICH EXPRESSION OF NITROGENASE IS INDEPENDENT OF GLUTAMINE SYNTHETASE CONTROL

F. M. AUSUBEL, R. F. MARGOLSKEE and NANCY MAIZELS

Biological Laboratories, Harvard University, Cambridge, Massachusetts 02138, U.S.A.

INTRODUCTION

Ammonia represents the synthesis of nitrogenase in *Klebsiella pneumoniae* as in other nitrogen fixing organisms. In a closely related species, *K. aerogenes*, ammonia also controls several other enzymes involved in the utilization of nitrogenous compounds, in particular glutamine synthetase and the enzymes of the histidine utilization (*hut*) operon (Prival and Magasanik, 1971). Recently, Magasanik and his collaborators proposed that the intracellular level of glutamine synthetase mediates ammonia repression (Magasanik *et al.*, 1974). According to this model, glutamine synthetase acts as a positive transcriptional activator by binding specifically to the promoters of several nitrogen yielding operons. This model derives from the observations that glutamine auxotrophs of *K. aerogenes* can not derepress the *hut* operon during ammonia limited growth, while mutants which produce glutamine synthetase constitutively synthesize derepressed levels of the *hut* enzymes even in the presence of high levels of ammonia (Prival *et al.*, 1973).

It seemed likely that glutamine synthetase might play a similar role in the regulation of nitrogenase synthesis since expression of the *nif* operon in *K. pneumoniae* is affected by ammonia levels in a manner analogous to the *hut* operon in *K. aerogenes*. Streicher *et al.*, 1974, and Tubb, 1974, have shown that glutamine synthetase appears to act as a positive control element in the regulation of nitrogenase expression: glutamine auxotrophs (*gln*A^-) do not derepress nitrogenase during ammonia limited growth; complementation of *K. pneumoniae gln*A^- mutants with an F'*gln*$^+$ episome from *E. coli* simultaneously restores glutamine synthetase and nitrogenase activities; and mutants which synthesize glutamine synthetase constitutively in high levels of ammonia (*gln*C$^-$) also synthesize nitrogenase constitutively.

Since all the elements of the *nif* operon appear to be coordinately controlled (St John *et al.*, 1975; Davis *et al.*, 1972), it is likely that a single chromosomal site responds to glutamine synthetase activation. A prediction of this model is that we should be able to define this site by mutation; there should exist mutations in the *nif* promoter region which allow transcription of all of the *nif* genes independently of glutamine synthetase activation. These would be analogous to the *E. coli* lactose operon mutation, UV5 (Arditti *et al.*, 1968), in which *lac* transcription is independent of catabolite activator protein.

In this paper we report the isolation of mutants in *K. pneumoniae* in which derepression of nitrogenase is independent of glutamine synthetase. These mutations map near *nif* and expression of nitrogenase in these mutants is still sensitive to ammonia repression. This latter observation implies the existence of a separate *nif* repressor which responds to ammonia.

EXPERIMENTAL PLAN

Although *gln*A$^-$ auxotrophs of *K. pneumoniae* are phenotypically *nif*$^-$, one cannot select directly for *nif*$^+$ revertants which are glutamine synthetase-independent since the amount of exogenous glutamine the cells require to supplement the glutamine auxotrophy is sufficient to serve as a sole nitrogen source. We devised the following the strategy to isolate *nif*$_{\text{GSind}}$ mutants which circumvents this difficulty. Glutamine synthetase has at least two important roles in *K. pneumoniae*, as an essential catalyst in the biosynthesis of glutamine and as an activator in stimulating transcription of *nif* genes. We hoped to mutate these two activities independently, and to isolate mutants which lacked the regulatory activity but retained the catalytic activity. We propose to adopt the nomenclature *gln*A$^{\pm}$ and *gln*R$^{\pm}$ to designate the biosynthetic and transcriptional activation roles of glutamine synthetase respectively, so the mutants genotype would be designated *gln*A$^+$R$^-$ and the phenotype *gln*$^+$*nif*$^-$*hut*$^-$. These mutants could then be used directly for selecting phenotypic *nif*$^+$ revertants.

Among the *nif*$^+$ revertants we expect: 1) second site mutations in glutamine synthetase which restore transcriptional activator activity (*gln*A$^+$R$^+$); 2) promoter mutations in the *nif* operon which render *nif* transcription independent of glutamine synthetase activation (*gln*A$^+$R$^-$*nif*$_{\text{GSind}}$). The first type of revertants can be identified by their simultaneous reversion to both *nif*$^+$ and *hut*$^+$; their genotype can be verified by cotransduction of the *nif*$^+$ phenotype with markers linked to *gln*A. The second type of revertants should be *nif*$^+$*hut*$^-$; the *nif*$^+$ phenotype in these strains should be linked to the *nif-his* region of the *K. pneumoniae* genome.

BACTERIAL STRAINS AND MEDIA

Bacterial Strains: All *Klebsiella pneumoniae* strains were derived from M5al (Streicher *et al.*, 1971) and are listed in Table I.

Media: Basal salts (BS) medium contained 1.05% KH_2PO_4, 0.45% KH_2PO_4, 0.2% $(NH_4)_2SO_4$, 0.01% $MgSO_47H_2O$, 0.001% NaCl, 0.0001% $CaCl_2$, 0.4% glucose. Nitrogen free (NF) medium was the same as BS except that it contained no $(NH_4)_2SO_4$; nitrogen sources such as histidine were added at 0.2%. Nitrogen free medium for the induction of nitrogenase (NIF) contained 0.625% Na_2HPO_4, 0.075% KH_2PO_4, 0.2% NaCl, 0.01% $MgSO_47H_2O$, 0.001% $FeSO_4$, 0.001% $NaMoO_4$, 2% glucose or 2% sucrose. LB broth was 1% tryptone, 0.5% yeast extract, 0.5% NaCl, pH 7.5. Solid media contained 1.5% Bacto-Difco agar except for NIF media which contained 1.2% Serva high gel strength agar. Tetrazolium (TZ) indicator plates contained 2.55% Difco antibiotic medium No. 2, 0.005% 2,3,5-triphenyl-2H-tetrazolium chloride (Eastman), 1% rhamnose and 0.05% glutamine.

RESULTS

Mapping the glnA *region of the* K. pneumoniae *genome*

To distinguish between mutations in the *nif* operon itself and mutations in glutamine synthetase which affect *nif* expression, we isolated mutants in genes linked to *gln*A. In *K. aerogenes* and in *E. coli*, *gln*A is linked to *rha* (rhamnose utilization) and *met*B (a methionine biosynthesis enzyme)(Streicher *et al.*, 1975; Magasanik *et al.*, 1974). The data presented in Table II show that, in *K. pneumoniae*, five independent *rha* mutants were 7 to 12% cotransducible with *gln*A in P1-mediated transductions. A single *met*B mutant was 4% cotransducible with *gln*A and 46% cotransducible with *rha*. In a three factor cross (P1.KP5022 x KP5106-3) a double selection for *gln*$^+$*met*$^+$ yielded 68% *gln*$^+$*met*$^+$*rha*$^+$. The only map order consistent with this result is *gln*A - *rha* - *met*B. The same order of genes is found in *E. coli* and *K. aerogenes* (Magasanik *et al.*, 1974).

Isolation of glnA$^+$*R*$^-$ *mutants of* K. pneumoniae

In order to obtain mutants with the genotype *gln*A$^+$R$^-$, we asked if the phenotypic *gln*$^+$ revertants of *gln*$^-$ auxotrophs were *nif*$^-$*hut*$^-$. We screened *gln*$^+$ revertants of eight independent *gln*A$^-$R$^-$ strains for aerobic growth on NFglucose-histidine agar (*hut*$^+$ phenotype) and for anaerobic growth on NIFglucose agar

Table 1

Bacterial Strains

Strain	Genotype	Relevant phenotype	Source and reference
KP5022	$hisD^- hsp^-$	$gln^+ nif^+$	derivative of M5al Streicher *et al.*, 1971
KP5060	$hisD^- hsp^-$	$gln^- nif^- hut^-$	derivative of M5al Streicher *et al.*, 1971
KP5106-1, 2,4,7	$hisD^- hsp^- glnA^-_{100}$	$gln^- nif^- hut^- rha^-$	EMS of KP5060 this work
KP5106-3	$hisD^- hsp^- glnA^-_{100}$ $rha^-_{101,102,104,107}$	$gln^- nif^- hut^-$ $rha^- met^-$	EMS of KP5060 this work
KP502-3 KP503-2	$hisD^- hsp^- glnA^+ R^-$	$gln^+ nif^- hut^-$	EMS of KP5060 this work
KP507-2,4	$hisD^- hsp^- glnA^-$ $glnR^+_{nif}$ $glnR^-_{hut}$	$gln^+ nif^+ hut^-$	NTG of KP502-3 this work
KP5160-3 KP5161-3	$hsp^- gln^- A^+ R^-_{nif,hut}$ nif_{GSind}	$gln^+ nif^+ hut^-$	NTG of KP502-3 this work

(*nif*$^+$ phenotype). This test divided the eight *gln*A$^-$R$^-$ strains into two categories: one class had a spontaneous reversion rate to *gln*$^+$ of approximately 10^{-6}; of several hundred revertants which we screened, all were *nif*$^+$*hut*$^+$. The other class had a spontaneous reversion rate to *gln*$^+$ of less than 10^{-9} and mutagenesis was required to obtain *gln*$^+$ revertants. Among these latter revertants, approximately 50% had the phenotype *gln*$^+$*hut*$^-$ *nif*$^-$. Two mutants with this phenotype, KP502-3 and KP503-2, were chosen for further study; Table III details the properties of these mutants. These data demonstrate that the *gln*A$^+$R$^-$ phenotype is due to a mutation in *gln*A, either at a single site or at two very closely linked sites.

Isolation of glnA$^+$*R*$^+$ nif^R hut$^-$ *mutants of* K. pneumoniae

Two different mutagenesis procedures produced phenotypic *nif*$^+$ revertants of strain KP502-3 (*gln*A$^+$R$^-$). In the procedure used to isolate mutants described in this section, an LB broth-grown culture was spread at a density of 5 x 10^7 cells per plate on NIFsucrose agar, and then a small crystal of nitrosoguanidine (NTG) was placed on the agar in the center of the plate. After incubation at 30°C for 5 days, only those plates containing a crystal of NTG contained *nif*$^+$ colonies. A total of 48 independent *nif*$^+$ revertants were purified and screened for their ability to grow on NFglucose-histidine agar (*hut*$^+$ phenotype). Of the 48 *nif*$^+$ revertants, 43 were *hut*$^+$. We assumed that these revertants had the genotype *gln*A$^+$R$^+$. The five remaining *nif*$^+$ revertants failed to grow on NFglucose-histidine agar and were tested further as presumptive *nif*$_{\text{GSind}}$ mutants. However, mapping localized these mutations to the *gln* region of the chromosome. The data in Table IV show that in two of the mutants (KP507-2 and KP507-4) the *nif*$^+$*hut*$^-$ phenotype is 100% cotransducible with *gln*A and 8-12% cotransducible with *rha*. The other three mutants gave similar results. These experiments indicate that the glutamine synthetase regulatory activities for *nif* and *hut* can be mutated independently.

Isolation of nif$_{GSind}$ *Mutants of* K. pneumoniae

It is not surprising that only revertants mapping in the *gln*A gene were obtained in the selection described above. The *gln*A gene is probably 2-3 orders of magnitude larger than the (hypothetical) *nif* promoter. Rather than screen several hundred *nif*$^+$ derivatives of KP502-3 in the hope of finding *nif* promoter mutants, we decided to enrich specifically for mutations linked to the *nif-his* region of the chromosome. We used a procedure developed by Oeschger and Berlyn (1974) which takes advantage of the fact that NTG causes closely linked multiple mutations at the replicating forks of the chromosome. In *K.*

Table II

P1 Mediated Transduction of the *gln*A region of *K. pneumoniae*

Cross	Donor strain (rha^+)	Recipient Strain	Recipient Genotype (all $glnA^-$)	Selected phenotype	No. transductants	Donor character % *rha*	Donor character % *gln*	Donor character % *met*
1	KP5022	5106-1	*rha*101	gln^+	298	9	-	-
2	KP5022	5106-2	*rha*102	gln^+	140	12	-	-
3	KP5022	5106-4	*rha*104	gln^+	96	3	-	-
4	KP5022	5106-7	*rha*107	gln^+	50	10	-	-
5	KP5022	5106-3	*rha*103*met*100	gln^+	200	9	-	-
6	KP5022	5106-3	*rha*103*met*100	met^+	100	46	4	-
7	KP5022	5106-3	*rha*103*met*100	$gln^+ met^+$	96	72	-	-

Strain KP5060 was mutagenized with ethyl methane sulfonate (EMS) as described by Prival and Magasanik (1971) and then plated on TZrhamnose agar at a density which gave isolated colonies. Dark red colonies (putative rha^- mutants) appeared at a frequency of approximately 0.1%. Five independent dark red mutants were purified and the presence of *rha* mutations in these strains was verified on BSrhamnose agar. However, these mutants proved to be leaky on BSrhamnose agar and thus could not be used as selective markers. Similar results have been reported in *Klebsiella aerogenes* and *E. coli* by Streicher *et al.*, 1975. Lysates of phage P1clr100Km on strain KP5022 were prepared and transductions using these lysates were performed as described by Goldberg *et al.*, 1974. One rha^- strain (KP5106-3) was found to contain an unselected auxotrophic marker linked to *gln*A and *rha*103. This auxotrophy could be satisfied by addition of methionine to BSglucose agar and was subsequently shown by R. Goldberg to be a mutation in *met*B.

Table III

Characterization of Glutamine Synthetase Regulatory Mutants

Strain	Relevant genotype	% Linkage of *gln*R phenotype to *gln*A	Relative nitrogenase activity	Growth on: NIFglucose	Growth on: NFgluhis
KP5022	*gln*A$^+$R$^+$	100	1.00	+	+
KP5060	*gln*A$^-$R$^-$	100	0.004	-	-
KP502-3	*gln*A$^+$R$^-$	100	0.013	-	-
KP503-2	*gln*A$^+$R$^-$	not tested	0.004	-	-

A 5 ml LB culture of KP5060 was mutagenized with EMS as in Table II and 0.05 ml portions of mutagenized cells were inoculated into several 5 ml lots of NIFsucrose media supplemented with 0.001% histidine and 0.01% glutamine in 18 x 150 mm test tubes. The tubes were fitted with serum stoppers, flushed with N_2 gas, and incubated with shaking at 30°C. At two days each culture was diluted 100-fold into fresh media and incubated as above. This procedure was repeated 4 times. Finally, a clone was purified from each culture. No *nif*$^+$ revertants were found but 6 independent mutants with the genotype *gln*A$^+$R$^-$ were obtained.

The linkage of *gln*R$^-$ to *gln*A and *rha* was determined by growing phage P1clr100Km on strain KP502-3 and transducing strain KP5106-1 (as described in Table II), selecting *gln*$^+$ transductants on BSglucose agar, and then screening the *gln*$^+$ transductants for *rha*, *hut*, and *nif* on TZrhamnose agar, NFglucose-histidine agar, and NIFglucose agar respectively. Of 250 *gln*$^+$ transductants screened, 17 were *rha*$^+$, 250 were *hut*$^-$, and 250 were *nif*$^-$. Nitrogenase assays were performed by growing cells in 5 ml of NIFsucrose media at 30°C in Klett tubes fitted with serum stoppers and sparged with N_2. At 100 Klett units (green filter), 1 ml of acetylene was added, the tubes were incubated at 30°C with shaking, and gas samples were removed periodically for ethylene determination.

Table IV

Characterization of Mutations affecting Glutamine Synthetase Regulation of Nitrogenase Expression

Strain	Genotype	Phenotype	Relative nitrogenase activity		Linkage of nif^+ to:		Growth on NFgluhis
			$+NH_4^+$	$-NH_4^+$	*rha*	*his*	
KP5022	$glnA^+R^+$	$gln^+hut^+nif^+$	0.0	1.0	0%	20-80%	+
KP502-3	$glnA^+R^-$	$gln^+hut^-nif^-$	0.0	0.011			-
KP507-2	$glnA^+R^+_{nif}R^-_{hut}$	$gln^+hut^-nif^+$	0.0	0.52	12%	0%	-
KP507-4	$glnA^+R^+_{nif}R^-_{hut}$	$gln^+hut^-nif^+$	0.0	0.28	8%	0%	-
KP5160-3	$glnA^+R^-nif_{GSind}$	$gln^+hut^-nif^+$	0.0	0.72	0%	42%	-
KP5161-3	$glnA^+R^-nif_{GSind}$	$gln^+hut^-nif^+$	0.0	0.73	0%	30%	-

Strains KP507-2 and KP507-4 were isolated as described in the text. Strains KP5160-3 and KP 5161-3 were isolated using the procedure of Oeschger and Berlyn (1974) with some modifications: Strain KP502-3 was grown at 37°C in LB to 100 Klett units (green filter). A fresh acetone solution of NTG at 10 mg/ml was added to the cells to give a final NTG concentration of 0.25 mg/ml. The cells were incubated for 10 min at 37°C without aeration, washed twice in 0.01 *M* $MgSO_4$ and concentrated 25 fold. A 0.01 ml portion of mutagenized cells was spread on each of several NIFglucose agar plates and incubated anaerobically (0.3% CO_2, 99.7% N_2 atmosphere) at 30°C for five days. Phage P1clr100Km mediated transductions and nitrogenase assays were performed as in Tables II and III.

pneumoniae, *his* and *nif* are closely linked whereas *gln* is unlinked to both *his* and *nif*. Therefore, by selecting simultaneous $his^+ nif^+$ revertants of NTG mutagenized KP502-3, we can enrich for nif^+ revertants which are linked to *his*. Two independent $nif^+ his^+ hut^-$ revertants obtained in this manner (KP5160-3 and KP5163-3) were chosen for further study. The data in Table IV show that the nif^+ phenotype in these strains is between 30-40% cotransducible with *his*D, yet that nitrogenase activity in these strains is completely repressed by 0.2% ammonia. We conclude from these data that the mutants have the genotype nif_{GSind}.

DISCUSSION

Two important conclusions can be drawn from the data presented in this paper: 1) The hypothesis is that glutamine synthetase plays a direct role in the transcriptional regulation of the *hut* and *nif* operons is supported by the isolation of glutamine synthetase mutants which exhibit different regulatory phenotypes with respect to *nif* and *hut* control. 2) The *nif* operon in *K. pneumoniae* is probably subject to direct negative control by an ammonia-sensing repressor since two nif_{GSind} mutants are sensitive to ammonia repression. This second conclusion contradicts the theory that glutamine synthetase is the sole regulatory element for *nif* expression (Streicher *et al.*, 1974).

Although it is likely that a single chromosomal site in the *nif* operon responds to glutamine synthetase activation, our results do not rule out the possibility that there are two or more independent operons all of which are independently regulated by glutamine synthetase. Our results are also not inconsistent with a regulatory mechanism in which glutamine synthetase interacts with a specialized control protein which in turn interacts with *nif* promoters. If this were the case, the nif_{GSind} mutants which we isolated could be mutations in such a control protein.

We are currently performing complementation tests to determine whether the nif_{GSind} mutants are cis-dominant as expected for promoter mutations or whether they can act in trans as expected for mutations in a *nif* specific control protein. We are also searching for mutations in the proposed *nif* repressor and in its proposed operator in the *nif* operon.

ACKNOWLEDGEMENTS

N.M. wishes to thank Professor Bob Pritchard and the Genetics Department of the University of Leicester for their hospitality during the initiation of this work. We would also like to thank Rich Goldberg for characterizing our *met*B mutant and Frank Cannon for many helpful discussions. N.M. is a Junior Fellow of the Society of Fellows, Harvard University. This work was supported by grants to F.M.A. from Research Corporation and the National Science Foundation.

REFERENCES

Arditti, R.R., Scaife, J.G. and Beckwith, J.R. (1968). *J.Molec. Biol.*, 38: 421.

Davis, L., Shah, V., Brill, W. and Orme-Johnson, W. (1972). *Biochem.Biophys.Acta*, 256: 512.

Goldberg, R., Bender, R. and Streicher, S. (1974). *J.Bact.*, 118: 810.

Magasanik, B., Prival, M., Brenchley, J., Tyler, B., DeLeo, A., Streicher, S., Bender, R. and Paris, G. (1974). *Current Topics in Cellular Regulation*, 8: 119.

Oeschger, M. and Berlyn, M. (1974). *Molec.gen.Genet.*, 134:77.

Prival, M.J. and Magasanik, B. (1971). *J.Biol.Chem.*, 246: 6288.

Prival, M.J., Brenchley, J.E. and Magasanik, B. (1973). *J.Biol. Chem.*, 248: 4334.

St John, R., Johnston, M., Seidman, C., Garfinkel, D., Gordon, J., Shah, V. and Brill, W.J. (1975). *J.Bact.*, 121: 759.

Streicher, S., Gurney, E. and Valentine, R. (1971). *Proc.Nat. Acad.Sci. USA*, 68: 1174.

Streicher, S., Shanmugam, K., Ausubel, F., Morandi, C. and Goldberg, R. (1974). *J.Bact.*, 120: 815.

Streicher, S., Bender, R. and Magasanik, B. (1975). *J.Bact.*, 121: 320.

Tubb, R. (1974). *Nature*, 251: 481.

CLONING OF *HIS* AND *NIF* GENES FROM *KLEBSIELLA PNEUMONIAE*

F.M. AUSUBEL*, F.C. CANNON** and R. RIEDEL*

INTRODUCTION

A cluster of *nif* genes maps between the operator end of the *his* operon and *shi*A (shikimate permease) on the *Klebsiella pneumoniae* chromosome (Streicher *et al.*, 1971; Dixon and Postgate, 1971; Shanmugam *et al.*, 1974; St John *et al.*, 1975). Most *nif* mutations are 20-80% cotransducible with *his*D (Streicher *et al.*, 1972; St John *et al.*, 1975). It is likely that all of the *nif* structural genes are located in this cluster since transfer of the *his-nif* region to *E. coli* by conjugation conferred nitrogen-fixing ability upon the *E. coli* recipient (Dixon and Postgate, 1972).

The transfer of *K. pneumoniae nif* genes to *E. coli* (Dixon and Postgate, 1972) has facilitated the construction of plasmids carrying nitrogen fixing genes. Cannon *et al.* (1976) isolated an F' factor (FN68) in *E. coli* K12 carrying *K. pneumoniae his* and *nif* genes, and Dixon *et al.* (1976) constructed a promiscuous *nif-his* plasmid (RP41) derived by recombination between FN68 and the P-type plasmid RP4. Although FN68 and RP 41 are useful genetic tools, they are unsuitable for experiments which require the large-scale isolation of purified *nif* DNA. The molecular weight of FN68 was 136±3 megadaltons which made its quantitative purification virtually impossible; and RP41, which had a molecular weight of 59 megadaltons, was more difficult to purify by conventional techniques than naturally occurring plasmids of comparable size (F. Cannon, unpublished results).

We have constructed, *in vitro*, three classes of small amplifiable plasmids carrying DNA fragments from RP41 which include

*Biological Laboratories, Harvard University, Cambridge, Massachusetts 02138, U.S.A.

**Unit of Nitrogen Fixation, University of Sussex, Brighton BN1 9QJ, England.

his and *nif* genes. Plasmids in the molecular weight categories of 5.3 and 6.2 megadaltons complemented *his*D mutations, while plasmids with a molecular weight of 14 megadaltons complemented *his*D and a closely linked *nif* mutation.

RESTRICTION ENDONUCLEASE DIGESTION OF RP41 DNA

Figure 1D shows that Sal 1, a restriction endonuclease from *Streptomyces albus garcia,* cleaved RP41 DNA into at least 4 DNA fragments which were separable on 0.65% agarose gels. In a comparable digest of RP4 DNA (not shown), two DNA fragments were generated, one of which comigrated with an RP41 fragment (the third band from the top) suggesting that this fragment was common to both plasmids.

At least 11 DNA bands were resolved on 0.75% agarose gels after digesting RP41 DNA with EcoR1, a restriction enzyme from *E. coli* (Figure 1B). In a comparable digest of RP4 DNA (not shown), one DNA fragment was generated which migrated at the limiting mobility of the gel indicating a molecular weight greater than 30 megadaltons.

CLONING OF *HIS* AND *NIF* GENES ON PMB9

Cotransduction frequencies of *nif* mutations with *his*D indicate that *nif* genes may span a region of the *K. pneumoniae* chromosome which could accommodate 15-20 genes (Streicher *et al.*, 1972). Since most of the EcoR1 fragments in terminal digests of RP41 DNA were less than 6.0 megadaltons (Figure 1B), it is likely that the cluster of *nif* genes lying between *his*D and *shi*A contains more than one EcoR1 regognition site. Thus, to increase the probability of cloning all of the *nif* genes on one fragment, in cloning experiments *in vitro*, we used partial EcoR1 digests of RP41 DNA generated in the presence of distamycin A, an antibiotic which binds to A-T rich DNA sequences and blocks cleavage by EcoR1 (Braga *et al.*, in press). As a cloning vehicle we chose pMB9, a 3.5 megadalton amplifiable plasmid carrying tetracycline (TC) resistance genes, the replicator region from colE1, and a single recognition site for Eco R1 (Boyer *et al.*, in preparation).

In a typical cloning experiment, RP41 DNA was digested with EcoR1 in the presence of 12 ug/ml distamycin A. These conditions resulted in a partial digest in which six fragments, all larger than 5 megadaltons were discernable above a background smear of DNA. This DNA was concentrated and then ligated to EcoR1 linearized pMB9 DNA using T4 polynucleotide ligase. The ligation reaction was incubated at 12.7°C for 16 h. The reaction concentrations of pMB9 DNA (25 ug/ml) and RP41 DNA (84 ug/ml) were chosen according to the formulae of Dugaiczyk *et al.*

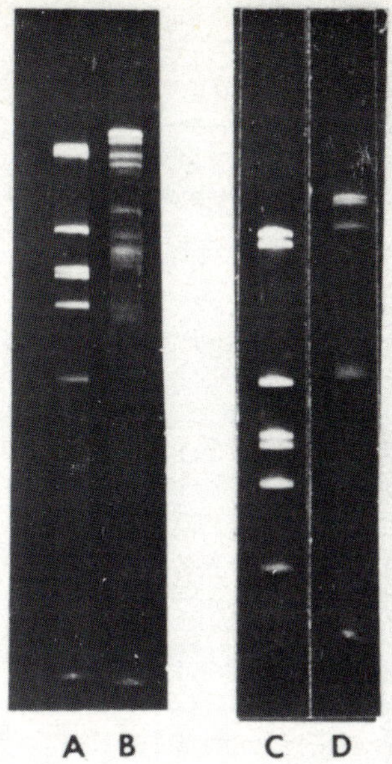

Fig. 1. *Agarose gel electrophoresis of RP41 and phage lambda DNA digested with restriction endonucleases EcoR1 and Sal 1. Supercoiled RP41 DNA, obtained from ethidium bromide cesium chloride gradients, was digested with EcoR1 (gel B) and Sal 1 (gel D). Phage lambda DNA was prepared from C1857S7 by the method of Pirrotta* et al. *(1971) and restricted with EcoR1. Restriction reaction mixtures contained: EcoR1 (five-fold excess), 50 mM Tris HCl, pH 7.6, 10 mM* $MgCl_2$, *0.15% Triton X-100, 1 mM dithiothreitol and 40 mM NaCl; Sal1 (five-fold excess), 140 mM NaCl, 6 mM* $MgCl_2$, *and 50 ug/ml bovine serum albumin. Digestions were carried out for one h (EcoR1) and 3 h (Sal 1) at 37°C.*

Agarose gels (10.5 cm) were prepared by the method of Hayward and Smith (1972) using the modifications described by Hamer and Thomas (1975). DNA samples in 10 to 100 1 containing 5% glycerol, 0.005% bromophenol blue, and 10 mM EDTA, pH 8.0 were layered under the electrophoresis buffer (40 mM Tris, 5 mM sodium acetate. (1 mM EDTA, pH 7.8) as described by Bedbrook and Ausubel (in preparation). Electrophoresis was for 12 h on 0.75% agarose (EcoR1) or 16 h on 0.65% agarose (Sal 1) at 1 mA/tube. The gels were stained and photographed as described by Hamer and Thomas (1975).

A and C, lambda DNA digested with R1; RP41 DNA digested with R1; D, RP41 DNA digested with Sal 1.

(1975) to increase the probability of end to end ligation between RP41 fragments and pMB9. The reaction was then diluted 50 fold and reincubated with ligase for 16 h at 12.7°C to promote circularization of linear molecules. The ligated DNA was then used to transform a *his*D,nif_{213}*hsp*,str^r derivative of *K. pneumoniae* in order to select chimaera plasmids carrying *his*, *nif* and TC resistance genes.

The recipient strain used in transformations, KP5058, was chosen because: 1) The *hsp* mutation prevented restriction of the transforming DNA which originated from *E. coli*. 2) The *his*D mutation allowed direct selection for his^+, circumventing problems in selecting directly for a Nif^+ phenotype (Dixon and Postgate, 1972). 3) The close linkage of *his*D and nif_{213} (80% cotransducible in P1 mediated transductions), increased the probability of identifying plasmids carrying only a portion of the *his*-linked *nif* genes. These His^+ transformants were then screened for Nif^+ using the acetylene reduction test.

Of 140 His^+TC^r transformants obtained, 9 were Nif^+. Supercoiled plasmid DNA was isolated from the 9 Nif^+ transformants and from 14 His^+Nif transformants, and analyzed by agarose gel electrophoresis. Plasmids from His^+Nif^- transformants fell into two size categories of 5.3 and 6.2 megadaltons. All 9 His^+Nif^+ transformants yielded plasmids of 14 megadaltons. One plasmid from each size category was chosen for further study: pCRA10, pCRA13 and pCRA37 had molecular weights of 5.3 (±0.3), 6.2 (±0.3), and 14.0 (±0.5) megadaltons respectively. The molecular weights of the plasmids were determined by using supercoiled pMB9 multimers as size markers (J. Bedbrook and F. Ausubel, in preparation).

Figure 2G,H,I,J show the EcoR1 digestion patterns of pCRA37, 13,10 and pMB9 DNA. Two fragments, one of which corresponded to pMB9, were common to all three plasmids. A third band was only present in pCRA37 and pCRA13. Two additional bands were found only in pCRA37. (The second band from the top in Figure 2G is a doublet which is resolved on lower percentage gels.)

Figure 2B,C,D,E show the Sal 1 digestion patterns of pCRA37, 13,10 and pMB9 DNA. Figure 3 shows the only order of the EcoR1 and Sal 1 fragments in pCRA37 which is consistent with the combined digestion patterns shown in Figure 2.

All three plasmid DNAs transformed strain KP5058 for His^+ and TC^r. Although purified pCRA37 DNA contained a single molecular species as determined by agarose gel electrophoresis, only 20% of the His^+TC^r transformants of KP5058 were also Nif^+. Preliminary evidence suggests that the Nif^- phenotype of the majority of transformants is due to instability of the plasmid rather than a failure of the *nif* genes on the plasmid to express. In addition to this inherent instability of pCRA37, we

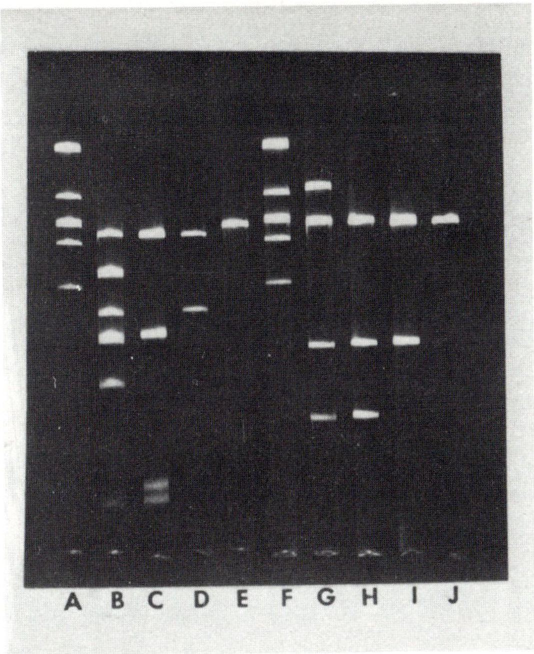

Fig. 2. *Agarose gel electrophoresis of plasmid DNA digested with EcoR1 and Sal1. (For DNA preparation and restriction details, see Fig. 1). EcoR1 digest of lambda DNA; A and F. Sal 1 digest of: pCRA 37, B; pCRA 13, C; pCRA 10, D; pMB9, E. EcoR1 digest of: pCRA 37, G; pCRA 13, H; pCRA 10, I; pMB9, J. Electrophoresis was for 11.5 h on 1% agarose at 1 mA/tube. The second and fourth bands from the top in B and the second band from the top in G are doublets which can be resolved on lower percentage gels.*

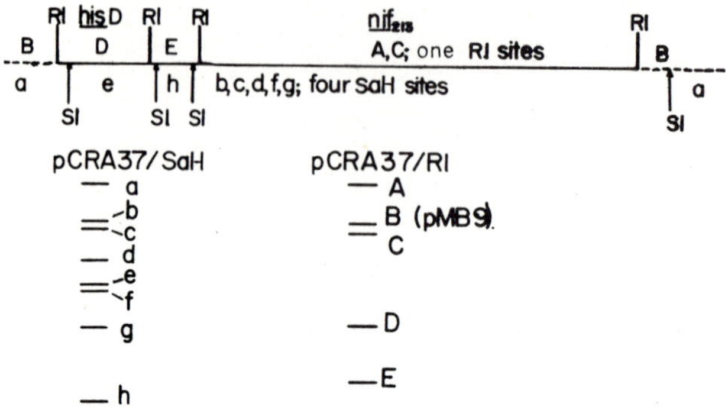

Fig. 3. *Order of EcoR1 and Sal 1 restriction fragments in pCRA37. Above: Restriction map of pCRA37 drawn to scale. Below: Diagrammatic representation of the gels shown in Figs 2B and 2G.*

observed that all three plasmids segregated rapidly during growth in the absence of stringent selection conditions.

DISCUSSION

Since the *nif* mutation in KP5058 (*nif*$_{213}$) which is complemented by pCRA37 is 80% cotransducible with *his*D (Streicher *et al.*, 1972), pCRA37 contains *nif* genes which are proximal to *his*. Experiments are now in progress to determine whether pCRA 37 will complement *nif* mutations which are distal to *his*.

The DNA fragment cloned in pCRA10 is 1.8 megadaltons and since it complements *his*D, it must contain a piece of the *his* operon which includes at least the operator and promoter and *his*GD genes.

We envisage several potential uses for small amplifiable plasmids carrying nitrogen fixing genes: 1) DNA fragments carrying *nif* promoters generated by restriction endonucleases could be used in *in vitro* studies of *nif* transcriptional control. 2) In a linked transcription-translation system, *nif* DNA fragments could be used in conjunction with antibodies to purified nitrogenase components to identify specific *nif* genes. Fragments found to carry genes of interest could then be cloned separately. 3) By using appropriate gene manipulation techniques, eukaryotic DNA replicons and promoters could be associated with *nif* genes in plasmids that would be potentially useful in studies on the uptake, replication and expression of *nif* genes in plant cells.

ACKNOWLEDGEMENTS

We would like to thank N. Maizels for useful suggestions about the use of distamycin A. and J. Bedbrook for many helpful discussions. This work was supported by a National Science Foundation grant to F.M.A.

REFERENCES

Braga, E.A., Nosikov, V.V., Tanyashin, V.I., Zhuze, A.L. and Polyanovski, O.L. (1976). *Dokladi of Acad. of Sci. USSR*, in press.

Cannon, F.C., Dixon, R.A. and Postgate, J.R. (1976). *J.Gen. Microbiol.*, 93: 111-125.

Dixon, R.A. and Postgate, J.R. (1971). *Nature*, 234: 47-48.

Dixon, R.A. and Postgate, J.R. (1972). *Nature*, 237: 102-103.

Dixon, R., Cannon, R. and Kondorosi, A. (1976). *Nature*, 260: 268-271.

Dugaiczyk, A., Boyer, H.W. and Goodman, H.M. (1975). *J.Mol.Biol.*, 96: 171-184.

Hamer, D.H. and Thomas, C.A. (1975). *Chromosoma*, 48: 243-267.

Hayward, G.S. and Smith, M.H. (1972). *J.Mol.Biol.*, 63: 383-395.

Shanmugam, K.J., Loo, A.S. and Valentine, R.C. (1974). *Biochim. Biophys.Acta*, 338: 545-553.

St John, R.T., Johnston, H.M., Seidman, C., Garfinkel, D., Gordon, J.K., Shah, V.K. and Brill, W.J. (1975). *J.Bact.*, 121: 759-765.

Streicher, S.L., Gurney, E.G. and Valentine, R.C. (1971). *Proc. Nat.Acad.Sci. USA*, 68: 1174-1175.

Streicher, S.L., Gurney, E.G. and Valentine, R.C. (1972). *Nature*, 239: 495-499.

Pirrotta, V., Ptashne, M., Chadwick, P. and Steinberg, R. (1971). *In:* Procedures in Nucleic Acid Research (Cantoni, G. and Davies, D.R., eds), pp. 703-715. Harper and Row, New York.

THE CORRELATION OF NITRATE REDUCTASE AND NITROGENASE IN *RHIZOBIUM* SYMBIOSIS

T. SIK and ILONA BARABAS

Institute of Genetics, Biological Research Centre, Hungarian Academy of Sciences, Szeged, Hungary.

INTRODUCTION

The correlation between nitrate reductase and nitrogenase enzyme activities in symbiotic dinitrogen fixation has long been known (Cheniae and Evans, 1960). The simultaneous appearance of these two reducing enzymes was not well understood. Evans and Russel (1971) suggested that the two enzyme systems share a common structural component which included the Mo atoms. This hypothesis was supported by the structural similarity of the Mo-containing components and by the enzyme complementation *in vitro* of a defective assimilatory nitrate reductase from *Neurospora crassa* mutant with the MoFe component of nitrogenase from *Rhizobium japonicum* bacteroids (Ketchum *et al.*, 1970).

A mutation in the suggested common component should have a pleiotropic effect on both nitrate reductase and nitrogenase. Exploiting this principle, nitrate reductase mutants of *Rhizobium meliloti* were obtained by Kondorosi *et al.* (1973) which were symbiotically ineffective.

Later it was shown that a special type of nitrate reductase was missing from the mutants. This nitrate reductase is formed in cells grown on amino acids with vigorous aeration and it cannot be classified unambiguously either as respiratory (dissimilatory) or as assimilatory type (Sik *et al.*, 1976).

In order to characterize the isolated mutants and to find an explanation for the pleiotropy, the appearance of nitrate reductase activity was studied in *Rhizobium meliloti* strain 41 and in three pleiotropic mutants, 461,69 and 77, under different conditions. A regulatory rather than structural connection between nitrate reductase and nitrogenase was found.

NITRATE REDUCTASE IN *RHIZOBIUM*

Rhizobium meliloti strain 41 can utilize ammonium, nitrate or amino acids equally well as nitrogen sources. Different

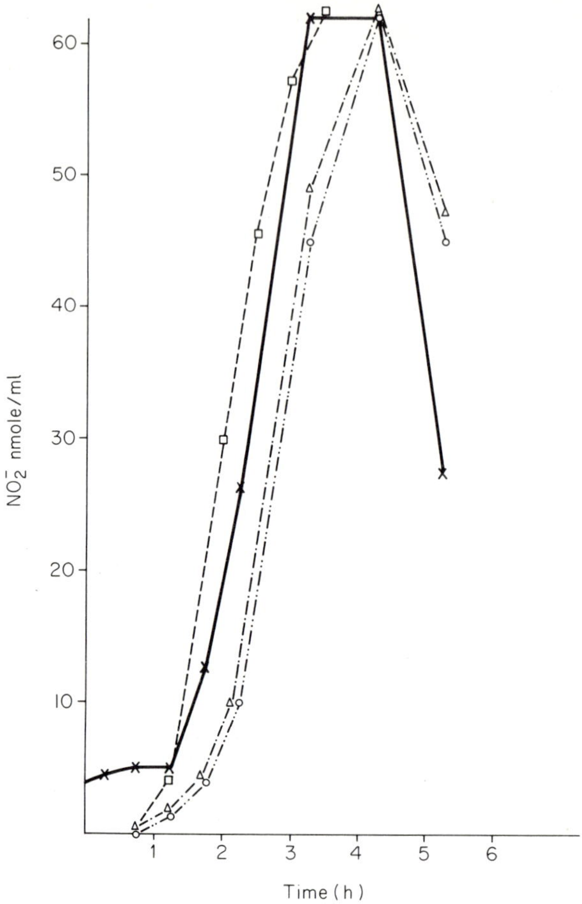

Fig. 1. *Assimilatory Nitrate Reductase in* Rhizobium meliloti *and its mutants.* Rh. meliloti *41 and the mutants 461, 69 and 77 were cultured overnight on amino acid or glucose ammonium medium. The culture was sedimented, washed and diluted 10-fold into vitamin-glucose and 10 mM NO_3-containing medium. The cultures were aerated at 30°C, and in 0.5 ml samples the NO_2 was determined with Griess-Ilosvay reagent (sulphanilic acid and α-naphtylamine in acetic acid). The diazo dye formed was measured in photometer and given as n mole NO_2/ml.* ×——× R. meliloti *41 (wild type); pleiotropic mutants:* o-·-·-o *461;* Δ-·-·-Δ *69 and* □- - - -□ *77.*

physiological conditions were chosen to distinguish the detectable nitrate reductase as assimilatory, respiratory or assignable to other functions.

Assimilatory nitrate reductase was measured when nitrate was the only nitrogen source in the culture medium. Respiratory nitrate reductase was detected in non aerated culture in the presence of nitrate despite other nitrogen compounds. A third type of nitrate reductase was also demonstrated in a culture grown on amino acids both as nitrogen and carbon sources (Sik *et al.*, 1976).

With *Rhizobium meliloti*, aeration was necessary to maintain growth with a generation time of about two hours. Under respiratory nitrate-reducing conditions (without aeration) only residual cell division could be detected.

The nitrate reductase activity was measured *in vivo* as nitrite formed in 10 min from nitrate added or present in the medium. From such measurements, conclusions on induction and enzyme synthesis are necessarily indirect.

Assimilatory Nitrate Reductase

This enzyme is formed in bacteria growing with nitrate as the only nitrogen source. Fig. 1 shows the induction of assimilatory nitrate reductase activity both in the wild type and in the three pleiotropic mutants (461, 69 and 77).

The cells were precultured on amino acid or ammonium-glucose medium, washed and diluted into nitrate-glucose medium. Nitrate reduction started with a lag period, attributable to the nitrogen pool accumulated in the cells during pre-growth. After the induction of nitrate reductase, increasing amounts of nitrite accumulated until nitrite reductase, the second enzyme of the assimilatory chain was also induced and nitrite was reduced to ammonia.

The induction of nitrite reductase separately is shown on Fig. 2. Wild type and mutant Rhizobia were precultured as described and transferred to nitrite-glucose medium. The activity of this enzyme was measured as the amount of nitrite reduced. Nitrite reduction also started after a lag period, the length of delay depending on pregrowth conditions.

Both enzymes of nitrate assimilation could be detected in *Rhizobium meliloti* and its mutants. Only the mutant 77 showed longer lag period in the induction of nitrite reductase. No considerable difference could be measured in their growth on nitrate. Consequently, the nitrate reductase defects in the mutants can not be determinants for either component of the assimilatory nitrate reductase system.

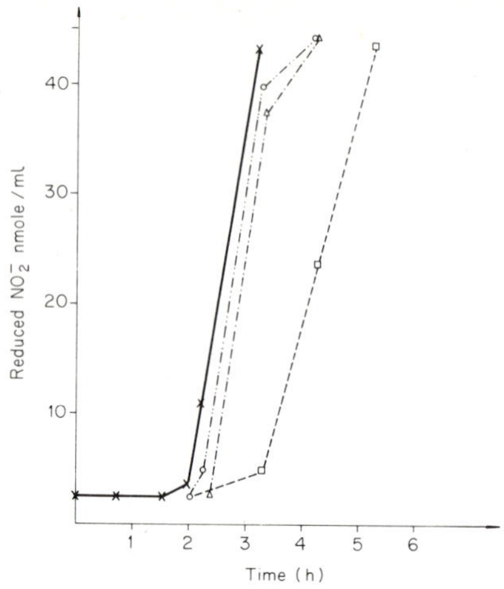

Fig. 2. *Nitrite reductase in* Rh. meliloti *and its mutants. Bacteria were cultured as for* Fig. 1. *but diluted into vitamin-glucose medium with 50 μM NO_2 as nitrogen source and aerated at 30°C. In samples from the culture the remaining nitrite was determined. Activity given as reduced amount of nitrite.* x——x R. meliloti *41; pleiotropic mutants:* o-·-·-o *461;* Δ-·-·-Δ *69 and* □---□ *77.*

Respiratory Nitrate Reductase

The activity in *Rhizobium meliloti* 41 and in the mutants was measured in nitrate-glucose medium without aeration. There was no cell division under these conditions.

The induction of the respiratory nitrate reductase, (Fig. 3) started immediately cells were transferred to non aerated nitrate-glucose medium. The higher enzyme activity in the wild type was due to some nitrate reductase already synthesized during the pregrowth on amino acids.

The respiratory enzyme in the strain 41 was also induced in the presence of amino acids or ammonium (Fig. 4). Ammonium repressed nitrate reductase activity in the mutants, amino acids only in the mutants 69 and 77. Mutant 461 reduced nitrate anaerobically in the presence of amino acids.

Clearly both wild type and mutant rhizobia were capable of respiratory nitrate reductase activity, though the repressibility of the enzyme by other nitrogen compounds differed. Thus the nitrate reductase mutations did not alter the structure of the enzyme, but its regulation was changed.

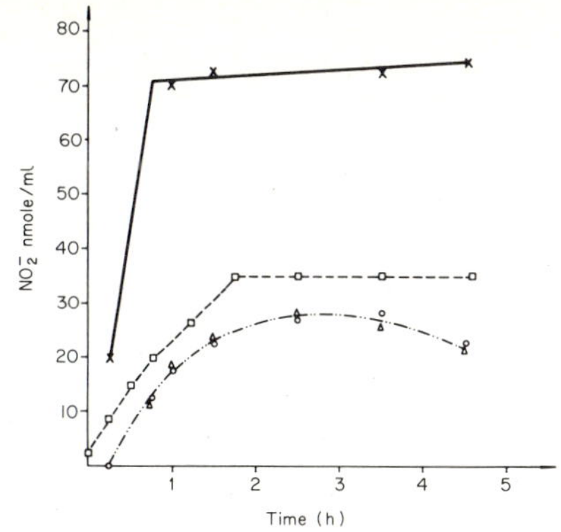

Fig. 3. *Respiratory Nitrate Reductase in* Rhizobium meliloti *and its mutants. Bacteria were cultivated and diluted 10-fold into nitrate-containing medium as for* Fig. 1., *but kept at 30°C without aeration.* NO_2 *was determined in withdrawn samples.* x——x R. meliloti *41;* o—··—o *461;* Δ—·—Δ *69 and* □-----□ *77.*

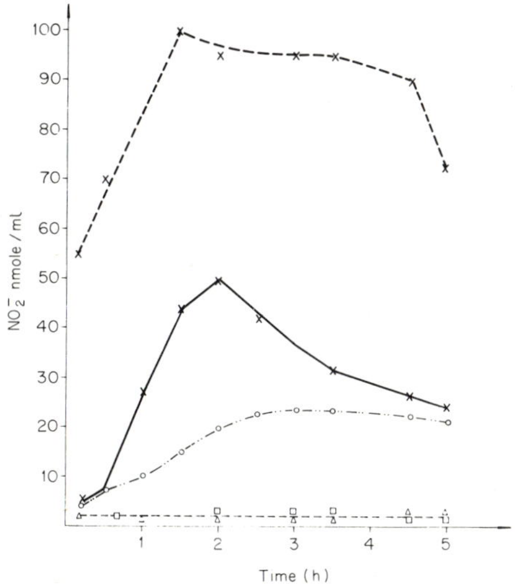

Fig. 4. *Respiratory Nitrate Reductase of* Rhizobium meliloti *and its mutants in the presence of* NH_4^+ *or amino acids. Bacteria were cultured as for* Fig. 1., *but diluted into vitamin-glucose-ammonium or vitamin-amino acid medium with 10 mM nitrate and kept at 30°C without aeration. The* NO_2 *formed was determined in samples.* x——x R. meliloti *41 in the presence of* NH_4^+; x----x R. meliloti *41 in the presence of amino acids;* o—··—o *mutant 461 in the presence of amino acids;* Δ—·—Δ *mutant 69 and* □-----□ *77 in the presence of ammonia or amino acids.*

Regulatory Nitrate Reductase

The nitrate reductase synthesized in wild type *Rhizobium meliloti* grown with aeration on amino acid medium, was called regulatory. Its synthesis was insensitive, but activity sensitive, to oxygen. The activity was determined as the amount of nitrite formed from added nitrate in a sample without aeration over 10 min. Nitrate reductase activity increased in parallel with the cell multiplication (Fig. 5). The activity was 25-30 nmole NO_2^-/10 min/10^9 cells and this level did not change during growth.

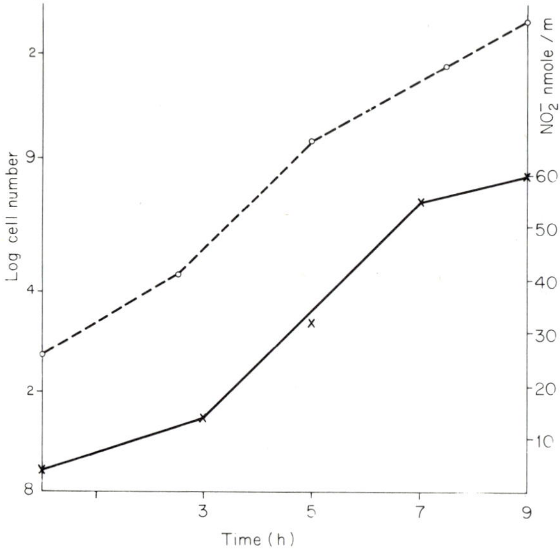

Fig. 5. *Regulatory Nitrate Reductase in* Rhizobium meliloti. R. meliloti *41 was grown in amino acid-vitamin medium at 30°C with vigorous aeration. 0.5 ml samples were withdrawn at different times from the culture, 10 mM NO_3 added, kept without aeration for 10 min and the amount of NO_2 determined. The cell number was also determined by plating at different times.* x——x *nitrate reductase activity;* o------o *cell growth.*

This type of nitrate reductase could not be detected in the three mutants, so its peculiar induction was studied in detail (Sik *et al.*, 1976).

The enzyme was formed in *R. meliloti* 41 growing on amino acids without externally added nitrate. Nitrite could be detected in the culture after stopping the aeration. This observation is evidence for endogenous formation of nitrate, which can then be reduced by the nitrate reductase.

During the growth of *Rhizobium* on amino acids the pH increased from 7.0 to 7.8. Tris or phosphate buffer inhibited the synthesis of nitrate reductase. In cultures of the mutants the increase of pH was slightly higher.

To explain these phenomena, we suggest that ammonia is accumulated when amino acids are deaminated during their utilization both as carbon and nitrogen source. This ammonia increases the pH and becomes oxidized to nitrate. The induction of the regulatory nitrate reductase is connected to this process.

In support of this hypothesis up to 10 mM ammonium chloride or potassium nitrate added to the amino acid-medium enhanced rather than inhibited the formation of regulatory nitrate reductase, proving that ammonia from the deamination of amino acids or internal nitrate formation cannot be inhibitory. In contrast, 10 mM glucose decreased enzyme activity and 10 mM glucose with 5 mM ammonium together caused further repression, suggesting that the deamination of amino acids was under catabolite repression and no nitrate reductase was formed without ammonia accumulation.

The regulatory enzyme was extracted from disrupted cells grown on amino acid medium. The purified enzyme was not sensitive to oxygen but only methyl viologen or benzyl viologen nitrate reductase activity could be detected (G. Kiss, personal communication).

Tungsten is known to inhibit competitively synthesis of the Mo-Fe-nitrogenase component (Benemann *et al.*, 1973; Nagatani and Brill, 1974). The regulatory nitrate reductase of *Rhizobium meliloti* like the assimilatory enzyme, was inhibited by 5 μM W and could be restored by 50 to 100 μM Mo, indicating that the enzyme contains a Mo-component.

In summary, the assimilatory enzyme was detectable in both the wild type and mutant bacteria; the respiratory enzyme was also present in the mutants, but repressible by ammonia or amino acids; the regulatory activity was missing from the mutants.

PLEIOTROPY

Plant tests showed that mutants 461, 69 and 77 can form nodules and bacteroids on the roots of the host plant, but these are ineffective. The reason might be a certain connection between the nitrogenase enzyme system and the mutated regulatory nitrate reductase system.

The possibility of a common Mo-Fe-component being responsible can be excluded because assimilatory nitrate reductase was determined in the mutants. This enzyme is known to contain a molybdoprotein component and this was also demonstrated by the inhibition with W. The regulatory enzyme was also sensitive to W, but the mutation could have occurred in Mo-Fe-component of this enzyme if a second Mo-Fe-component were presumed. This would contradict the one common component hypothesis.

The reason for pleiotropy may be functional or regulatory rather than structural. Functional relation between nitrate reductase and nitrogenase has been suggested and demonstrated (Daniel and Appleby, 1972; Rigaud *et al.*, 1973). A possible regulatory connection between the two enzyme systems can be suggested from the characteristic induction of the regulatory nitrate reductase. The nitrate present at the induction of this enzyme was thought to have been oxidized from accumulated ammonium in vegetative growth of the bacteria. Ammonium accumulation may also happen in bacteroids during nitrogen fixation when the product of nitrogenase is not readily bound or transported by the plant. An eventual oxidation to nitrate could eliminate ammonia repression of the nitrogenase. The appearance of nitrate reductase seems to be an indirect consequence of this reaction.

The mutants were unable to eliminate ammonia, therefore their ineffectiveness should be the result of nitrogenase repression. This type of ineffectiveness ought to be phenotypically reversible by stimulating the binding, transport or utilization of ammonia by the plant.

PHENOTYPIC REVERSION

Based on this assumption, 5 to 20 μg/ml of different compounds were added to the agar substrate of plants nodulating with the mutants. Purines and their nucleosides, riboflavin as well as glutamate, aspartate, arginine and ornithine were active. The nitrogenase activity of the plants nodulated by the mutants was partially or completely restored to the level of the wild type. Complete reversion was achieved by purines, especially with adenine and hypoxanthine. Medium activity was restored with purine nucleosides and riboflavin. Among the amino acids tested, glutamate resulted higher than medium reversion, with the others only partial, low nitrogenase activity could be measured.

The regulatory nitrate reductase of the bacteria was not restored by any of the compounds.

Since so many different compounds were suitable for phenotypic reversion, general stimulation of ammonia binding, plant transport or ammonia utilization is most probably a satisfactory for the reversion.

Stimulatory effect of the adenine was already presumed by studying the nodulation of auxotrophs (Pankhurst and Schwinghamer, 1974). The other two purines and riboflavin may have the same effect. The reverting amino acids, especially glutamate and aspartate, probably simply facilitate the binding of ammonia produced by the nitrogenase.

Thus the reversion demonstrated was actually due to the release of nitrogenase from ammonia repression. Revertible mutants cannot be structural. The pleiotropy of these *Rhizobium meliloti* nitrate reductase mutants is based on an indirect connection between nitrate reductase and nitrogenase, which is a new aspect of correlation.

REFERENCES

Benemann, J.R., Smith, G.M., Kostel, P.J. and McKenna, C.E. (1973). *FEBS Lett.*, 29: 219-221.

Cheniae, G.M. and Evans, H.J. (1960). *Plant Physiol.*, 35: 454-462.

Daniel, R.M. and Appleby, C.A. (1972). *Biochim.Biophys.Acta*, 275: 347-354.

Evans, H.J. and Russel, S.A. (1971). Physiological Chemistry of Symbiotic Nitrogen Fixation *In:* The Chemistry and Biochemistry of Nitrogen Fixation (J.R. Postgate, ed.), pp. 191-244. Plenum Press, London, N.Y.

Ketchum, P.A., Cambier, H.Y., Frazier, W.A., Madansky, C. and Nason, A. (1970). *Proc.Natl.Acad.Sci.U.S.A.*, 66: 1016-1023.

Kondorosi, A., Barabas, I., Svab, Z., Orosz, L., Sik, T. and Hotchkiss, R.D. (1973). *Nature New Biol.*, 246: 153-154.

Nagatani, H.H. and Brill, W.J. (1974). *Biochim.Biophys.Acta*, 362: 160-166.

Pankhurst, C.E. and Schwinghamer, E.A. (1974). *Arch.Microbiol.*, 100: 219-238.

Rigaud, J., Bergersen, F.J., Turner, G.L. and Daniel, R.M. (1973). *J.Gen.Microbiol.*, 77: 137-144.

Sik, T., Kondorosi, A., Barabas, I. and Svab, Z. (1976). Proc. 1st Int. Symp. on Nitrogen Fixation (W.E. Newton and C.J. Nyman, eds) Washington State Univ. Press, 2: 374-382.

SOME EFFECTS DERIVED FROM THE PRESENCE OF EXTRACHROMOSOMAL DNA IN *RHIZOBIUM MELITOTI*

J. OLIVARES, E. MONTOYA and A. PALOMARES

Department of Microbiology, Estación Experimental del Zaidín, C.S.I.C. Granada, Spain.

INTRODUCTION

The enzyme polygalacturonase has been implicated in the rhizobia-legume roots association for the last fifteen years. Fahraeus and Ljunggren (1959) and Ljunggren and Fahraeus (1961) reported evidence on the fact that the production of such enzyme by plants might be induced by the presence of the corresponding bacteria. However, this thesis has been debated in several papers by different authors who have found in rhizobia no correlation between infectiveness and polygalacturonase induction ability (for example: Munns, 1968; Lillich and Elkan, 1968; MacMillan and Cooke, 1969).

Fahraeus and Ljunggren found a close relationship between polygalacturonase induction and the presence of polysaccharides in the bacterial capsule. Later, Ljunggren (1969) reported new data on the role of polysaccharides in the infection mechanism. Although different authors (Hume, 1966; Lillich and Elkan, 1968; MacMillan and Cooke, 1969; Solheim and Raa, 1971) disagree with Ljunggren's opinion because of their failure to perform comparable experiments successfully, Hubbell (1970) and Solheim and Raa (1973) conclude that the extracellular polysaccharides of *Rhizobium* may be one of at least two factors involved in both root hair curling and infection thread formation.

Several authors have reported the presence of extrachromosomal DNA in different species and strains of *Rhizobium*. In most studies its presence was deduced from the effects of treatments with chemicals that, at subbacteriostatic concentrations, selectively act on the expression of different characters. The response of infectiveness was studied by Higashi (1967), Dunican and Cannon (1971), Parisjskaya (1973) and Zurkowski *et al.* (1973) using acridine orange and sodium dodecyl sulphate as eliminating agents and mitomycin C as inducing one. After the treatment of infective *Rhizobium* with curing chemicals the authors found a sharply decreasing ability of bacteria to form nodules in the corresponding legume roots.

Higashi (1967) reported genetical transfer of host specificity, one of the symbiotic properties of the rhizobia legume roots association from *R. trifolii* to *R. phaseoli*. Dunican and Tierney (1974) presented evidence of the transfer of the nitrogen fixation character from *R. trifolii* to *Klebsiella aerogenes*; *nif* gene transfer occurred at a higher frequency than that expected for genetic material included in the chromosome, consistent with the view that such information might be located on extrachromosomal elements.

Klein *et al.* (1975) presented physical evidence of the existence of a plasmid in *R. japonicum*. Luyindula *et al.* (1975) and Tshitenge *et al.* (1975) found similar genetic elements in *R. phaseoli* (*vigna*) and several strains of *R. trifolii*.

POLYGALACTURONASE INDUCTION ABILITY LINKED TO PLASMID DNA

We have studied the ability of different strains of *R. melitoti* to form polygalacturonase in response to induction (polygalacturonase induction ability). This character shows clear instability related with that recorded for Inf^+ and Eff^+ characters in many investigations (Fig. 1).

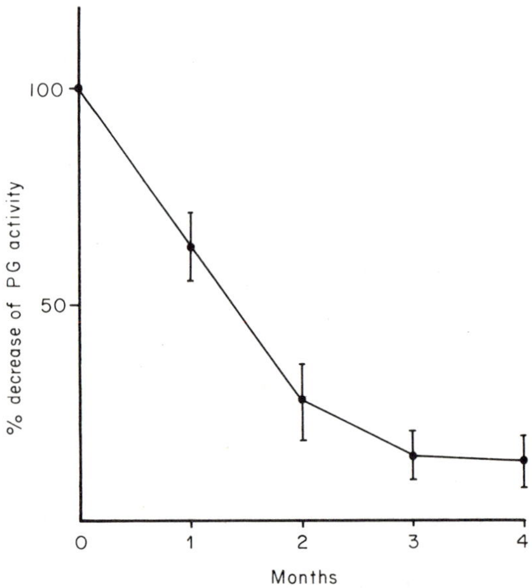

Fig. 1. *Decrease of polygalacturonase (PG) induction ability of a strain of* Rhizobium melitoti *(Rm 4) along its maintenance passage. The value obtained after its isolation was considered as 100.*

Polygalacturonase activity was determined following the technique of Ljunggren and Fahraeus (1961) with slight modifications which increased the sensitivity (Palomares, 1975).

This instability suggested that the character could be linked to extrachromosomal DNA which was gradually repressed or lost during passage of the culture.

The existence of a plasmid in *R. melitoti* has been investigated by Palomares (1975). Experiments using acridine orange, as plasmid-eliminating agent, have shown that the majority of the subcultures obtained from colonies of a *R. melitoti* strain with high polygalacturonase induction ability showed extremely low activity after treatment. On the other hand, wild strains with naturally low induction ability, showed a high one under

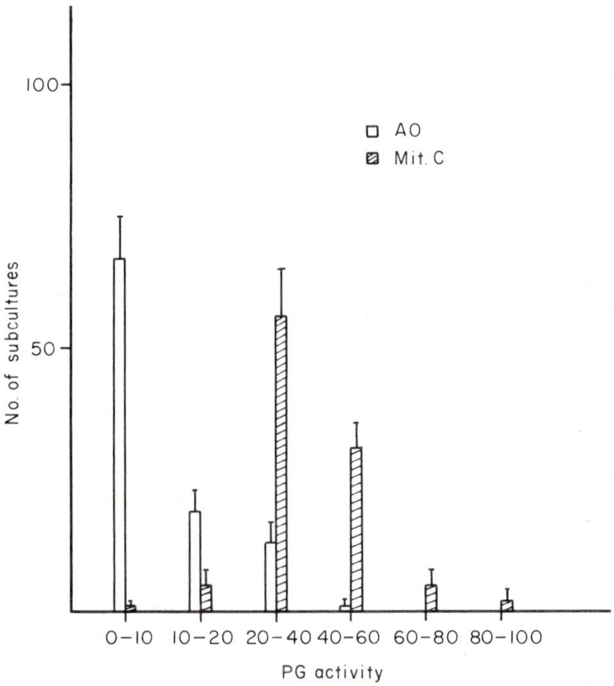

Fig. 2. *Percentage of subcultures obtained from colonies coming from a wild strain* Rhizobium melitoti *(Rm4), showing a high polygalacturonase inducing ability, after its treatment with acridine orange, and another strain (Rm6), with natural low enzyme induction ability, after its treatment with mitomycin C. (Polygalacturonase (PG) activity expressed as % decrease in viscosity of 1% (w/v) solution of low methoxyl pectin in 24 h.)*

adequate conditions in presence of mitomycin C, a plasmid-inducing agent (Fig. 2). Thus the plasmid is gradually repressed and not lost in natural conditions.

Consistent with this conclusion, strains previously cured with acridine orange and then treated with mitomycin C did not recover polygalacturonase induction ability.

From these results it is possible to conclude that the pectin enzyme induction character is linked to extrachromosomal DNA. A biophysical study of this genetic element was therefore carried out.

DNA of wild and cured strains was labelled with 6-^{3}H thymidine by shaking a culture of the bacteria in minimal medium (Ferry *et al.*, 1959) supplemented with 6-^{3}H thymidine and 2 deoxyadenosine. After 16 h at 28°C, cells were washed with TES buffer and lysed with lysozyme and sodium dodecyl sulphate. The cleared lysate was used for dye-buoyant equilibrium centrifugation.

The lysate was adjusted to a density of 1.625 with cesium chloride and ethidium bromide was added as prescribed. After centrifugation at 130,000 g for 30 h at 20°C the polycarbonate

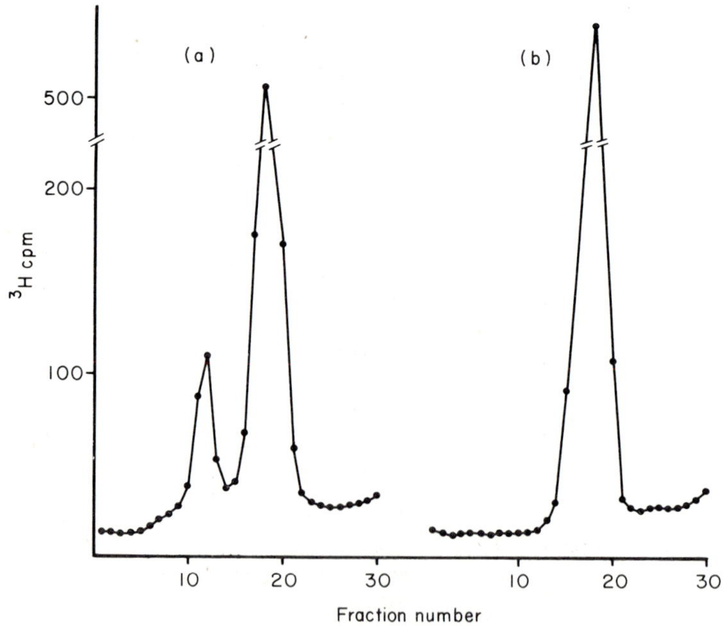

Fig. 3. *Profile of the dye-buoyant density centrifugation of ^{3}H labelled cell lysates of two* Rhizobium melitoti *strains. (a) wild strain (Rm11) and (b) the corresponding cured one (Rm11c).*

tubes were punctured at the bottom and the centrifugate collected in fractions of 0.5 ml. Two drops of each fraction were put on Whatman No. 3 filter discs, treated as usual and assayed for ^{3}H radioactivity.

Fig. 3 shows the profiles of ^{3}H labelled extrachromosomal DNA of both a wild and a cured strain of *R. melitoti*. The rapidly sedimenting DNA peak attributable to a possible plasmid, is clear in the lysate of the wild strain.

Fig. 4 shows the profile of ^{3}H-labelled satellite DNA from *R. melitoti* analysed by velocity centrifugation in neutral sucrose gradient. Supercoiled DNA, purified by dye-buoyant density centrifugation, was dialysed overnight and then layered on sucrose gradient with ^{14}C labelled Col El DNA as a sedimentation marker. After centrifugation for 60 min at 120,000 g at 20°C fractions were collected on Whatman No. 3 filter discs, precipitated with cold trichloroacetic acid and assayed for ^{3}H and ^{14}C radioactivity.

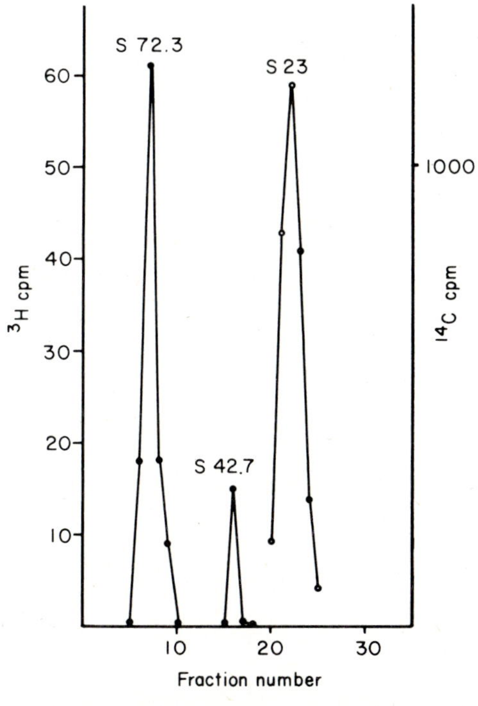

Fig. 4. *Profile of neutral sucrose centrifugation of the* ^{3}H *labelled satellite DNA from* Rhizobium melitoti *(Rml1) (•).* ^{14}C *labelled Col El DNA has been taken as marker for molecular weight determination (∘).*

The extrachromosomal DNA from *R. melitoti* showed a sedimentation coefficient of 72.3 S. A small peak of 42.7 S corresponding to open circular DNA could be expected. Taking the relationship between sedimentation coefficient and molecular weight given by Bazaral and Helinski (1968) and having as reference the values for Col E1 DNA (23 S and $4.2 \cdot 10^6$ daltons) the value 72.3 S corresponds to a molecular weight of $59.6 \cdot 10^6$ daltons.

Tshitenge *et al.* (1975) found for plasmid DNA of *R. phaseoli* (*vigna*) a molecular weight higher than 26.10^6 corresponding to *Escherichia coli* R6K plasmid.

PLASMID DNA AND THE PHENOTYPES Inf^+ AND Eff^+

Different authors have tried to associate the presence of extrachromosomal DNA with infectiveness and effectiveness in rhizobia. Higashi (1967) reported that infectiveness of *R. trifolii* on clover roots decreased when bacteria were cultivated in presence of acridine orange. Dunican and Cannon (1971) correlated effectiveness with a plasmid in *R. trifolii* too, since a change of viomycin sensitivity and resistance occurred after treatment of bacteria with plasmid-eliminating agents. Vio^s has been considered related to effectiveness (Schwinghamer, 1964; Hendry and Jordan, 1969). There is evidence that viomycin sensitivity is a plasmid linked character in *R. trifolii* (Cannon, 1972). Dunican and Tierney (1974) and Dunican *et al.* 1976) reported in a similar way.

Tshitenge *et al.* (1975) concluded that all effective strains of *R. phaseoli* (*vigna*) and *R. trifolii* studied contained plasmid DNA, while non-infective strains might or might not contain plasmid DNA. In contrast, the results of our experiments are quite different (Table 1).

All cured strains, although having low polygalacturonase induction ability, showed an infectiveness and effectiveness as high as the wild strains. More cured strains were assayed and were Inf^+ and Eff^+ but their plasmid DNA had not been biophysically characterized so they have not been included in this table.

STUDIES ON EXTRACELLULAR POLYSACCHARIDES

The composition of the extracellular polysaccharides of rhizobia has been widely studied. However, the chemical, physico-chemical and immunological methodology used do not establish any significant correlation between chemical composition and the different bacterial species and strains.

Palomares (1975) found qualitative and quantitative differences between extracellular polysaccharides from wild and cured

Table I

Results of an experiment on inoculation of *Medicago sativa* plants with different strains of *Rhizobium melitoti* in pots under adequate greenhouse conditions.

Strain[a]	Inf[b]	Eff[c]	PG[d]
Rm 4	100	168	70
Rm 4c	102	140	10
Rm 8	100	140	70
Rm 8c	101	160	10
Rm 11	100	170	70
Rm 11c	97	165	10
Control	-	100	-

a) c strains derived from the corresponding parent wild strain after treatment with acridine orange.

b) The nodules/plant ratio for the wild strains is taken as 100.

c) The weight of plant/pot has been taken as effectiveness index. The weight control is considered as 100.

d) Polygalacturonase activity is expressed as % decrease in viscosity of a 1% (w/v) solution of low methoxyl pectin in 24 h.

strains of *R. melitoti*. Fig. 5, taken from his experiments, shows the amount of extracellular polysaccharides, assayed following to Amarger *et al.* (1967), produced in 100 ml of culture medium during several days by a couple of *R. melitoti* strains, representative of those studied.

Although the growth is the same in both cultures, the level of production by the two types of strains is clearly different. The polygalacturonase inducing ability shown by both preparations in presence of *Medicago sativa* seedlings was also different. According to Palomares, a higher concentration of extracellular polysaccharide from the cured strain is necessary to reach acceptable levels of polygalacturonase induction than of the polysaccharide produced by the corresponding wild strain. This could mean that a qualitative difference exists between both preparations.

Three major fractions were isolated from each preparation by gel filtration through Sephadex G-25. Only the first one,

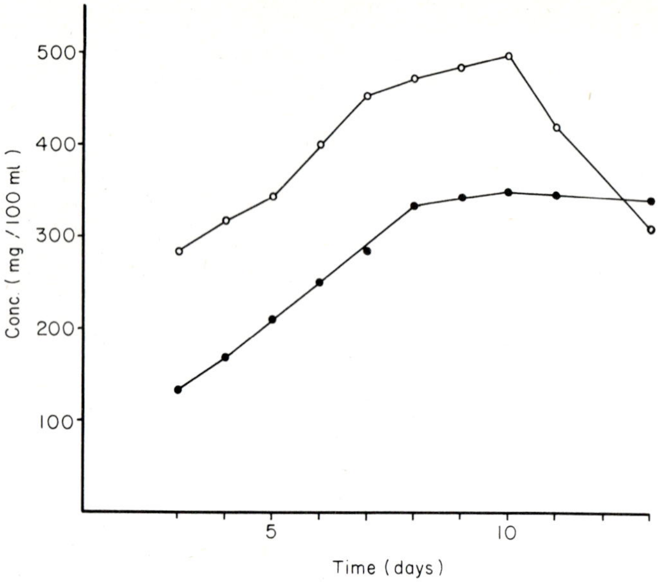

Fig. 5. *Extracellular polysaccharide production by a wild strain (○) and a cured one (•) of* Rhizobium melitoti.

independently of its origin, showed polygalacturonase inducing ability, although in a lower degree when it originated from cured strains.

The active fractions were passed through the gel again to improve their purity and hydrolysed with boiling hydrochloric acid. Chromatographic analysis of the hydrolysates gave a different monomer composition for each fraction. Active polysaccharide from wild strains, showing a high polygalacturonase inducing ability, contained glucose, galactose, mannose, uronic acid and deoxyose in the proportion of 2:2:1:1:1, while that from the cured strains consisted of glucose, galactose and mannose in the ratio 4:1:1.

In comparison with these results, Amarger *et al.* (1967)did not find mannose, while Zevenhuizen (1971) found it together with galactose, but in very low proportion, and glucose in a really high proportion (76-83%). Using electrophoresis on DEAE dextran, Dudman and Bishop (1968) and Dudman (1972) separated the acidic polysaccharides containing glucuronic acid. Stevenson *et al.* (1970) found traces of substances which seem to be residues of cell wall. In general, the presence of small amounts (about 1 per cent) of different components may be considered as contaminants of other structures. This can also be true of DNA detected in extracellular polysaccharide preparations and that was supposed to be involved in the induction of polygalacturonase (Ljunggren, 1961 and Keele *et al.*, 1974). DNA precipitates at the same conditions than polysaccharides do.

Hepper (1972) found a ratio glucose:galactose of 4:1, similar to that reported by Palomares in preparations from cured strains.

Although Chaudhari *et al.* (1973) have proposed a defined structure for polysaccharides from a strain of *R. trifolii* that may be related to the host specificity, it is possible to affirm with Bjorndal *et al.* (1971) that there is no correlation between the composition of these substances and the infectivity or effectivity characters in rhizobia.

As one sees after consulting the bibliography on this subject (Dudman, 1964, 1968, 1972, 1976; Dudman and Heidelberger, 1969; Heidelberger *et al.*, 1970; Bailey *et al.*, 1971; as well as the papers above quoted) there are differences in the results reported that may be due to the strain variability and to the lack of standardization of the methodology followed.

CONCLUSIONS

Polygalacturonase induction ability depends on the extracellular polysaccharides produced by *Rhizobium*. These polysaccharides are of different nature and enzyme-inducing ability according to whether wild or cured strains produce them.

The role of the extrachromosomal DNA in their biosynthesis has not been clarified. Infectiveness and effectiveness in *R. melitoti* seem not to be linked to plasmid DNA.

ACKNOWLEDGEMENTS

We wish to thank J.R. Postgate and F.C. Cannon for the facilities given to study the plasmid DNA.

REFERENCES

Amarger, N., Obaton, M. and Blanchère, H. (1967). *Can.J.Microbiol.*, 13: 99-105.
Bailey, R.W., Greenwood, R.M. and Craig, A. (1971). *J.gen.Microbiol.*, 65: 315-324.
Bazaral, M. and Helsinki, D.R. (1968). *J.Mol.Biol.*, 36: 185-194.
Bjorndal, H., Erbing, C., Lindberg, B. and Ljunggren, H. (1971). *Acta Chem.Scand.*, 25: 1281-1286.
Cannon, F.C. (1972). Ph.D. Thesis, University of Galway, Ireland.
Chaudhari, A.J., Bishop, C.T. and Dudman, W.F. (1973). *Carbohyd.Res.*, 28: 221-231.
Dudman, W.F. (1964). *J.Bacteriol.*, 88: 640-645.
Dudman, W.F. (1968). *J.Bacteriol.*, 95: 1200-1201.

Dudman, W.F. and Bishop, C.T. (1968). *Can.J.Chem.*, 46: 3097-3104.
Dudman, W.F. and Heidelberger, M. (1969). *Science*, 164: 954-955.
Dudman, W.F. (1971). *Appl.Microbiol.*, 21: 973-985.
Dudman, W.F. (1972). *Anal.Biochem.*, 46: 680-690.
Dudman, W.F. (1976). *Carbohydr.Res.*, 46: 97-110.
Dunican, L.K. and Cannon, F.C. (1971). *Plant Soil* (Spec. Vol.), 73-79.
Dunican, L.K. and Tierney, A.B. (1974). *Biochem.Biophys.Res. Comm.*, 57: 62-72.
Dunican, L.K., O'Gara, F. and Tierney, A.B. (1976). *In:* Symbiotic nitrogen fixation in plants (P.S. Nutman, ed.), pp. 77-90. Cambridge University Press.
Fahraeus, G. and Ljunggren, H. (1959). *Physiol.Plant.*, 12: 145-154.
Ferry, P., Blanchère, H. and Obaton, M. (1959). *Ann.Agron.*, 10: 219-233.
Heidelberger, M., Dudman, W.F. and Nimmich, W. (1970). *J.Immunol.*, 104: 1321-1328.
Hendry, G.A. and Jordan, D.C. (1969). *Can.J.Microbiol.*, 15: 671-675.
Hepper, C.M. (1972). *Ant. van Leeuwenhoek*, 38: 437-445.
Higashi, S. (1967). *J.gen.appl.Microbiol.*, 13: 391-403.
Hubbell, D.H. (1970). *Bot.Gaz.*, 131: 337-342.
Hume, D.J. (1966). *Diss.Abstr.*, 27: 22-23.
Keele, B.B., Wheat, M.R.W. and Elkan, G.H. (1974). *J.appl.Microbiol.*, 20: 187-196.
Klein, G.E., Jenison, P., Haak, R.A. and Matthysse, A.G. (1975). *Experientia*, 31: 352-353.
Lillich, T.T. and Elkan, G.H. (1968). *Can.J.Microbiol.*, 14: 617-625.
Ljunggren, H. (1961). *Nature*, 191: 623.
Ljunggren, H. and Fahraeus, G. (1961). *J.gen.Microbiol.*, 26: 521-528.
Ljunggren, H. (1969). *Physiol.Plant.*, Suppl. V. 1-82.
Luyindula, N., Tshitenge, G., Lurquin, P. and Ledoux, G. (1975). *Arch.Int.Physiol.Biochim.*, 83: 199-200.
MacMillan, J.D. and Cooke, R.C. (1969). *Can.J.Microbiol.*, 15: 643-645.
Munns, D.N. (1968). *Plant Soil*, 28: 129-146.
Palomares, A.J. (1975). Ph.D. Thesis, Universidad de Granada, Spain.
Parisjskaya, A.N. (1973). *Mikrobiologiya*, 42: 119-121.
Schwinghamer, E.A. (1964). *Can.J.Microbiol.*, 10: 221-233.
Solheim, B. and Raa, J. (1971). *Plant Soil*, 35: 275-280.
Solheim, B. and Raa, J. (1973). *J.gen.Microbiol.*, 77: 241-248.
Stevenson, F.J., Clapp, C.E. and Molina, J.A.E. (1970). *Soil*

Sci.Soc.Am.Proc., 34: 759-762.
Tshitenge, G., Luyindula, N., Lurquin, P.F. and Ledoux, L. (1975). *Biochim.Biophys.Acta*, 414: 357-361.
Zevenhuizen, L.P.T.M. (1971). *J.gen.Microbiol.*, 68: 239-243.
Zurkowski, W., Hoffman, W. and Lorkiewicz, Z. (1973). *Acta Microbiol.Polon.*, Ser. A 22: 55-60.

NITROGEN FIXATION IN PLANTS: AN EXPANDING HORIZON?

A.H. GIBSON, W.R. SCOWCROFT and JANET D. PAGAN

Division of Plant Industry, CSIRO, P.O. Box 1600, Canberra City, A.C.T., 2601, Australia.

INTRODUCTION

The realization that nitrogen is one of the principal limitations to increasing world food production, that resources for present methods of producing nitrogen fertilizers are not inexhaustible, and that it is feasible to rectify many of the problems by biological means, has resulted in a major resurgence in research into the means by which plants and N_2-fixing bacteria form, and function, as symbiotic associations. As the major single supplier of biologically-fixed nitrogen, the legume-*Rhizobium* symbiosis is at the forefront of these investigations. Elucidation of the various physiological processes involved, and of their genetic regulation, has significance in seeking ways to increase N_2 fixation by legumes; it also opens the way to exploring the feasibility of extending this type of symbiosis to other agriculturally-important plants. At the same time, the investigation of many other diazotrophic symbiotic associations has been initiated or expanded, including those involving non-leguminous Angiosperms, the Cycads, the Bryophytes, fungi (lichens) and even marine diatoms. Of great potential significance is the symbiosis between various graminaceous species and *Spirillum*, first announced at the Pullman Meeting (Döbereiner and Day, 1976). There is also renewed interest in many N_2-fixing bacteria and blue-green algae, both in their "free-living" form and in loose association with the roots, and leaves, of many plants.

From all these studies, a better understanding of the significance of biological N_2 fixation in both global productivity and in natural habitats is evolving. In addition, a closer insight into the development, regulation and function of many symbiotic associations is evolving. In this paper, various methods of investigating those associations will be reviewed, including some of our own studies with *Rhizobium*, and a brief examination made of approaches to extending N_2 fixation by plants. Emphasis will be given to those approaches involving a true symbiotic situation, although the transfer of nitrogen fixation genes to the plant genome will be considered.

SYSTEMS

Various systems and approaches have been developed to examine the fine details of symbiotic associations. At the genetic level, factors affecting the nodulation of legumes have been described (reviewed by Holl and LaRue, 1976), but follow-up studies at the physiological level have not been productive. The genetic control over the N_2-fixing association between *Spirillum lipoferum* and some family Gramineae species (von Bulow and Döbereiner, 1975; Döbereiner and Day, 1976) offers scope for examining the critical factors involved in the development of such associations and for the extension of this type of symbiosis to other species and cultivars.

For this review, plant science is defined as the physiological and biochemical studies designed to elucidate critical factors in the symbiosis between prokaryotes and their symbiotic partners. Most of the studies relate to legumes, due to their agricultural importance and the relative lack of information for other associations. Recent developments in studies of infection and nodule development will be considered, but the main emphasis will be given to tissue culture-bacteria associations and the observation of N_2 fixation by rhizobia in plant-free culture.

Infection

Since 1973, when Hamblin and Kent suggested that lectins, such as phytohaemaglutinin, could be responsible for binding rhizobia to the root hairs, a number of papers on the topic have appeared. Lectins from certain hosts react only with those rhizobia able to nodulate that host; however, not all rhizobia capable of nodulating the host will react (Bohlool and Schmidt, 1974; Dazzo and Hubbell, 1975a,b). Dazzo and Hubbell (1975b) proposed that the root hairs of the legumes and their nodulating rhizobia have common surface antigenic determinants, and that the multivalent lectins act as a bridge between the bacteria and the host; cellulose microfibril formation by rhizobia may also be involved (Napoli *et al.*, 1975). An alternative hypothesis is that the cell wall lipopolysaccharides (O-antigens), rather than the exopolysaccharides, are the bacterial component reacting with the lectin (Wolpert and Albersheim, 1976).

However, attachment is only the first step towards symbiosis. Root hair invasion, infection thread growth and nodule initiation have received little attention in the past 2 years.

The difficulties in studying invasion processes are exemplified in the work of Napoli and Hubbell (1975) and the failure, again, to involve pectic and cellulolytic enzymes in the infection process (Hunter and Elkan, 1975). Even so, knowledge of the processes in legumes heavily outweighs that from the nodulating non-legumes. Although the endophyte of *Alnus* nodules gains entry to the host through the root hairs and the development of the nodules has been described (Angulo *et al.*, 1976), further detailed studies with other non-legumes are needed. It is with these species, and with the *Spirillum*-grass associations, that the most valuable leads may come for the extension of symbiotic associations to other plants.

Nodule Development

The cytology and anatomy of nodule development has been recently reviewed for legumes (Libbenga and Bogers, 1974; Dart, 1976) and non-legumes (Angulo *et al.*, 1976; Gardner, 1976). At the biochemical level there is increasing evidence that glutamine synthetase (GS)(Dunn and Klucas, 1973; Brown and Dilworth, 1975; Kurz *et al.*, 1975; Robertson *et al.*, 1975; McParland *et al.*, 1976) may be responsible for the assimilation of the ammonia released by bacteroids (Bergersen, 1965; Kennedy, 1966) rather than the more common enzyme of ammonia assimilation, glutamate dehydrogenase (GDH). The enzyme(s) responsible for the synthesis of asparagine, the common and economical transport form in nodulating legumes, remain to be characterized. In non-legume nodules, the transport product is citrulline, formed by the action of ornithine carbamyl transferase in the host cell mitochondria and endoplasmic reticulum (Gardner, 1976).

Legume-Tissue Culture Associations

As a means of attempting to elucidate the factors controlling infection, nodule formation, and the induction of nitrogenase activity, Holsten *et al.* (1971) developed a soybean cell tissue culture-*Rhizobium japonicum* system in which they detected nitrogen fixation. Close examination of the tissue revealed structures akin to infection threads between the cells and that 1-10% of the plant cells were filled with bacteria which, at times, were enclosed in vesicles.

Since then, there have been a number of reports showing C_2H_2-dependent C_2H_4 production in tissue culture associations. Child and LaRue (1974) and Phillips (1974a) found that soybean cells induced nitrogenase activity in some strains of *Rhizobium japonicum*, although in Phillips' investigation, strains of cowpea rhizobia (*Rhizobium* sp.) showed highest activity. Host genotypes were confirmed as important in determining the level of activity developed, and although there was a broad correlation between the activity in nodules and in tissue culture associations, this has not been investigated further. Further investigation resulted in enhanced nitrogenase activity when various nitrogen sources, especially glutamine, and citric acid cycle intermediates were included in the agar medium (Phillips, 1974b, 1976; Child and LaRue, 1976). In all cases, the cultures were grown under air, and 0.2 atm O_2 in the assay gas mixture promoted highest activity. The conclusion apparent to most was that the tissue culture cells were providing respiratory protection to the rhizobia.

In only one of these five papers was comment made about infection of the host cells, and this was to the effect that only a low proportion of cells contained bacteria, that these bacteria were not enclosed in a membrane envelope, and that there was no evidence for true infection threads (Child and LaRue, 1976). However, Ranga Rao *et al.* (1974) then reported results similar in most respects to those of Holsten *et al.* (1971), except that they used *Pisum sativum* cells inoculated with *R. leguminosarum*. Furthermore, Reporter *et al.* (1975) observed that soybean cells in liquid suspension became infected when inoculated with *R. japonicum*, but as with a previous study (Davey *et al.*, 1973), proof of symbiotic N_2 fixation in this situation has yet to be presented.

In the past two years, development of the system has occurred in three directions. Firstly, associations have been shown for a range of legumes and the rhizobia normally capable of nodulating these legumes - cowpea (Scowcroft and Gibson, 1975), lupins (Werner and Oberlies, 1975), *Stylosanthes gracilis* and red clover (Rango Rao, 1976). Child (1975), using strains of cowpea rhizobia and different sources of host cells (peas, sweet clover, *Vicia hajastana*), confirmed that the ability of a strain to nodulate a host was not essential for the induction of nitrogenase in the bacteria. Of interest also was the observation of an N_2-fixing association between *Azotobacter* and cultured carrot cells in which each "symbiont" was dependent on the other (Carlson and Chaleff, 1974).

The second development concerned the induction of nitrogenase activity in 6 strains of *R. japonicum* separated from suspensions of cultured soybean cells by 0.2 to 0.4 micron filters (Reporter and Hermina, 1975). Levels of activity comparable to those obtained with bacteroids from nodules were found when the bacteria were grown adjacent to a pot containing medium from which the plant cells had been filtered. Hopefully, this work will permit definition of any essential factors excreted by the cultured plant cells.

The third development involved tissue-cultured cells of non-legumes and strains of cowpea rhizobia. Due to the repeated failure of many researchers to show N_2 fixation by rhizobia in pure culture, Dilworth and Parker (1969) proposed that the host plant synthesized the protein, or part of the protein, needed for complete nitrogenase synthesis in the bacteria. Alternatively, host mRNA transcribed for nitrogenase may have been transmitted to the bacterial ribosomes. Thus it was logical that tissue cultured cells of non-legumes be examined to ascertain whether they may have the essential factors.

Successful induction of nitrogenase activity in strain 32H1, in association with cultured cells of non-legumes, was described simultaneously by Child (1975) and Scowcroft and Gibson (1975), the former working with wheat, brome grass and rapeseed, the latter with tobacco. $^{15}N_2$ incorporation into the callus, but particularly into the bacteria underlying the supporting filter paper, was shown in the tobacco associations. As with the legume cell associations, maximum activity was observed with 0.2 atm O_2 in the gas mixture (Gibson *et al.*, 1976a). Cultures of strain 32H1 grown alone on the agar medium failed to develop nitrogenase activity. The results of this study indicated that these 4 non-legumes all possessed the factor(s) necessary for the induction of nitrogenase, and that these factor(s) were not the unique property of legumes. They also indicated that if infection of non-legume roots by rhizobia could be achieved, the environment would not be unfavourable for the growth of the bacteria, and possibly for N_2 fixation. Nitrogen fixation by rhizobia in *Trema cannabina* nodules (Trinick, 1973) confirms that this optimism is warranted.

In these studies, N_2 fixation by rhizobia remaining on the agar continued after the callus was removed. Furthermore, nitrogenase activity could be induced in strain 32H1 by culturing the bacteria adjacent to, but not in contact with, tobacco (Scowcroft and Gibson, 1975) and rice (LaRue *et al.*, 1975) cells. This indicated that the "inducing" factor was diffusible and therefore more likely to be nutritional than regulatory in nature.

Various organic nitrogen sources proved adequate to induce nitrogenase activity in the tobacco-strain 32H1 association, with best activity at 5 mM N (Gibson *et al.*, 1976a). At this concentration, both NO_3^- and NH_4^+ inhibited nitrogenase induction. Apart from strain 32H1, a number of strains of cowpea rhizobia and *R. japonicum* showed nitrogenase activity in association with tobacco cells. However, the majority of strains tested, including *R. japonicum* 61A76 which had been induced previously (Holsten *et al.*, 1971; Child and LaRue, 1976; Phillips, 1974a) were nitrogenase inactive.

N_2 Fixation in Cultured Rhizobia

In parallel with the tissue cultures studies, circumstantial evidence was accumulating which indicated that rhizobia possessed all the genes necessary for nitrogenase synthesis and function. Dunican and Tierney (1974) had reported the probable transfer of "nif" genes from *Rhizobium trifolii* to a non-fixing strain of *K. aerogenes*, using an R plasmid from *P. aeruginosa*. Phillips *et al.* (1973) and Whiting and Dilworth (1974) found, on the basis of several criteria, close similarity between MoFe protein fractions of nitrogenase from bacteroids of the same strain but taken from nodules on genetically different legumes. When different cowpea strains were used to nodulate cowpeas, the amino acid composition of the two MoFe proteins differed. Furthermore, Bishop *et al.* (1975) detected cross-reactivity between antisera prepared against MoFe protein and extracts of *R. japonicum* cells grown and assayed under various conditions.

With this background of genetical, biochemical and tissue culture evidence, it was not surprising that five laboratories should simultaneously produce evidence for N_2 fixation in cultured rhizobia. In three cases, the bacteria were grown under air on agar (Kurz and LaRue, 1975; McComb *et al.*, 1975; Pagan *et al.*, 1975), whilst the other cases involved liquid suspensions, under a low O_2 tension (Keister, 1975; Tjepkema and Evans, 1975). The Canberra and Saskatoon groups' observations arose from a consideration of nutritional conditions, in particular the supply of a carbon source more favourable for growth than the sucrose in the tissue culture medium, and the need for a low concentration of combined nitrogen. The Perth group found activity in their "no plant cell" controls (inoculated with 32H1 only), and after much worry and checking, concluded that nitrogenase had been induced in the bacteria.

The only common ingredients to all systems were strain 32H1, the inorganic salts, inositol and vitamins. However, strain 32H1 is not unique as 22 other strains (9 *Rhizobium* sp., 12 *R. japonicum*, 1 *R. leguminosarum* - Keister, 1975; Kurz and LaRue, 1975; Pagan *et al.*, 1975) also show nitrogenase activity in culture; subsequently we have shown additional strains with an inducible nitrogenase. Neither inositol nor vitamins are essential for activity in strain 32H1 (Gibson *et al.*, 1976b). Most of the systems contained two carbon sources, with pentoses and citric acid cycle intermediates predominant, but one carbon source is adequate (Tjepkema and Evans, 1975; Gibson *et al.*, 1976b). As a range of sources of combined nitrogen can be used, from inorganic NO_3^- and NH_4^+ salts to casamino acids, no specific requirement for a particular N-source can be invoked. Glutamine, at 1-2 mM, appears to promote the highest activity (Gibson *et al.*, 1976b). High phosphate enhanced activity in 32H1 (Keister, 1975; Gibson *et al.*, 1976b); this is probably a pH buffer effect, extending the period of activity.

All agar cultures were grown under air, and where tested, showed a sharp peak in maximum nitrogenase activity at pO_2 = 0.2 atm. In liquid cultures, microaerophilic conditions were required. In chemostat cultures of strain 32H1, activity was not observed until the O_2 concentration in solution fell to about 1 μM (Bergersen *et al.*, 1976). Furthermore, the highest levels of activity were obtained when the chemostat-grown cells were assayed in the presence of leghaemoglobin, indicating the importance of maintaining the low pO_2 in the immediate presence of the cells. This aspect has been discussed previously (Bergersen, 1976). The best explanation for the activity in the agar grown cultures is that there is a zone within the mass within which the bacterial cells have nitrogenase activity (Kurz and LaRue, 1975; Pagan *et al.*, 1975). This is supported by the pO_2 concentration data, by the loss of activity when the culture is disturbed, and by the sharp temperature optimum for activity, such optimum being the growth temperature. Hence it would appear that the cultures provide a degree of respiratory protection for some bacteria within the mass. The state of the polysaccharide formed by the bacteria could be crucial in determining O_2 diffusion into the culture.

The necessity for combined nitrogen in the initial agar medium has, so far, been absolute. Removal of glutamine from the influent medium in chemostat culture led to increase in activity, but this was not sufficient for continued exponential growth (Bergersen, 1976). However, when active cultures growing on Millipore filter discs on glutamine-containing agar were transferred to N-free agar containing 100 mM arabinose, activity could be maintained at a reasonable rate for 6 weeks (20-40 nmole C_2H_4/culture/h). The protein level rose from 1 mg/culture to 4.3 mg/culture during that period, suggesting a C_2H_2/N_2 ratio of 1.5:1, but with wide limits due to variability between assayed discs.

The principal advantage gained in showing that strains of rhizobia can fix nitrogen in pure culture is that the bacteria can be studied independently of the host. Hopefully, this will lead to better definition of the relationship between plant and bacteria in the symbiotic situation. With this information attention can be directed towards the plant or the bacteria, as is appropriate, to achieve higher, and more efficient, rates of nitrogen fixation. Furthermore, in association with studies concerning the infection of the plant and nodule initiation, the information can be used to explore the feasibility of extending the symbiosis to agriculturally-important non-legumes. Similar considerations will apply to the symbiosis between non-legumes and their microendophyte when techniques have developed to a stage where the latter can be readily cultured. The studies of Lalonde and Knowles (1975) suggest this is imminent; the next step will be to seek N_2 fixation in culture.

SOME CURRENT INVESTIGATIONS

The investigations in the Canberra laboratories have continued in several directions. Apart from the studies required to define the conditions required for N_2 fixation by rhizobia, using the chemostat (Bergersen, 1976), an attempt has been made to study the regulation and control of nitrogenase synthesis in strain 32H1. Various nitrate reductase deficient mutants have been used to investigate the relationship between nitrogenase and nitrate reductase, and of the inhibition of nodulation and N_2 fixation by nitrate.

Regulation and Control

A considerable body of evidence implicates glutamine synthetase (GS) as having a key role in the synthesis of nitrogenase in *Klebsiella pneumoniae* (Streicher *et al.*, 1974; Tubb, 1974; Bishop, McParland and Evans, 1975; Shanmugan and Valentine, 1975a,b) and possibly in *Azotobacter vinelandii* (Gordon and Brill, 1974). The state of adenylylation of the enzyme is critical, with the deadenylylated form (normally found when the NH_4^+ level in the cells is low) possibly acting as the promoter of the "nif" operon (Shanmugan and Valentine, 1975b; Valentine, 1976). An understanding of the control of nitrogenase synthesis in rhizobia is essential if these bacteria are to be used to extend the range of plants able to form symbiotic N_2-fixing associations. The experimental approach adopted has been to study NH_4^+ repression of nitrogenase and to try and ascertain the relationship between nitrogenase and GS.

As stated previously, a source of combined nitrogen is essential for nitrogenase synthesis in rhizobia, and relatively high levels of nitrogenase activity can be achieved with concentrations of NH_4^+ in the medium that would be expected to inhibit nitrogenase synthesis in other diazotrophs (Gibson *et al.*, 1976b). However, nitrogenase activity in *A. vinelandii* and *K. pneumoniae* was found in this system (10 mM NH_4^+) and along with the observations on strain 32H1, there was reason to believe that NH_4^+ levels *within* the cultures were less than that required to repress nitrogenase synthesis. In order to study repression of nitrogenase in active cultures, a system was developed in which rhizobia were grown in soft agar on Millipore filter discs placed on CS8 medium (100 mM arabinose, 12.5 mM succinate, 2 mM glutamine, 30 mM KH_2PO_4 and the other nutrients as in CS7 - Pagan *et al.*, 1975; Gibson *et al.*, 1976b). When the disc cultures showed reasonable activity (15 nmoles C_2H_4/

culture/h, or higher) they were transferred to CS10, containing 100 mM arabinose and inorganic salts, and with pH 7.0 (Scowcroft *et al.*, 1976). Various inhibitors of nitrogenase synthesis, and of the enzymes of NH_4^+ assimilation, were added to CS10.

(i) Ammonia, glutamine and nitrate all inhibited nitrogenase activity, with the concentrations required for 50% inhibition in a one hour assay being 1.5 mM, 0.1 mM, and 0.15 mM respectively.

(ii) Ammonia inhibited nitrogenase immediately after transfer. The level of inhibition remained constant with time and was dependent on NH_4^+ concentration (Fig. 1). Activity was not

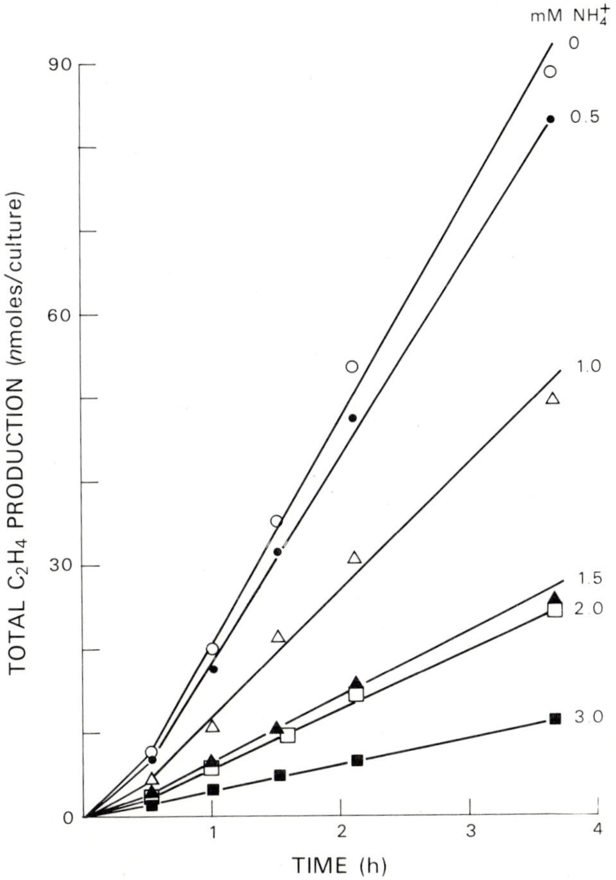

Fig. 1. *Effect of NH_4^+ concentration (mM) on C_2H_4 production (nmole) by disc cultures of strain 32H1 during a 220 minute assay period following transfer of discs to CS10 medium.*

restored on transfer to N-free CS10.

(iii) Methionine sulfoximine and methionine sulfone are inhibitors of the GS and GOGAT enzymes involved in NH_4^+ assimilation. They also allow derepression of nitrogenase synthesis in *K. pneumoniae* and *A. vinelandii* in the presence of NH_4^+ (Gordon and Brill, 1974). With nitrogenase-active cultures of strain 32H1, neither compound was able to prevent NH_4^+ inhibition of nitrogenase activity, even at concentrations that were slightly inhibitory in themselves. Exposure of the discs to methionine sulfoximine for 2 hours prior to transfer to NH_4^+-media did not prevent inhibition.

(iv) Disc cultures of strain 32H1 showing inhibition by NH_4^+ of nitrogenase activity were removed from the discs, and extracts made for the determination of GS activity, in both the presence and absence of Mg^{2+}, by the γ-glutamyl transferase assay (Stadtman *et al.*, 1970; Shapiro and Stadtman, 1972). No change in total GS activity was detected, nor was there any detectable change in the state of adenylylation. NADH-dependent and NADPH-dependent GDH activity was assayed (Meers *et al.*, 1970; Brown and Dilworth, 1975). No GDH activity could be detected in the cultures, even after 24 h exposure to 2, or 4 mM NH_4^+.

In summary the nature of the inhibition by NH_4^+ (constant with time), the inability to mitigate the inhibition with methionine sulfoximine and methionine sulfone, the failure to detect a change in state of adenylylation of GS, and the absence of GDH, indicated that control of nitrogenase synthesis and activity in strain 32H1 was not the same as that observed in wild-type *Klebsiella pneumoniae* (Scowcroft *et al.*, 1976).

As earlier observations indicated that the optimum pO_2 for nitrogenase activity in cultured 32H1 was 0.15-0.20 atm (Pagan *et al.*, 1975) or 0.20-0.25 atm (Gibson *et al.*, 1976b), all assays were done with 0.2 atm O_2 in the gas mixture. Based on the hypothesis that nitrogenase activity in these cultures may be restricted to a zone in which the O_2 concentration was optimal for the development of activity, it was considered that NH_4^+ and glutamine may alter the oxygen demand of the entire culture so that the O_2 concentration in the zone was no longer adequate for maximum activity. When this was tested, it was found that the apparent inhibition by NH_4^+ of nitrogenase was a consequence of a changed relationship between activity and gas phase O_2 concentrations. The inhibition by NH_4^+ of nitrogenase activity could be alleviated by raising the pO_2 of the gas mixture from 0.2 to 0.3 atm (Fig. 2).

The results with NH_4^+ were mimicked by changing the carbon source in the CS10 medium. Previous experiments (Gibson *et al.*, 1976b) had shown that different Citric Acid Cycle intermediates varied in their support of nitrogenase activity when included

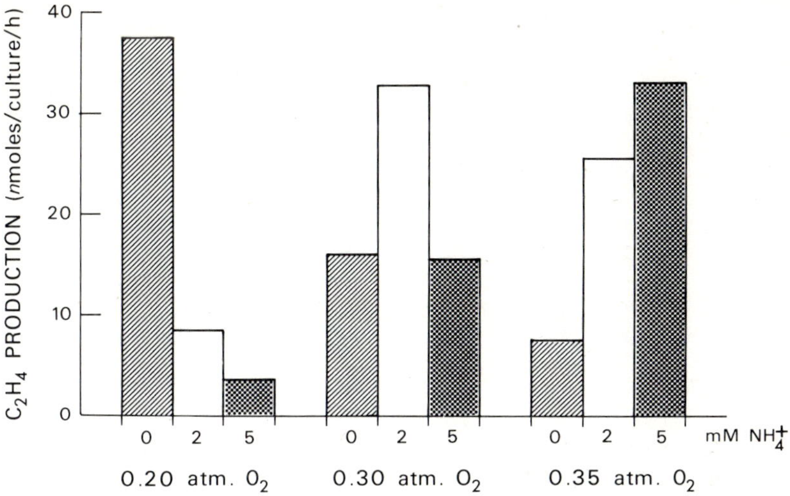

Fig. 2. *The interaction between NH_4^+ and pO_2 on nitrogenase activity in 7 day disc cultures of strain 32H1.*

in the medium on which cultures were grown. When disc cultures were transferred to CS10 containing various carbon sources, strong interactions between the level of nitrogenase activity and the gas phase O_2 concentration were observed (Table I).

Table I

Effect of various carbon sources (0.67 g/100 ml, equivalent to 25 mM succinate) on nitrogenase activity (nmole C_2H_4/disc/hr) during two assays with 0.2 and 0.3 atm.O_2 in the assay gas mixture. pH adjusted to 7.0.

Carbon source	pO_2 = 0.2 atm.		pO_2 = 0.3 atm.	
	0-140 min	140-320 min	0-70 min	70-220 min
Nil	52.4	15.5	49.3	9.7
Inositol	55.1	18.1	51.9	10.5
Glucose	53.5	22.2	57.6	11.2
Malate	52.2	73.8	55.4	40.3
Arabinose	31.2	48.5	61.3	52.3
Pyruvate	26.9	36.3	59.2	63.2
Succinate	14.2	18.9	60.2	73.0
2-Oxoglutarate	7.9	11.4	26.3	23.8
Lactate	5.4	3.5	36.6	37.0

These results are consistent with the suggestion that oxygen demand in the whole culture was stimulated and the O_2 concentration in the "nitrogenase active" zone was altered. Increasing the pO_2 from 0.20 to 0.275 atm in the assay gas mixture increased the level of nitrogenase activity in cultures inhibited at 0.2 atm O_2 (Table II), indicating that these cultures were O_2-deficient at the lower O_2 concentration. However, when cultures with low activity in assay with 0.35 atm O_2 in the gas mixture were provided with 0.275 atm O_2, no recovery was detected, indicating that irreversible O_2 inhibition had occurred. A simplified model attempting to explain these results is shown in Fig. 3. Based on results with the chemostat (Bergersen *et al.*, 1976), it is tempting to suggest that the O_2 concentration favouring expression of nitrogenase activity is about 0.1 μM.

Table II

Effect of transferring disc cultures of strain 32H1 to media containing different citric acid cycle intermediates, and assaying with 3 pO_2 levels in the gas mixture. After the stated periods, some discs were regassed with 0.275 atm O_2 in the gas mixture (nmole C_2H_4/culture/h).

Treatment	Malate	Pyruvate	Lactate	Succinate
pO_2 = 0.20 0-100 min	45.4	27.8	10.9	9.1
100-170 min	67.2	33.8	14.0	14.4
Transfer to pO_2 = 0.275 atm (70 min)	26.7	53.9	59.2	41.6
pO_2 = 0.275 atm 0-110 min	46.2	63.1	41.7	39.0
pO_2 = 0.35 atm 0-110 min	7.4	15.5	-	-
100-170 min	10.4	22.8	-	-
Transfer to pO_2 = 0.275 atm (230 min)	7.6	16.8	-	-

Returning to a consideration of the inhibition of nitrogenase activity by NH_4^+, there is good reason to believe that strain 32H1 is little affected by concentrations of NH_4^+ that would repress nitrogenase synthesis in other N_2-fixing microorganisms. Keister (1975) was unable to detect NH_4^+-induced repression in strain 32H1 grown in liquid culture, while Bergersen *et al.* (1976) found that NH_4^+ caused inhibition when added to chemostat cultures under conditions where high local concentrations of O_2 existed; under assay conditions where the O_2

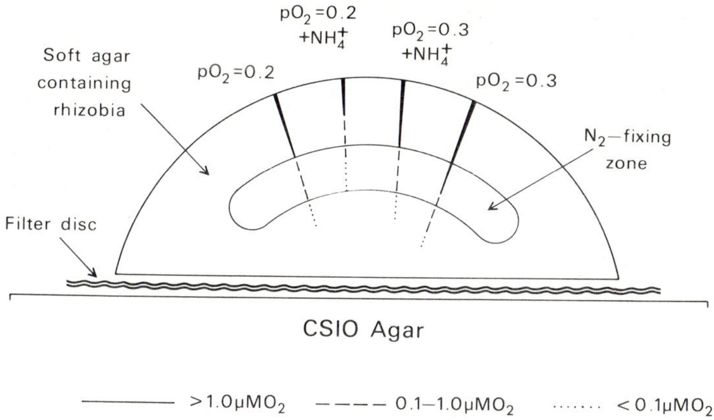

Fig. 3. *Proposed model for the effects of* NH_4^+ *and* O_2 *concentration on* O_2 *supply to the zone of* N_2*-fixing bacteria within disc cultures of strain 32H1.*

concentration and supply was stabilized by leghaemoglobin, 5 mM NH_4^+ had no effect on nitrogenase activity in an 80 minutes assay. One possible explanation of the results (Scowcroft *et al.*, 1976) is that strain 32H1 is similar to those mutant strains of *K. pneumoniae* which synthesise GS constitutively and in which GDH formation is repressed (Brenchley *et al.*, 1973; Shanmugam and Valentine, 1975a). Another possibility is that strain 32H1 may be defective in its ability to adenylylate GS (Foor *et al.*, 1975) so that there is non-repressible synthesis of the biologically-active unadenylylated GS. Numerous other possible explanations can be offered to explain the control of nitrogenase in strain 32H1, but these must await further developments in techniques, based on a better understanding of the conditions needed for synthesis and function of the enzyme. At present we know little of the turnover rates of nitrogenase and distinction between repression and inhibition of activity is difficult.

Nitrate Reductase Deficient Mutants

Kondorosi *et al.* (1973) and Sik *et al.* (1976) used nitrate reductase deficient mutants of *Rhizobium meliloti* to examine the possible commonality of a component of nitrate reductase and nitrogenase. A proportion of their mutants showed low, or no, nitrogenase activity in symbiosis in *Medicago sativa*.

Using resistance to chlorate as the basis of the selection procedure, a number of nitrate reductase deficient mutants of strain 32H1 were isolated (Pagan *et al.*, 1976) and used to examine the relationship between nitrate reductase and nitrogenase, as well as the nature of nitrate inhibition of nitrogenase in culture. Of 283 mutants selected, 148 had low, or undetectable nitrate reductase (NR) activity. Forty six of these isolates showed levels of nitrogenase activity comparable with the parent strain 32H1 when grown on CS8 medium. Of the other two examined, one (NR293) had very low activity (less than 2% of 32H1) and failed to nodulate both *Macroptilium atropurpureum* (Siratro) or *Vigna unguiculata* (cowpea, cv. Meringa). The other, NR165, had low nitrogenase activity in culture (less than 20% of 32H1), and although it nodulated Siratro promptly, the nodules were 4-5 days slower in providing adequate nitrogen for plant growth than those formed by 32H1 and 3 other mutants. Single colony isolates from the NR165 nodules also had low nitrogenase activity, and very low NR activity (less than 1% of 32H1). The three other nitrate reductase deficient mutants tested formed a fully effective symbiosis with Siratro, and these mutants, plus NR165, and NR293, gave a reaction of identity with 32H1 in gel-immune diffusion analysis. Hence in strain 32H1, and the mutants derived from this strain, there was a poor correlation between nitrate reductase deficiency and low nitrogenase activity in culture, indicating that the mutations were not concerned with a factor common to nitrogenase and NR. However, as found by the Hungarian workers, some NR deficient mutants were also altered in their symbiotic characteristics, and such mutants also had low nitrogenase activity. Of interest to the question as to the location of genes responsible for N_2 fixation and symbiosis is the fact that the non-nodulating mutant (NR293) contains a plasmid, as does parent strain 32H1 (J.B. Langridge, personal communication).

Nitrogenase activity in disc cultures of the mutants transferred to CS10 medium containing nitrate, was unaffected up to 5 mM NO_3^-, and in some cases 25 mM NO_3^-. Parent strain 32H1 was 50% inhibited by 0.1 mM nitrate. Nitrite severely inhibited nitrogenase activity in both parent 32H1 and the mutant strains. Hence nitrate inhibition of nitrogenase activity in cultured 32H1 is due to an effect of nitrite, similar to that described for *R. japonicum* and *R. lupini* bacteroids and crude extracts of these bacteroids (Rigaud *et al.*, 1973; Kennedy *et al.*, 1975).

Inhibition of Nodulation

Inhibition of nodule initiation by nitrate has been attributed to the catalytic destruction of indolelacetic acid (IAA) by nitrite, with both IAA and nitrite being produced by the rhizobia (Tanner and Anderson, 1964). IAA is considered essential for infection and nodule initiation (Kefford *et al.*, 1969; Libbenga and Torrey, 1973) and support for the Tanner-Anderson hypothesis is found in the partial alleviation of nitrite inhibition by supplemental IAA (Valera and Alexander, 1965; Munns, 1968). To test this hypothesis, Siratro plants were inoculated with strain 32H1 and each of 4 nitrate reductase deficient mutants, and various levels of NO_3^- were added to the medium. Nitrate delayed the time of initial nodulation by the nitrate reductase deficient mutants (Gibson and Pagan, 1976). Similarly, the nodulation of *Trifolium subterraneum* by two nitrate reductase deficient mutants of *Rhizobium trifolii* strain TA1 (nitrate reductase in this strain is inducible by nitrate) was delayed in the presence of NO_3^-. These results suggest that the Tanner-Anderson hypothesis requires modification. If the delayed nodule initiation is due to destruction of IAA by NO_2^-, the NO_2^- must be produced by the plant. A more likely explanation is that initiation is retarded by a nitrate-induced carbohydrate insufficiency in the roots, a possibility recognised by Tanner and Anderson (1964).

Inhibition of Symbiotic Nitrogen Fixation by Nitrate

The inhibitory effects of nitrate on nodulation and nitrogen fixation are attributed to "photosynthate deprivation", occasioned by the use of photosynthate in the NO_3^--assimilatory processes depriving the nodules of their carbohydrate requirements (e.g. Oghoghorie and Pate, 1971). However, it is feasible that NO_3^- entering the nodules will be reduced to NO_2^-. Virtanen (1950) suggested that NO_2^- could form a complex with leghaemoglobin, and thus destroy its role as a supplier of O_2 to the bacteroids. Alternatively, NO_2^- could inhibit nitrogenase activity (Rigaud *et al.*, 1973; Kennedy *et al.*, 1975).

To test the hypothesis that NO_2^- could reduce the role of the nodule as a sink for photosynthate, Siratro plants nodulated by strain 32H1 and each of 4 NR deficient mutants, were supplied with 1 and 3 mg NO_3^--N 23 days after inoculation (Gibson and Pagan, 1976). Acetylene reduction assays 24 and 48 h later showed a decline in nitrogenase activity by all plants supplied with KNO_3, regardless of the strain forming the nodules (Fig. 4). Nodule weight determinations 48 h after NO_3^- addition indicated that the immediate effect of NO_3^- was to inhibit further their nodule growth. There was also a decline in the nitrogenase specific activity of the nodules. Nitrate reductase activity was high in the 32H1-nodules but could not be detected in nodules formed by the mutant strains. Similar results were attained when nitrate was applied to *Trifolium subterraneum* plants nodulated by strain TA1 and the two nitrate reductase deficient mutants of this strain.

The failure to prevent inhibition by NO_3^- of nitrogen fixation in nodules produced by nitrate reductase deficient mutants suggests that the effect of NO_3^- is due to factors other than

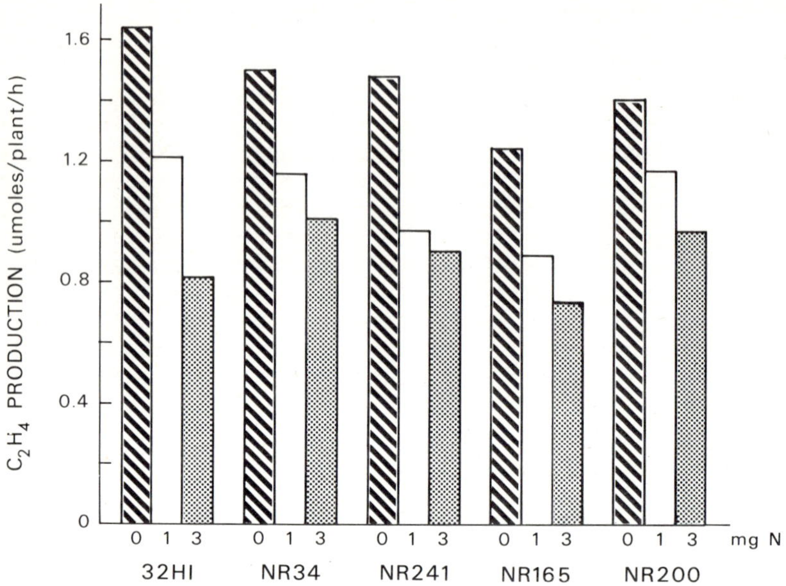

Fig. 4. *Nitrate effects on* C_2H_4 *reduction by* Macroptilium atropurpureum *plants nodulated by strain 32H1 and each of 4 nitrate reductase deficient mutants derived from 32H1. Nitrate added 48 h prior to assay.*

NO_2^- produced by the bacteria. The immediate effect on nodule growth could be used as evidence to support the photosynthate deprivation hypothesis. Another objective in using the nitrate reductase deficient mutants was to examine the feasibility of this approach in selecting strains for commercial inoculants, such strains being able to nodulate, and fix nitrogen, in the presence of moderate levels of nitrate. The results obtained indicate that this approach would have no sound basis.

Variation Within Cultures

Variability between single colony isolates with regard to nitrogenase activity in culture, and in symbiotic performance with Siratro, has been examined, using 20 single colony isolates of the strains of cowpea rhizobia 32H1 and CB756. With both strains, there were significant differences between the 20 isolates in the level of nitrogenase activity after 6 days on CS7 medium, although the growth of all cultures (dry weight) was similar. When the same isolates were tested for symbiotic effectiveness on Siratro grown with only the roots inside test-tubes and the shoots freely exposed above an Al-foil cap (Gibson

1965; Vincent, 1970), all isolates formed an effective symbiosis. No significant differences (P = 0.05) were found between the isolates. These results may be contrasted with those of Herridge and Roughley (1975) who observed differences in symbiotic effectiveness between 15 single colony isolates of CB756, with the more effective isolates being those isolated as small colonies. On CS7 medium, those isolations made from small colonies were not significantly different from each other in the level of nitrogenase activity achieved (Table III). Those from the medium and large colonies showed greater variation within the groups. A lower level of nitrogenase activity was observed among isolates from large colonies. This is similar to the effects observed by Herridge and Roughley (1975) for differences in symbiotic effectiveness.

Table III

Variation in nitrogenase activity (nmole C_2H_4/culture/h) between single colony isolates of strain CB756 grown on CS7 agar for 6 days. The isolates were from small, medium and large colonies. At inoculation, all cultures received same number of cells.

Small		Medium		Large	
Isolate	Rate	Isolate	Rate	Isolate	Rate
3	44.7	13	50.4	27	34.5
2	38.0	11	40.6	24	28.9
1	37.8	12	32.3	26	28.2
5	34.7	17	31.0	22	22.7
6	33.9	14	30.5	21	22.0
7	32.0	16	26.9	25	10.2
		15	21.0	23	0.2
$\bar{x}$ =	36.9	$\bar{x}$ =	33.2	$\bar{x}$ =	21.0

Variation in symbiotic effectiveness between single colony isolates of rhizobia is a feature of some strains (Labandera and Vincent, 1975; Roughley, 1976; Gibson *et al.*, 1976c), and from the results presented here, variation exists in the ability to fix N_2 in culture. The factors responsible for this variation have not been ascertained, but attention to the composition of the stock culture medium is one aspect that should be considered.

In this context, it is of interest to note that strains 29C2 and 32Z3 that fixed N_2 in tissue culture associations involving soybeans, peas, and non-legumes (Phillips, 1974a; Child and LaRue, 1974; Child, 1975), were not active in tobacco cell cultures (Gibson *et al.*, 1976a), nor when grown on CS7 medium

(Pagan *et al.*, 1975). Phillips (1974a) reported a decline in the activity of strain 32H1 over a period of time, whilst Dr. J.J. Child (personal communication) lost all activity with the *R. japonicum* strain 61A76 on two occasions and had to obtain new stock cultures. All our attempts to induce nitrogenase activity in *R. leguminosarum* strain TA101 have been unsuccessful, even after plant passage and various modifications to the media and physical conditions; this contrasts with the results of Kurz and LaRue (1975).

These points are raised *only* to indicate the undefined subtleties which exist in the culture of rhizobia, particularly in trying to induce nitrogenase activity. Until the factors responsible for this variation in, and loss of, nitrogenase activity have been defined, anomalous situations are likely to arise with different laboratories reporting different results. Urgent attention to this aspect of nitrogen fixation by rhizobia in culture is required.

Purity of Cultures

While this situation exists, and even after it is clarified, great care must be exercised in ensuring that the cultures being handled are indeed pure strains of rhizobia. The principal criterion for recognizing *Rhizobium* is its ability to nodulate a legume, ideally the one from which it was isolated or from the same cross-inoculation group (Fred *et al.*, 1932). As a group, there are no distinguishing physiological or morphological characters, and attempts to devise a selective medium (Graham, 1969) have not met universal acceptance. Serological characteristics (Dudman, 1976) and the use of antibiotic resistance, phage resistance, and both biochemical and nutritional markers (Schwinghamer, 1976) can be used for identifying strains once their purity is established. The slow-growing nature of most rhizobia - even *R. trifolii* and *R. leguminosarum* have generation times of the order of 4-6 hours and single colonies of *Rhizobium* sp. and *R. japonicum* usually take 5-7 days to appear - creates special problems in maintaining bacteriological control over cultures. Furthermore, there are considerable difficulties in adequately surface-sterilizing seeds to ensure that all N_2-fixing bacteria are excluded (Evans *et al.*, 1972; unpublished results).

Variation within strains discussed under the previous heading make the task of ensuring the purity of a culture very difficult; by the same reasoning, it is vital that the task be done properly. The same general considerations apply to all N_2-fixing endophytes. By comparison, the task with rhizobia is simple compared with the *Alnus*-type microsymbiont (*Frankia* sp.) and blue-green algae.

THE EXPANDING HORIZONS - IN WHICH DIRECTIONS?

As stated at Pullman (Gibson, 1976a), the greatest barrier to increased biological dinitrogen fixation is the relatively poor yield of legumes, relative to that of grasses (including sugar cane), cereals, and such crops as sunflower and safflower. In terms of protein yield, legumes are the equal of most crops, but without a heavy premium on protein, and while N-fertilizers are reasonably priced and available, legumes will not attract the attention they deserve.

Concentrating research efforts on symbiotic associations, such as those found with bacteria in the nodules on legumes, and *Alnus*-type non-legumes (see Bond, 1976), with blue-green algae in *Gunnera* (Silvester, 1976), *Macrozamia* (Nathaniels and Staff, 1975; Halliday and Pate, 1976), and *Azolla* (Peters, 1975, 1976), and with *Spirillum* in the roots of graminaceous species, appears to offer the greatest scope for increasing biological N_2 fixation. In these systems there is a direct relationship between the host and microsymbiont, the former providing carbohydrate and receiving fixed nitrogen in return. The associative systems, such as found with N_2-fixing bacteria in the rhizosphere of many plants, is less direct, depending on root exudation as a supply of energy, an uncertain passage of fixed nitrogen back into the plants, and the existence of conditions favourable for N_2 fixation. There is also the thought that nitrogen-fixing genes can be incorporated into the plant genome, or into endophytes able to invade plant tissues.

Legumes and Nodulating Non-legumes

Various approaches to increasing symbiotic nitrogen fixation by legumes were considered at Pullman (Gibson, 1976a). Rather than reiterate these approaches, which are still valid and viable, consideration here is given to approaches based more directly on the title of this session, Genetics and Plant Science.

Efficiency of N_2 Fixation The energy costs of assimilating nitrogen, be it N_2 by biological fixation or NO_3^- by reduction, are similar and considerable (e.g. Hardy and Havelka, 1975). Based on Minchin and Pate's (1973) data for peas, which have been confirmed in the essential details for lupins (Pate, 1976), 12 to 17% of the net gain of carbon from the atmosphere (i.e. excluding shoot respiration) may be utilized in nitrogen assimilation by legumes. For nitrogen fixation *per se*, the cost (excluding that required for nodule growth, and that returned to the plant in amines and amino acids) can be estimated as 3.8 mg carbohydrate per mg N_2 fixed. This is in close agreement with the theoretical value proposed by Bergersen (1971) of 3.2 mg/mg N_2, and lower than the value of 6.4 mg/mg N_2 which can be calculated from Hardy and Havelka's estimate (1975) of 24 ATP's (or their equivalent)/molecule N_2 fixed.

Various examples of increased efficiency of N_2 fixation can be cited, involving "super-efficient" strains of *Rhizobium japonicum* that form nodules with twice the rate of N_2 fixation to that found with "normal" strains (Döbereiner *et al.*, 1970); differences in efficiency between *R. trifolii* strains (Gibson, 1966) which in some cases is due to 4-aminobutyrate accumulation in the nodules (Freney and Gibson, 1975); high rates of N_2 fixation in nodules on plants with increased rates of photosynthesis (Streeter, 1974; Gibson, 1976b; Havelka and Hardy, 1976); differences in the rates of N_2 fixation and H_2 evolution from nodules on a range of legumes and non-legumes, with some nodules apparently utilizing as much energy in H_2 evolution as in C_2H_4 production (Schubert and Evans, 1976). Comparison of the growth of *Spirillum* cultures in agar suggests that N_2 fixation, relative to NH_4^+ assimilation, is only 60% efficient in terms of glucose utilized (Day and Döbereiner, 1976) but the situation in plants may be more favourable.

The full significance of these observations has not been evaluated but with N_2 fixation consuming such a high proportion of net photosynthate production, any efforts to increase the efficiency of N_2 fixation could be beneficial to overall plant growth.

Legumes with Low Nitrogen Concentrations Legumes generally have a higher concentration of N in their vegetative tissues, and certainly in their seeds, than most graminaceous species and other non-legumes. They also have lower growth rates and dry matter yields. Penning de Vries (1975) and others have shown that protein synthesis and metabolism is very demanding in terms of energy. Is this a reason for the lower yield of legumes? If so, is it possible to develop legumes that do not have such high concentrations of N in the tissue, and divert the energy so conserved into higher growth rates? As a consequence, such plants may fix even greater amounts of N_2. Or would it be easier to try and get nodules onto grasses and cereals!

Why do Legumes Nodulate? Information as to the physiological processes involved in infection, nodule initiation, and nodule development is accumulating, albeit rather slowly. For any systematic attempt to confer on non-legumes the ability to nodulate, there should be a full inventory of the major physiological processes involved, with an understanding of the role of the host and the rhizobia at each step. With this information, non-legumes can be examined to ascertain which physiological reactions associated with symbiotic development can occur. For example, if lectins are important in the initial stage of infection, what type of lectins, or their equivalent, do non-legumes produce? Is it possible to modify the antigenic surface of rhizobia to recognize these lectins? There is also the problem of root hair invasion, and the question arises as to the uniqueness of this feature to legumes, and to those nodulating non-legumes. The observation of infected cells in suspension tissue cultures (Holsten *et al.*, 1971) raises the possibility that root hairs are not essential, but caution should be exercised in extrapolating from these observations to the whole-plant situation. Another problem is that of nodule initiation. In legumes, either inner (temperate-legumes) or outer (tropical-legumes) cortical cells divide to provide a nodule meristem, possibly as a consequence of the release of cytokinins and indoleacetic and by the rhizobia (Libbenga *et al.*, 1973). In *Alnus* there is also cortical proliferation, but the principal development initiates in the pericycle (Angulo *et al.*, 1976). The question that is raised here is to what extent can the cortical cells of non-legumes be stimulated to provide a nodule-like structure?

The information gained from an examination of *Alnus* and other nodulating non-legumes, and from the *Spirillum* associations, may be more relevant to the extension of symbiotic nodule formation to other plants than the corresponding information from legumes.

How do Nodules Function? Although the state of knowledge of the biochemistry of nitrogenase in rhizobia is well advanced, many questions remain unanswered about its regulation and function. Why is ammonia released from the endophyte when in most N_2-fixing organisms it is used for further growth? Defective synthesis of the NH_4^+ assimilating enzymes GOGAT and GDH, as found in certain mutants of *K. pneumoniae* (Shanmugam and Valentine, 1975a), may be characteristic of some microsymbionts, although most rhizobia examined have GOGAT or GDH (e.g. Brown and Dilworth, 1975). Alternatively, the fall in the level of DNA in *Rhizobium* bacteroids (Dilworth and Williams, 1967; Cannon, 1974; Agarwal and Mehta, 1974; Sutton, 1974) may result in a loss in the ability to synthesize enzymes essential for growth; again, this is not universal (Sutton, 1974; Reijnders *et al.*, 1975). These are but two possibilities.

Although nitrogenase can be induced in some strains of rhizobia grown on defined medium, this has not been possible with other strains. This raises the question of the necessity of a plant factor for nitrogenase synthesis, as suggested by Dilworth and Parker (1969), not only for rhizobia but for the endophyte of *Alnus*-type nodules, and other N_2-fixers such as *Anabaena azollae* in *Azolla* fronds (Peters and Mayne, 1974). At least with *Spirillum*, no host factor appears necessary to induce nitrogenase (Day and Döbereiner, 1976).

Physical conditions within the host appear to be critical for nitrogenase induction and function, particularly with regard to O_2 tension and supply (Bergersen and Goodchild, 1973; Appleby *et al.*, 1976; Bergersen and Turner, 1975). However, little is known of the form of protection against excess O_2, or of the mechanism by which O_2 is supplied to the endophyte, in systems other than legume nodules. The N_2-fixing nodules on *Trema cannabina*, which contain cowpea rhizobia as the endophyte, do not contain leghaemoglobin but do possess active polyphenol oxidases (Coventry *et al.*, 1976). This and the observation that the polyphenol oxidase, tyrosinase, can be reversibly oxygenated, prompted Coventry *et al.* (1976) to suggest this as a possible means by which oxygen protection and oxygen supply are maintained in the nodules of non-legumes. Another alternative is that the mass of bacteria in the nodule provides a degree of oxygen protection for a limited proportion of the population that are engaged in N_2 fixation. This would be similar to the situation described for *Derxia* cultures (Hill, 1971) and postulated for cultures of *Rhizobium* on agar.

The supply of photosynthate has emerged as a major factor controlling rates of N_2 fixation in legumes (e.g. Havelka and Hardy, 1976; Hardy and Havelka, 1975), *Alnus* (Wheeler and Lawrie, 1976), *Gunnera* (Silvester, 1976), and *Azolla* (Peters and Mayne, 1974). However, relatively little is known of the metabolism, if any, of the sucrose before it reaches the endophyte, or of subsequent carbon metabolism in nodules. While bacteroids from soybean nodules respond to some Citric Acid Cycle intermediates (e.g. succinate), but not others (Bergersen and Turner, 1967), their natural substrate has yet to be defined. In soybean nodules (Ching, 1976) and in suspensions of *R. japonicum* bacteroids (Appleby *et al.*, 1975), the level of ATP, the energy charge of the system and the O_2 supply to the system, are important in regulating nitrogenase activity. Of interest is the recent observation that trehalose is present in soybean nodules, but not in other parts of the plant (Streeter, 1976), but the significance of this observation remains to be ascertained.

Tolerance of Nitrogen Fixation to Combined Nitrogen A view is commonly expressed that it would be advantageous to seek strains of rhizobia able to fix N_2 in the presence of combined nitrogen. This approach may be questioned on several grounds. All available information suggests that the inhibitory effects of combined nitrogen on N_2 fixation are caused by a deficiency of photosynthate in the nodules. Hence the selection of strains able to synthesize high levels of nitrogenase in the presence of combined nitrogen, could be a fruitless exercise. Furthermore, the existing data on nitrogenase activity in the cultured cowpea strain, 32H1 (Keister, 1975; Bergersen *et al.*, 1976; Gibson *et al.*, 1976b; Scowcroft *et al.*, 1976) suggests that it is able to tolerate higher levels of NH_4^+ and other sources of combined nitrogen than either *K. pneumoniae* or *A. vinelandii*. A similar situation exists with *Anabaena azollae* (Peters, 1975), while the nitrogenase activity of *Spirillum* in association with *Digitaria* and *Pennisetum* is not inhibited by NO_3^--applications that would retard activity in legume nodules (Döbereiner *et al.*, 1975). A more logical approach may be to select hosts that are less susceptible to inhibition of N_2 fixation in the nodules by NO_3^-, possibly on the basis of lower nitrate reductase activities in the roots, or in the plants (Gibson, 1976a,b).

Mutant strains of rhizobia, or other diazotrophs, able to fix N_2 in the presence of combined nitrogen have considerable value in genetic and biochemical studies, and the comments on not seeking such mutants are only intended to dissuade those with intentions of seeking better strains for field inoculation. Their attentions would be better directed towards the host.

Alternative Strategies

Inherent in the foregoing is the belief that expansion of the range of plants capable of forming symbiotic N_2-fixing associations will come from a thorough knowledge of existing systems. In this section, several alternative approaches are considered, including the incorporation of the nitrogen-fixing genes (*nif*) into the genome of plants and into micro-organisms currently not known to possess these genes.

The ability to fix N_2 is restricted to the prokaryotes. In attempting to ascertain whether eukaryotes could support nitrogen-fixing genes, consideration has been given to the evolutionary development of nitrogenase (most recently by Postgate, 1974a, and Dilworth, 1974), but this has been inconclusive. Apart from the problem of introducing the *nif* genes into the plant genome, and maintaining them there, the crucial factors appear to be whether or not protection against destruction by O_2 can be achieved in the host cytoplasm or its organelles, whether electron transport pathways can lead into the nitrogenase complex, and how will regulation of nitrogenase be effected. Gallon *et al.* (1975) consider that nitrogenase in the blue-green alga *Gloeocapsa* may be associated with the thylakoid membranes of the chloroplasts, and consequently use reductant directly from photosynthesis (photosystem I). This appears an improbable site for nitrogenase, due to high rates of O_2 evolution from the associated photosystem II, at least in higher plants. A site *within* the mitochondria may be more protective, particularly in non-photosynthesizing cells such as in the roots. The electron transport pathway associated with nitrogenase in N_2-fixing bacteria appears to be unique and unlikely to occur naturally in higher plants. Although the precise pathways remain to be elucidated, ferredoxin and flavodoxin are involved (Evans and Phillips, 1975). Finally there is the problem of regulating nitrogenase, not only so it will be synthesized and function, but that its activity will be integrated with the other enzymes in the host cells. Difficulties have been encountered in achieving expression of *K. pneumoniae nif* genes transferred on a plasmid to *Agrobacterium tumefaciens* and *Rhizobium meliloti;* evidently synthesis of the protein occurred to a stage where it would cross-react with *Klebsiella* molybdoprotein antiserum but no further (Dixon *et al.*, 1976).

One problem associated with the expression of the *nif* genes is that bacterial mRNA's are characteristically translated by 70S ribosomes whereas the eukaryotic cytoplasmic mRNA's are translated by 80S ribosomes. The 70S ribosomes in eukaryotes appear to be confined to the mitochondria and chloroplasts, and it is possible that the *nif* genes would have to be associated with the mitochondrial DNA (see also above) if they are to be expressed.

However, these theoretical difficulties will not dissuade geneticists from attempting to get the expression of *nif* genes in plant material. The problems of determining whether transfer has occurred, or the degree to which the protein has been synthesized towards a functional level, are considerable but not insurmountable. The approaches vary, but one depends on the incorporation of a plasmid containing the *nif* genes, and possibly genes connected with regulation and electron transport, into protoplasts of the recipient host (Shanmugam and Valentine, 1975b). In another approach, incorporation of the *nif* genes into a DNA molecule which can be transcribed in plant cells, e.g. Cauliflower Mosaic Virus, may alleviate the translation problem.

A more pragmatic approach, which avoids transfer of *nif* genes, is to attempt protoplast fusion, using cells of a legume (or nodulating non-legume) and a non-nodulating plant of agricultural importance. By this technique, it is hoped that all the genes required for the establishment and function of a symbiotic association will persist through to the plant developed from the resulting callus.

Postgate (1974b) has suggested that *nif* genes be incorporated into benign endocellular pathogens which would proliferate harmlessly in the plant tissue. This approach utilizes the infective capability of the pathogen, and depends on blocking the toxin production from such organisms (*Pseudomonas, Erwinia)*. The permanence of the blockage would be a source of concern, for fear of back mutation to permit pathogenicity on the selected, or even unselected, hosts. *Agrobacterium tumefaciens* is another logical candidate for such treatment, while Postgate (1974b) also suggests that endotrophic mycorrhizae would be a useful carrier. Again, problems of oxygen protection could be paramount. The consequences of these approaches could be exciting - in more ways than one.

CONCLUSIONS

From experience, the nodule-type of symbiotic association, preferably that found with legumes rather than those found with *Alnus*-type non-legumes or possibly the *Spirillum*-grass associations, appeal as the most beneficial type of N_2-fixing association. However, there are many problems in attempting to extend the ability to form such associations into "non-nodulating" plants. The tremendous potential of the *Spirillum*-based associations and their apparent genetic base (Bulow and Döbereiner, 1975) present hope that such associations can be extended to many members of the Family Gramineae. The other approaches, involving the transfer of *nif* genes, or the ability to form symbiotic associations, into plant cells has many problems from both the technical and ecological viewpoint, but in these days where Science Fiction tends to become reality, there is little point in discounting the likelihood of success. However, I believe that existing systems have much to offer, both in the improvements that can be made to them, and in providing essential background information towards a more logical extension of symbiotic associations into plants without this capability at present.

ACKNOWLEDGEMENTS

The senior author appreciates the generosity of the Charles F. Kettering Foundation, Consejo Superior de Investigaciones Cientificas and the University of Salamanca for the opportunity to attend this meeting. The discussions with many colleagues, especially F.J. Bergersen, W.F. Dudman, T.J. Higgins and E.A. Schwinghamer and G.L. Turner, are sincerely appreciated.

REFERENCES

Agarwal, A.K. and Mehta, S.L. (1974). *Biochem.Biophys.Res.Comm.*, 60: 257-265.

Angulo, A.F., Dijk, C. van and Quispel, A. (1976). *In:* Symbiotic Nitrogen Fixation in Plants (P.S. Nutman, ed.), pp. 475-484, Cambridge University Press, Cambridge.

Appleby, C.A., Turner, G.L. and Macnicol, P.K. (1975). *Biochim. Biophys.Acta*, 387: 461-474.

Appleby, C.A., Bergersen, F.J., Macnicol, P.K., Turner, G.L., Wittenberg, B.A. and Wittenberg, J.B. (1976). *In:* Proc. 1st Int.Symp. on Nitrogen Fixation (W.E. Newton and C.J. Nyman, eds), Vol. 1, pp. 274-292. Washington State University Press, Pullman.
Bergersen, F.J. (1965). *Aust.J.Biol.Sci.,* 18: 1-9.
Bergersen, F.J. (1971). *In:* Biological Nitrogen Fixation in Natural and Agricultural Habitats (T.A. Lie and E.G. Mulder, eds), pp. 511-524. Martinus Nijhoff, The Hague.
Bergersen, F.J. (1976). *In:* Proc.2nd Int.Symp. on Nitrogen Fixation (W.E. Newton, J.R. Postgate and C. Rodriguez Barrueco, eds), p. - . Academic Press, New York.
Bergersen, F.J. and Goodchild, D.J. (1973). *Aust.J.Biol.Sci.,* 26: 741-756.
Bergersen, F.J. and Turner, G.L. (1967). *Biochim.Biophys.Acta,* 141: 507-515.
Bergersen, F.J. and Turner, G.L. (1975). *J.Gen.Microbiol.,* 91: 345-354.
Bergersen, F.J., Turner, G.L., Gibson, A.H. and Dudman, W.F. (1976). *Biochim.Biophys.Acta,* 44: 164-174.
Bishop, P.E., Evans, H.J., Daniel, R.M. and Hampton, R.O. (1975). *Biochim.Biophys.Acta,* 381: 248-256.
Bishop, P.E., McParland, R.H. and Evans, H.J. (1975). *Biochem. Biophys.Res.Comm.,* 67: 774-781.
Bohlool, B.B. and Schmidt, E.L. (1974). *Science,* 185: 269-271.
Bond, G. (1976). *In:* The Biology of Nitrogen Fixation (A. Quispel, eds.), pp. 499-520. North Holland, Amsterdam, and Oxford.
Brenchley, J.E., Prival, M.J. and Magasanik, B. (1973). *J.Biol. Chem.,* 248: 6122-6128.
Brown, C.M. and Dilworth, M.J. (1975). *J.Gen.Microbiol.,* 86: 39-48.
Bulow, J.F.W. von and Döbereiner, J. (1975). *Proc.Nat.Acad.Sci., USA,* 72: 2389-2393.
Cannon, F.C. (1974). *Heredity,* 33: 133-134.
Carlson, P.S. and Chaleff, R.S. (1974). *Nature,* 252: 393-395.
Child, J.J. (1975). *Nature,* 253: 350-351.
Child, J.J. and LaRue, T.A. (1974). *Plant Physiol.,* 53: 88-90.
Child, J.J. and LaRue, T.A. (1976). *In:* Proc.1st Int.Symp. on Nitrogen Fixation (W.E. Newton and C.J. Nyman, eds), Vol. 2, pp. 447-455. Washington State University, Pullman.
Ching, T.M. (1976). *Life Sciences,* 18: 1071-1076.
Coventry, D.R., Trinick, M.J. and Appleby, C.A. (1976). *Biochim. Biophys.Acta,* 420: 105-111.
Dart, P.J. (1976). *In:* A Treatise on Dinitrogen Fixation (R.W. F. Hardy and W.S. Silver, eds), Vol. 2A (in press). Wiley-Interscience, New York.

Davey, M.R., Cocking, E.C. and Bush, E. (1973). *Nature*, 244: 460.
Day, J.M. and Döbereiner, J. (1976). *Soil Biol.Biochem.*, 8: 45-50.
Dazzo, F.B. and Hubbell, D.H. (1975a). *Plant and Soil*, 43: 713-717.
Dazzo, F.B. and Hubbell, D. H. (1975b). *Appl.Microbiol.*, 30: 1017-1033.
Dilworth, M.J. (1974). *Ann.Rev.Plant Physiol.*, 25: 81-114.
Dilworth, M.J. and Parker, C.A. (1969). *J.Theoret.Biol.*, 25: 208-218.
Dilworth, M.J. and Williams, D.C. (1967). *J.Gen.Microbiol.*, 48: 31-36.
Dixon, R., Cannon, F. and Kondorosi, A. (1976). *Nature*, 260: 268-271.
Döbereiner, J. and Day, J.M. (1976). *In:* Proc.1st Int.Symp. on Nitrogen Fixation (W.E. Newton and C.J. Nyman, eds), Vol. 2, pp. 518-538. Washington State University Press, Pullman.
Döbereiner, J., Day, J.M. and Bulow, J.F.W. von (1975).Proc.Int. Wheat Conference, Zagreb.
Döbereiner, J., Franco, A.A. and Guzman, I. (1970). *Pesqui. Agropecuar.Brasil*, 5: 155-161.
Dudman, W.F. (1976). *In:* A Treatise on Dinitrogen Fixation (R. W.F. Hardy and A.H. Gibson, eds), Vol. 2B (in press), Wiley Inter-science, New York.
Dunican, L.K. and Tierney, A.B. (1974). *Biochem.Biophys.Res. Comm.*, 57: 62-72.
Dunn, S.D. and Klucas, R.V. (1973). *Can.J.Microbiol.*, 19: 1493-1499.
Evans, H.J. Campbell, N.E.R. and Hill, S. (1972). *Can.J.Microbiol.*, 18: 13-21.
Evans, H.J. and Phillips, D.A. (1976). *In:* Nitrogen Fixation by Free-living Micro-organisms (W.D.P. Stewart, ed.), pp. 389-420. Cambridge University Press, Cambridge.
Foor, F., Janssen, K.A. and Magasanik, B. (1975). *Proc.Nat.Acad. Sci., USA*, 72: 4844-4848.
Fred, E.B., Baldwin, I.L. and McCoy, E. (1932). Root Nodule Bacteria and Leguminous Plants. University of Wisconsin Press, Madison.
Freney, J.R. and Gibson, A.H. (1975). *Aust.J.Plant Physiol.*, 2: 663-668.
Gallon, J.R., Kurz, W.G.W. and LaRue, T.A. (1975). *In:* Nitrogen-Fixation in Free-living Micro-organisms (W.D.P. Stewart, ed.), pp. 159-176. Cambridge University Press, Cambridge.
Gardner, I.C. (1976). *In:* Symbiotic Nitrogen Fixation in Plants (P.S. Nutman, ed.), pp. 485-496. Cambridge University Press, Cambridge.

Gibson, A.H. (1965). *Aust.J.Biol.Sci.*, 18: 295-310.
Gibson, A.H. (1966). *Aust.J.Biol.Sci.*, 19: 499-515.
Gibson, A.H. (1976a). *In:* Proc.1st Int.Symp. on Nitrogen Fixation (W.E. Newton and C.J. Nyman, eds), Vol. 2, pp. 400-428. Washington State University Press, Pullman.
Gibson, A.H. (1976b). *In:* Symbiotic Nitrogen Fixation in Plants (P.S. Nutman, ed.), pp. 385-404. Cambridge University Press, Cambridge.
Gibson, A.H. and Pagan, J.D. (1976). In preparation.
Gibson, A.H. Child, J.J., Pagan, J.D. and Scowcroft, W.R. (1976a). *Planta (Berl.)*, 128: 233-239.
Gibson, A.H., Scowcroft, W.R., Child, J.J. and Pagan, J.D. (1976b). *Arch.Microbiol.*, 108: 45-54.
Gibson, A.H., Date, R.A., Ireland, J.M. and Brockwell, J. (1976c). *Soil Biol.Biochem.*, in press.
Gordon, J.K. and Brill, W.J. (1974). *Biochem.Biophys.Res.Comm.*, 59: 967- 971.
Graham, P.H. (1969). *Appl.Microbiol.*, 17: 769-770.
Halliday, J. and Pate, J.S. (1976). *Aust.J.Plant Physiol.*, 3: 349-358.
Hamblin, J. and Kent, S.P. (1973). *Nature*, 245: 28-30.
Hardy, R.W.F. and Havelka, U.D. (1976). *Science*, 188: 633-643.
Havelka, U.D. and Hardy, R.W.F. (1975). *In:* Proc.1st Int.Symp. on Nitrogen Fixation (W.E. Newton and C.J. Nyman, eds), Vol. 2, p. 456-475. Washington State University Press, Pullman.
Herridge, D.F. and Roughley, R.J. (1975). *J.Appl.Bacteriol.*, 38: 19-27.
Hill, S. (1971). *J.Gen.Microbiol.*, 67: 77-83.
Holl, F.B. and LaRue, T.A. (1976). *In:* Proc.1st Int.Symp. on Nitrogen Fixation (W.E. Newton and C.J. Nyman, eds), Vol. 2, pp. 391-399. Washington State University Press, Pullman.
Holsten, R.D., Burns, R.C., Hardy, R.W.F. and Hebert, R.R. (1971). *Nature*, 232: 173-176.
Hunter, W.J. and Elkan, G.H. (1974). *Can.J.Microbiol.*, 21: 1254 -1258.
Kefford, N.P., Brockwell, J. and Zwar, J.A. (1960). *Aust.J. Biol.Sci.*, 13: 456-467.
Keister, D.L. (1975). *J.Bacteriol.*, 123: 1265-1268.
Kennedy, I.R. (1966). *Biochim.Biophys.Acta*, 130: 295-303.
Kennedy, I.R., Rigaud, J. and Trinchant, J.C. (1975). *Biochim. Biophys.Acta*, 397: 24-35.
Kondorosi, A., Barabas, I., Svab, Z., Orosz, L., Sik, T. and Hotchkiss, R.D. (1973). *Nature*, 246: 153.
Kurz, W.G.W. and LaRue, T.A. (1975). *Nature*, 256: 407-409.
Kurz, W.G.W., Rokosh, D.A. and LaRue, T.A. (1975). *Can.J.Microbiol.*, 21: 1009-1012.
Labandera, C.A. and Vincent, J.M. (1975). *J.Appl.Bacteriol.*, 39: 209-211.

Lalonde, M. and Knowles, R. (1975). *Can.J.Microbiol.*, 21: 1901-1920.
LaRue, T.A., Kurz, W.C.W. and Child, J.J. (1975). *Can.J.Microbiol.*, 21: 1884-1886.
Libbenga, K.R. and Bogers, R.J. (1974). *In:* The Biology of Nitrogen Fixation (A. Quispel, ed.), pp. 430-472. North Holland, Amsterdam and Oxford.
Libbenga, K.R. and Torrey, J.G. (1973). *Amer.J.Bot.*, 60: 293-299.
Libbenga, K.R., Iren, F. van, Bogers, R.J. and Schraag-Lamers, M.F. (1973). *Planta (Berl.)*, 114: 29-39.
McComb, J.A., Elliott, J. and Dilworth, M.J. (1975). *Nature*, 256: 409-410.
McParland, R.H., Guevara, J.G., Becker, R.R. and Evans, H.J. (1976). *Biochem.J.*, 153: 597-606.
Meers, J.L., Tempest, D.W. and Brown, C.M. (1970). *J.Gen.Microbiol.*, 64: 187-194.
Minchin, F.R. and Pate, J.S. (1973). *J.Exptl.Bot.*, 24: 259-271.
Munns, D.N. (1968). *Plant Soil*, 29: 257-262.
Napoli, C.A. and Hubbell, D.H. (1975). *Appl.Microbiol.*, 30: 1003-1009.
Napoli, C., Dazzo, F. and Hubbell, D.H. (1975). *Appl.Microbiol.*, 30: 123-131.
Nathaniels , C.P. and Staff, I.A. (1975). *Amer.J.Bot.*, 62: 232-235.
Oghoghorie, C.G.O. and Pate, J.S. (1971). *In:* Biological Nitrogen Fixation in Natural and Agricultural Habitats (T.A. Lie and E.G. Mulder, eds), pp. 185-202. Martinus Nijhoff, The Hague.
Pagan, J.D., Child, J.J., Scowcroft, W.R. and Gibson, A.H. (1975). *Nature*, 256: 406-407.
Pagan, J.D., Scowcroft, W.R. and Gibson, A.H. (1976). In preparation.
Pate, J.S. (1976). *In:* Transport and Transfer Processes in Plants (I.F. Wardlaw and J.B. Passioura, eds), in press. Academic Press, New York.
Penning de Vries, F.W.T. (1975). *Ann.Bot.*, 39: 77-92.
Peters, G.A. (1975). *Arch.Microbiol.*, 103: 113-122.
Peters, G.A. (1976). *In:* Proc.1st Int.Symp. on Nitrogen Fixation (W.E. Newton and C.J. Nyman, eds), Vol. 2, pp. 592-610. Washington State University Press, Pullman.
Peters, G.A. and Mayne, B.C. (1974). *Plant Physiol.*, 53: 820-824.
Phillips, D.A. (1974a). *Plant Physiol.*, 53: 67-72.
Phillips, D.A. (1974b). *Plant Physiol.*, 54: 654-655.
Phillips, D.A. (1976). *In:* Proc.1st Int.Symp. on Nitrogen Fixation (W.E. Newton and C.J. Nyman, eds), Vol. 2, pp. 367-373. Washington State University Press, Pullman.

Phillips, D.A., Howard, R.L. and Evans, H.J. (1973). *Physiol. Plant.*, 28: 248-253.
Postgate, J.R. (1974a). Symp.Soc.Gen.Microbiol. XXIV, 263-292.
Postgate, J.R. (1974b). *J.Appl.Bact.*, 37: 185-202.
Ranga Rao, V. (1976). *Plant Sci.Letts.*, 6: 77-83.
Ranga Rao, V., Sopory, S. and Subba Rao, N.S. (1974). *Curr.Sci. (India)*, 43: 503-505.
Reijnders, L., Visser, L., Aalbers, A.M.J., Kammen, A. van and Houwers, A. (1975). *Biochim.Biophys.Acta*, 414: 206-216.
Reporter, M. and Hermina, N. (1975). *Biochem.Biophys.Res.Comm.*, 64: 1126-1133.
Reporter, M., Raveed, D. and Norris, G. (1975). *Plant Sci.Letts*, 5: 73-76.
Rigaud, J., Bergersen, F.J., Turner, G.L. and Daniel, R.M. (1973). *J.Gen.Microbiol.*, 77: 137-144.
Robertson, J.G., Farnden, K.J.F., Warburton, M.P. and Banks, J. M. (1975). *Aust.J.Plant Physiol.*, 2: 265-272.
Roughley, R.J. (1976). *In:* Symbiotic Nitrogen Fixation in Plants (P.S. Nutman, ed.), pp. 125-136. Cambridge University Press, Cambridge.
Schubert, K.R. and Evans, H.J. (1976). *Proc.Nat.Acad.Sci.USA*, 73: 1207-1211.
Schwinghamer, E.A. (1976). *In:* A Treatise on Dinitrogen Fixation (R.W.F. Hardy and W.S. Silver, eds), Vol. 2A (in press). Wiley-Interscience, New York.
Scowcroft, W.R. and Gibson, A.H. (1975). *Nature*, 253: 351-352.
Scowcroft, W.R., Gibson, A.H. and Pagan, J.D. (1976). *Biochem. Biophys.Res.Comm.*, in press.
Shanmugam, K.T. and Valentine, R.C. (1975a). *Proc.Nat.Acad.Sci. USA*, 72: 136-139.
Shanmugam, K.T. and Valentine, R.C. (1975b). *Science*, 187: 919-924.
Shapiro, B.M. and Stadtman, E.R. (1972). *In:* Methods of Enzymology (H. Taber and C.W. Taber, eds), Vol. 17A, pp. 910-922. Academic Press, New York.
Sik, T., Kondorosi, A., Barabas, I. and Svab, Z. (1976). *In:* Proc. 1st Int. Symp. Nitrogen Fixation (W.E. Newton and C. J. Nyman, eds), Vol. 2, pp. 374-382. Washington State University Press, Pullman.
Silvester, W.B. (1976). *In:* Symbiotic Nitrogen Fixation in Plants (P.S. Nutman, ed.), pp. 521-538. Cambridge University Press, Cambridge.
Stadtman, E.R., Ginsburg, A., Ciardi, J.E., Yeh, J., Henning, S.B. and Shapiro, B.M. (1970). *In:* Advances in Enzyme Regulation (G. Weber, ed.), Vol. 8, pp. 99-118. Pergamon Press, Oxford.
Streeter, J.G. (1974). *J.Exptl.Bot.*, 25: 189-198.

Streeter, J.G. (1976). Plant Physiol., A.M. Supp., 57: 79.
Streicher, S.L., Shanmugam, K.T., Ausubel, F., Morandi, C. and Goldberg, R.B. (1974). *J.Bacteriol.*, 120: 815-821.
Sutton, W.D. (1974). *Biochim.Biophys.Acta*, 366: 1-10.
Tanner, J.W. and Anderson, I.C. (1964). *Plant Physiol.*, 39: 1039-1043.
Tjepkema, J. and Evans, H.J. (1975). *Biochem.Biophys.Res.Comm.*, 65: 625-628.
Trinick, M.J. (1973). *Nature*, 244: 459-460.
Tubb, R.S. (1974). *Nature*, 251: 481-485.
Valentine, R.C. (1976). *In:* Proc.2nd Int. Symp. on Nitrogen Fixation (W.E. Newton, J.R. Postgate and C. Rodriguez Barrueco, eds), pp. - . Academic Press, New York.
Valera, C.L. and Alexander, M. (1965). *Nature*, 206: 326.
Vincent, J.M. (1970). A Manual for the Practical Study of Root-Nodule Bacteria. p. 82, IBP Handbook No. 15. Blackwell Scientific Publications, Oxford and Edinburgh.
Virtanen, A.I. (1950). *In:* Proc. 7th Int.Bot.Congr., Stockholm, pp. 156-159.
Werner, D. and Oberlies, G. (1975). *Die Naturwiss.*, 62: 350-351.
Wheeler, C.T. and Lawrie, A.C. (1976). *In:* Symbiotic Nitrogen Fixation in Plants (P.S. Nutman, ed.), pp. 497-509. Cambridge University Press, Cambridge.
Whiting, M.J. and Dilworth, M.J. (1974). *Biochim.Biophys.Acta*, 371: 337-351.
Wolpert, J.S. and Albersheim, P. (1976). *Biochem.Biophys.Res. Comm.*, 70: 729-737.

THE PHYSIOLOGY OF ACETYLENE REDUCTION IN PURE CULTURES OF RHIZOBIA

D.L. KEISTER* and V. RANGA RAO

Charles F. Kettering Research Laboratory, Yellow Springs, Ohio 45387, U.S.A.

INTRODUCTION

Since the initial reports on nitrogen fixation (acetylene reduction) by pure cultures of rhizobia appeared (Keister, 1975; Kurz and LaRue, 1975; McComb *et al.*, 1975; Pagan *et al.*, 1975; Tjepkema and Evans, 1975), a number of laboratories have begun utilizing this discovery to further the study of nitrogen fixation in these symbiotic bacteria. Work of this nature will be enhanced by a thorough study of the physiology of nitrogen fixation in this important group of micro-organisms. We recently described the development of nitrogenase activity in stationary (non-shaking) liquid cultures of *Rhizobium sp.* 32H1 and *R. japonicum* 3I1b83 (Evans and Keister, 1976) and their oxygen and nutritional requirements in shake cultures (Keister and Evans, 1976). The major determinant of nitrogenase synthesis and activity was the oxygen concentration. The oxygen requirement, perhaps better described as oxygen tolerance, is lower than for other known aerobic nitrogen-fixing micro-organisms[1]. By appropriately adjusting the oxygen concentration of the gas phase and the cell density of the liquid phase, high specific rates of acetylene reduction can be obtained with several strains of rhizobia.

This paper will describe some physiological factors involved in the induction and control of nitrogenase activity of *R. japonicum* 61A76, as well as further studies of *R. sp.* 32H1 and *R. japonicum* 3I1b83.

---

* To whom inquiries should be addressed: Contribution No. 576 of the Charles F. Kettering Research Laboratory, Yellow Springs, Ohio 45387.

METHODS AND MATERIAL

Stock cultures of rhizobia were maintained and grown as described by Keister (1975). All acetylene reduction experiments were conducted as follows: aerobic cells in the logarithmic phase of growth were harvested, washed and resuspended in 3 mM KH_2PO_4, 0.4 mM $MgSO_4$, 30 μM $NaMoO_4$, 0.1 mM iron citrate and a zwitterionic buffer, 50 mM (except 32H1 in which 10 or 50 mM phosphate was used). The pH was adjusted with NaOH. The experiments with *R. japonicum* 83 and *R. sp.* 32H1 were done in 1 litre flat bottom reagent bottles (Fisher Scientific #2-940-5E) with 20 ml of medium. The experiments with 61A76 were done in 500 ml reagent bottles (Wheaton Scientific #291039) with 10 ml medium and both were sealed with a rubber sleeve stopper (Wheaton #224097) wired tightly shut. Oxygen was removed by repeated evacuation and filling with argon (usually three cycles) until the concentration was less than 0.005%. The bottles were then filled with argon: C_2H_2:CO_2(89:10:1) and oxygen injected to the appropriate level. The cultures were shaken on a rotary shaker at 170 rpm at 27°C. The oxygen concentration was monitored at approximately 12 h intervals and pure oxygen was added to maintain the desired concentration. Oxygen and ethylene were determined by gas chromatography as described by Keister and Evans, 1976.

RESULTS

Effect of Carbon Dioxide and pH on Acetylene Reduction

Several years ago, Lowe and Evans (1962) demonstrated that several species of rhizobia required CO_2 for aerobic growth. Table I illustrates that CO_2 stimulated acetylene reduction markedly with strain 83. However, we observed a large differential in the pH of the medium caused by the buffering capacity of dissolved CO_2. A much smaller stimulatory effect and pH shift was observed with 32H1.

Clearly pH affected acetylene reduction. Fig. 1 illustrates a rather sharp pH optimum for strain 83 at pH 6. The growth rate was much less dependent on pH. The pH profile of 32H1 and 61A76 (not shown) showed a rather broad pH tolerance between pH 5.5 and 7.5. A reinvestigation of the effect of CO_2 in a buffered medium illustrated that CO_2 had only a small stimulatory effect at pH 6 (Table I, expt. 2). The pH effect provides an interpretation of the reports (Keister, 1975; Keister and Evans, 1976) that a high phosphate concentration was somewhat stimulatory for strain 32H1. The pH rises rapidly in a medium in which sodium gluconate is the carbon source and nitrogenase activity was low at pH greater than 7.5 (Fig. 1).

Table I

Effect of CO_2 on acetylene reduction with *R. japonicum* 83

Expt.	Additions	Final[a] cell concentration	Final pH[b]	nmoles C_2H_4/mg protein/hr
1.	---	39	7.7	150
	CO_2, 1%	45	7.1	304
	CO_2, 2.5%	45	6.9	343
	CO_2, 5%	44	6.7	444
2.	MES, pH 6.0	42	6.1	536
	MES, 1% CO_2	38	6.0	572

The experiment was conducted as described in the Materials and Methods Section. CO_2 and O_2 (0.12%) were added to the gas phase. Succinate (0.2%) and histidine (0.1%) were the carbon and nitrogen sources. The specific activity was determined between 41 and 52 hours. MES, (2[N-Morpholino]-ethane sulfonic acid). [a]Initial cell concentration was 17 g protein/ml for experiment 1 and 15 for 2. [b]Initial pH was 7.2 for experiment 1.

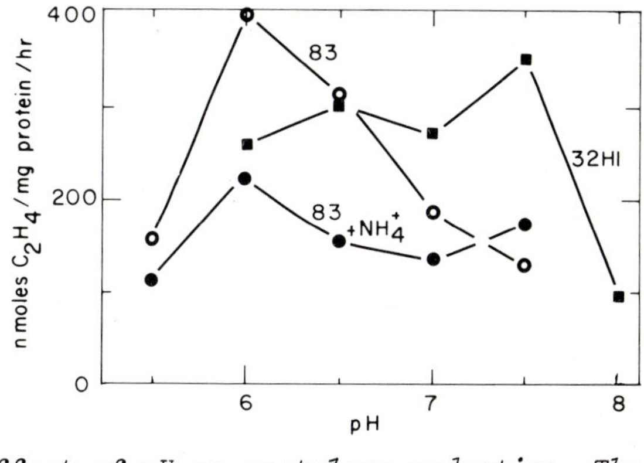

Fig. 1. *Effect of pH on acetylene reduction. The following buffers were used at 50 mM: MES, pH 5.5 and 6; PIPES (piperazine-N, N'-*bis*-2-ethanesulfonic acid), pH 6.5 and 7; HEPES [4-(2-hydroxyethyl)-1-piperazine-ethanesulfonic acid), pH 7.5; Tricine (N-[tris-(hydroxymethyl)-methyl] glycine), pH 8. Succinate (0.2%) were used for 83; succinate and glutamate were used for 32H1. Specific activities were determined between 65 and 76 h with an average oxygen concentration of 0.13%.*

We do not know whether CO_2 is required by these strains of rhizobia since CO_2 is produced by respiration. However, we now routinely add 1% CO_2 to the gas phase and use a well buffered medium to prevent large shifts in the pH. These are necessary conditions for obtaining maximum specific rates.

The Oxygen Requirement of R. japonicum *61A76*

The oxygen requirement for 61A76 is presented in Fig. 2. As for 32H1 and 83, a very low oxygen concentration in the gas

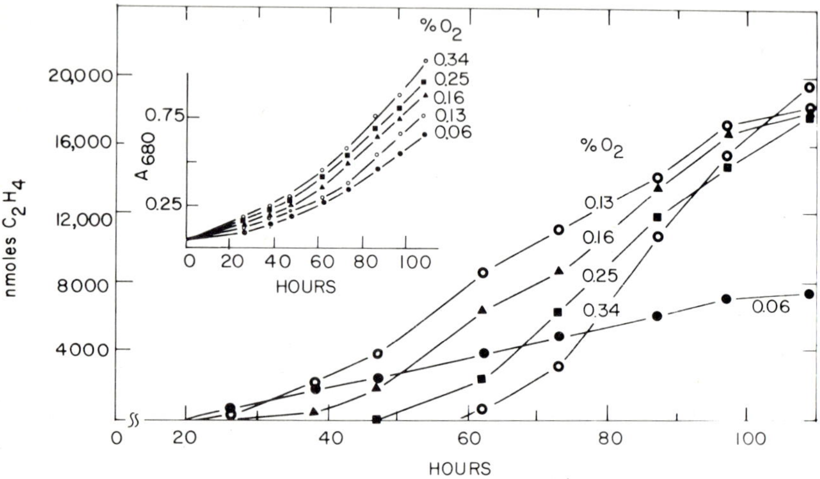

Fig. 2. *The effect of oxygen on the growth and acetylene reduction with* R. japonicum *61A76. Gluconate (0.3%) and glutamate (0.1%) were the carbon and nitrogen sources and the medium was buffered at pH 7 with MOPS [3-(N-morpholino)propanesulfonic acid]. The oxygen concentrations given are the average during approximately 12 h periods. The ranges were 0.06% (0.04-0.08); 0.13% (0.08-0.16); 0.16% (0.14-0.2); 0.25% (0.15-0.3); 0.34% (0.21-0.4). The ordinate is total nmoles C_2H_4 per 500 ml flask containing 10 ml of culture.*

phase is best for initiation of nitrogenase synthesis while slightly higher concentrations (0.13%) gave higher specific rates of C_2H_2 reduction. The lag period was longer with higher oxygen concentrations.

The specific rate profile is shown in Fig. 3. Oxygen at 0.13% was generally best and high rates were found for at least 24 h. Note that even 0.34? O_2 gave a high specific rate after a long lag but the period of optimal activity was very short and the specific rates dropped rapidly as the cell density increased. This probably represents a short period during which an optimum

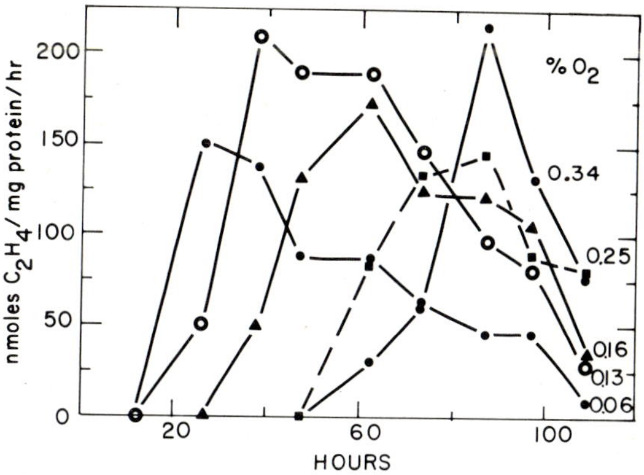

Fig. 3. *Specific rates of acetylene reduction with 61A76. The specific rates were calculated from the data presented in Fig. 1.*

balance was obtained between the oxygen-solution and the oxygen-consumption rates at higher cell densities.

We have not measured the dissolved oxygen concentration during the active acetylene-reducing phase, but it is almost certain that the cell density increased to a point where the O_2 is consumed rapidly enough to maintain the dissolved O_2 concentration at a level that the nitrogenase can tolerate.

Results similar to those obtained with 61A76 have been obtained with *R.sp.* 32H1 and *R. japonicum* 83 and are slightly different from those reported earlier for these strains Keister and Evans, 1976). In the earlier study, good specific rates were never observed with O_2 above 0.2%. This change in response to oxygen is the result of improved cultural conditions due mainly to better pH control and therefore a slightly more rapid cell growth.

Nutritional Requirements for Acetylene Reduction with 61A76

As with 32H1 and 83, the nutritional requirements of 61A76 are simple; a metabolizable carbon source, a combined nitrogen source and inorganic salts. Glutamate was the most effective compound (Fig. 4) and we have found it best for several other rhizobial strains as well. Nitrate, histidine and β-alanine were ineffective with 61A76. Nitrate (or nitrite) appears to be a general inhibitor of nitrogenase in several strains.

The effect of three carbon sources at various O_2 concentrations is shown in Fig. 5. A rather interesting inversion in

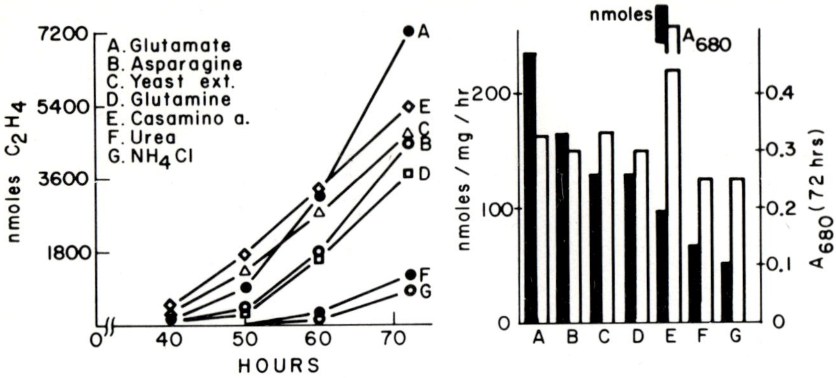

Fig. 4. *Effect of various nitrogen sources on growth and ethylene formation in 61A76. Gluconate was the carbon source and the nitrogen sources were added at 10 mM except yeast extract and casamino acids (0.1%). The average oxygen was 0.13% and specific rates were calculated between 60 to 72 hours.*

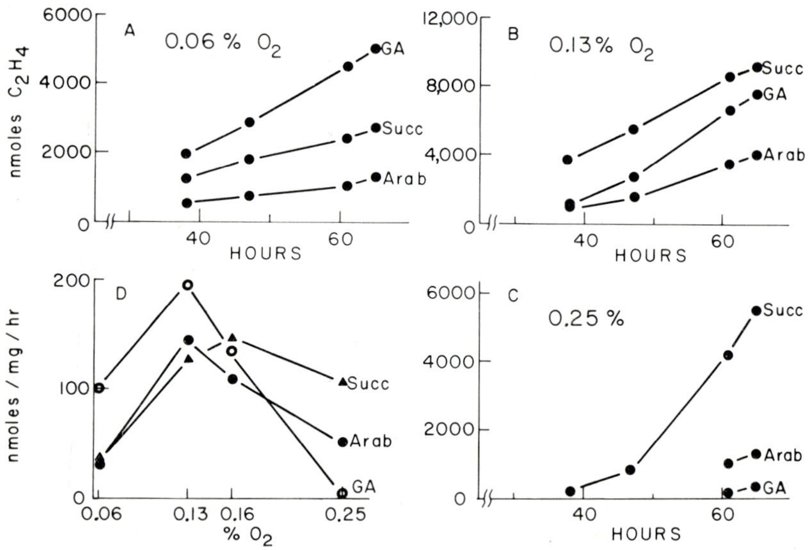

Fig. 5. *The effect of the carbon source on acetylene reduction in 61A76. Arabinose, gluconate (GA) and succinate were 10 mM and glutamate (0.1%) was the nitrogen source. Specific rates were calculated between 47 to 61 hours.*

effectiveness is observed when gluconate is compared with succinate at 0.06% and 0.13% O_2. Gluconate was more effective at the low O_2 concentration whereas succinate was better at higher oxygen. These results emphasize the importance of investigating several parameters of each assay system of each rhizobial strain.

Table II

Specific rates of acetylene reduction with 32H1

Additions	Protein µg/ml	nmoles C_2H_4/ mg protein/h
NH_4Cl, 10 mM	54	227
Glutamate, 0.1%	50	343
Glutamate + NH_4^+	59	179
Glutamine, 0.1%	52	249
Glutamine + NH_4^+	49	278
Phenylalanine	40	51
Phenylalanine + NH_4^+	46	227
Asparagine	51	260
Asparagine + NH_4^+	47	288

See legend to Fig. 6 for details

Effect of NH_4Cl

Keister and Evans (1976) found that NH_4Cl had no inhibitory or repressive effect on nitrogenase in strains 32H1 and 83 in dilute shaken cultures under low oxygen concentration. In contrast, 10 mM NH_4 Cl was inhibitory for 32H1 and 61A76 in stationary cultures (Evans and Keister, 1976; Evans *et al.*, 1976). Kurz and LaRue (1975) found no inhibition of C_2H_2 reduction by NH_4Cl with 61A76 grown on solid media and likewise Gibson *et al.* (1976) found little inhibition by ammonia with 32H1 at 10 mM but about 50% inhibition b6 20 mM when grown on solid media.

We have continued studying the NH_3 effect in an attempt to understand whether ammonia is exerting a controlling effect on nitrogenase synthesis as in other nitrogen-fixing micro-organisms.

R. japonicum 83 In our earlier studies (see Fig. 5, Keister and Evans, 1976) NH_4^+ was a better nitrogen source than glutamate and did not cause repression even at 30 mM. More recent results reveal that 10 mM NH_4Cl derepressed the activity in the presence of histidine by up to 50%. With less effective nitrogen sources the effect of NH_4^+ was minimized. The degree of inhibition was very pH-dependent and the maximum inhibition was found at the optimum pH. At pH 7.5, NH_4^+ was stimulatory. Our earlier results were obtained in a lightly buffered medium and are easy to interpret in terms of pH shifts resulting from NH_4 assimilation or utilization of succinate or gluconate.

R. japonicum 61A76 NH_4^+ was a poor nitrogen source for the development of nitrogenase (Fig. 4) and severely depressed acetylene reduction when added with glutamate or glutamine (Fig. 6). Thus this strain appears to be more sensitive to NH_4^+ under these conditions than either 32H1 or 83.

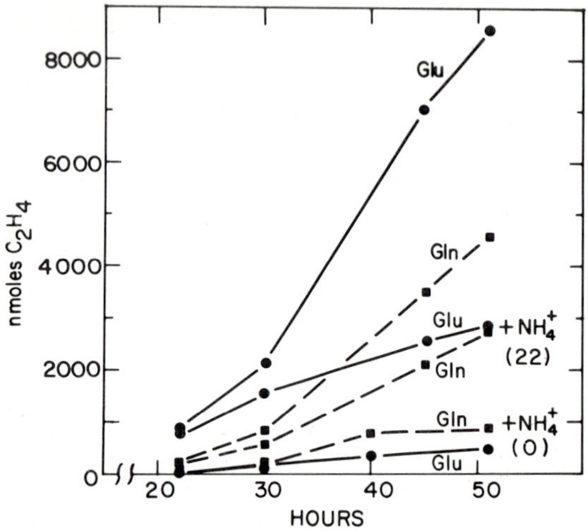

Fig. 6. *Effect of NH_4+ on* R. japonicum *61A76. Gluconate (10 mM) was the carbon source. Glutamate (Glu, 0.1%) or glutamine (Gln, 0.1%) was the nitrogen source. $(NH_4)_2SO_4$ (10 mM) was added at 0 time (0) or after development of the activity at 22 h (22). Oxygen was 0.13%.*

R. sp. 32H1 The effect of ammonium chloride on 32H1 at pH 6 with various nitrogen sources is shown in Fig. 7. There is a small (ca. 25%) depression of activity with glutamate as the nitrogen source. Higher concentrations had no greater effect. With glutamine, asparagine or phenylalanine, which are not as good nitrogen sources as glutamine, there was no inhibitory effect of ammonia; ammonia actually stimulated.

Likewise, when arabinose, a carbon source less effective than gluconate, was used, ammonia had no repressive effect (Fig. 8).

The modest effects of ammonia on C_2H_2 reduction with 32H1 was reflected in the growth of the organism. With gluconate or glutamate, NH_4^+ stimulated growth by about 20%, but with gluconate and glutamine or aspartate, ammonia inhibited by about 5 to 10%. Likewise, NH_4^+ inhibited growth slightly when arabinose

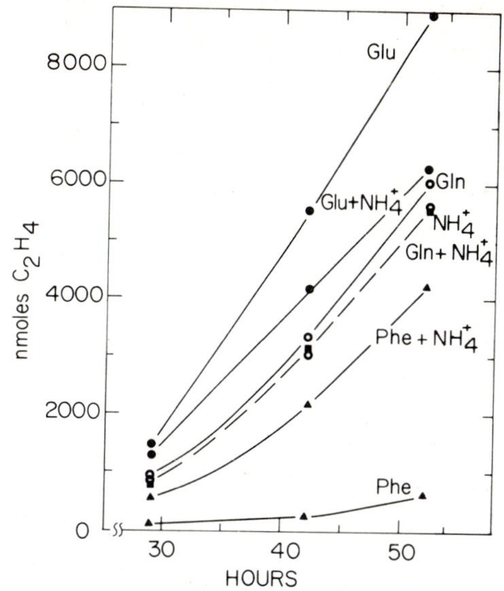

Fig. 7. *Effect of* NH_4^+ *on* R. sp. *32H1. Succinate (0.2%) was the carbon source and the medium was buffered at pH 6 with MES. Phosphate was 10 mM and the nitrogen sources (Glu, glutamate; Gln, glutamine; Phe, phenylalanine) were 0.1% except* NH_4Cl *was 10 mM. The oxygen averaged 0.1% for the first 29 h and then was increased to 0.13%. The specific rates, calculated between 42 and 53 h are given in Table II.*

was the carbon source. Thus we believe at least under some conditions, NH_4^+ does not cause repression of nitrogenase with 32H1.

With 83, which shows an intermediate effect of ammonia, it is possible that the inhibition is due to repression but it is also possible that it is a metabolic effect. With 61A76 in which C_2H_2 reduction is severely but not completely depressed by NH_4^+, the effect is most likely due to repression of nitrogenase synthesis.

DISCUSSION

We have now characterized three strains of rhizobia for their oxygen and nutritional requirements for acetylene reduction in liquid culture. The oxygen requirement is very similar for all three strains that so far have been found to give high specific rates of acetylene reduction. Very low oxygen in the gas phase (ca. 0.06%) gives the more rapid synthesis of nitrogenase but a slightly higher oxygen (0.13%) usually gives the

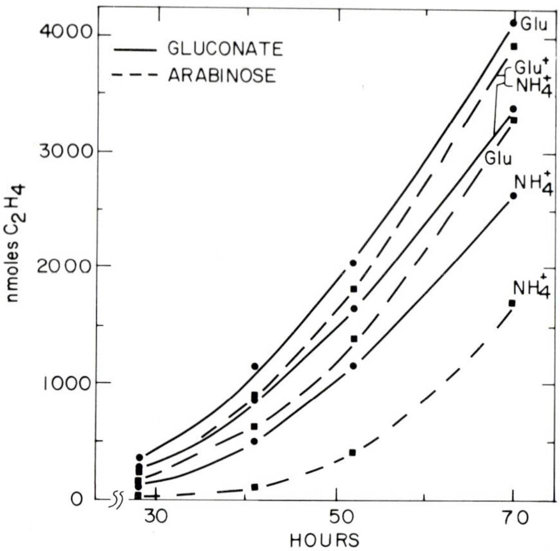

Fig. 8. *Effect of* NH_4^+ *on* R. sp. *32H1 with gluconate and arabinose as the carbon sources. Conditions were essentially as for Fig. 7. Glutamate (glu) and/or* NH_4Cl *was added as noted.*

best specific rates when low initial cell densities (5-15 x 10^7 cells/ml) are used. With the optimized cultural conditions described in this paper, good activity can be observed with oxygen up to 0.34% at very specific times. This temporal high activity undoubtedly represents that short period during which the oxygen solution rates and oxygen consumption rates are balanced at the maximum solution oxygen concentration from which the cells can protect the nitrogenase enzyme. These parameters are clearly shown in Fig. 9 which illustrates C_2H_2 reduction during changing oxygen conditions. Curve A is a control with 0.08% O_2 throughout the experiment. Curve B had 0.08% O_2 until the nitrogenase activity was established and then 0.3% O_2 was added (38 h). Observe that nitrogenase activity was immediately inhibited and was not re-established until after 62 h when the cell density had more than doubled (to 0.14 mg/ml). Curve D had 0.4% O_2 (0.34% average) for 40 h. The gas phase was then removed and the O_2 level lowered to 0.12%. Nitrogenase activity began almost immediately. The control for this curve is the 0.34% curve of Fig. 2 and it can be seen that when the oxygen concentration was maintained at 0.34%, no activity

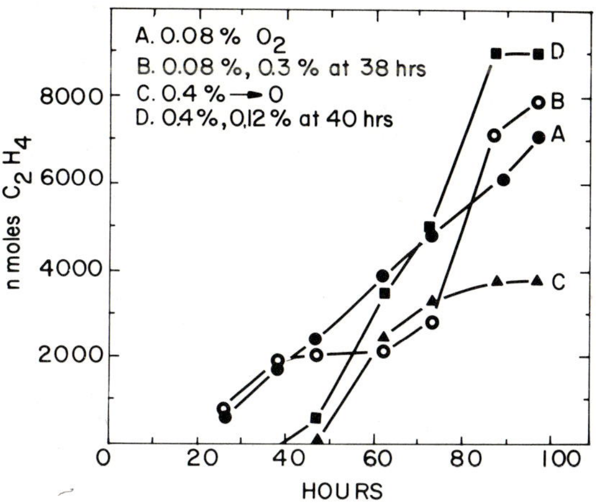

Fig. 9. *Effect of changing oxygen on 61A76. Conditions were as for Fig. 2. A. An initial oxygen concentration of 0.08% (average 0.06%) was maintained throughout the experiment. B. 0.08% O_2 was changed to 0.3% (average 0.25%) at 38 hours. C. 0.4% O_2 with no further injections. Intermediate levels were 0.23% at 47 h; 0.12% at 62 h; and 0.08% at 73 h. D. 0.4% O_2 for 40 h. The gas phase was then removed and replaced with 0.12% O_2 and increased to 0.2% at 73 h.*

was apparent until 62 h. Thus the cells have adapted to the low oxygen environment and only the establishment of a tolerable oxygen level is required to observe acetylene reduction. Curve C represents 0.4% oxygen initially which is allowed to fall to 0% oxygen during the course of the experiment.

The nutritional requirements for the development of nitrogenase activity in the several strains that we have studied are quite simple. An effective carbon source and a nitrogen source (other than nitrate) which the cells can use in addition to the usual inorganic compounds. A combined nitrogen source is required in our experimental system since no gaseous N_2 is present which could produce NH_3. However, in a few experiments we have tried to grow the cells on N_2 without success and at this point we are not sure whether this is possible or not. O'Gara and Shanmugam (1976) recently demonstrated that some species of Rhizobium exported NH_3 into the medium when presented with $^{15}N_2$.

The control of nitrogenase synthesis in some rhizobia does not appear to be as sensitive to ammonia repression as in other nitrogen-fixing bacteria. The nitrogenase activity in *R. sp.* 32H1 is slightly depressed by NH_4^+ but our data indicate that this may be a metabolic effect rather than repression. Under low oxygen, the cells are certainly energy-limited and therefore any disturbances or changes in their metabolism may be expected to have an effect on nitrogenase activity. Even with 61A76 which was severely depressed by NH_4^+, there was significant activity developed in the presence of NH_4^+. With 83 only about 50% depression was found. Whether the effects noted are repression or not awaits more definitive experiments.

NOTES

[1]However, other papers in this volume on *Spirillum lipoferum* by J. Döbereiner and by R. H. Burris and a recent publication by Berndt *et al.* (1976) on *Corynebacterium autotrophicum* show that a group of micro-organisms is becoming apparent which reduce nitrogen only at very low oxygen partial pressures.

REFERENCES

Berndt, H., Ostwal, K.-P., Lalucat, J., Schumann, C., Mayer, F. and Schlegel, H.G. (1976). *Arch.Microbiol.*, 108: 17-26.

Evans, W.R. and Keister, D.L. (1976). *Can.J.Microbiol.*, 22: 949-952.

Evans, W.R., Keister, D.L. and Hall, G. (1976). *Plant Physiol.*, 57: 103 Suppl.

Gibson, A.H., Scowcraft, W.R., Child, J.J. and Pagan, J.D. (1976). *Arch.Microbiol.*, 108: 45-54.

Keister, D.L. (1975). *J.Bacteriol.*, 123: 1265-1268.

Keister, D.L. and Evans, W.R. (1976). *J.Bacteriol.*, 129: 149-153.

Kurz, W.G.W. and LaRue, T.A. (1975). *Nature*, 256: 407-408.

Lowe, R.H. and Evans, H.J. (1962). *Soil Science*, 94: 351-356.

McComb, J.A., Elliot, J. and Dilworth, M.J. (1975). *Nature*, 256: 409-410.

O'Gara, F. and Shanmugam, K.T. (1976). *Biochim.Biophys.Acta*, 437: 313-321.

Pagan, J.D., Child, J.J., Scowcroft, W.R. and Gibson, A.H. (1975). *Nature*, 256: 406-407.

Prival, M.J. and Magasanik, B. (1971). *J.Biol.Chem.*, 246: 6288-6296.

Smith, G.R., Halpern, Y.S. and Magasanik, B. (1971). *J.Biol. Chem.*, 246: 3320-3329.

Tjepkema, J.D. and Evans, H.J. (1975). *Biochem.Biophys.Res. Commun.*, 65: 625-628.

STABILITY OF LEGHAEMOGLOBIN IN YELLOW LUPIN NODULES

M.J. DILWORTH and D.R. COVENTRY

*Department of Soil Science and Plant Nutrition,
University of Western Australia
and
School of Environmental and Life Sciences,
Murdoch University, Western Australia.*

Leghaemoglobin, the red pigment of effective leguminous root nodules, is an essential feature of the nitrogen fixation system produced by interaction of *Rhizobium* and legume root cells, although no such pigment is apparently necessary when *Rhizobium* forms N_2-fixing nodules on roots of the non-legume *Trema cannabina* (Coventry *et al.*, 1976). The leghaemoglobin is confined to the bacteroid-containing cells of the legume nodule (Smith, 1949) and no traces of globin can be detected in other plant parts (Cutting and Schulman, 1968; Godfrey *et al.*, 1975).

Within the individual nodule cell, the site of leghaemoglobin has been sought by a number of techniques, with some confusion still existing as to its precise intracellular localization. Light microscope observations (Bergersen and Goodchild, 1973), electron microscope autoradiography of ^{59}Fe-labelled nodules (Dilworth and Kidby, 1968) and light and electron microscope studies of diaminobenzidine-stained material (Bergersen and Goodchild, 1973; Truchet, 1972) indicate a localization between bacteroid wall and peribacteroid membrane. However, other studies of diaminobenzidine-stained nodules (Dart, 1968) and recent work using ferritin-conjugated antibody to leghaemoglobin (Verma and Bal, 1976) indicate a site in the plant cytoplasm. Resolution of this conflict is obviously of some importance since the role currently suggested for leghaemoglobin function in the legume root nodule implicitly assumes the position immediately around the bacteroid. The crux of the problem is possible relocation of soluble leghaemoglobin during fixation and staining for microscopy.

Leghaemoglobin is a product specified by plant genes (Dilworth, 1969; Cutting and Schulman, 1971) and is translated on plant cytoplasmic ribosomes (Verma *et al.*, 1974; Godfrey *et al.*,

1975; Coventry and Dilworth, 1976). In at least two nodule systems - soybean and lupin - globin or leghaemoglobin production by the plant actually precedes nitrogenase development in the bacteroid (Godfrey *et al.*, 1975; Bergersen and Goodchild, 1973), though the nature of the signal from bacteroid to plant for leghaemoglobin synthesis is not known. A 9S messanger RNA fraction has been isolated from soybean root nodules and translated with a wheat-germ protein-synthesizing system (Verma *et al.*, 1974); the production of two of the leghaemoglobin components in a ratio atypical of the one found in the nodule may explain how the overall ratio between different components varies with the stage of development of the nodule (Fuchsman *et al.*, 1976). The primary amino-acid sequences of the leghaemoglobins (Ellfolk and Sievers, 1971; Lehtovaara and Ellfolk, 1974; Richardson *et al.*, 1975; Egorov *et al.*, 1976; Burrell *et al.*, 1976) shows a high degree of variability between species, and some polymorphisms even within an individual component (Ellfolk and Sievers, 1971; Richardson *et al.*, 1975). The three-dimensional arrangement of the molecule is very similar to that of myoglobin (Vainshtein *et al.*, 1974).

While globin synthesis is a plant function, haem synthesis appears to be a bacteroid property (Cutting and Schulman, 1969; Godfrey and Dilworth, 1971; Godfrey and Dilworth, 1975); the controls over this process either from the plant or bacteroid side are at present unknown. It appears probable, however, that globin production in the plant plays a role in regulating haem biosynthesis in the bacteroid by removing haem bound to the bacteroids (Cutting and Schulman, 1972; Godfrey *et al.*, 1975). The non-participation of the plant in nodular haem biosynthesis depends on the apparent lack of a biosynthetic route to 5-amino-laevulinic acid in the plant fraction. If current work on aminolaevulinic acid synthesis in plants (Beale *et al.*, 1975; Troxler and Brown, 1975) is successful in establishing an alternative route to aminolaevulinic acid to that catalyzed by aminolaevulinic acid synthetase, the sites of haem synthesis in the nodule may need to be re-investigated. Rates of haem synthesis in the bacteroids are, however, quite adequate to account fully for the rate of haem synthesis needed for leghaemoglobin production (Cutting and Schulman, 1969; Godfrey and Dilworth, 1971).

A role for leghaemoglobin in oxygen transport into the nodule has long been postulated, since nodules are known to fix no nitrogen anaerobically. The work of Appleby, Bergersen and others has shown, with systems *in vitro*, that leghaemoglobin acts to facilitate diffusion of oxygen across the leghaemoglobin-filled spaces, and to deliver the oxygen at the bacteroid surface at a buffered concentration optimal for the oxidase-driven phosphorylation system supplying ATP for nitrogenase action (Appleby *et al.*, 1975; Bergersen and Turner, 1975a,1975b; Wittenberg, 1976). The experiments *in vitro* model a system where the leghaemoglobin is located immediately around the bacteroid itself, and therefore assume that leghaemoglobin is indeed localized there in the intact nodule. The degree of oxygenation of leghaemoglobin in the nodule (Appleby, 1962) agrees well with the concentration of oxygen favouring optimal oxidative phosphorylation in leghaemoglobin-bathed bacteroid suspensions (Bergersen and Turner, 1973b) and therefore fits well with the assumed location. If the leghaemoglobin is in fact located only in the plant cytoplasm there still remains a formidable diffusion barrier to oxygen movement in the liquid inside the peri-bacteroid membrane.

Leghaemoglobin can be rapidly degraded if actively fixing nodulated plants are simply placed in the dark (Virtanen and Miettinen, 1949; Roponen, 1970); in the case of peas, most of the leghaemoglobin can be lost in 24 h (Roponen, 1970). Normally, nitrogenase activity is lost before extensive loss of leghaemoglobin occurs in the normal process of nodule senescence (Klucas, 1974). However, there is little information relating to actual turnover of the leghaemoglobin during normal nodule function.

Simple darkening experiments with yellow lupins held at 20° failed to show any sign of loss of leghaemoglobin even after four days, indicating that even deprivation of fresh photosynthate for that period did not initiate extensive degradation in our system.

During work on D-*threo*-chloramphenicol inhibition of protein and leghaemoglobin synthesis in lupins, we developed techniques which allowed us to label the leghaemoglobin and then examine the stability of the pigment under conditions where active fixation was occurring.

The major problem involved in studying protein turnover by labelling techniques is producing a pure protein in which radioactivity can be measured specifically. With leghaemoglobin, isolation of highly purified material can be achieved rapidly with some nodule systems - in our case the yellow lupin (*Lupinus luteus* L.). Lupin nodules are ground in barbiturate buffer (ionic strength, 0.1; pH, 8.6) at 4° and the macerate filtered and centrifuged to remove bacteroids and other debris. Trial experiments showed that gel filtration of the supernatant through a 90 x 1.8 cm column of Sephadex G-75 in the same buffer produced four coloured fractions, two of them with significant absorption at 405 nm (Fig. 1). The second had an absorption spectrum typical of leghaemoglobin and, after electrophoresis in 7.5% polyacrylamide gel in Tris-glycine and scanning at

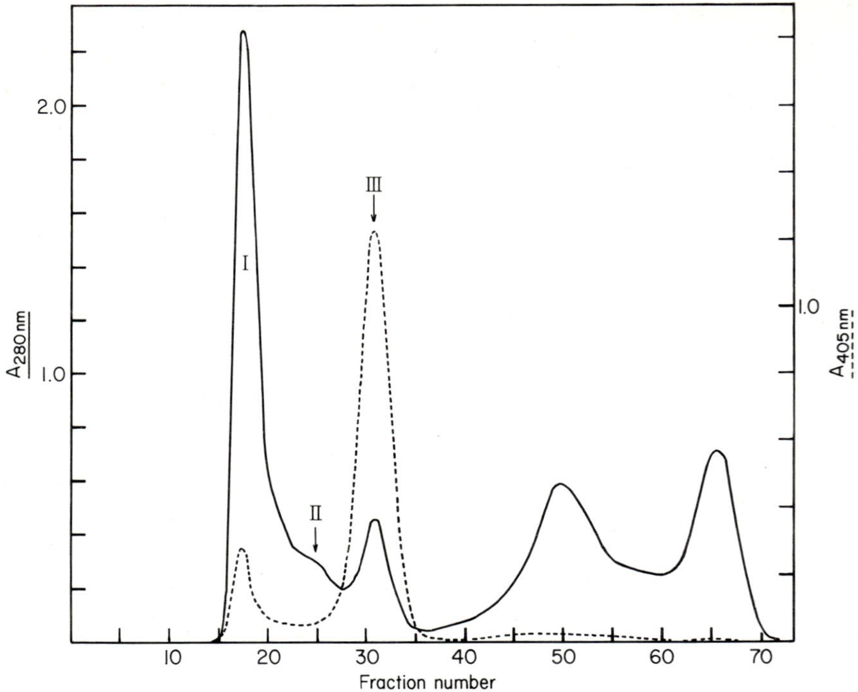

Fig. 1. *Elution profile for a lupin nodule supernatant on a 90 x 1.8 cm diameter column of Sephadex G-75 in barbiturate buffer, pH 8.6. Fraction volume, 3.0 ml.*

———————— *absorbance at 280 nm;*

- - - - - - - - - - - *absorbance at 405 nm.*

280 nm, proved to be better than 95% pure leghaemoglobin. Ouchterlony double diffusion against a rabbit antibody to purified leghaemoglobin showed that the first fraction contained no detectable leghaemoglobin, while the second reacted strongly. The other lower molecular weight fractions were also devoid of leghaemoglobin. Gel filtration on columns of Sephadex G-75 was therefore used throughout this work to isolate leghaemoglobin of this purity. The technique is not applicable to other nodules - broad bean, for example, gives a leghaemoglobin preparation not more than about 50% pure under comparable conditions.

To label the leghaemoglobin we administered labelled material via two different routes - ^{14}C as $^{14}CO_2$ to be photosynthesized in tops and translocated downwards, and 4, 5- ^{3}H leucine, administered via the roots in a nutrient solution.

For ^{14}C labelling, yellow lupins were germinated in sterile sand inoculated with strain WU8 of *Rhizobium lupini* and later transferred to polythene slats over a nutrient solution lacking nitrogen. The plant tops were sealed into a 55 dm^3 plastic chamber for exposure to 200 uCi $^{14}CO_2$ (8.3 $\mu Ci.mmol^{-1}$, 1% by volume) for eight hours. After exposure, $^{14}CO_2$ was absorbed from the chamber with NaOH, and the plants returned to the normal nutrient solution for periods up to twelve days. For ^{3}H-labelling, soil-grown lupins were transplanted into nutrient solution and maintained there for at least seven days until nitrogenase activity was adequately re-established. The roots were then aerated for eight hours at 20° in a nutrient solution to which had been added 0.5 $\mu Ci.cm^{-3}$ L-4, 5- [^{3}H]leucine, specific activity 53 $Ci.mmol^{-1}$. The roots were then thoroughly washed, and replaced in the normal nutrient solution for sampling every two days for 14 days.

Leghaemoglobin was then isolated by gel filtration as indicated, and haem removed from it by adding to a solution at 4° and pH 2 an equal volume of cold ethyl methyl ketone and shaking (Teal, 1959). Haem removal eliminated the possibility that measured radioactivity might only represent incorporation into haem. Soluble plant protein in the first peak from the G-75 columns was precipitated with trichloroacetic acid, washed with trichloroacetic acid, and finally with absolute ethanol before being dissolved in 1 M Hyamine hydroxide for counting.

Bacteroids were isolated by centrifugation, washed free of plant proteins and sonicated. Bacteroid soluble proteins were also precipitated with trichloroacetic acid and counted in a similar way (Coventry and Dilworth, 1975).

Leghaemoglobin quantities were measured using radial immunodiffusion methods (Mancini *et al.*, 1965), but incorporating ampicillin, streptomycin and cycloheximide into the gels so that they could be allowed to diffuse aerobically at 37° in a humid dessicator rather than anaerobically under paraffin oil. Gel diffusion plates were photographed by indirect light and the zones within the precipitin rings cut out of prints and weighed. Standard leghaemoglobin samples were included on each diffusion plate.

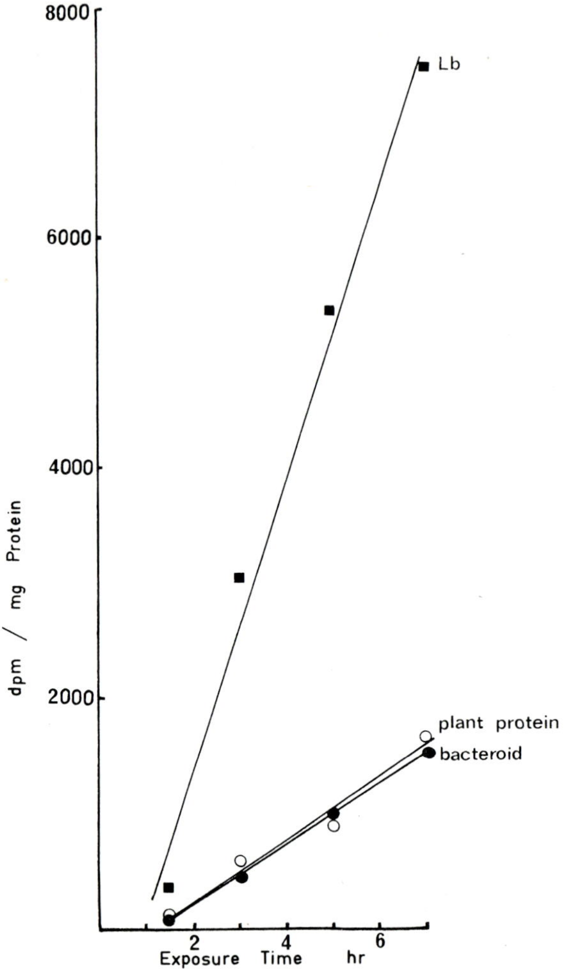

Fig. 2. *Incorporation of* ^{14}C *from* $^{14}CO_2$ *into various protein fractions of yellow lupin root nodules as a function of labelling.* ■ *leghaemoglobin;* ○ *soluble plant protein;* ● *soluble bacteroid protein.*

Incorporation of ^{14}C into plant protein, Lb and bacteroid protein was linear with time for all fractions, after a lag of about one hour for solute translocation from leaves to nodules (Fig. 2). It is clear from the data that a considerable part of the protein radioactivity is in the leghaemoglobin fraction, though leghaemoglobin represents only about 8-12% of lupin nodule soluble protein. When the decay of specific radioactivity in the various fractions was measured, the data conformed to a logarithmic loss of specific radioactivity with time after pulse labelling (Fig. 3).

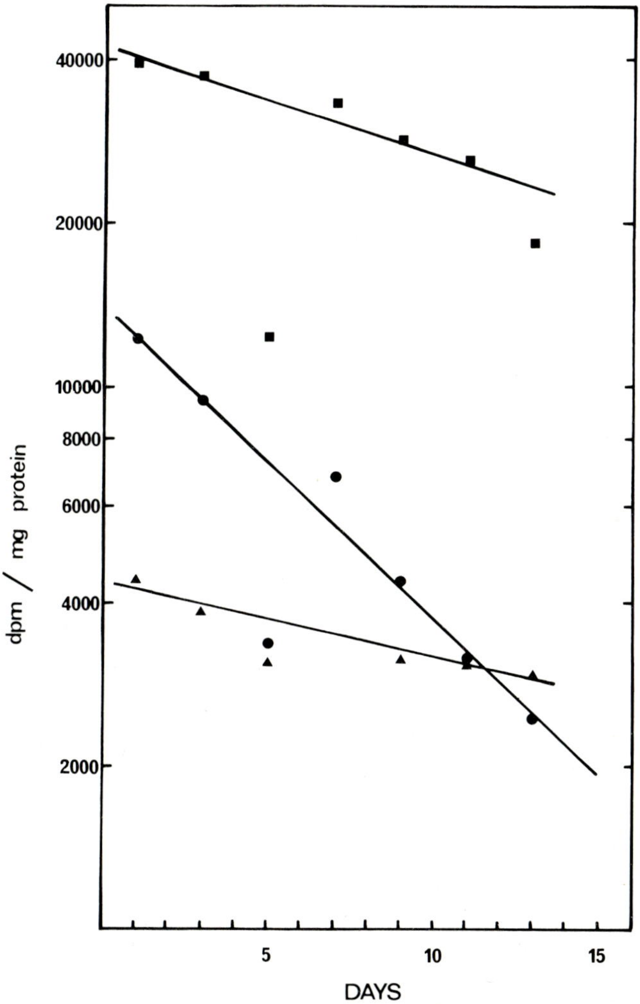

Fig. 3. *Decay of specific radioactivity of various protein fractions of yellow lupin root nodules with time after administration of an 8-hour pulse of* ^{14}C *as* $^{14}CO_2$. ■ *leghaemoglobin;* ▲ *soluble plant protein;* ● *soluble bacteroid protein.*

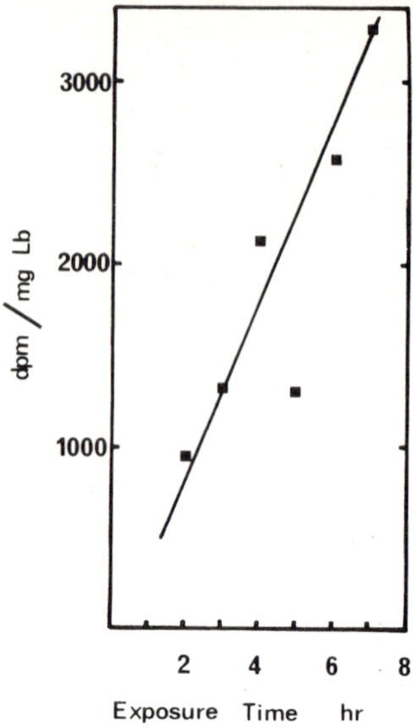

Fig. 4. *Incorporation of ^3H from L-4, 5 $[^3H]$ leucine into leghaemoglobin of yellow lupin root nodules as a function of time of labelling.*

Incorporation of label from ^3H-leucine is also linear with time after a lag of about the same length (Fig. 4). With leucine as the labelled precursor a relatively smaller amount of the total radioactivity was incorporated into Lb than was the case with $^{14}CO_2$ labelling. This can be clearly seen from a comparison of Figs 3 and 5; with $^{14}CO_2$ the specific radioactivity of Lb is always greater than that of general soluble plant protein, while the reverse is true for ^3H-leucine labelling. It must be assumed that the pools to which radioactivity from the two sources is allocated must be different, and that one particularly favours leghaemoglobin synthesis. When the data for specific radioactivity of the protein fractions are plotted against time, a similar logarithmic decay of specific radioactivity is obvious (Fig. 5). Calculations from the data of Figs 3 and 5 lead to the times for 50% reduction of specific radioactivity shown in Table I. Despite the very considerable differences between the distribution of label between the three protein fractions when different radioactive precursors are used, the decay rates for specific radioactivity are in good agreement.

In these experiments, the plant material which could be labelled was very limited and accordingly the absolute increase

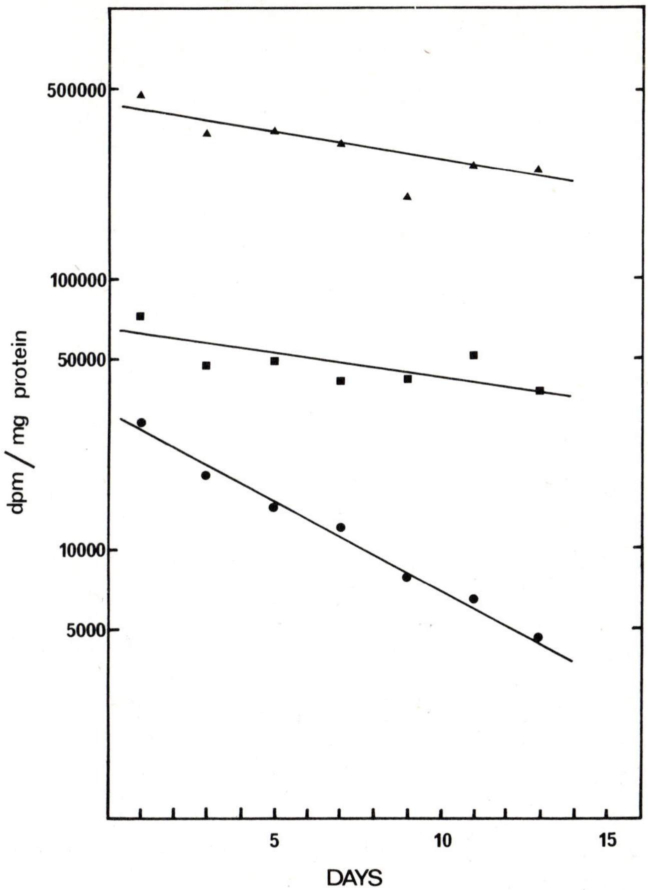

Fig. 5. *Decay of specific radioactivity of various protein fractions of yellow lupin nodules with time after administration of an 8-hour pulse of ^3H as L-4, 5- $[^3H]$ leucine.* ■ *leghaemoglobin;* ▲ *soluble plant protein;* ● *soluble bacteroid protein.*

in the protein fractions over the "chase" period was not measured. Any new synthesis of leghaemoglobin during this period will give a decay rate for specific radioactivity which represents dilution from new ynthesis. The actual rate of leghaemoglobin degradation in actively fixing nodules will therefore be less than that actually measured in Figs 3 and 5. We calculate that synthesis of 0.22 mg leghaemoglobin per plant in the lupins labelled with $^{14}CO_2$ during the chase period would mean that no loss of radioactivity occurred from the

Table I

Time (days) required for lupin nodule fractions to halve their specific radioactivity after pulse labelling

Protein fraction	Radioactive precursor	
	$^{14}CO_2$	^{3}H-leucine
Soluble plant protein	20	15
Leghaemoglobin	18	18
Bacteroid protein	5	4.5

leghaemoglobin fraction. In a similar calculation, 0.4 mg increase in leghaemoglobin per plant in the lupins labelled with ^{3}H-leucine would mean that no loss of label occurred.

A further problem is the re-utilization of labelled amino-acids resulting from any protein degradation, or from mobilization of labelled compounds for later protein synthesis. If this were considerable, our estimates of loss of radioactivity would be smaller than the actual rate of loss from the protein fractions. In view of the rapid flush of amino-acid nitrogen through the nodule from new fixation, this complication seems unlikely to be significant.

It can be concluded that leghaemoglobin is a relatively stable protein in N_2-fixing lupin nodules, with an apparent half-life of 18 days or greater, but that degradation during active N_2 fixation is not established.

Since bacteroid numbers and leghaemoglobin content per unit of nodule tissue remain relatively constant in yellow lupin nodules during the active N_2 fixation period (Dilworth and Williams, 1967), bacteroid protein turnover must be faster than the plant protein turnover, and the data of Fig. 3 and Fig. 5 must indicate actual turnover of bacteroid proteins.

REFERENCES

Appleby, C.A. (1962). *Biochim.Biophys.Acta,* 60: 226-235.

Appleby, C.A., Turner, G.L. and MacNicol, P.K. (1975). *Biochim. Biophys.Acta,* 387: 461-474.

Beale, S.I., Gough, S.P. and Granick, S. (1975). *Proc.Natl.Acad. Sci.USA,* 72: 2719-2723.

Bergersen, F.J. and Goodchild, D.J. (1973). *Austral.J.Biol.Sci.,* 26: 741-756.

Bergersen, F.J. and Turner, G.L. (1975a). *J.gen.Microbiol.,* 89: 31-47.

Bergersen, F.J. and Turner, G.L. (1975b). *J.gen.Microbiol.*, 91: 345-354.
Coventry, D.R. and Dilworth, M.J. (1975). *J.gen.Microbiol.*, 91: 69-75.
Coventry, D.R. and Dilworth, M.J. (1976). *Biochim.Biophys.Acta*, in press.
Coventry, D.R., Trinick, M.J. and Appleby, C.A. (1976). *Biochim. Biophys.Acta*, 420: 105-111.
Cutting, J.A. and Schulman, H.M. (1968). *Fed.Proc.*, 27: 768.
Cutting, J.A. and Schulman, H.M. (1969). *Biochim.Biophys.Acta*, 192: 486-493.
Cutting, J.A. and Schulman, H.M. (1971). *Biochim.Biophys.Acta*, 229: 58-62.
Cutting, J.A. and Schulman, H.M. (1972). *Biochim.Biophys.Acta*, 261: 321-327.
Dart, P.J. (1968). Proc.4th Europ.Reg.Conf.Electron Microscopy, Rome, 69-70.
Dilworth, M.J. (1969). *Biochim.Biophys.Acta*, 184: 432-441.
Dilworth, M.J. and Appleby, C.A. *In:* Dinitrogen Fixation. Wiley Publishing Co., in press.
Dilworth, M.J. and Kidby, D.K. (1968). *Exptl Cell Res.*, 49: 148-159.
Dilworth, M.J. and Williams, D.C. (1967). *J.gen.Microbiol.*, 48: 31-36.
Egorov, Ts.A., Feigina, M.Ya., Kazakov, V.K., Shakhparonov, M. I., Mitaleva, S.I. and Ovchinnikov, U.A. (1976). *Bioorganischeskaya Khimiya*, 2: 125-128.
Ellfolk, N. and Sievers, G. (1971). *Acta Chem.Scand.*, 25: 3532-3534.
Fuchsman, W.H., Barton, C.R., Stein, M.M., Thompson, J.T. and Willett, R.M. (1976). *Biochem.Biophys.Res.Commun.*, 68: 387-392.
Godfrey, C.A., Coventry, D.R. and Dilworth, M.J. (1975). *In:* Nitrogen Fixation by Free-Living Micro-organisms (W.D.P. Stewart, ed.), Vol. 6, pp. 311-332, International Biological Programme. Cambridge University Press, Cambridge.
Godfrey, C.A. and Dilworth, M.J. (1971). *J.gen.Microbiol.*, 69: 385-390.
Hurrell, J.G.R., Nicola, N.A. and Leach, S.J. (1976). Somiplan Symposium, Kuala Lumpur, August 1976.
Klucas, R.V. (1974). *Plant Physiol.*, 54: 612-616.
Lehtovaara, P. and Ellfolk, N. (1974). *FEBS Letters*, 43: 239-240.
Mancini, G., Carbonara, A.O. and Heremans, J.F. (1965). *Immunochem.*, 2: 235-254.
Smith, J.D. (1949). *Biochem.J.*, 44: 585-591.
Richardson, M., Dilworth, M.J.and Scawen, M. (1975). *FEBS Letters*, 51: 33-37.

Roponen, I. (1970). *Physiol.Plant.*, 23: 452-460.
Teale, F.W.J. (1959). *Biochim.Biophys.Acta*, 35: 543.
Troxler, R.F. and Brown, A.S. (1975). *Plant Physiol.*, 55: 463-467.
Truchet, G. (1972). *C.r.hebd.Seanc.Acad.Sci., Paris*, 274: 1290-1293.
Vainshtein, B.K., Arutyunyan, E.H., Kuranova, I.P., Borisov, V.V., Sosfenov, N.I., Pavlovskii, A.G., Grebenko, A.I. and Konerova, N.V. (1975). *Nature*, 254: 163-164.
Verma, D.P.S. and Bal, A.K. (1976). Personal communication.
Verma, D.P.S., Nash, D.T. and Schulman, H.M. (1974). *Nature*, 251: 74-77.
Virtanen, A.I. and Miettinen, J.K. (1949). *Acta Chem.Scand.*, 3: 17-21.
Wittenberg, J.B. (1976). *Colloquium on Oxygen and Physiological Function, Anaheim, Calif., April 1976.*

STUDY FRAMEWORKS FOR SYMBIOTIC NITROGEN FIXATION

P.S. NUTMAN

Rothamsted Experimental Station,
Harpenden, Herts, U.K.

What I shall try to do in the next quarter of an hour is to outline two frameworks that might provide points of reference. This is not easy for two reasons: firstly, because one person's framework is another person's straight jacket, and secondly, because the field of nitrogen fixation, even if one restricts oneself as I shall do to the nodulated legume, is so complex as almost to defy its orderly dissection so as to display all its components in their proper relationships, and it is the relationship of the parts to the whole that I wish to stress.

A time-honoured way of cracking a difficult problem is to strike it with a heavy object, and then painfully to piece together recognisable fragments - a method used with some success in nuclear physics, and even, using a biological example, in early electron microscopic section work which employed high speed rotary microtomy. This meeting has shown that *we* use far more subtle and refined methods, with demonstrable success, though at times we may seem to lose all connections with reality.

Let us then belabour the nub of our problem with a notional hammer, with consequences imagined in Fig. 1.

Some parts of the entangled complex seem to have separated and others remain closely linked, but not so much so as perhaps entirely to obscure some valid relationships.

Two spheres of fundamental interest are first identify the separate, discrete and probably immiscible genetic apparatus of the legume host and of its bacteria. In effective symbiosis their gene products become functionally interdependent giving rise to unique structures. Whether these require the actual interchange of genetic material between the symbionts is open to serious question, but they certainly depend on the transfer of some kinds of biochemical messages across the membranes, which always separate the bacteria and the plant.

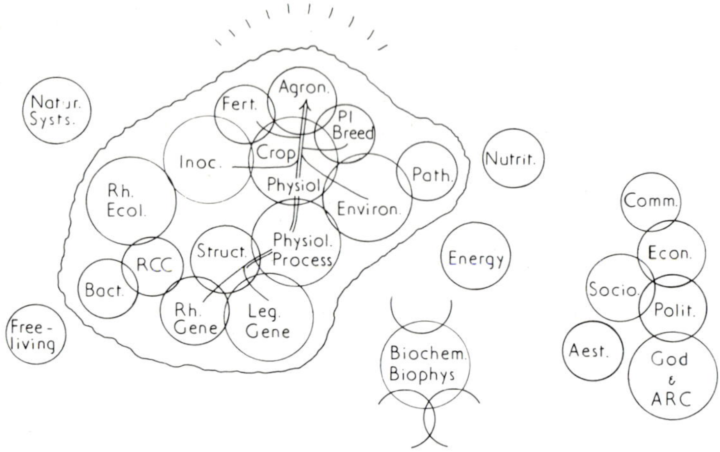

Fig. 1.

The symbiosis also interacts with the environment and the other components as indicated. To the agriculturalist these comprise crop physiology and agronomy of the nodulated plant.

Somewhat peripheral to these central fields of interest is the bacteriology of *Rhizobium*, the curating of *Rhizobium* collections, the unravelling of *Rhizobium* ecology and the practices of inoculation, fertiliser use and plant breeding.

Underlying or rather interpenetrating these areas of interest, are corresponding ones concerning biochemistry and biophysics.

Still more peripheral is the study of N_2-fixation by free-living organisms and by leguminous and non-leguminous symbioses in natural habitats. Also to be considered are energy constraints and nutritional caveats, the former dealt with in several papers, but the latter entirely neglected.

Still more peripheral and some decidedly not the concern of this meeting, are matters of economics (that dismal science), customs of farming and food preferences, sociology, politics, certainly aesthetics and even perhaps religion. The circle marked 'Comm' can signify communication such as we enjoy here or committee as you wish. To the uninitiated 'ARC' means the Agricultural Research Council, in the U.K.

Whether this way of looking at things has any heuristic value is for you to judge. Some of you would doubtless prefer the abstractions and disciplines of mathematical modelling, and others with whom I would include myself, may also need something more concrete to guide their understanding or stimulate their imagination.

Fig. 2 attempts to provide this more concrete alternative frame of reference in the form of an ideogram of a nodulated legume, on which is represented the principal events, factors, relationships and interactions known to affect the symbiosis - its structures, development, physiology and metabolism.

Outside the root the rhizobia live freely in the soil stimulated by root exudates, probably exchanging genetic information amongst themselves, and possibly also with other micro-

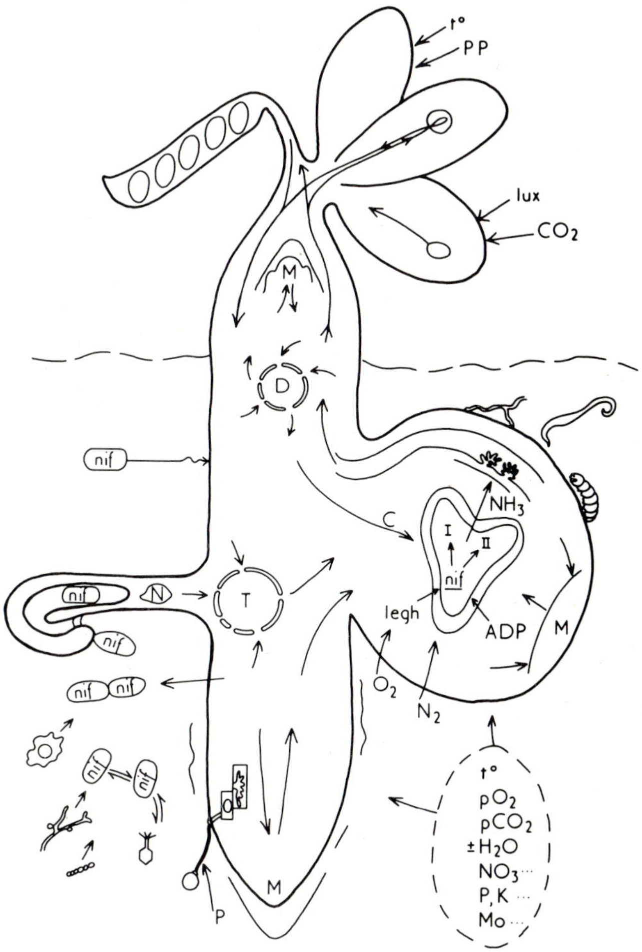

Fig. 2.

organisms and suffering, as do we all, the vissicitudes of community life - predation, competition, suppression, degradation, and even simple neglect.

But the very few rhizobia, the 'elect', succeed against all odds in penetrating the root, in associating with the nodule cells, in undergoing the miraculous conversion within them into the celibate bacteroid, and in the end fixing nitrogen, but not without the help of a peculiar assortment of enzymes, metabolic regulators, co-factors and so on. This N then enters the host's life stream, by a devious route, nourishing its roots and leaves and filling its storage organs with protein-rich materials.

Affecting this microcosm are a plentitude of internal factors, ultimately genetic, most of which seem to be linked to the multifarious processes in the host that govern its growth especially those centred in the plant's meristems. These are also affected directly or indirectly by the whole environment: by physical factors such as light intensity and duration; by water relations; by CO_2 and photosynthetic efficiency; and by nutritional inputs, such as phosphate and minor elements, and by biological factors such as the harmful attacks of fungi, insects and eelworms, or the beneficial symbiosis of mycorrhizal fungi.

Let me also draw your attention to the highly irregular and as yet unavailing attempts of the nif gene to insinuate itself into the plant by unorthodox means, or at any rate in any meaningful way.

Complex as it may appear, this ideogram is a simplification which contributors are amply confirming during this meeting. This is an important point and I would like to illustrate it by outlining what is at present known, or believed to be known, about the pre-infection processes - a focus of interest covered by only a very small part of this diagram.

The present state of knowledge, dogma and hypothesis about pre-infection is summarised in Fig. 3. Much is familiar to you and I do not propose to take you laboriously through its different parts. Suffice to remind you that this particular area of study is in a state of great flux and uncertainty. Data and ideas are being poured in and nearly as precipitately withdrawn. No longer may polygalacturonidase induction be thought to be a central feature and the attractive hypothesis based upon it for explaining specificity may be due for discarding although Olivares' work indicates a new function for PG, the stimulation of IAA formation. Perhaps the lectin story will also not be confirmed. Here, at any rate, we may be about to pass from a phase of almost frenetic elaboration to some simplification and rationalisation.

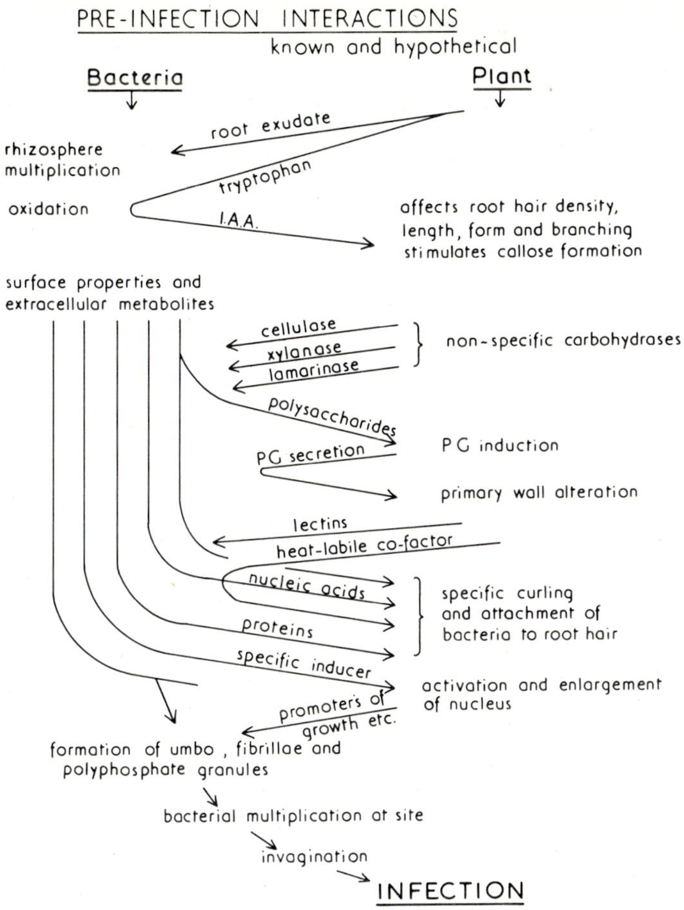

Fig. 3.

The correct conclusion is that out of confusion will come eventually a measure of clarification.

This excursion has been somewhat teleological, not to say anthropomorphic and it is now high time that we moved on to the film and the first serious paper.

FAO'S ACTIVITIES IN BIOLOGICAL NITROGEN FIXATION

S.A. KRASTANOV* and F. RIVEROS**

In 1973/74 it became increasingly evident that the world's raw material and energy resources were not unlimited and would have to be used more sparingly in future.

This situation called for a revision of the technological approach in general, and, in particular, a reconsideration of the use of organic materials as fertilizers including the use of legumes for nitrogen fixation.

A special consultancy was devoted to nitrogen fixation and guidelines for an action programme on the use of Rhizobia were established during 1975. This consultancy stressed the necessity to identify the potential and the factors limiting the widespread use of inoculants in the field. Furthermore, it was recommended that the zones or regions where they may be useful be determined and that the benefits derived from symbiotic nitrogen fixation be demonstrated as has already been shown in developed countries.

At the 18th Session of the FAO Conference, held in Rome in November 1975, special stress was placed on biological nitrogen fixation. The Conference noted the importance of Rhizobia fixing bacteria in increasing the yield of legumes and, in some cases, the ameliorative effect on soil. It was recommended that highest priority be given to the biological fixing of nitrogen through nitrogen fixing bacteria. Some of the positive features of this process are:

1. Legumes are very important as a prospective source of protein in the diet of the population in developing countries.

2. Considerable knowledge has been accumulated on biological nitrogen fixation, which is a pre-requisite for the success of its practical application.

A considerable amount of basic as well as applied research is being currently carried out in many countries in the tropical regions of Latin America, Asia and Australia. It would not

* Land and Water Development Division, FAO, Rome.

** Plant Production and Protection Division, FAO, Rome.

be exaggerated to say that the study of biological N fixation in many tropical countries is considered as a number one problem in the research policy on agrobiology.

3. The intensive nitrogen fixing ability of legumes in the tropical regions could contribute successfully in overcoming or easing the shortage of nitrogenous fertilizers in a number of developing countries.

Following the recommendation of its General Conference, FAO has recently initiated practical measures for transferring knowledge and experience accumulated in developed countries to the developing countries. This is reflected in a project document entitled "Development of a Coordinated International Programme on Biological Nitrogen Fixation by Micro-organisms" which is financially supported by UNEP and was approved at the beginning of 1976. The immediate objectives of the project are defined as follows:

1. To make an assessment of ongoing activities in biological nitrogen fixation carried out by various institutions (national, multinational, international) in developing countries in order to identify the gaps particularly at the application level.

2. To evolve a cooperative programme aiming at developing, assisting and promoting rational capabilities and facilities in developing countries for a wider use of biological nitrogen fixation, including *Rhizobium*.

3. To initiate demonstrations on beneficial effects of Rhizobia on grain and fodder legume crops in developing countries.

At present this project is in progress. It is too early to give any evaluation of the situation but, nevertheless, some of the gaps at the application level could be identified at the very beginning of the project implementation. One of these gaps is a lack of sufficient knowledge concerning the activity of the Rhizobia strains under the specific ecological conditions in the developing countries. The shortage of experience on the efficiency of the inoculation of the various crops and crop varieties is another obstacle to wide scale application of these techniques in developing agriculture.

Though studies on biological nitrogen fixation are being carried out at present by a number of national institutions in developing countries, it should be stressed that these programmes need further strengthening. Detailed studies need to be made of local ecological conditions and local and introduced Rhizobia strains have to be tested as well as the response of the local and improved crop varieties. In this connection, national institutes in developed countries could give valuable assistance to developing countries in support of improved agricultural techniques.

INVESTIGATIONS OF POSSIBLE LIMITATIONS OF NITROGEN FIXATION BY LEGUMES: (1) METHODOLOGY, (2) IDENTIFICATION, AND (3) ASSESSMENT OF SIGNIFICANCE

R.W.F. HARDY, J.G. CRISWELL and U.D. HAVELKA

E.I. du Pont de Nemours and Company,
Central Research and Development Department,
Experimental Station, Wilmington, Delaware 19898, U.S.A.

INTRODUCTION

Some of the recent work of our laboratory on identification and quantitative assessment of significance of limitations of N_2 fixation by mainly soybeans and in some cases peanuts and peas is summarized in this report. The need for additional nitrogen input in legumes is presented and the two general types of solutions - additional nitrogen fertilization and enhanced biological N_2 fixation - to achieve this goal are compared on the basis of current knowledge. A brief description of three methodology modifications used to measure N_2 fixation is included. Specific limitations investigated include: (1) photosynthate, (2) reproductive sink intensity, (3) rhizosphere temperature and (4) rhizosphere pO_2.

ASSESSMENT OF THE PROBLEMS

Need for Increased Nitrogen Input into Legumes

There is a widely recognized need to increase grain legume yields in both more developed and less developed countries. In the former, for example, the low-yielding but less profitable soybean crop is in direct competition with the high-yielding more profitable corn crop. Substantial improvement in soybean yield should cause the allocation of corn acreage to soybean acreage. Soybeans are the major feed protein in the United States and export of 50+% (Breedlove, 1976) of production makes them a major source of feed protein in countries where increasing affluence is expanding the demand for animal versus plant foods. In a recent report (Scrimshaw *et al.*, 1975) the need for U.S. soybeans is projected to increase by 73% from the mean for 1970-72 to that for 1985; more than that for any other major crop. In the less-developed countries a similar competition exists between the high-protein, low-yield grain legumes which serve as an important contributor of protein to the diet and the low-protein higher-yield cereal grains with the world non-U.S.-soybean legume: cereal grain ratio declining during the last decade (Fig. 1). Substantial increases in the yield of grain legumes used as food should reverse this nutritionally undesirable trend.

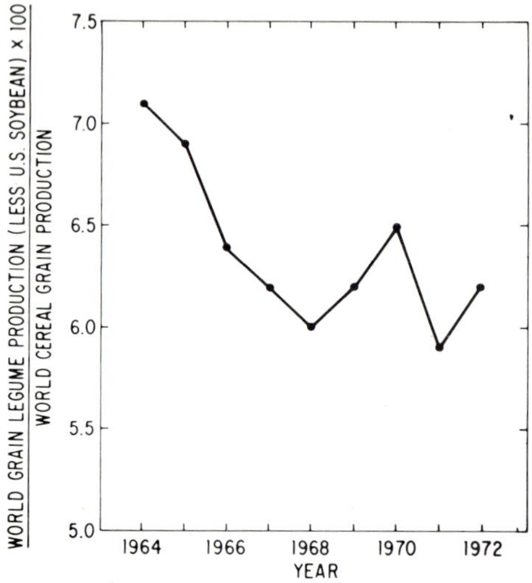

Fig. 1. *Declining ratio of non-U.S. soybean grain legume: cereal grain world production during recent years.*

The high protein content of both food and feed legumes makes the provision of substantially increased nitrogen input a necessity for higher yield. This additional nitrogen, in theory, may be provided by additional fertilizer nitrogen or enhanced biological N_2 fixation (Hardy, 1975; Hardy and Havelka, 1974, 1975b; Hardy *et al.*, 1975a, 1975b, 1976; Havelka and Hardy, 1976b; Ozbun *et al.*, 1976; Silver and Hardy, 1976).

Types of Information Needed

Five types of information are needed for yield improvement through development of logical solutions to overcome limitations in physiological processes or biochemical reactions such as utilization of fertilizer or symbiotic N_2 fixation by legumes. Arranged in chronological sequence these types are the following:

(1) identification of the specific physiological or biochemical limitation;

(2) quantitative assessment of the significance of the limitation, preferably in the field under normal crop production conditions rather than under the artificial conditions of the growth room or greenhouse where limitations probably differ from the field;

(3) development of specific but rapid and simple screens to select genetic, chemical, or other solutions that will overcome the limitation (plant breeders and plant growth regulator scientists need to examine thousands of treatments to select effective ones);

(4) discovery of practical solutions;

(5) implementation of the solutions.

Outside of the empirical research directed towards the selection of improved rhizobia and development of improved inoculation techniques such as, for example, granular inoculum, the majority of the research on symbiotic N_2 fixation in legumes has focused on the identification area utilizing either parts of plants or in some cases intact plants for the most part under the unnatural production conditions of the laboratory. In our opinion more effort is needed in areas 2, 3, 4 and 5 as defined above.

Comparison of Additional Nitrogen Fertilization and Enhanced Biological N_2 Fixation

Up until this past year, additional nitrogen fertilization either applied as a foliar or soil application at a single time or multiple times did not appear to be an effective approach since legumes gave no significant positive yield responses. The low rate of nitrogen fertilization of soybeans (16 Kg/ha) in the U.S. versus corn (128 Kg/ha), wheat (52 Kg/ha), and cotton (83 Kg/ha) (Havelka and Hardy, 1976b) further attests to the ineffectiveness of fertilizer nitrogen. This ineffectiveness has been attributed to the compensatory decrease in nitrogen-input from N_2 fixation that approximates to the increased nitrogen-input from fertilizer nitrogen (see e.g., Hardy *et al.*, 1973). Within the past year (Garcia and Hanway, 1976) significant yield increases have been reported in soybeans (600-1200 Kg/ha) by multiple foliar fertilization from early pod-filling to almost complete pod-filling with a mixture of N, P, K, and S where nutrient ratio is similar to that found in the mature seed (Table I). There is some question as to the practicality of the solution since the foliar application dictates the use of low-biuret urea and polyphosphates to avoid toxicity to the foliage and the cost of the chemicals and their multiple application has been estimated to exceed 50/ha.

In the case of biological N_2 fixation previous results from our laboratory (Hardy and Havelka, 1974, 1975a; Havelka

Table I

Foliar fertilization of soybeans (Garcia and Hanway, 1976)

Total Nutrient Application N (Kg/ha)	P	K	S	Applications No.	Yield (Kg/ha)
-	-	-	-	-	2980
80	8	24	4	4	3550
-	8	24	4	2	2810
80	-	24	4	4	3090
80	8	-	4	5	2840
80	8	24	-	4	3310
40	4	12	2	4	2940
120	12	36	6	5	3220

and Hardy, 1976a) and one other group (Phillips *et al.*, 1976) and the additional evidence included in this report demonstrate that increased net photosynthate production achieved by a substantial decrease of photorespiration through CO_2-enrichment of the canopy of field-grown legumes - soybeans, peanuts, peas - can produce large increases in N_2 fixation and of even greater importance, yield. For example, 1500 ppm CO_2 surrounding the canopy of field-grown peas increases the yield of seed by over 1800 Kg/ha; an effect greater than that reported for foliar fertilization. CO_2-enrichment is, of course, not economically attractive but routes to equivalent but practical solutions by chemical or genetic means to completely or partially eliminate photorespiration are being pursued in several laboratories in both the public and private sectors. No useful solutions have been reported.

We report here some information on other possible limitations of biological N_2 fixation. These include long exposures to low rhizosphere temperatures and even short exposures to high rhizosphere temperatures. This information for the rhizosphere supplements previous data of temperature effects on the whole plant (see e.g., Gibson, 1976a, 1976b; Dart *et al.*, 1975). Subatmospheric rhizosphere pO_2's previously believed to be extremely detrimental to N_2 fixation (Bergersen, 1962; Gibson, 1976a, 1976b) are shown to be of little concern since the N_2-fixing plant, at least in the case of soybean, adapts to altered pO_2. Exceptions are extremely low pO_2 where the adaptation is incomplete and low pO_2 in combination with low temperatures where the adaptation is slow. This O_2 control system appears to be used to husband carbohydrate in the substrate-depleted nodule. The significance of these limitations, unlike the limitation imposed by photosynthesis, has not been assessed in the field. Such an assessment will be needed to determine if they are of great significance to legume yield or merely interesting physiology of the legume system. Additional research is also needed to uncover additional limitations such as possibly nodular hormonal contents. Research of types 3, 4 and 5 should provide practical solutions to enhance biological N_2 fixation and yields of legumes.

RESULTS

Modifications in Methodology

Three modifications of the C_2H_2 to C_2H_4 assay used are described briefly in this section along with a summary of their respective advantages and disadvantages (Table II). More complete descriptions are available elsewhere (see Criswell *et al.*, 1976; Hardy and Holsten, 1976).

The first method is the excised nodulated root-plastic bag technique (Hardy and Holsten, 1971; Havelka and Hardy, 1976a). This method has been employed routinely in our laboratory and as its name indicates, uses excised nodulated roots rather than intact plants.

The second method is the intact plant-growth pot incubation technique which enables measurements on intact plants grown in pots in the laboratory or field (Criswell *et al.*, 1976; Quebedeaux *et al.*, 1975). The pot is sealed (Fig. 2) and becomes the incubation vessel with initiation of assay by C_2H_2 injection. Other gas-producing or gas-utilizing activities of the rhizosphere (respiration, H_2 evolution, etc.) can also be assessed with this technique. The $N_2[C_2H_2]$-fixing activity of the intact plant measured by this system is constant over at least several hours, in contrast to decapitation of the shoot which reduces activity significantly in the 30 min immediately following decapitation. The excised nodulated root techniques may thus underestimate $N_2[C_2H_2]$ fixation by 20-30% (Fig. 3). We have used this intact-plant-growth pot technique in recent years for experiments in the growth room on reproductive sink demand and rhizosphere pO_2 and temperature as discussed below.

The third method, the continuous flow field technique, *in situ*, is still undergoing development (Criswell and Hardy, 1976). An open cylinder (Fig. 4) whose diameter and length depend upon the diameter and depth of the nodulated root zone is driven into the soil to enclose the nodulated zone. The top of the cylinder

Table II

Advantages and Disadvantages of $N_2[C_2H_2]$ Fixation Assay Methods

Advantages	Disadvantages
Excised Nodulated Root-Plastic Bag Technique	
High capacity	Plant part
Simple equipment	Destructive - single assay on same plant
No corrections for moisture content	Measurement of gas phase volume
Mobile in field	Large number of assays with variability of individual plants
	Lower activity than intact plant, especially in incubations longer than 30 min
Intact Plant-Growth Pot Technique	
Intact plant	Time consuming
Multiple assays on same plant	Measurement of gas phase volume
Study kinetics	Modified pots
Separate treatment of shoot and root	Correction for moisture content
100% nodule recovery	Correction for degree of saturation by C_2H_2
Smaller number of assays with decreased variability of plants on defined conditions	Unnatural support media (sand)
Simultaneous measurement of other rhizosphere gases	
Continuous Flow *In Situ* Field Technique	
Intact plant	Extremely time consuming
In situ in field	Special chamber
Continuous measurement on same plant	Unrecognized errors
Relatively mobile in field	Lateral root shear
Study kinetics	Ethylene oxidation by soil bacteria but acetylene inhibits
Separate treatment of shoot and root	
Simultaneous measurement of other rhizosphere gases	

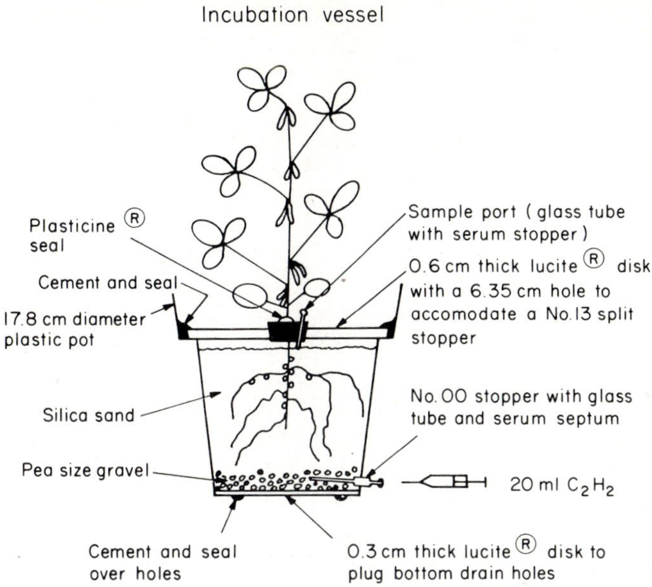

Fig. 2. *Diagram of intact plant-incubation pot assay technique for measurement of* N_2 *[C_2H_2] fixation (Criswell* et al., *1976).*

is sealed around the plant and an Ar:O_2 mixture, or air containing 0.1 atm C_2H_2, is continuously forced into the soil through metal probes whose exit ports are located well below the nodulated ozone and within the cylinder. Some of the introduced gas is forced upward through the nodulation zone where an amount of C_2H_2 proportional to the N_2-fixing activity is reduced to C_2H_4, and this C_2H_4 plus the unreduced C_2H_2 flow out the exit port. The following calculation is used: N_2[C_2H_2]-fixing activity/plant·min =

$$\frac{\text{exit flow rate (ml/min) x moles } C_2H_4\text{/ml exit gas x moles } C_2H_2 \text{ ml inlet gas}}{\text{moles } C_2H_2\text{/ml exit gas}}$$

This method enables the continuous measurement of N_2[C_2H_2] fixation by minimally disturbed field-grown plants. We are still in the stage of comparing profiles of N_2 fixation obtained by this method with those obtained by the excised nodulated root-plastic bag technique. None of the data reported below were obtained with this technique.

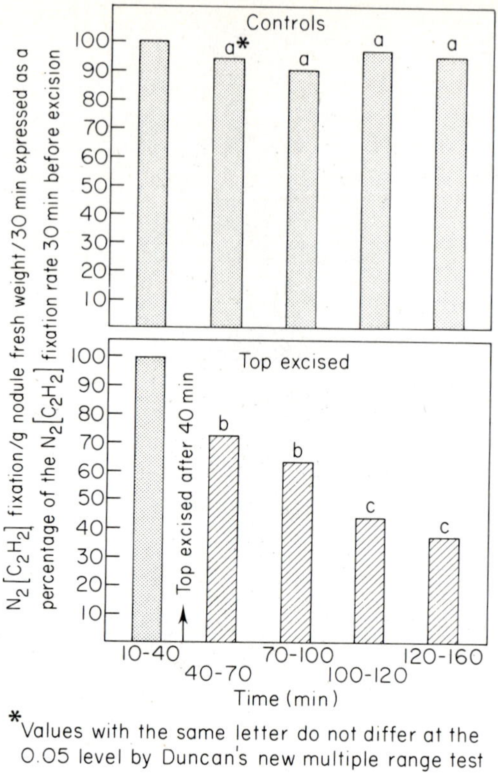

Fig. 3. *N_2 [C_2H_2] fixation measured by the intact plant-incubation pot assay over 3 h and comparison with plants decapitated at 40 min.*

Additional Evidence on Photosynthate Limitation

The number of factors that influence photosynthate available to the nodule and their parallel effect on N_2 fixation in the soybean has expanded to an impressive list (Table III) during the past three years. A recent addition is the demonstration that some of the decline in N_2 fixation at low water potential arises from a primary decrease in photosynthesis and that overcoming this decrease by CO_2 enrichment around the canopy reverses the decline (Huang *et al.*, 1975).

We have reported a multifold enhancement of $N_2[C_2H_2]$ fixation by field-grown soybeans in which the canopy was enriched to 800 to 1200 ppm CO_2 during the day from about 40 days of age to maturity (Hardy and Havelka, 1975a; Havelka and Hardy, 1976a) A special open-top, side-enclosed chamber enabled defined enrichment of canopy CO_2 of field-grown plants without alteration

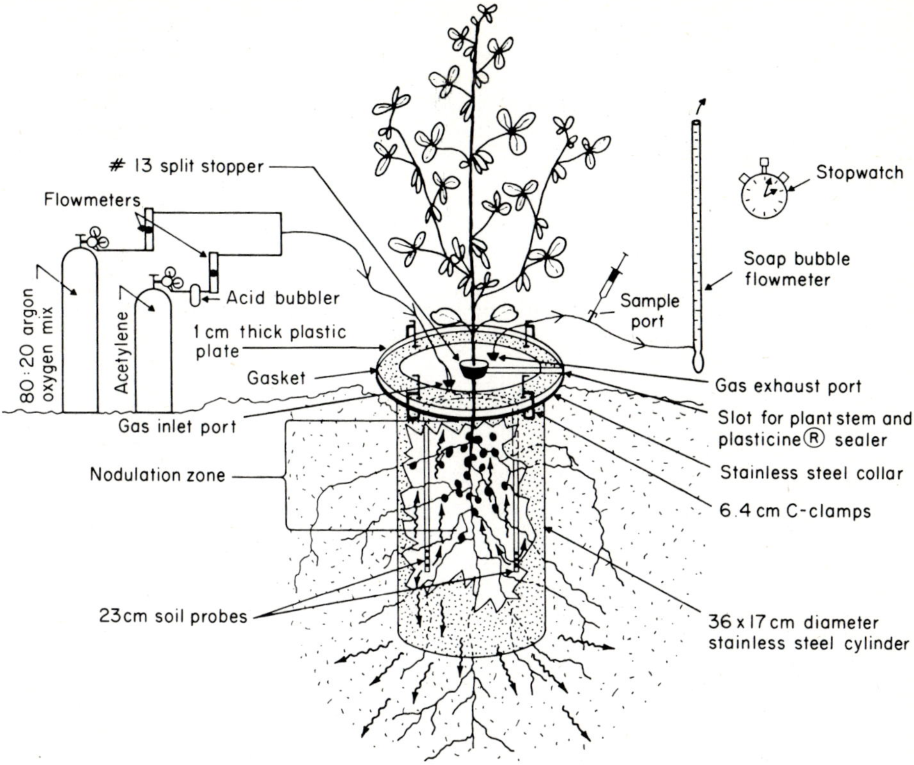

Fig. 4. *Diagram of the continuous flow* in situ *field technique for measurement of* $N_2[C_2H_2]$ *fixation.*

in other environmental factors such as temperature or light intensity. The increase in soybean $N_2[C_2H_2]$ fixation was a product of increased specific activity of the nodule, increased average nodule mass, and delayed loss of exponential phase of N_2 fixation. We attributed the increase to the enhanced availability of photosynthate to the nodule to support the high energy consumption of N_2 fixation and proposed that photosynthate was a major factor limiting N_2 fixation in at least field-grown soybeans and possibly other legumes.

We suggested that the major effect of increased CO_2 was to decrease photorespiration. Two experiments summarized below provide further support for this hypothesis. The first examined the relationship between CO_2 concentrations from 300 to 1500 ppm around the canopy and $N_2[C_2H_2]$ fixation by field-grown soybeans. The total amount of $N_2[C_2H_2]$ fixed per plant·season increased in a linear manner with CO_2 from 300 to 1000

Table III

Photosynthesis and Symbiotic N_2 Fixation by Legumes (Silver and Hardy, 1976)

Factor	Effect on N_2 Fixation	
	Increase	Decrease
Light quantity	Day Long days Supplemental light	Night Short days Shading
Source size	Additional foliage Low planting density	Defoliation High planting density
CO_2/O_2 ratio in canopy	Increased pCO_2 Decreased pO_2	Decreased pCO_2 Increased pO_2
Photosynthetic type	C_4 ( )	C_3
Competitive sinks	Removal of reproductive structures	Development of reproductive structures
Translocation		Girdling
Water		Low water potential

ppm and was curvilinear from 1000 ppm to 1500 ppm as shown in Fig. 5 (Havelka and Hardy, 1974). This relationship is consistent with that expected for the CO_2/O_2 competition characteristic of photorespiration. In the other experiment the CO_2/O_2 ratio around the canopy was altered by either CO_2 enrichment at a constant pO_2 of 0.21 atm as in the previous experiments or by O_2 depletion or O_2 enrichment at a constant CO_2 concentration of 300 ppm (Quebedeaux *et al.*, 1975). During vegetative growth including the early flowering stage there was a close correlation between CO_2/O_2 ratio and $N_2[C_2H_2]$ fixation providing strong support that the effect of CO_2 enrichment was mediated by decreased photorespiration (Fig. 6). The relationship was less well correlated at maturity and this is attributed to the arresting of reproductive development by subatmospheric O_2 during the reproductive growth phase.

We have extended our observations on CO_2 enrichment and N_2 $[C_2H_2]$ fixation of field-grown legumes to include peanuts and peas (Havelka and Hardy, 1976c). Both crops were grown in open-top, side-enclosed chambers and the treated plots received 1500 ppm CO_2 from early vegetative growth to maturity while the

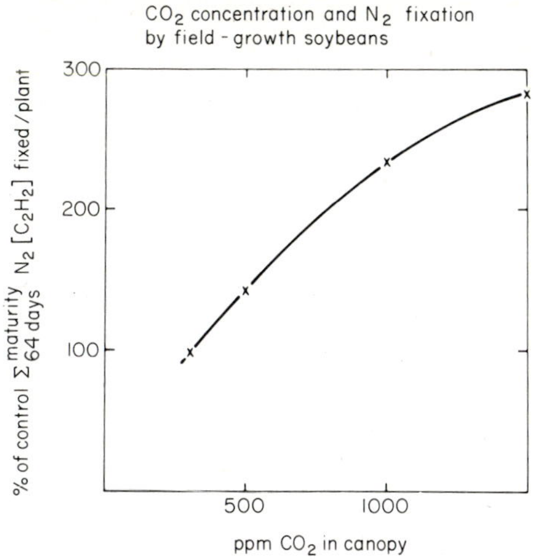

Fig. 5. *Relationship of $N_2[C_2H_2]$ fixed by field-grown soybeans and pCO_2 around canopy.*

controls received air (300 ppm CO_2). With peanuts the amount of $N_2[C_2H_2]$ fixed per plant was increased by 59% from 157 mg/plant to 248 mg/plant while on an area basis it was increased by 76% from 58 Kg/ha to 102 Kg/ha with the greater mature plant population contributing the additional increase similar to that previously observed for soybeans. This increase was composed of an increased mass of nodules and a delayed senescence of nodules with no increase in nodule specific activity. The normal specific activity of peanut nodule is 2 to 3 times that of soybean nodules. Yield of the aerial part of the peanuts was increased by 44% and yield of nuts was increased by 30%. Plant senescence was caused by cool temperatures so that it was not possible to demonstrate a delay similar to that found in soybeans. In the case of peas the amount of $N_2[C_2H_2]$ fixed was increased by 46% from 87.5 mg/plant (87.5 Kg/ha) to 128 mg/plant (128 Kg/ha). This increase came from an increased specific activity of pea nodules, increased mass of nodules, and delayed senescence of nodules. Yield of dried peas was increased by 53% from 3446 Kg/ha to 5262 Kg/ha. Plant senescence was delayed by 4 to 7 days by CO_2 enrichment. $N_2[C_2H_2]$ fixation was calculated to provide about 45% of total N in both air grown and CO_2-enriched plants.

Reproductive Sink Demand

Subatmospheric concentrations of O_2 around the canopy produce dramatic decreases in reproductive growth of all major

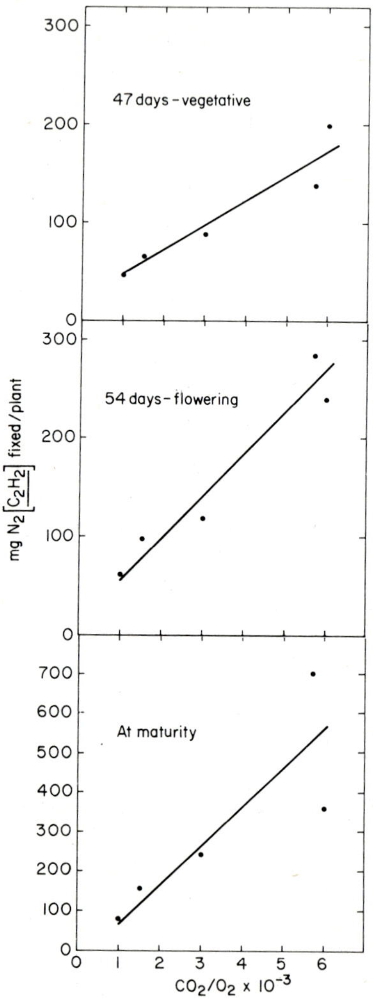

Fig. 6. *Relationship of* $N_2[C_2H_2]$ *fixed by soybeans and* CO_2/O_2 *ratio around canopy up to indicated stage of development.* CO_2/O_2 *ratio altered by* O_2 *depletion or enrichment at constant* pCO_2 *and* CO_2 *enrichment at constant* pO_2.

crops that have been tested (Quebedeaux and Hardy, 1973, 1976). In the case of soybeans 5% O_2 almost completely arrests seed development. The rate of $N_2[C_2H_2]$ fixation decreases within a few days after arrest of reproductive growth by 5% O_2(Fig. 7), suggesting that decreased reproductive sink demand will limit N_2 fixation even in the presence of adequate photosynthate (Quebedeaux *et al.*, 1975). The nitrogen content of the seeds at maturity is decreased by 5% O_2 while that in leaves is increased from 2 to 6% and that in stems from 1 to 3%. It is not known if increased reproductive sink demand will increase N_2 fixation.

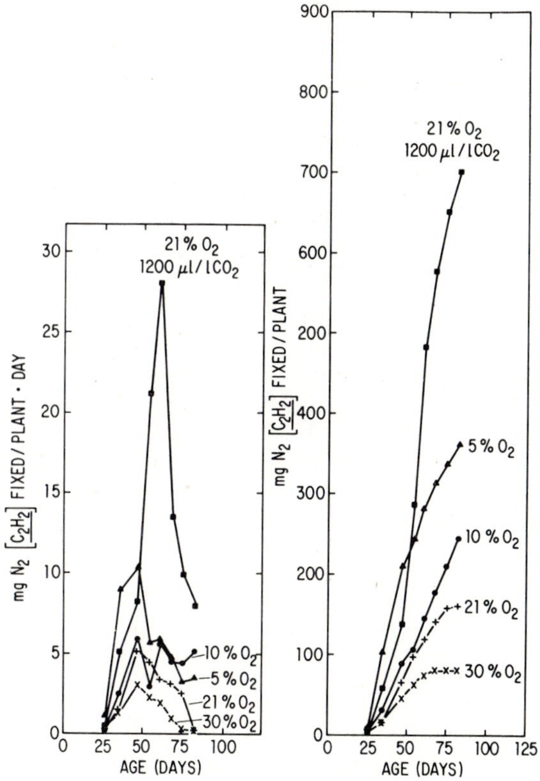

Fig. 7. *$N_2[C_2H_2]$ fixation rate of soybeans where the canopy gas phase was 5, 10, 15, 21 and 30% O_2 at 300 ppm CO_2 and 1200 ppm CO_2 at 21% O_2. Note decrease in $N_2[C_2H_2]$-fixing activity under subatmospheric pO_2 at a time coincident with the arrest of reproductive growth (Criswell* et al., *1976).*

Rhizosphere Temperature

Equilibrium and kinetic responses of soybean $N_2[C_2H_2]$ fixation to altered temperature of the rhizosphere were measured in the growth room (Hardy and Criswell, 1976). For previous information on temperature effects see Gibson (1976a,b) and Dart *et al.* (1975). The equilibrium response of soybean $N_2[C_2H_2]$ fixation to altered temperature over the range of 14 to 34°C is quadratic, $Y = 114 + 20X - 0.45\ X^2$, with optima at 22 to 24°C and greater decreases are produced by heating than cooling. The ability of the plant to respond rapidly to altered environmental conditions such as temperature may be of equal or greater importance than equilibrium responses since in the field situation continuous change rather than constancy is the normal situation. The kinetic response of soybean $N_2[C_2H_2]$ fixation to altered rhizosphere temperature shows a rapid decline and recovery following cooling (24 to 14°C) and rewarming (14 to 24°C) and a slow decline and very slow recovery following heating (24 to 34°C) and recooling (34 to 24°C). These temperature effects suggest that long exposures to low rhizosphere temperatures may reduce N_2 fixation substantially while even short exposures to high rhizosphere temperatures may be detrimental.

Rhizosphere pO_2

Excised nodulated roots of soybeans have been reported over the years (e.g. Bergersen, 1962) to show substantial increases in N_2 fixation with supra-atmospheric pO_2 and large decreases with subatmospheric pO_2. These observations suggest that the soybean nodule may be limited by the pO_2 in air and extremely limited by the subatmospheric pO_2 found in poorly aerated soils such as occur with water logging.

In a comprehensive study of the effect of pO_2 on $N_2[C_2H_2]$ fixation by excised nodulated roots from field-grown soybeans measured at six times during the complete growth cycle, stimulations of up to 91% occurred from ≥ 26 to $\leq$ 41% O_2 with the pO_2 for maximum stimulation varying with the age of the plant while ≥ 65 to $\leq 90\%$ O_2 inhibited $N_2[C_2H_2]$ fixation; inhibitions of up to 87% occurred from ≥ 6 to $\leq$ 21% O_2 (Criswell *et al.*, to be published). The effects of altered rhizosphere pO_2 on intact plants measured immediately after alteration was less than that with excised nodulated roots with decreases in the intact plants always occurring at pO_2's $< 21\%$ and usually at pO_2's $> 21\%$. These qualitative and quantitative differences demonstrate the need to utilize intact rather than excised plants.

Longer-term studies of the effect of altered rhizosphere pO_2 led to the discovery of the ability of the soybean N_2-fixing

Table IV

Adaptation of Soybean $N_2[C_2H_2]$ Fixation to Altered Rhizosphere pO_2 (Criswell *et al.*, unpublished).

Time hr	Altered pO_2[(1)] atm	N_2 Fixation by Altered pO_2
		g $N_2[C_2H_2]$ fixed/g nodule fr wt.hr as % of ambient control
0	0.06	55
4	0.06	58
24	0.06	122
96	0.06	121
99	0.21	58

system to adapt to modified rhizosphere pO_2's (Criswell *et al.*, to be published). After 4 to 24 hours of a continuous exposure of the rhizosphere to O_2 concentrations between 6 and 33% there was a complete recovery of $N_2[C_2H_2]$ fixation. Return of the rhizosphere to ambient pO_2 produced an initial decrease in $N_2[C_2H_2]$ fixation which re-adapted to control rates after 4 to 24 h (Table IV). The adaptation was incomplete at 2 to 3% O_2. The adaptation was slower and less complete at cool (14 to 19°C) than at ambient (24°C) or hot (29 to 34°C) temperatures. These observations show a capability of the nodule to contend with a broad range of pO_2 values and thus eliminates rhizosphere pO_2 as a major practical limitation in soybean N_2 fixation except for extreme subatmospheric pO_2 values or combinations of low temperature and low pO_2. We suggest that the adaptation results from the development of an optimal internal pO_2 in the nodule for N_2 fixation over a wide range of pO_2 values. There are many mechanisms that can be suggested but at this time no specific mechanism has been identified. The increasing recognition of the sophistication of the legume-rhizobia symbiosis in areas such as the ability to adapt to fluctuating pO_2's makes it difficult to visualize the construction of man-made symbioses that will be useful for nitrogen input for high yield under the fluctuating environmental conditions of crop production.

The initial response of substrate-depleted intact soybean plants to altered pO_2 differs from that observed with normal plants. The inhibition by low pO_2 is much greater and an elevated pO_2 of about 30% substantially increases $N_2[C_2H_2]$ fixation. We suggest that the nodule may be utilizing its proposed ability to regulate internal pO_2 in order to husband limited substrate reserves of the depleted plants. Thus, the pO_2 adaptation capability may be a control system used for situations in addition to altered rhizosphere pO_2.

REFERENCES

Bergersen, F.J. (1962). *J.Gen.Microbiol.*, 29: 113-125.
Breedlove, S.E. (1976). *Foreign Agriculture*, 14(34): 8.
Criswell, J.G. and Hardy, R.W.F. (1976). *Agron.Abstr.*,
Criswell, J.G., Havelka, U.D., Quebedeaux, B. and Hardy, R.W.F. (1975). *Agron.Abstr.*, p. 131.
Criswell, J.G., Hardy, R.W.F. and Havelka, U.D. (1976). *In:* Proceedings of the World Soybean Research Conference (L. D. Hill, ed.), pp. 108-124. Interstate Printers and Publishers Inc., Danville, Illinois.
Dart, P., Day, J., Islam, R. and Dobereiner, J. (1975). *In:* Symbiotic Nitrogen Fixation in Plants (P. Nutman, ed.), Vol. 7, pp. 361-383. Cambridge University Press, London.
Garcia, R. and Hanway, J.J. (1976). *Agron.J.*, 68: 653-656.
Gibson, A.H. (1976a). *In:* A Treatise on Dinitrogen Fixation. Section IV. Agronomy and Ecology (R.W.F. Hardy and A.H. Gibson, eds), pp. 393-450. Wiley-Interscience, New York.
Gibson, A.H. (1976b). *In:* Proceedings of 1st International Symposium on Nitrogen Fixation (W.E. Newton and C.J. Nutman, eds), Vol. 2, pp. 400-428. Washington State University Press, Pullman, Washington.
Hardy, R.W.F. (1975). *In:* Research for Food in Century III, Proceedings and Minutes of Twenty-Fourth Annual Meeting of the Agricultural Research Institute, pp. 115-130. Agricultural Research Institute, Washington, D.C.
Hardy, R.W.F. and Criswell, J.G. (1976). *Agron.Abstr.*,
Hardy, R.W.F. and Havelka, U.D. (1974). *Crops Soils*, 26(5): 10-13.
Hardy, R.W.F. and Havelka, U.D. (1975a). *In:* Symbiotic Nitrogen Fixation in Plants (P. Nutman, ed.), Vol. 7, pp. 421-439. Cambridge University Press, London.
Hardy, R.W.F. and Havelka, U.D. (1975b). *Science*, 188: 633-643.
Hardy, R.W.F. and Holsten, R.D. (1971). *In:* Proceedings of the Fourth Australian Legume Nodulation Conference. CSIRO, Canberra, Australia, Paper No. 34.
Hardy, R.W.F. and Holsten, R.D. (1976). *In:* A Treatise on Dinitrogen Fixation. Section IV. Agronomy and Ecology (R.W. F. Hardy and A.H. Gibson, eds), pp. 451-486. Wiley-Interscience, New York.
Hardy, R.W.F., Burns, R.C. and Holsten, R.D. (1973). *Soil Biol. Biochem.*, 5: 47-81.
Hardy, R.W.F., Filner, P. and Hageman, R.H. (1975a). *In:* Crop Productivity . . . Research Imperatives (A.W.A. Brown,

T.C. Byerly, M. Gibbs and A. San Pietro, eds), pp. 133-176. Michigan Agricultural Experiment Station, East Lansing, Michigan and Charles F. Kettering, Yellow Springs, Ohio.

Hardy, R.W.F., Frederick, L.R., Harper, J.E., Sloger, C. and Weber, D.F. (1975b). *In:* Enhancing Biological Nitrogen Fixation (H.J. Evans, ed.), pp. 37-43. The National Science Foundation, Washington, D.C.

Hardy, R.W.F., Havelka, U.D. and Quebedeaux, B. (1976). *In:* Genetic Improvement of Seed Proteins, pp. 196-228. National Academy of Sciences, Washington, D.C.

Havelka, U.D. and Hardy, R.W.F. (1974). *Agron.Abstr.*, p. 33.

Havelka, U.D. and Hardy, R.W.F. (1976a). *In:* Proceedings of the 1st International Symposium on Nitrogen Fixation (W.E. Newton and C.J. Nyman, eds), Vol. 2, pp. 456-475. Washington State University Press, Pullman, Washington.

Havelka, U.D. and Hardy, R.W.F. (1976b). *In:* Protein Resources and Technology: Status and Research Needs. Vol. II. Common Issues Affecting Availability of Protein Resources and Their Research Needs (N.S. Scrimshaw, D.I.C.Wang and M. Milner, eds), pp. 327-399. U.S. Government Printing Office, Washington, D.C.

Havelka, U.D. and Hardy, R.W.F. (1976c). *Agron.Abstr.*,

Huang, C.-Y., Boyer, J.S. and Vanderhoef, L. (1975). *Plant Physiol.*, 56: 228-232.

Ozbun, J.L., Carlson, P.S., Hardy, R.W.F., Johnson, H.J., Robinson, K.L., Rutger, J.N., Stuber, D.W. and Wallace, D.H. (1976). Potential Increases in Food Supply Through Research In Agriculture - Research Areas which have Potential for Increasing Crop Production. The Agricultural Experiment Station, New York State College of Agriculture and Life Sciences, Cornell University, Ithaca, New York.

Phillips, D.A., Newell, K.D., Hassell, S.A. and Felling, C.S. (1976). *Am.J.Bot.*, 63: 356-362.

Quebedeaux, B. and Hardy, R.W.F. (1973). *Nature*, 243: 477-479.

Quebedeaux, B. and Hardy, R.W.F. (1976). *In:* CO_2 Metabolism and Productivity (R.H. Burris and C.C. Black, eds), pp. 185-204. University Park Press, Baltimore, Maryland.

Quebedeaux, B., Havelka, U.D., Livak, K.L. and Hardy, R.W.F. (1975). *Plant Physiol.*, 56: 761-764.

Scrimshaw, N.S., Wang, D.I.C. and Milner, M. (eds)(1975). Protein Resources and Technology: Status and Research Needs, Vol. 1. Summary and Research Recommendations. U.S. Government Printing Office, Washington, D.C.

Silver, W.S. and Hardy, R.W.F. (1976). *In:* The Role of Biological Nitrogen Fixation in Forage-Livestock Systems (C.S. Hoveland, ed.), pp. 1- . American Society for Agronomy, Madison, Wisconsin.

THE RELATION OF HYDROGEN REACTIONS TO NITROGEN FIXATION IN NODULATED SYMBIONTS

K.R. SCHUBERT and H.J. EVANS

Department of Botany and Plant Pathology, Oregon State University, Corvallis, Oregon 97331, U.S.A.

INTRODUCTION

The increasing world population and rapid depletion of fossil fuel supplies have stimulated renewed interest in developing methods of increasing agricultural productivity while minimizing the consumption of non-renewable energy resources. Aside from water, the most common limiting nutrient for plant growth is fixed nitrogen (Evans, 1975), a substantial portion of which is supplied to the agricultural industry as synthetically produced anhydrous ammonia. The availability of natural gas for the industrial synthesis of an adequate supply of anhydrous ammonia may be prohibitive, especially for some of the developing nations. Since biological nitrogen-fixing systems utilize photosynthetically stored energy instead of fossil fuels, a thorough understanding of factors limiting the biological process are essential. Hardy and Havelka (1975) have postulated that one limitation of N_2 fixation in legumes is the supply of photosynthate. If we assume that this is correct, then any source of inefficient utilization of photosynthate could theoretically reduce the quantity of N_2 fixed. A recent paper by Schubert and Evans (1976) implicated the ATP-dependent evolution of H_2 catalyzed by nitrogenase as a significant pathway whereby energy may be lost during the N_2-fixation process. The factors that affect H_2 evolution by nodulated symbionts and influence the relative efficiency of transfer of energy to the N_2-reducing system are being investigated in our laboratory. The purpose of this paper is to review briefly the status of our knowledge relevant to these areas.

METHODOLOGY

Several different techniques have been employed to measure ATP-dependent H_2 evolution and hydrogenase activity. The methods used in the experiments described here include either the amperometric or the gas chromatographic techniques for measurement of H_2 evolution or uptake (Schubert and Evans, 1976). An advantage of the amperometric technique is its adaptability for continuous assay of H_2 evolution or H_2 uptake by reaction mixtures containing purified enzyme preparations or by excised nodules. Other methods that may be applicable are listed in Table I.

NITROGENASE-CATALYZED HYDROGEN EVOLUTION

Studies In Vitro

The ATP-dependence of hydrogen evolution from reactions containing N_2-fixing extracts of *Azotobacter* was first established by Bulen, Burns and LeComte (1965). Subsequently, it was demonstrated that dithionite would serve as a reductant and both components of *Azotobacter* nitrogenase were essential for the reduction of protons (Bulen and LeComte, 1966). Similar findings have been reported for cell-free preparations of nitrogenase from soybean root nodules (Koch *et al.*, 1967; Bergersen and Turner, 1968), *Clostridium* (Winter and Burris, 1968), *Rhodospirillum* (Burns and Bulen, 1966), and all other sources examined (Burns and Hardy, 1975). The ratio of the number of ATP molecules hydrolyzed per pair of electrons transferred to an acceptor is independent of the substrate according to Hadfield and Bulen (1969). The values of this ratio range from less than 2 to 20 (Ljones and Burris, 1972b; Kelly, 1969; Winter and Burris, 1968; Burns and Hardy, 1975) and it has been postulated that the variation in the ratio of moles of ATP consumed per 2 electrons transferred during nitrogenase catalysis may be caused in part by a failure to account for H_2 evolution (Hwang and Burris, 1972). The ATP per 2 electron ratio increased with increasing temperature from 20°C to 40°C, but changes in pH within the range 6 to 8 apparently had no consistent effect (Hadfield and Bulen, 1969; Winter and Burris, 1968; Jeng *et al.*, 1970).

During catalysis by nitrogenase, protons and N_2 compete for electrons. In the absence of N_2 the entire energy flux available for nitrogenase activity is expended for the production of H_2 as shown in Figs 1 and 2 (Winter and Burris, 1968; Hwang *et al.*, 1973; Burns and Hardy, 1975). In the presence of an atmosphere of N_2,H_2 evolution occurs concomitantly with ammonia formation and during this process 25 to 35 percent of the total equivalents of reductant available to nitrogenase is diverted to the reduction of protons to H_2 (Hadfield and Bulen, 1969; Hwang and Burris, 1972; Koch *et al.*, 1967)(Figs 1 and 2). Kinetic analysis by Rivera-Ortiz and Burris (1975) indicated that H_2 evolution would still account for 13 to 23 percent of

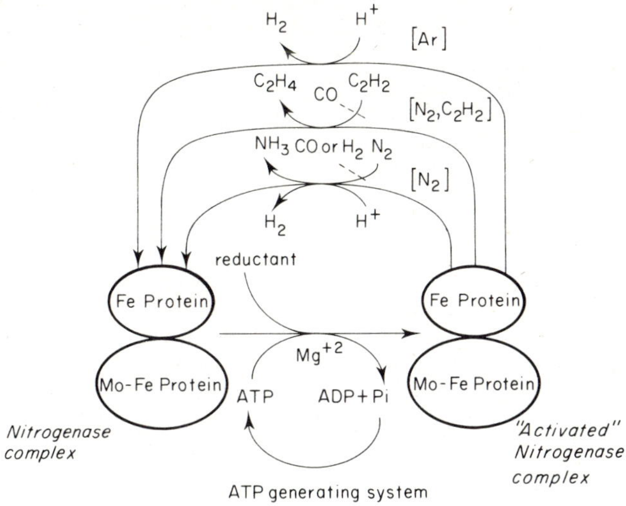

Fig. 1. *Reactions of nitrogenase based on results from cell-free preparations of nitrogenase under atmospheres shown in brackets. Inhibition by CO or* H_2 *is indicated by a dashed line.*

the total reductant consumed via nitrogenase even at an infinite concentration of N_2. Allocation of electrons between substrates is dependent to some extent on the ratio of the Fe protein component to the Mo-Fe protein component of nitrogenase (Ljones and Burris, 1972; Shaw *et al.*, 1975; Burns and Hardy, 1975). Increasing the ratio of the Mo-Fe to the Fe protein increases the proportion of electrons consumed in the hydrogen evolving reaction (Davis *et al.*, 1975; Schubert and Evans, 1976) Similarly the ATP concentration and the ATP to ADP ratio have been found to alter the allocation of electrons to acceptors such as H^+, C_2H_2 or N_2 (Bulen, 1976; Silverstein and Bulen, 1970; Davis *et al.*, 1975).

Unlike N_2, a saturating concentration of C_2H_2 essentially eliminates both H_2 evolution and N_2 reduction (Watt and Bulen, 1976; Rivera-Ortiz and Burris, 1975; Burns and Hardy, 1975) (Fig. 1). The other alternate substrates for nitrogenase including azide, cyanide, and N_2O compete for "activated" electrons and thus serve as inhibitors of H_2 evolution (Hwang *et al.*, 1973; Rivera-Ortiz and Burris, 1975). Cyanide seems to inhibit H_2 production by a mechanism that appears more complex than simple competitive inhibition (Hardy and Knight, 1967). Of all the substrates reduced by nitrogenase, N_2 is the only one that is inhibited by H_2 (Hwang *et al.*, 1973; Rivera-Ortiz and Burris, 1975), but the presence of H_2 in a reaction

Table I

Methods of measuring H_2 evolving or utilizing reactions

Method	ATP-dependent H_2 evolving activity catalyzed by nitrogenase	Hydrogenase activity	References
Manometry	+	+	Peck & Gest (1956); Koch, Evans & Russell (1967); Hyndman, Burris & Wilson (1953); Krasna & Rittenburg (1956).
Mass spectrometry	+	+	Dixon (1967); Hoch, Schneider & Burris (1960).
Gas chromatography	+	+	Schubert & Evans (1976).
Spectrophotometry	-	+	Peck & Gest (1954); Burris (1972); Kleiner & Burris (1970).
Dithionite oxidation assay	+	-	Ljones & Burris (1972a).
Hydrogen-deuterium (tritium) exchange	-	+	Kleiner & Burris (1970); Anand & Krasna (1965).
Ortho/para H_2 interconversion	-	+	Krasna & Rittenberg (1954).
Radioisotopic technique	-	+	Marayama (1975).
Amperometric technique	-	+	Schubert & Evans (1976).

+ Method applicable for assay of activity

- Method not suitable as an assay.

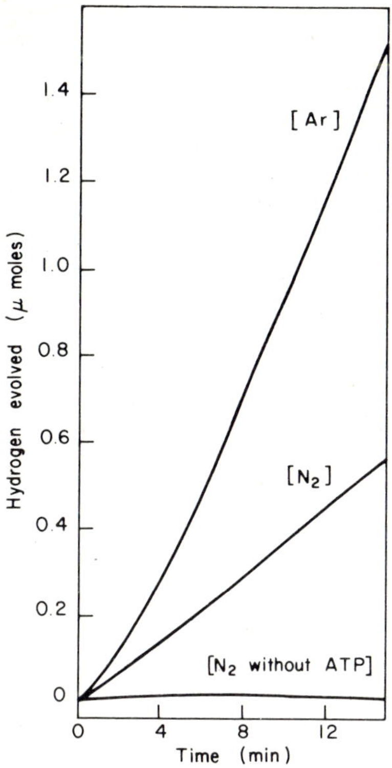

Fig. 2. *Continuous amperometric measurements of H_2 evolution from reaction mixtures containing cell-free preparations of nitrogenase from soybean nodules. The complete assay mixture contained 50 μ moles of creatine phosphate, 7.5 μ moles of $Na_2S_2O_4$, 10 μ moles of $MgCl_2$ and 37.5 moles of TES (N tris(hydroxymethyl)-methyl-2-amino-ethane sulfonic acid) buffer at pH 7.5 in a final volume of 1.5 ml. The assay temperature was 25°. Reactions were initiated by the addition of 50 μl (0.64 mg) of partially purified nitrogenase. Recorded tracings are shown for reactions under N_2 or Ar and also for a reaction without ATP under N_2. Tracings similar to that recorded where ATP was omitted were obtained when either dithionite or enzyme was left out or when oxygen was added to the reaction mixture. Elimination of the gas phase in the electrode chamber improved the linearity of the rate of H_2 evolution after an initial lag.*

containing the necessary components for the nitrogenase activity does not inhibit H_2 evolution (Rivera-Ortiz and Burris, 1975; Hwang *et al.*, 1973)(Fig. 1).

Carbon monoxide, as an inhibitor of nitrogenase, is unique in that it is the only known compound that noncompetitively inhibits the reduction of all nitrogenase substrates except H^+ (Winter and Burris, 1968; Hwang *et al.*, 1973; Rivera-Ortiz and Burris, 1975; Burns and Hardy, 1975)(Fig. 1). This fact has significant mechanistic implications and has led to the postulation that proton reduction occurs at a site distinct from that of N_2 reduction (Hwang *et al.*, 1973; Rivera-Ortiz and Burris, 1975). Likewise, the N_2-stimulated hydrogen-deuterium exchange reaction, which also is noncompetitively inhibited by CO, reportedly occurs at a site distinct from the H_2-evolving site (Bulen, 1976; Jackson *et al.*, 1968; Turner and Bergersen, 1969; Bergersen, 1963).

Studies In Vivo

Excised nodules Hydrogen production was observed for soybean nodules (Hoch *et al.*, 1957) and other legumes such as cowpeas (*Vigna unguiculata*) and lupines (*Lupinus* sp.)(Dart and Day, 1971a,b), however, not all nodulated symbionts exhibit net H_2 production. Exceptions include *Pisum sativum* (Dixon, 1967, 1969), *Alnus glutinosa, Hippophae rhamnoides* and *Casuarina cunninghamiana* (Dart and Day, 1971b). Hoch *et al.* (1960) established that the evolution of H_2 was catalyzed by the nitrogen-fixing system. Oxygen is required for both nitrogen fixation and nitrogenase-dependent H_2 evolution (Hoch *et al.*, 1960; Bergersen and Turner, 1968; Bergersen, 1963). This requirement undoubtedly is associated with the role of O_2 as the terminal electron acceptor necessary for the production of ATP via electron transport and oxidative phosphorylation (Bergersen and Turner, 1968). In the absence of N_2, the total electron flux available to nitrogenase apparently is used to reduce protons (Burns and Hardy, 1975)(Fig. 3A). According to Hoch *et al.* (1960) a pN_2 of 0.3 atm caused a 45% decrease in the rate of H_2 evolution in the absence of N_2 by soybean nodules. When saturating concentrations of C_2H_2 are used H_2 formation is completely inhibited and consequently, rates of acetylene reduction may be used to estimate the total electron flux through the nitrogenase system (Burris, 1975). Concentrations of carbon monoxide which do not inhibit respiration, but totally inhibit N_2 fixation either do not alter or stimulate H_2 evolution by nodules (Hoch *et al.*, 1960; Bergersen and Turner, 1968). Since the process of ATP-dependent H_2 evolution might compete for the energy available for N_2-fixation, a survey was conducted to determine the magnitude and generality of occurrence of H_2 evolution by nodulated symbionts. The amounts of H_2 and C_2H_4 produced by nodule samples were used as a basis for an estimation of the relative efficiency of energy utilization for

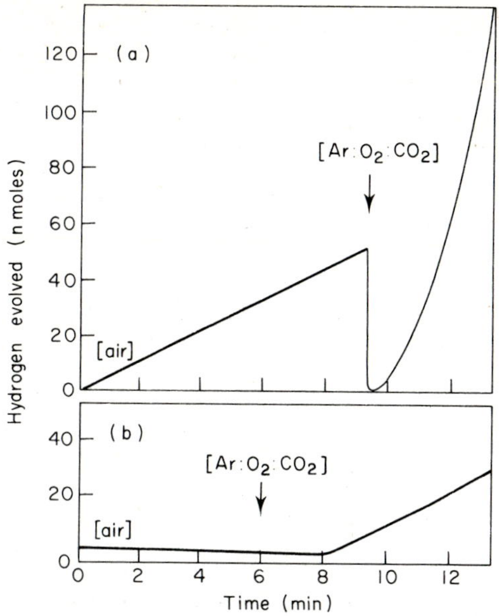

Fig. 3. *Amperometric measurements of H_2 evolution from excised soybean nodules (*Glycine max *cv. Anoka) inoculated with specific* Rhizobium japonicum *strains. Reactions were initiated in air and as indicated by the arrow, the chamber was flushed for 30 secs with 79.96% Ar, 20% O_2, 0.04% CO_2. Nodules used in experiments A and B were from plants inoculated with USDA strains 138 and 110 respectively. Excised nodules (0.58 g in A, 0.51 g in B) were utilized in each experiment. Other experimental details were described previously (Schubert and Evans, 1976).*

nitrogen reduction. Relative efficiency was defined as follows:

$$\text{Relative efficiency} = \frac{\text{Energy used for nitrogen reduction}}{\text{Total energy flux available for nitrogenase activity}}$$

In the calculation of relative efficiency the following assumptions concerning energy flux through nitrogenase were made:

(1) Under an atmosphere of air the total energy used by nitrogenase catalyzed reactions is proportional to the rate of N_2 reduction (N_2 reduction), plus the rate of H_2 evolution in air [H_2(air)].

(2) Under an atmosphere in which N_2 is replaced by argon the total energy consumption is proportional to the rate of

proton reduction [H_2(Ar)].

(3) In the presence of a saturating level of C_2H_2, the total energy used is proportional to the rate of C_2H_2 reduction [C_2H_2].

(4) The energy flux available for nitrogenase activity is not altered by changes in the gaseous environment. The derivation below can then be applied:

$$\text{Relative efficiency} = \frac{[N_2 \text{ reduction}]}{[N_2 \text{ reduction}]+[H_2(\text{air})]} = \frac{[N_2 \text{ reduction}]}{[H_2(\text{Ar})]}$$

$$\text{Relative efficiency} = \frac{[H_2(\text{Ar})]-[H_2(\text{Air})]}{[H_2(\text{Ar})]} = 1 - \frac{[H_2(\text{air})]}{[H_2(\text{Ar})]}$$

This should be equivalent to:

$$\text{Relative efficiency} = 1 - \frac{[H_2(\text{air})]}{[C_2H_2]}$$

In the event that nodules possess a mechanism for reutilizing part of the H_2 evolved, the relative efficiency calculated on the basis of H_2 evolved under argon as the total electron flux will be lower than the relative efficiency calculated where C_2H_2 reduction is used as the total electron flux.

Table II

Relative efficiencies of energy utilization by nitrogenase in nodulated symbionts[+a]

Species	Inoculant	Relative efficiency[b] $1-\frac{[H_2(\text{air})]}{[H_2(\text{Ar})]}$	$1-\frac{[H_2(\text{air})]}{(C_2H_2}$
Legumes			
Alfalfa (*Medicago sativa*	Commercial[c]	0.54±0.04	0.51±0.04
Alfalfa (*M. sativa*)	Native	0.51±0.03	0.53±0.03
Sweet clover (*Melilotus alba* and *M. officinalis*)	Native	0.59±0.03	0.59±0.03

Strawberry clover (*Trifolium fragiferum*)	Commercial	0.59 ± 0.01	0.55 ± 0.04
Alsike clover (*T. hybridum*)	Commercial	0.50 ± 0.04	0.68 ± 0.04
White clover (*T. repens*)	Commercial	0.50 ± 0.02	0.45 ± 0.04
Red clover (*T. pratense*)	Commercial	0.47 ± 0.08	0.49 ± 0.01
Subterranean clover (*T. subterraneum*)	Commercial	0.56 ± 0.02	0.41 ± 0.18
Lupine (*Lupinus* sp.)	Native	0.58 ± 0.05	0.69 ± 0.03
Soybean (*Glycine max*)	Commercial	0.57 ± 0.03	0.52 ± 0.07
Garden peas (*Pisum sativum*)	*R. leguminosarum* 128 C75	0.45 ± 0.03	0.46 ± 0.07
Scotch broom (*Cytisis scoparius*)	Native	0.52 ± 0.04	0.78 ± 0.05
Cowpeas (*Vigna unguiculata*) (formerly *V. sinensis*)	R. 32H1	0.99 ± 0.00	0.99 ± 0.00
Nonlegumes			
Red alder (*Alnus rubra*)	Native	0.94 ± 0.01	0.99 ± 0.001
Purshia tridentata	Native	0.88 ± 0.05	0.97 ± 0.01
Russian olive (*Elaeagnus angustifolia*)	Native	0.87 ± 0.05	0.97 ± 0.01
Ceanothus velutinus	Native	0.33 ± 0.23	0.77 ± 0.04
Myrica californica	Native	0.98 ± 0.0	0.99 ± 0.00

[a]This table is taken in part from Schubert and Evans (1976).

[b]Commercial inocula were obtained from The Nitragin Co. (Milwaukee, Wisc.).

[c]In the calculation of relative efficiency [H_2(air)] represents the rate of H_2 evolution in air, [H_2(Ar)] signifies the rate of H_2 evolution in 79.96% Ar: 20% O_2: 0.04% CO_2, and [C_2H_2] represents the rate of C_2H_2 reduction. Rates are expressed in units of equivalents/h/g.fr.wt. Values for the relative efficiency are expressed as means $\pm$ SEM.

According to the results of the survey (Table II), H_2 evolution may account for 30 to 60% of the energy utilized via nitrogenase in excised nodules of many agronomically important legumes including soybeans, *Glycine max;* alfalfa, *Medicago sativa;* and clovers, *Trifolium* sp. Since a few nodulated legumes (i.e. cowpeas, *Vigna unguiculata*) and most of the nonleguminous symbionts tested exhibited relative efficiencies that approached unity, it is plausible that genetic or other means might be employed to obtain plant-endophyte combinations with minimal losses of energy through H_2 evolution. Evidence presented in Figs 3A and 3B shows that the extent of H_2 evolution by nodules on soybean cultivar Anoka was dependent upon the *Rhizobium* strain used for inoculation. An investigation designed to select the most efficient combinations of two soybean cultivars and four *Rhizobium* strains is reported in Table III. Excised nodules from the soybean cultivar, Anoka, inoculated with *R. japonicum* strain USDA 110 exhibited negligible rates of H_2 production in air (Fig. 3B) and the relative efficiency value approached unity (Table III). The other three strains (USDA 31, USDA 117 and USDA 138) were less efficient on soybean cultivars Amsoy 71 or Anoka (Table III).

Table III

Relative efficiency as a function of host-rhizobium strain combination

Soybean cultivar	*Rhizobium*[a] *japonicum* strain	Relative efficiency[b]	
		$1 - \frac{[H_2(air)]}{[H_2(Ar)]}$	$1 - \frac{[H_2(air)]}{[C_2H_2]}$
Anoka	USDA 31	0.66 ± 0.07	0.64 ± 0.07
Anoka	USDA 117	0.71 ± 0.05	0.69 ± 0.08
Anoka	USDA 138	0.75 ± 0.04	0.67 ± 0.05
Anoka	USDA 110	0.99 ± 0.01	0.99 ± 0.01
Amsoy 71	USDA 31	0.68 ± 0.04	0.63 ± 0.04
Amsoy 71	USDA 117	0.68 ± 0.05	0.55 ± 0.04
Amsoy 71	USDA 138	0.66 ± 0.03	0.44 ± 0.05
Amsoy 71	USDA 110	0.91 ± 0.03	0.91 ± 0.03

[a]Soybean cultivars and *Rhizobium japonicum* strains were kindly provided by Dr. George Ham.

[b]Relative efficiencies are expressed as means ± SEM for 5 replicated treatments sampled at 21, 28 and 35 days of age. Hydrogen evolution from excised nodules was measured amperometrically and the rate of C_2H_2 reduction was determined by gas chromatographic analysis as described previously (Schubert and Evans, 1976). Plants were inoculated with single strains as indicated and cultured under controlled conditions in a growth chamber. The day temperature was 27°C and night temperature was 22°C. The light intensity was 17,600 lux (1600 foot candles) and the photoperiod was 16 h.

Intact Plants Schubert and Evans (1976) reported measurements of H_2 production by nodulated root systems of soybean plants *in situ*. Results from these experiments were consistent with those obtained for excised nodules. As shown in Fig. 4 intact cultures of nodulated cowpeas also evolved H_2. Results

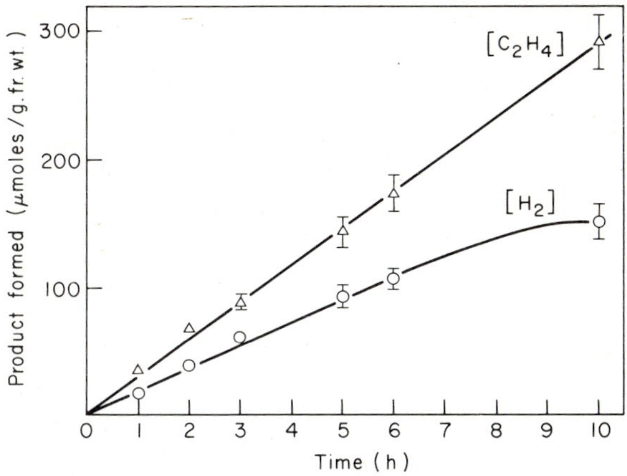

Fig. 4. *Measurement* in situ *of H_2 evolution and C_2H_2 reduction by greenhouse grown cowpeas (*Vigna unguiculata *cv. Whippoorwill) inoculated with* Rhizobium *strain 176A27 obtained from The Nitragin Co. Intact plants in pots were sealed in Saran bags as described previously (Schubert & Evans, 1976). Gas samples were removed for gas chromatographic analysis of H_2 (o) evolved from plants exposed to air and C_2H_4 (Δ) formed from plants exposed to 0.1 atm C_2H_2 in air. Mean rates of product formed ± SEM are given for five replicates.*

presented previously (Schubert and Evans, 1976) are consistent with the assumptions made in this report that the evolution of H_2 from nodules was not detectable when a saturating concentration of C_2H_2 was provided, and the rates of H_2 evolution by nodules in an atmosphere containing argon instead of N_2 was not measurably different from the rate of C_2H_2 reduction. Factors such as the rate of photosynthesis and water stress may affect the allocation of energy to the alternate substrates, H^+ or N_2. These factors have not been thoroughly investigated.

Energy charge which would be expected to be influenced by the supply of photosynthate to nodules *in vivo* (Ching *et al.*, 1975) has been implicated as a parameter controlling electron distribution between H^+ and N_2 (Bulen, 1976; Davis *et al.*, 1975; Silverstein and Bulen, 1970). These postulations are consistent with the results of experiments in which the effect of exposure of intact plants to light during incubation of the nodulated root system in air or air plus 0.1 atm C_2H_2 were studied

(Schubert and Evans, 1976). Under these conditions the relative efficiency was about 0.6 compared to 0.4 for the case in which the entire plant system was enclosed in the container during incubation (Schubert and Evans, 1976).

HYDROGENASE REACTIONS

The presence of hydrogenase in pea root nodules was first reported by Phelps and Wilson (1941). These findings were not confirmed by Shug *et al.* (1956), but more recently, Dixon (1967) has observed utilization of H_2 by excised pea root nodules. Although nodules from pea roots failed to exhibit net H_2 evolution, Dixon (1967) established that H_2 was evolved and re-utilized. This conclusion is in agreement with the observation that increasing the pN_2 from 0 to 1 atm stimulated the uptake of D_2 by pea root nodules and by washed bacteroids. Presumably, this stimulation in D_2 uptake was caused by an inhibition by N_2 of the internal formation of H_2 by nitrogenase (Dixon, 1967). In these experiments O_2 was required for H_2 uptake and the optimal pO_2 was about 0.04 atm (Dixon, 1968). In experiments with washed bacteroids, the addition of D_2 stimulated O_2 consumption, but decreased CO_2 formation which suggested that H_2 was metabolized via the electron transport system (Dixon, 1968). A ratio of two moles of D_2 utilized per mole of O_2 consumed is consistent with this conclusion (Dixon, 1968). Furthermore, the addition of pyruvate and succinate inhibited H_2 uptake by bacteroids presumably by competing for electron carrier sites in the oxidative phosphorylation pathway (Dixon, 1968). Measurements of H_2-dependent ATP synthesis as determined by hexokinase-catalyzed glucose-6-phosphate formation suggested that hydrogenase is coupled to oxidative phosphorylation apparently at the same site as succinate oxidation (Dixon, 1968).

There seems to be at least two classes of hydrogenase which differ in their physical and chemical properties. The first group which includes the enzyme from *Clostridium* is characterized by its apparent reversible nature, its capability to catalyse a rapid rate of hydrogen-deuterium exchange and its acceptance of a variety of electron donors or acceptors (Peck and Gest, 1954; Nakos and Mortenson, 1971; Chen and Mortenson, 1974). The second class includes the hydrogenase from pea nodules and *Azotobacter*. The enzymes from these sources are quite similar (Dixon, 1972) in that they fail to evolve H_2, catalyze a very slow exchange between H_2 and D_2O, and seem to be closely associated with particulate membrane fractions containing succinate oxidizing enzymes (Dixon, 1972). In addition, the hydrogenase from pea nodules and *Azotobacter* catalyze reactions with electron acceptors such as methylene blue and ferricyanide, but slowly catalyze reactions in which benzyl viologen is the acceptor (Dixon, 1972). Neither enzyme will reduce methyl viologen, pyridine nucleotides, NAD or NADP (Dixon, 1972). Recent evidence by Smith, Hill and Yates (Personal communication; Yates, this volume) suggests that C_2H_2 inhibits both classes of hydrogenases. Consideration of these findings is essential in the interpretation of results of experiments in which H_2 evolution or uptake are measured in the presence of C_2H_2. In agreement with Dixon (1972), our results indicate that the expression of hydrogenase activity in nodules is dependent upon the specific combination of host and endophyte. It is not clear whether the genetic information for hydrogenase is of plant or rhizobial origin. Recent investigations by Roberta McCrae, Karel R. Schubert and Harold J. Evans indicate that pure culture of certain strains of *Rhizobium japonicum* and "cowpea rhizobia" exhibit hydrogen-dependent methylene blue reduction. In these experiments, cells were cultured on a yeast extract mannitol medium and then exposed to 1% hydrogen before use. Hydrogenase activity has been measured in nodules of several of the more efficient symbionts including soybeans, cowpeas, and red alder, and it is clear that O_2 was required for H_2 uptake (Fig. 5). A

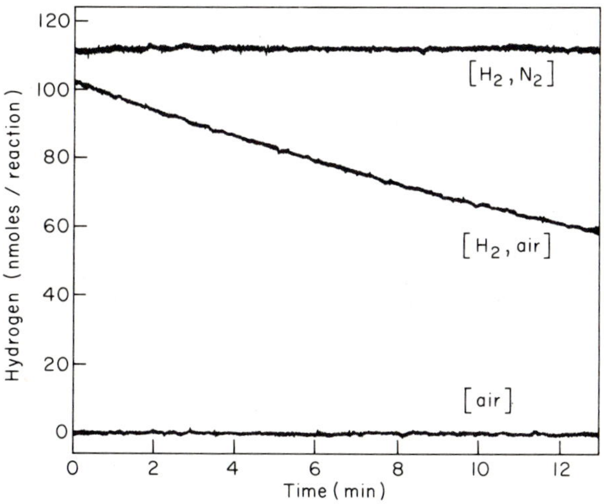

Fig. 5. *Amperometric measurement of hydrogen utilization by excised cowpea (*Vigna unguiculata *cv. Whippoorwill) nodules (0.4 g) inoculated with* Rhizobium *strain 176A28 obtained from The Nitragin Co. Tracings are shown for nodules in gas mixtures containing air; H_2 plus air; and H_2 plus N_2.*

typical amperometric tracing for H_2 uptake by excised cowpea nodules is shown in Fig. 5. The fact that the rates of C_2H_2

reduction were at least several times greater than the rates of H_2 production in $Ar:O_2:CO_2$ for many of the more efficient symbionts listed in Table II supported the conclusion that nodules from these symbionts contained hydrogenase.

GENERAL CONSIDERATIONS

The failure to evaluate the magnitude of H_2 evolution by N_2-fixing organisms may explain the wide range of values reported for the ratio of moles of C_2H_2 reduced per mole of N_2 fixed (Bergersen, 1970; Hardy *et al.*, 1973). The fraction of energy consumed for ATP-dependent H_2 production in air is not considered in the theoretical value for the ratio of C_2H_2 to N_2 of 3, which is often used to estimate N_2 fixation. For example, assuming that 50% of the total energy flux generated for nitrogenase reactions is lost as H_2 (relative efficiency of 0.5), then the ratio of moles of C_2H_2 reduced to moles of N_2 fixed would be 6 rather than 3. Therefore the ratio of C_2H_2 reduced to N_2 fixed needs to be determined experimentally. Likewise, extrapolations of rates of C_2H_2 reduction measured under subsaturating concentrations of C_2H_2 to maximal rates of C_2H_2 reduction are subject to error. A valid measure of nitrogenase activity must take into account concurrent rates of reduction of alternate substrates such as protons.

Several researchers have proposed that the evolution of H_2 by N_2-fixing organisms may be a source of inefficiency in the N_2-fixing process (Dixon, 1972; Bergersen *et al.*, 1973). Now that an assessment of the magnitude of this loss has been obtained (Schubert and Evans, 1976) a consideration of the effect of H_2 evolution on the utilization of energy in the nodule, schematically shown in Fig. 6 is necessary. In many respects, the loss of energy through the reduction of protons by nitrogenase is analogous to energy losses through photorespiration in plants with C3 type metabolism (Zelitch, 1975). To our knowledge neither process serves a useful purpose.

Dixon has discussed the possible role of hydrogenase as a means of reutilizing H_2 from nitrogenase in the more efficient symbionts (see also Yates, this volume). The energy cost of proton reduction via nitrogenase is extremely high in terms of both reducing power and ATP consumption (Fig. 6). On the basis of Dixon's (1972) experiments (Dixon, 1972) the two moles of ATP produced per mole of H_2 metabolized via hydrogenase could meet at least part of ATP demand for N_2-fixation (Fig. 6), but whether nitrogen-fixing symbionts have the capability to utilize hydrogenase as a catalyst to supply reductant to nitrogenase is not known. Whether low rates of H_2 evolution and high relative efficiencies of reductant utilization by nodulated

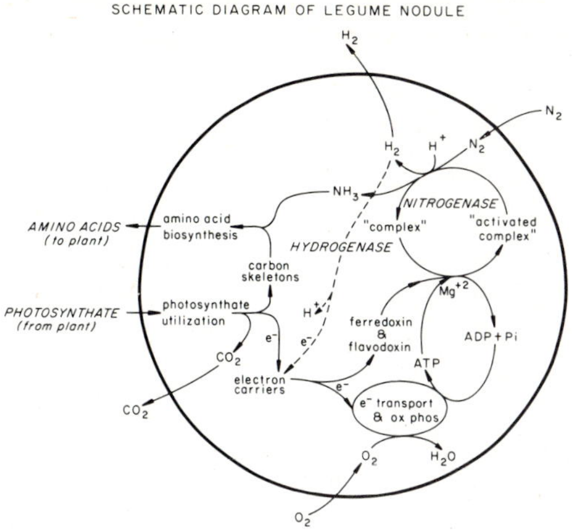

Fig. 6. *Schematic diagram of reactions involved in the utilization of energy for N_2 fixation within a nodule.*

symbionts is actually associated with increased N_2 fixation rates remains to be determined. Results of experiments now in progress suggest that the energy conserved in the more efficient symbionts may be stored or expended for other processes.

ACKNOWLEDGEMENTS

We acknowledge Mrs. Jean Engelke and Mr. Sterling Russell for technical assistance in assaying nodules and culturing plants. Gifts of *Rhizobium* strains and seeds from Dr. Joe Burton and Dr. George Ham are greatly appreciated. This research was supported by grants from The National Science Foundation (BMS-7417812), The Rockefeller Foundation (GA-AS7522) and by the Oregon Agricultural Experiment Station (Paper No. 4332).

REFERENCES

Anand, S.R. and Krasna, A.I. (1965). *Biochemistry*, 4: 2747-2753.

Bergersen, F.J. (1963). *Aust.J.Biol.Sci.*, 16: 669-680.

Bergersen, F.J. (1970). *Aust.J.Biol.Sci.*, 23: 1015-1025.

Bergersen, F.J. and Turner, G.L. (1968). *J.Gen.Microbiol.*, 53: 205-220.

Bergersen, F.J., Turner, G.L. and Appleby, C.A. (1973). *Biochim. Biophys.Acta*, 292: 271-282.

Bulen, W.A. (1976). *In:* Proceedings of 1st International

Symposium on Nitrogen Fixation (W.E. Newton and C.J. Nyman, eds), Vol. 1, pp. 177-186, Washington State University Press, Pullman, Washington.

Bulen, W.A. and LeComte, J.R. (1966). *Proc.Nat.Acad.Sci.*, 56: 979-986.

Bulen, W.A., Burns, R.C. and LeComte, J.R. (1965). *Proc.Nat. Acad.Sci. USA*, 53: 532-539.

Burns, R.C. and Hardy, R.W.F. (1975). Nitrogen Fixation in Bacteria and Higher Plants. Springer-Verlag, New York.

Burris, R.H. (1972). *In:* Manometric and Biochemical Techniques (W.W. Umbreit, R.H. Burris and J.F. Stauffer, eds), pp. 293-294, Burgess Publ. Co., Minneapolis, Minnesota.

Burris, R.H. (1975). *In:* The Biology of Nitrogen Fixation (A. Quispel, ed.), pp. 9-33. North-Holland Publ. Co., Amsterdam.

Chen, J.S. and Mortenson, L.E. (1974). *Biochim.Biophys.Acta*, 371: 283-298.

Ching, T.M., Hedtke, S., Russell, S. and Evans, H.J. (1975). *Plant Physiol.*, 55: 796-798.

Dart, P.J. and Day, J.M. (1971a). *Plant and Soil Special Vol.19*, 71: 167-184.

Dart, P.J. and Day, J.M. (1971b). Rothamsted Expt. Sta. Report for 1970, part I, p. 86.

Davis, L.C., Shah, V.K. and Brill, W.J. (1975). *Biochem.Biophys. Acta*, 403: 67-78.

Dixon, R.O.D. (1967). *Ann.Bot., N.S.*, 31: 179-188.

Dixon, R.O.D. (1968). *Arch.Mikrobiol.*, 62: 272-283.

Dixon, R.O.D. (1969).*Ann.Rev.Microbiol.*, 23: 137-158.

Dixon, R.O.D. (1972). *Arch.Mikrobiol.*, 85: 193-201.

Evans, H.J. (ed.)(1975). Enhancing Biological Nitrogen Fixation (Division of Biological and Medical Sciences of The National Science Foundation, Washington, D.C., USA).

Hadfield, L.K. and Bulen, W.A. (1969). *Biochemistry*, 8: 5103-5108.

Hardy, R.W.F. and Havelka, U.D. (1975). *Science*, 188: 633-643.

Hardy, R.W.F. and Knight, E., Jr. (1967). *Biochim.Biophys.Acta*, 139: 69-90.

Hardy, R.W.F., Burns, R.C. and Holsten, R.D. (1973). *Soil Biol. Biochem.*, 5: 47-81.

Hoch, G.E., Little, H.N. and Burris, R.H. (1957). *Nature*, 179: 430-431.

Hoch, G.E., Schneider, K.C. and Burris, R.H. (1960). *Biochim. Biophys.Acta*, 37: 273-279.

Hwang, J.C. and Burris, R.H. (1972). *Biochim.Biophys.Acta*, 283: 339-350.

Hwang, J.C., Chen, C.H. and Burris, R.H. (1973). *Biochim.Biophys.Acta*, 292: 256-270.

Hyndman, L.A., Burris, R.H. and Wilson, P.W. (1953). *J.Bact.*, 65: 522-531.
Jackson, E.K., Parshall, G.W. and Hardy, R.W.F. (1968). *J.Biol. Chem.*, 243: 4952-4958.
Jeng, D.Y., Morris, J.A. and Mortenson, L.E. (1970). *J.Biol. Chem.*, 245: 2809-2813.
Kelly, M. (1969). *Biochim.Biophys.Acta*, 171: 9-22.
Kleiner, D. and Burris, R.H. (1970). *Biochim.Biophys.Acta*, 212: 417-427.
Koch, B., Evans, H.J. and Russell, S. (1967). *Proc.Nat.Acad.Sci. USA*, 58: 1343-1350.
Krasna, A.I. and Rittenberg, D. (1954). *J.Am.Chem.Soc.*, 76: 3015-3020.
Krasna, A.I. and Rittenberg, D. (1956). *Proc.Nat.Acad.Sci.USA*, 42: 180-185.
Ljones, T. and Burris, R.H. (1972a). *Anal.Biochem.*, 45: 448-452.
Ljones, T. and Burris, R.H. (1972b). *Biochim.Biophys.Acta*, 275: 93-101.
Nakos, G. and Mortenson, L. (1971). *Biochim.Biophys.Acta*, 227: 576-583.
Marayama, Y. (1975). *In:* Nitrogen Fixation and Nitrogen Cycle (H. Takashaski, ed.), Vol. 12, pp. 35-40. Japanese Committee for the International Biological Program, Tokyo.
Peck, H.D. and Gest, H. (1954). *Biochim.Biophys.Acta*, 15: 587-588.
Peck, H.D. and Gest, H. (1956). *J.Bact.*, 71: 70-80.
Phelps, A.S. and Wilson, P.W. (1941). *Proc.Soc.Exp.Biol.(N.Y.)*, 47: 473-476.
Rivera-Ortiz, J.M. and Burris, R.H. (1975). *J.Bact.*, 123: 537-545.
Schubert, K.R. and Evans, H.J. (1976). *Proc.Nat.Acad.Sci.USA*, 73: 1207-1211.
Shah, V.K., Davis, L.C. and Brill, W.J. (1975). *Biochim.Biophys. Acta*, 384: 353-359.
Shug, A.L., Hamilton, P.B. and Wilson, P.W. (1956). *In:* Inorganic Nitrogen Metabolism (W.D. McElroy and B. Glass, eds), The John Hopkins Press, Baltimore.
Silverstein, R. and Bulen, W.A. (1970). *Biochemistry*, 9: 3809-3815.
Turner, G.L. and Bergersen, F.J. (1969). *Biochem.J.*, 115: 529-535.
Watt, G.D. and Bulen, W.A. (1976). *In:* Proceedings of 1st International Symposium on Nitrogen Fixation (W.E. Newton and C.J. Nyman, eds), Vol. 1, pp. 248-256, Washington University Press, Pullman, Washington.
Winter, H.C. and Burris, R.H. (1968). *J.Biol.Chem.*, 243: 940-944.
Zelitch, I. (1975). *Science*, 188: 626-633.

A SYNTHESIS PAPER ON NONLEGUMINOUS N_2-FIXING SYSTEMS

R.H. BURRIS

Department of Biochemistry and Center for Studies of N_2 Fixation, College of Agricultural and Life Sciences, University of Wisconsin-Madison, Madison, Wisconsin 53706, U.S.A.

The assignment given to me is to present a synthesis lecture on plant associations. As we have heard already about the association between leguminous plants and the rhizobia, the remarks in this section will be confined to other types of plant-micro-organism associations. The micro-organisms with which we will be concerned are bacteria, actinomycetes, and blue-green algae. These micro-organisms form associations with a variety of plants and achieve biological N_2 fixation together with them. The micro-organisms frequently are capable of N_2 fixation when growing by themselves, but it is the associations which will concern us particularly. Table I lists a number of these microbial-plant associations, which are involved in N_2 fixation.

Table I

Microbial-plant Associations that Fix N_2
Leguminous plants and *Rhizobium sp.*
Angiospermous non-leguminous plants and actinomycetes
Blue-green algae with: lichens, mosses, liverworts, cycads, *Gunnera*, *Azolla-Anabaena azollae*
N_2-fixing organisms in phyllosphere and rhizosphere
Non-nodulating, N_2-fixing organisms inside roots

The blue-green algae are perfectly capable of fixing N_2 by themselves, but they also form some interesting associations (Balandreau *et al.*, 1975). N_2 fixation in blue-green algae is generally associated with the heterocystous forms, but some nonheterocystous blue-greens can fix N_2 anaerobically or microaerophilically (Carpenter and Price, 1976).

Millbank (1974) discussed the associations of blue-green algae in some detail. Perhaps the most interesting and widespread of these associations is represented by the lichens. This association between algae and fungi, operating in close proximity, is the most widely studied of the blue-green associations. In the structure of the lichen, the fungus is the dominant member. Lichens take many forms, and thousands of species have been described. Millbank reports that 8% of the species are characterized by having blue-green algae in the association. Eight genera of cyanophyceae have been reported as the phycobiont in the association. By using ^{15}N as a tracer, it has been shown that blue-green algae of the genera *Nostoc* and *Calothrix* are active N_2 fixers in lichens. Others have been demonstrated with the acetylene reduction technique, e.g. the genus *Scytonema* has been shown to be active in this way.

The blue-green algae usually constitute a small percentage of the biomass of lichens. The evidence suggests, however, that they are highly active N_2 fixers in such an association. Apparently, they fix N_2 and then release fixed nitrogen rather readily for use by the fungus. Time course studies indicate that the release of fixed nitrogen is essentially linear for a couple of weeks. The release or non-release of fixed nitrogen could control growth of the lichen. The nature of the compounds furnished by the algae to the fungi is not established.

The presence or absence of water is crucial to lichen activity. The organisms remain viable even when dessicated, and they recover activity with remarkable speed when water is added; in less than an hour metabolic activity is restored after rehydration.

The saturation by light in the lichen association is rather similar to that for photosynthesis. Some lichens cease fixation of N_2 rather quickly when darkened, whereas fixation persists in others for some time. Activity is temperature sensitive, and metabolic activity is demonstrable from somewhat above freezing to 35°C. It has been reported that the optimal temperature for N_2 fixation is around 15°C, but this probably is quite variable with species.

The association of algae and fungi in the lichens shows remarkable versatility and ability to survive unfavourable conditions. It has been adapted to survive under conditions ranging from the antarctic to hot deserts. When conditions become favourable, the association quickly responds and becomes active again.

Although most interest in the past has centred on the algal association in lichens, there are a number of other associations of blue-green algae which have been studied less extensively. For example, certain liverworts contain a *Nostoc* species that can fix N_2 in association with the liverwort. This organism has been isolated and has been shown to be an active N_2 fixer when living separately.

Some cycads have root nodules which form near the soil surface and contain an abundance of blue-green algae. These nodules have a coralloid form, and the algae occur between the epidermal cells. Fixation of N_2 occurs in the light, and the fixed nitrogen is transferred to the cycads where it can be utilized effectively.

Among the angiosperms, certain species in the genus *Gunnera* also house *Nostoc*, in this case in an intracellular location. There is good evidence for active N_2 fixation by these blue-greens in association with the plant (Silvester, 1976).

Currently there is great interest in the association of blue-greens with the water fern *Azolla* (Peters and Mayne, 1974a, b; Peters, 1975). This fresh water fern has *Anabaena azollae* normally associated with it. This association has been exploited for many years in Southeast Asia as a means of adding fixed nitrogen in paddies. It seems clear that the subsequent growth of rice has benefited substantially from the N_2 fixation of this association. *Anabaena* has not been cultured apart from the water fern; however, it is possible to free the fern from the *Anabaena* and to use the algae-free fern as a control in experimentation. It has not been possible to re-establish the association to the satisfaction of most investigators. Remarkably high rates of N_2 fixation have been reported for the azollae. In some cases, over 300 kg N/(ha x yr) has been reported as a fixation level; Becking (1976) believes 62-125 kg N is more realistic. A particularly interesting feature of the association is the high frequency of heterocysts in the algal filaments. Commonly, 5 to 6% of the cells are heterocysts in *Anabaena*, but this may rise to above 30% in *Anabaena* in *Azolla*. As N_2 fixation is localized in the heterocysts, it is apparent that this high frequency has the potential to increase the N_2 fixation markedly. There is no hint what substance the plant supplies that induces this high frequency of heterocyst formation. The azolla system is under intensive investigation in several laboratories, so we can expect continuing improvement in our information concerning it.

Although it is peripheral to the interests of plant investigators, it is interesting that certain algae grow in association with marine animals. The living material of corals, the polyps, and the mantle of marine clams frequently carry high concentrations of green algae, which are referred to as zooxanthellae. These organisms photosynthesize actively and furnish photosynthate to the animal. On the Great Barrier Reef of Australia, Newcomb and Pugh (1975) discovered an ascidian (sea animals also known as sea squirts) which carried blue-green algae in its tissue rather than the usual green algae. I tested the organisms but found no evidence of acetylene reduction.

Another group of associations which deserve our attention are the associations between micro-organisms and leaf surfaces. This research area is a province of Ruinen (1974, 1975), and she has reviewed the subject in great detail. She introduced the term phyllosphere to describe the habitat of the organisms proliferating on leaf surfaces. In this ecosystem, moisture is furnished primarily by dew, and nutrients are furnished from plant exudates. Two kinds of syntrophy may take place. In one, the organisms growing on the leaf surface may furnish soluble nitrogenous compounds which can be absorbed through the leaf surface; this would constitute a closed cycle of foliar feeding between the leaf and the micro-organisms on its surface. Secondly, there may be transfer of N to the soil and root system when rains wash the organisms and materials from the leaf surface. In addition to carbohydrates, a variety of other compounds are present on leaf surfaces such as organic acids, amino acids, minerals and growth substances. If material is washed off, it may be recovered and re-cycled through the roots.

Leaf surfaces carry a mixed population of micro-organism, and among them are a number capable of N_2 fixation. In general, the population of micro-organisms increases as the leaf matures and declines as the leaf senesces. There is a potential for N_2 fixation in this system, but if readily available fixed nitrogen is present, the micro-organisms will use it rather than fixing N_2. Ruinen found that *Beijerinckia* is particularly abundant on leaf surfaces, and she recovered it from 95% of leaves examined in Indonesia. Certain species of the *Azotobacter* also are found in substantial numbers.

It is difficult to quantitate N_2 fixed in the phyllosphere, because such fixation is highly variable. It depends upon the period that the leaf remains moist and upon how persistent the flora is on the leaf before it is washed off by rain. It is suggested that the rapid growth of tropical forests may be quite dependent upon nitrogen which has been fixed in the phyllosphere. Again, in a quantitative sense, it is difficult to distinguish N_2 fixed on the leaf surface from that fixed in the soil or from that recycled through foliage decomposed at the soil surface. Although N_2 fixation in the phyllosphere poses difficult problems of quantitation, it deserves further attention.

Nodulated leaves constitute another curious bacterial-plant association often cited as having potential for N_2 fixation (see reviews by Becking (1974) and van Hove (1976). Early work with *Ardisia* indicated that infection occurred in young leaves, and it was apparent that the plants grew poorly in the absence of the bacteria (the bacteria can be destroyed by heating the seed to 40°C). Although there are positive claims for N_2 fixation in this association, careful tests in at least three laboratories with $^{15}N_2$ and C_2H_2 have been completely negative.

The *Psychotria* species also have attracted attention, because they have nodules distributed over much of the lower surface of their leaves. Silver *et al.* (1963) and Centifanto and Silver (1964) isolated an organism from the leaves, which they described as a *Klebsiella*. The pure culture fixed N_2 actively under anaerobic conditions, and they reported that if they rubbed it on leaves together with an abrasive they could establish nodules. Becking checked this report and was unable to produce nodules with the strain that had been deposited in the American Type Culture Collection. Under natural conditions the infection appears on young leaves and the bacteria are located intercellularly; with aging they appear to break down the plant cells. Becking (1974) and van Hove (1976) have been unable to demonstrate any fixation of ^{15}N or reduction of acetylene by leaves of *Psychotria*. Becking attributes positive growth responses observed to elaboration of growth substances by the micro-organisms.

There have been claims for N_2 fixation in the leaf glands of *Dioscorea* species. The claims are similar to those for *Psychotria*, but N_2 fixation never has been verified with sensitive modern techniques.

Now I would like to turn your attention to plant-bacterial associations in the soil. The rhizosphere usually is defined as the soil volume which is influenced by roots. It is apparent that there is a gradient of effects from directly adjacent to the roots to some distance from them. We will not worry about distinguishing the rhizoplane and other subunits of the rhizosphere volume.

It is axiomatic to say that bacteria and other microorganisms are abundant in soil. Bacteria retain their viability and physiological characteristics in soil over extended periods. Furthermore, the bacteria commonly multiply in soil if supplied a suitable energy source. The inorganic nutrient elements generally are available in sufficient quantity in the soil to support growth, and in a reasonably rich soil secondary factors such as growth substances also are available in sufficient quantity. Soil is an unusually plastic medium that might be characterized as a friendly home for micro-organisms. From a single soil sample one can isolate strictly aerobic and strictly anaerobic organisms, and thermophiles and psychrophiles exist together in soil. The question could be posed: If soil is such a favourable medium for micro-organisms, why is N_2 fixation ever limiting for crop growth? There are many species of soil bacteria capable of N_2 fixation. Why don't they supply all the needs of our crop plants? The answer: usually there is a limitation of substrates to furnish energy to support growth and N_2 fixation by the organism. Addition of sugar to the soil will elicit a substantial growth of free-living N_2-fixing organisms. Skepticism is justified in evaluating claims of improvement in N_2 fixation following inoculation of soil with free-living N_2 fixers. Inoculation of legumes with the rhizobia is a special case, because these organisms associate with specific plants and they may be absent from soils in which the leguminous symbiont is not native. On the other hand, if the pH of the soil is correct and an energy source is available, free-living N_2 fixers generally will be present and growing already. Energy sources usually are limiting factors, as it takes a great deal of energy to support microbial growth and N_2 fixation. This energy requirement often is overlooked by those who promise to nodulate corn with N_2 fixers but infer that there will be no loss in yield. Even bacteria and plants never get something for nothing, and a substantial amount of photosynthate must be channelled toward N_2 fixation in any plant-bacterial association that reduces N_2 to ammonia.

The question arises: Where is sufficient energy available to support vigorous N_2 fixation in bacterial-plant associations? Obviously this is available for the legume-rhizobia association, because the rhizobia have direct access to photosynthate. The same applies to the angiospermous nonleguminous plants, as they also have a direct pipeline to photosynthate. Plants such as sugarcane release substantial amounts of sugar into the soil directly or by way of leaf exudates that are washed into the soil. *Spirillum lipoferum* invades the roots of a number of plants and inside the root system has direct access to photosynthate. We already have cited the case of the association in *Azolla*, in which the blue-green algae lives in a favourable environment and can capture its energy directly from sunlight. It is apparent that the free-living blue-green algae are another group able to utilize photosynthetic energy directly.

The root and bacterial association in leguminous plants already has been covered thoroughly. During the present sessions Dr. Bond and others will discuss the angiospermous nonleguminous plants. Although we need not elaborate, this is a fascinating but often neglected group that effects a vigorous fixation of N_2 over extensive land surfaces.

There is a long history of experimentation on inoculation of soil with various strains of the azotobacter. These are vigorous aerobic soil organisms and many have believed that all one needed to do was to add them to soil and they then would fix N_2 and benefit the associated crops by supplying fixed nitrogen to the roots. Numerous claims of benefits from such inoculation have been recorded and for some years the practice of inoculation was endorsed in Russia. However, in recent years the practice of inoculation with azotobacter seems to have been de-emphasized in Russia and elsewhere, because of the difficulty of establishing that any benefit accrues from the practice (Mishustin and Shilinikova, 1971).

As indicated, a readily available source of energy is crucial in establishing N_2 fixation by any type of organism. If the plant excretes available substrates through its root system, it is apparent that micro-organisms will take advantage of this to grow and fix N_2 under conditions which do not repress the nitrogenase. Examination of such situations, reveals that certain types of plants excrete sufficient carbohydrate to support a sizeable microbial population (Döbereiner *et al.*, 1972a). Not only the azotobacter group benefits, but strictly anaerobic organisms and facultative organisms also may thrive in association with roots.

Döbereiner (1974) and Döbereiner and Day (1975) have investigated rhizosphere associations under tropical conditions. They point out that N_2 fixation in the rhizosphere may be particularly active under tropical conditions because of the abundance of different types of N_2-fixing organisms there and because some of them are tolerant to acid conditions. The temperature is favourable for the growth of the organisms year-round, and there is a reasonable abundance of highly efficient C_4 type of plants to support rhizosphere associations. Plants that grow over an extended period (e.g. sugarcane) will furnish a continuing supply of substrate to associated organisms. *Beijerinckia* and *Azotobacter paspali* are common in tropical regions. Döbereiner *et al.* (1972b) have studied the association between *Azotobacter paspali* and *Paspalum notatum*. Although *Azotobacter paspali* in laboratory culture shows an optimum pH near 6.7 and a marked sensitivity to acid conditions, when growing in the rhizosphere of *Paspalum notatum* it appears to grow well even in somewhat acid soils. The organism has a rather specific association with this particular plant.

Beijerinckia is another tropical organism with good capacity for N_2 fixation. Grasses with which it is associated show benefit, and this suggests that active N_2 fixation occurs. *Beijerinckia* establishes an active association with sugarcane, and may be abundant under conditions unfavourable to the azotobacter (Döbereiner *et al.*, 1972a).

Let us turn our attention now to a discussion of the association between *Spirillum lipoferum* and higher plants. This interesting organism was first described by Beijerinck (1925), and he indicated that it was capable of fixing N_2. However, he did not obtain it in pure culture. When Schröder (1932) recovered a pure culture, she was unable to demonstrate N_2 fixation by *S. lipoferum*. It may well be that the associated organism in the mixed culture studied by Beijerinck reduced the O_2 concentration so that *S. lipoferum* was able to fix N_2. When Schröder incubated the pure culture at a normal concentration of O_2, apparently its N_2-fixation activity was suppressed. Becking (1963) re-isolated this organism and demonstrated N_2 fixation in pure culture. His tests were made with $^{15}N_2$ as a tracer, and he produced unequivocal evidence of its capability for N_2 fixation. Again the organism was largely ignored until new observations were made by Döbereiner and Day (1976).

They observed that *Digitaria decumbens* grew well and appeared green on a nitrogen-deficient soil. An examination of the root system revealed that it housed an abundance of bacteria, and the roots yielded an organism which was characterized as *S. lipoferum*. The organism grew under microaerophilic conditions when supplied with organic acids as a substrate, and it proved to be capable of fixing N_2.

Day *et al.* (1975) extended the observations from *Digitaria decumbens* to other tropical grasses and found that a number of them could form active associations with the organism. Interest was stirred particularly by the report of Von Bülow and Döbereiner (1975) that *S. lipoferum* could associate with *Zea mays*, and that roots removed from the plant could reduce acetylene vigorously after preincubation at a low partial pressure of O_2. Examination of the roots indicated that the organism was inside the root tissue as well as on the surface. Because the organism is microaerophilic in habit when fixing N_2, it appeared that the lowered partial pressure of O_2 inside the root was favourable for N_2 fixation by the organism. This location also carried the obvious benefit that the bacteria were in close proximity to the source of photosynthate which was translocated from the top to the root of the plant. The potential of the system for partially supplying the needs for fertilizer nitrogen was evident, and many people have joined the quest to unravel the secrets of the spirillum-plant interactions.

We now turn to specifics of our studies on *S. lipoferum*, as grown by itself and in association with higher plants. In considering *S. lipoferum*, we find this organism entirely capable of fixing N_2 by itself. If one supplies $^{15}N_2$ to the organism growing on a suitable medium under microaerophilic conditions, it is easy to show that it assimilates ^{15}N. Table II

Table II

Fixation of $^{15}N_2$ by *S. lipoferum*
20 ml culture of *S. lipoferum* in
60 ml serum bottle + $p^{15}N_2$ 0.30 atm
(about 80 atom % ^{15}N excess), pO_2 0.01 atm, pA 0.69 atm
Shake 30 min, 30°C.
Recover and digest cells.
Mass analysis showed 0.50 atom % ^{15}N excess.
About 15 nmoles N_2 fixed/(h x ml culture).

indicates the ^{15}N levels that accumulate in the organism. It is equally easy to demonstrate that *S. lipoferum* is capable of reducing acetylene to ethylene.

S. lipoferum can be grown on a relatively simple medium as described by Döbereiner and Day (1976). We have modified their medium somewhat by adding micronutrient elements and a considerably higher concentration of phosphate to supply increased buffering capacity (Okon *et al.*, 1976a). The increased buffering capacity is particularly helpful, because the organism becomes inactive above a pH of about 7.8. *S. lipoferum* tolerates 10 g of phosphate salts per litre without any apparent toxicity. A small amount of yeast extract gets the organism underway with decreased lag, probably because it furnishes growth substances. Although malate generally has been used as the energy source in the medium, it is satisfactory to substitute succinate or lactate, as they also support rapid growth of the organism. Pyruvic acid is a good substrate, but is not convenient for general use. We start growth at a pH of 6.8 and keep the pH below 7.8.

Control of the partial pressure of O_2 is critical in growing *S. lipoferum* under conditions requiring N_2 fixation. If ammonia is supplied as a source of nitrogen, the organism grows very well at a pO_2 of 0.2 atm (Okon *et al.*, 1976c). Under such conditions *S. lipoferum* grows as a typical aerobic organism, and an adequate supply of O_2 hastens its growth. However, when the organism is fixing N_2, it operates as a microaerophilic organism.

There are a variety of ways to determine the optimal pO_2 for *S. lipoferum*. Several experimental approaches have verified the observation of Döbereiner and Day (1976) that best growth on N_2 occurs near 0.5% O_2. The most satisfactory method we have employed has been to supply O_2 through an oxygenstat. This can be used with an exponentially growing culture which is constantly increasing its total demand for oxygen. The apparatus employs a sterilizable O_2 electrode immersed in the culture medium, and this electrode (Borkowski and Johnson, 1967) continuously feeds a signal to the oxygenstat to indicate the dissolved O_2 concentration. As the dissolved O_2 decreases below a set level, the oxygenstat responds to the electrode signal by opening a solenoid valve to supply supplemental air to the continuous gas stream (tank N_2 plus a limited amount of air). When the dissolved O_2 concentration returns to a set level, the signal from the O_2 electrode causes the oxygenstat to close the solenoid valve to the supplemental air line. The dissolved O_2 thus cycles somewhat above and below a set point. As the culture grows and demands an increased supply of O_2, the oxygenstat meets the demand by more frequent addition of supplemental air, but the concentration of dissolved O_2 still remains at the original set level. With such an apparatus one can demonstrate (Okon, Houchins, Albrecht and Burris, 1976) that an exponentially growing culture of *S. lipoferum* has an optimal pO_2 of 0.005 to 0.007 atmospheres as shown in Table III.

Table III

Effect of pO_2 on doubling time and C_2H_2 reduction by *S. lipoferum*

pO_2 (atm)	doubling time (h)	nmoles C_2H_2 red. / h x ml culture
0.002-0.003	9 -10	250-280
0.005-0.007	5.5- 7	280-300
0.009-0.011	7.5- 8.5	80-100
0.014-0.018	8.5- 9.5	50- 75

It has been a general observation that aerobic N_2-fixing organisms vary in their efficiency of fixation with a change in the partial pressure of O_2. *S. lipoferum* is no exception, as is shown in Table IV. The efficiency is approximately 10 mg of cell nitrogen produced per g of malate consumed at the optimal pO_2. At partial pressures of O_2 below the optimal level, efficiency is slightly greater, and at higher pO_2's the efficiency is distinctly less.

The rates of growth are quite different when the organism is supplied ammonia as a nitrogen source and when it must fix N_2. As shown in Fig. 1, the doubling time for growth of *S.*

Table IV

Effect of pO_2 on efficiency of conversion of malate to cells

pO_2 (atm)	mg N assimilated / g malate consumed
0.002 to 0.003	11.9 to 12.5
0.005 to 0.007	8 to 10
0.009 to 0.011	2.5 to 4
0.014 to 0.018	1.2 to 2.1

lipoferum on the medium with ammonia is about 1 h. If a culture growing on ammonia is used for inoculating (about 10% by volume inoculum) the nitrogen-free medium, there is an initial rapid growth on the new medium and the doubling time is about 2 h. This rapid growth undoubtedly is at the expense of combined nitrogen carried over with the heavy inoculum. After the fixed nitrogen is exhausted, the organism quickly adapts to the use of N_2 and establishes a growth rate characterized by a doubling time of 5.5 to 7 h. If one wishes to grow a N_2-fixing culture of *S. lipoferum* in quantity, the most advantageous technique is to transfer a heavy ammonia-grown inoculum and allow the culture to grow until it exhausts the ammonia and establishes vigorous N_2 fixation. Such cultures can be brought to an optical density approaching 1.0 in considerably less than a day. Organisms cultured in this way have a perfectly respectable N_2-fixing system.

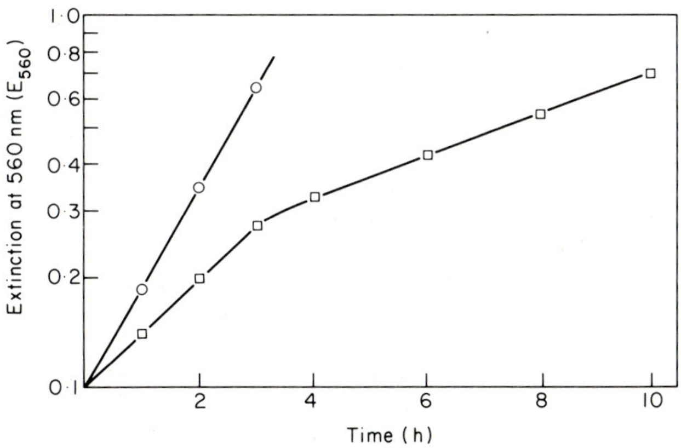

Fig. 1. *Growth of* S. lipoferum *on N_2 (□) and on NH_4^+ (○). The initial growth of the N_2 culture is supported by fixed N supplied with the inoculum, and the final slope is representative of growth on N_2.*

As indicated, *S. lipoferum* grows very well on malate as a substrate. As anticipated, they metabolize this substrate vigorously (Okon *et al.*, 1976b). The respiratory activity of washed cells of *S. lipoferum* can be tested by observing how much various substrates increase the respiratory rate above the endogenous rate. Table V shows that they respire the organic acids

Table V

Uptake of O_2 by intact cells of *S. lipoferum*

Grown on 0.04 M	$Q_{O_2}(N)$ measured on 0.04 M			
	malate	succinate	lactate	pyruvate
Malate	870	840	458	342
Succinate	725	875	520	385
Lactate	285	250	663	730
Pyruvate	242	276	519	630

very rapidly. Malate, succinate, lactate and pyruvate are particularly well used. Organisms grown on malate metabolize rapidly and vice versa, and organisms grown on lactate metabolize pyruvate very well and vice versa.

In contrast to its rapid metabolism of organic acids, *S. lipoferum* utilizes sugars with a rate of O_2 uptake just above the endogenous rate. As reported, they grow very poorly with sugars as a substrate.

If the cells of *S. lipoferum* are disrupted and the soluble extract is tested for its metabolic activity, the same picture emerges as with the intact cells; i.e., the cell-free extract oxidizes organic acids vigorously but oxidizes the sugars very slowly (Table VI). Thus, the lack of oxidation of the sugars by intact cells cannot be attributed to any permeability barriers that block access of sugars to the enzyme systems of *Spirillum lipoferum*.

There have been few reports on the physiology of *S. lipoferum* and little effort to isolate nitrogenase from the organism. We have found it possible, however, to disrupt the organisms and to recover a cell-free extract capable of reducing acetylene to ethylene. Disruption by osmotic shock is rather ineffective, but the organisms can be disrupted by passage through a French pressure cell under anaerobic conditions. If the cells after harvest and the extracts from the cells are handled anaerobically, it is possible to recover active nitrogenase.

Table VI

Uptake of O_2 by cell-free extracts from *S. lipoferum*. Rates corrected for resp. without added substrate.

Substrate	$Q_{O_2}(N)$
0.04 M succinate	100-200
0.04 M malate, citrate, α-ketoglutarate, oxaloacetate, fumarate, isocitrate, lactate or pyruvate	10- 30
0.04 M fructose 1,6-diphosphate, glyceraldehyde 3-phosphate or phosphoenol pyruvate	1- 5
0.04 M glucose, galactose, fructose, glucose 6-phosphate, gluconate 6-phosphate or fructose 6-phosphate	0-0.5

Purification has not yet yielded homogeneous Mo-Fe and Fe proteins, but specific activities in the region of 500 nanomoles of acetylene reduced per mg protein per min have been reached. Purification to this level has been achieved by anaerobic manipulation of the extracts with methods which have been successful in purifying the nitrogenase from *R. rubrum*. As indicated in Table VII, the requirements for enzymatic activity are exactly the same as for other nitrogenases, i.e., there must be a supply of MgATP and a strong electron donor.

Table VII

Requirements for nitrogenase activity in a cell-free extract from *S. lipoferum*

Reaction mixture	nmoles C_2H_4 produced / min x mg protein
Complete	3.5
- Mn^{++}	0.020
- Mg^{++}	0.025
- $Na_2S_2O_4$	0.000
- ATP generating system	0.000

One interesting property of the Fe protein of the nitrogenase from *S. lipoferum* is its requirement for activation. To date, the only nitrogenase components for which a need for activation has been demonstrated are the Fe proteins from *R. rubrum* and *S. lipoferum*. The Fe protein is inactive as isolated, but if it is incubated with the activating factor from *R. rubrum* (Ludden and Burris, 1976) it is converted to an active form. Likewise, extracts from *S. lipoferum* can effect some activation of the inactive Fe protein from *R. rubrum*.

The cross reactions of the Mo-Fe protein and the Fe protein from various nitrogenases has indicated that closely homologous organisms give enzymatically active crosses, whereas highly diverse organisms, such as *C. pasteurianum* and *A. vinelandii*, give inactive crosses. Extracts from *S. lipoferum* have been crossed with components of other organisms, and as shown in Fig. 2, it crosses with nitrogenase components from most other organisms. However, its Fe protein does not cross with the Mo-Fe protein of *C. pasteurianum*, and its Mo-Fe protein is inactive with the Fe protein of *C. pasteurianum*, *Bacillus polymyxa* and *Chromatium vinosum*.

FE PROTEIN \ MO-FE PROTEIN	A.v.	K.p.	R.r.	S.l.	C.v.	B.p.	C.p.
Azotobacter vinelandii		+	+	+	+	+	−
Klebsiella pneumoniae	+		+	+	+	+	+
Rhodospirillum rubrum	+	+		+	+	+	−
Spirillum lipoferum	+	+	+		+	+	−
Chromatium vinosum	+	+	+	−		−	−
Bacillus polymyxa	+	+	+	−	−		+
Clostridium pasteurianum	−	+	−	−	−	+	

Fig. 2. *Cross reactions between Fe proteins and Mo-Fe proteins derived from various micro-organisms. The diagonal represents homologous crosses, all of which are active. The symbol + indicates that the cross generates a catalytically active nitrogenase; a − represents an inactive cross.*

Great interest attaches to the association between *S. lipoferum* and a variety of grasses. The observation by Döbereiner and Day (1976) and Day and Döbereiner (1976) that *Digitaria decumbens* carries *S. lipoferum* in its root system and grows vigorously on soils of low nitrogen content elicited widespread interest in this previously neglected organism. Day *et al.* (1975) also demonstrated that the association is not confined to *D. decumbens* but embraces a wide variety of grasses and other types of plants as well.

We have investigated the infection of maize roots under bacteriologically controlled conditions. *Zea mays* seed was surface-sterilized with alcohol and hypochlorite. The seeds then were germinated in contact with nutrient medium so that if any contaminating organisms were present they would proliferate and indicate their presence. Non-contaminated seedlings were placed in jars with sterilized soil. The jars were stoppered with cotton to exclude contaminating organisms. Some of the plant cultures were inoculated with *S. lipoferum* and others were maintained as aseptic controls. After a period of growth, the plants were harvested aseptically; the controls lacked in *S. lipoferum* or other micro-organisms. In contrast, the roots of plants which had been inoculated with *S. lipoferum* carried the organism. Not only was the organism present on the surface of the roots, but when the roots were surf ce sterilized and then crushed, a large number of *S. lipoferum* were liberated from inside the tissue. Upon culture, the organisms showed the typical characteristics of *S. lipoferum.* Thus, it is possible to state that Koch's postulates have been fulfilled as far as infection and reisolation of the organism are concerned. Surface-sterilized roots yield very few organisms when shaken in sterile buffer, but upon crushing, substantial numbers of recoverable organisms are released. This test verifies the observations of Döbereiner and Day (1976) that the organism can be found inside the roots. Our tests of root systems of plants removed from soil inoculated with *S. lipoferum* indicate that the organism is both inside and outside the roots. Most probable number counts indicate that the number of *S. lipoferum* on the surface of the root is substantially greater than the number inside the root, but there is no difficulty in showing that the organisms can in fact penetrate and remain viable inside the root (Okon *et al.*, 1976c).

It has been customary in recent years to determine the N_2-fixing capacity of systems based upon their ability to reduce acetylene. For example, Von Bülow and Döbereiner (1975) have reported vigorous acetylene reduction by excised roots which had been preincubated at a low pO_2 overnight before being supplied acetylene. We have made numerous tests with this technique and likewise have observed high rates of acetylene reduction; our highest rates with maize roots were about 1/7th the highest rates reported by Von Bülow and Döbereiner (1975).

As a different approach to evaluating potential N_2 fixation, we have placed maize plants in large plastic cans which enclose their root systems but leave their stems and leaves exposed. Then we have supplied acetylene to the enclosed root system and have tested for its reduction. Under these conditions, the reduction of the acetylene by the intact plant is very limited. The dilution of ethylene formed in such a large volume container decreases the sensitivity of the method, but calculations show that gas chromatography should be of adequate sensitivity to detect acetylene reduction at a rate considerably less than that observed with the preincubated detached roots. In fact, we can measure the ethylene formed although the amount is small. Likewise, our tests with soil cores, which have included the root system directly around the base of the maize plant, have shown low rates of acetylene reduction. As shown in Fig. 3 the rates

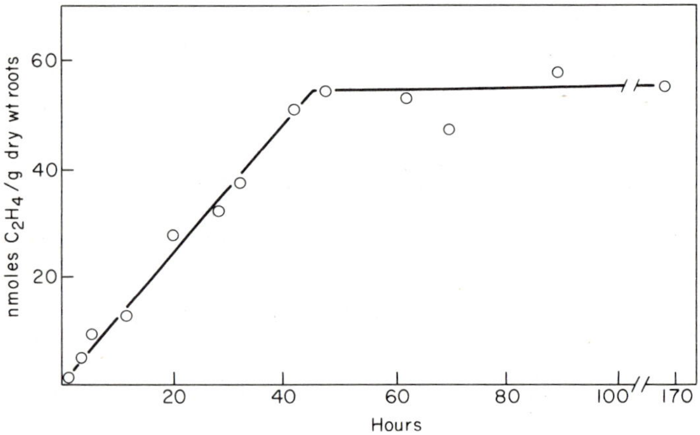

Fig. 3. *Time course of C_2H_4 production in a soil core that included the central root system of a maize plant. The nmoles of C_2H_4 shown are cumulative with time.*

are very low; however, if these cores are incubated at warm temperatures, the rates are rather steady and production of C_2H_4 persists for about 45 h and then ceases. If these rates are used to estimate the amounts of N_2 fixed by maize during the entire growing season, rates of fixation would be about 2 kg of nitrogen fixed in a growing season/hectare.

The question has been raised, "why are the rates of acetylene reduction higher in excised roots than in intact plants or soil cores?" *S. lipoferum* is a rather sensitive organism, and if one removes it carefully from an exponentially growing liquid culture to test its acetylene reduction capabilities, there is a substantial lag before it resumes acetylene reduction. This seems to be associated with shock in handling or exposure to O_2. Adding acetylene to intact plants in large containers or to roots in soil cores should involve minimum disturbance and damage to the organisms active in the root system. In our experience these systems never have supported high rates of acetylene reduction even after allowing them to stand for hours after addition of acetylene. The possibility remains that shock may damage the acetylene-reducing capacity under these conditions, however, one must entertain the other possibility that preincubation at a low pO_2 yields an inaccurately high acetylene reduction rate. That this possibility must be considered seriously is supported by our observation of the proliferation of micro-organisms during the preincubation of roots under low O_2 concentration. We have placed excised roots in closed containers, lowered the O_2 concentration, and then have followed the gas exchange by mass spectrometric measurements (Table VIII).

Table VIII

Gas metabolism of roots of *Zea mays* inoculated with *S. lipoferum*. Roots were removed from the plant, placed in a bottle, and the gas was sampled periodically for analysis with a mass spectrometer

Hours	Gas composition % O_2	Gas composition % CO_2
0.0	19.8	0.6
3.75	9.7	5.2
9.0	5.9	10.8
24.75	2.5	14.4

Under these conditions, O_2 is utilized and the system goes anaerobic rather rapidly. Then a fermentative metabolism is established and a high concentration of CO_2 is produced. The fermentation probably produces lactic acid, which is an excellent substrate for the growth of *S. lipoferum*. With a readily metabolizable substrate available, the organisms may proliferate considerably more rapidly than they do in normal plant tissue and this in turn would lead to more rapid acetylene reduction.

We have measured the total number of *S. lipoferum* associated with freshly harvested and preincubated maize root systems by the most probable number (MPN) method (Okon *et al.*, 1976c). In applying this technique, the root tissue was ground, diluted, and aliquots of the diluted suspension were added to

7 ml bottles containing semisolid malate medium. During incubation, the bottles containing *S. lipoferum* form a typical layer of growth a few mm below the surface where the pO_2 is optimal. Each bottle showing such growth was then supplied acetylene and tested for ethylene production. Characteristic growth plus acetylene reduction are presumptive evidence of the presence of *S. lipoferum*. With results of this test and the use of MPN tables, we demonstrated that *S. lipoferum* may increase in numbers a hundred-fold or more in excised maize roots incubated for a day at 30°C at a pO_2 initially adjusted to 0.01 atm. These data are shown in Table IX. A hundred-fold multiplication implies

Table IX

Reduction of C_2H_2 and proliferation of *S. lipoferum* under microaerophilic conditions in detached roots from inoculated, field-grown maize

C_2H_2 reduction and proliferation of *S. lipoferum* in maize roots (inoc. field) under microaerobic conditions				
Variety	nmoles C_2H_4 produced / min x g dry wt		MPN of *S. lipoferum* / g dry wt	
	0 h	24 h	0 h	24 h
Wis. 900	0	425	10,000	2,000,000
	0	275	1,500	3,200,000
Field maize	0	9	13,000	320,000
	0	14	5.700	190,000
Golden cross bantam	0	12	1,500	1,400,000

a very short generation time for the organism under these conditions. As indicated earlier, the organism has the potential for doubling in about an hour under optimal conditions and a generation time between 3 and 4 h would be short enough for a hundred-fold multiplication in 24 h.

The data on rates of acetylene reduction in excised roots versus intact plants and the rapid proliferation of organisms in excised root tissue suggest that the measurement of acetylene reduction by preincubated excised roots should be used as evidence for the presence of N_2-fixing organisms and their potential for N_2 fixation, but such data should not be interpreted in quantitative terms as a measure of the amount of N_2

fixation actually accomplished by the plant as it grows in the field.

We have tested the association of *S. lipoferum* with various plants in a greenhouse and in growth chambers. Table X shows the response observed under greenhouse conditions. The maize

Table X

Yield of maize in the greenhouse with and without *S. lipoferum* inoculum; Hoagland's solution -N was supplied

	Inoculated	Control
Nitrogenase activity, nmoles C_2H_4/(h x g dry wt)	381	0.0
Weight of roots, g	3.67	2.52
Weight of shoots, g	11.33	12.36
% total N in roots	0.62	0.55
% total N in shoots	0.52	0.40

plants were grown on sterilized soil and the seeds were surface sterilized. Some were inoculated and uninoculated controls also were retained. As indicated, there was very little response in terms of yield or total nitrogen in the plants. We have done comparable tests in growth chambers and likewise have found minimal increase in growth or total nitrogen content of the maize plants.

Suggestions have been made that the association between maize and *S. lipoferum* is better under tropical than under temperate conditions. To test the influence of light intensity and temperature we have grown maize inoculated and uninoculated with *S. lipoferum* under 16 different conditions of temperature and light in a crossed-gradient room in the University of Wisconsin Biotron. The room is arranged in such a fashion (Fig. 4) that one can apply 4 different temperature conditions and 4 different light intensities to give 16 combinations of light and temperature. The light intensities ran from 500 to 3000 foot candles and the temperatures during the daytime were 28, 32, 36 and 40°C. The night temperatures were 10° lower than daytime temperatures. Maize seeds were surface sterilized and placed in sterile sand after pregermination under aseptic conditions. The plants were supplied nitrogen-free Hoagland's solution and at intervals were given increments of combined nitrogen; the increments finally totalled about 30% the level supplied in normal Hoagland's solution. Fig. 5 shows the results of the experiment. It is apparent that there was very

40° 500	40° 1300	40° 2400	40° 3000
36° 500	36° 1300	36° 2400	36° 3000
32° 500	32° 1300	32° 2400	32° 3000
28° 500	28° 1300	28° 2400	28° 3000

Fig. 4. *Temperatures and light intensities in the 16 treatment blocks in the Biotron crossed-gradient room. The temperatures in °C and the light intensities in foot candles are approximate.*

little difference between inoculated and uninoculated plants. After a 100-day growing period in the crossed-gradient room, the yield based on dry weight or total nitrogen of the inoculated and uninoculated plants were scarcely distinguishable under any of the temperature of light conditions. The best response in the ly treatments corresponds to about 8 kg of nitrogen fixed per hectare in a growing season. However, it should be emphasized that with the small number of plants that could be accommodated, this 8 kg increase is not statistically valid, and it represents the most favourable response rather than a typical response. The typical responses showed minimal benefits that were well within the statistical errors of the experiment.

Table XI

Field trials, Wisconsin River, 1975
Maize ± *S. lipoferum*

Cultivar	Inoc.		Uninoc.	
	grams	% N	grams	% N
W182E	100	1.15	110	1.68
A654	111	1.05	134	1.53
A629	101	1.09	82	1.69
Gld. Cross	120	1.38	169	1.23
Wis. 900	64	1.28	93	1.56
W153R	105	1.12	74	1.51
Field corn	191	1.03	153	1.14
Gold Rush	162	0.94	83	1.65
Gld. Beauty	137	1.27	60	1.57

What is of primary concern is whether or not the use of *S. lipoferum* in association with certain plants can increase the yield of the plants in the field. We have made field trials in 1975 and 1976, and Table XI shows the results of field trials which were made on a sandy soil in the Wisconsin River valley. A slightly higher percentage of the uninoculated plants gave a better total yield than the inoculated plnats, but the differences were well within the statistical variations of the experiment.

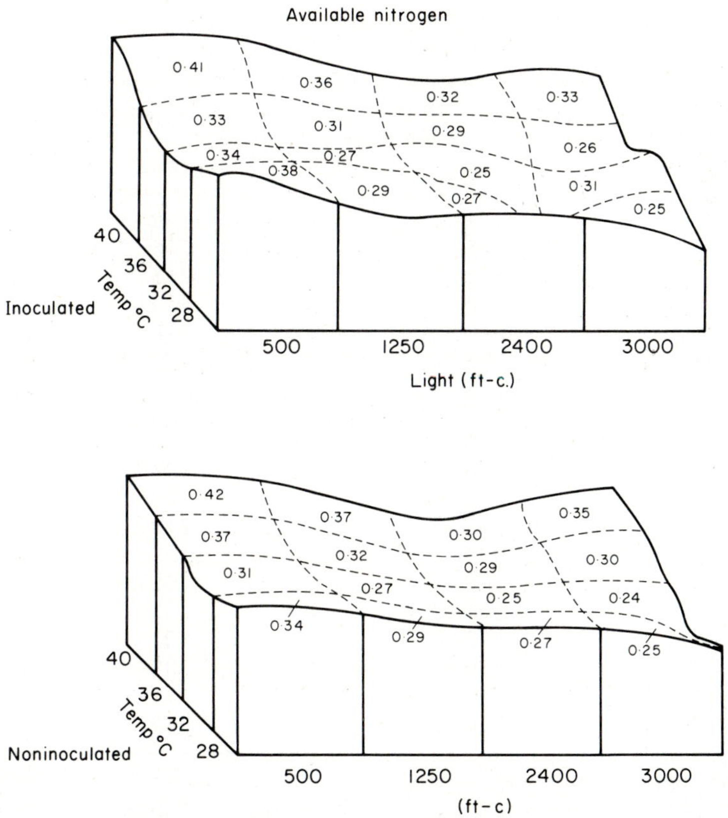

Fig. 5. *Comparison between inoculated and uninoculated maize plants grown under the 16 conditions of temperature and light intensity supplied in the crossed-gradient room of the Biotron. "Available nitrogen" is the product of the £ N x dry weight of the maize plants.*

On a somewhat heavier soil near Madison, Wisconsin, we tested 19 different varieties or species of plants (Table XII). In this experiment, 14 of the 19 showed a positive response and 5 showed a negative response to inoculation with *S. lipoferum*. Again the results are not statistically significant but they definitely were in a positive direction toward benefit from *S. lipoferum* inoculation. Use of these data to calculate yields and increased nitrogen on a hectare basis indicates a possible increase in total nitrogen of 2 to 5 kg per hectare during a growing season.

Table XII

Field trials, Charmany Farms, 1975
Maize etc. ± *S. lipoferum*

Cultivar	Inoc.		Uninoc.	
	grams	% N	grams	% N
W59E	163	0.99	169	1.02
W153R	225	1.25	246	1.16
W64A	240	1.19	243	1.23
Wis 900	376	1.12	422	1.07
Field corn	519	1.09	318	1.07
Sorghum	385	0.37	478	0.40
Sorghum-Sudan	592	0.56	356	0.45
W729D	223	1.23	223	1.10
W182E	205	1.13	178	1.08
W629A	113	1.39	103	1.45
W513	205	1.28	145	1.25
W61	235	1.19	218	1.13
A629	237	1.15	223	1.10
W438	289	1.11	266	1.12
A654	211	1.26	223	1.19
Gold Rush	372	1.18	302	1.12
Gld. Cross	439	1.15	390	0.96
Gld. Beauty	329	1.10	286	1.12
White Cloud	320	1.06	262	1.06

What can we say regarding the potential of *S. lipoferum* for N_2 fixation in association with various plants? Clearly, the organism is an active N_2 fixer by itself. In addition, there is evidence that the organism can invade various plants and set up an association both inside and outside the roots. This association is capable of reducing acetylene, as demonstrated by Döbereiner and Day (1976) and subsequently in many other laboratories. The quantitative benefits of the association are less clear, and to date the field trials have shown rather limited increases in N_2 fixation and yield. In the association the plant is capable of supplying photosynthate, including organic acids which furnish an excellent substrate for the growth of *S. lipoferum*. Although available yield data are not particularly convincing of benefit, nevertheless we should remember that field trials have been limited to a year or two. This time has been inadequate to select well adapted strains or to manipulate *S. lipoferum* genetically to improve its performance in association with higher plants. Likewise, time has been inadequate to select the most effective plant cultivars. There are experimental suggestions that the association between *S. lipoferum* and higher plants has a high potential for N_2 fixation. This should encourage investigators to continue and to expand their studies of this interesting interaction between plants and bacteria in the hope that the association can be improved to the point where it will furnish a substantial portion of the fixed nitrogen required by nonleguminous plants.

ACKNOWLEDGEMENTS

These investigations were supported by the College of Agricultural and Life Sciences, University of Wisconsin-Madison, by Public Health Service Grant AI-00848 from the National Institute of Allergy and Infectious Diseases, by National Science Foundation Grant PCM74-17604, and by Grant 161209 from the University of Wisconsin Graduate School Research Committee. We thank J. Döbereiner for the cultures used in many of the experiments reported.

REFERENCES

Balandreau, J., Rinaudo, G., Fares-Hamad, I. and Dommergues, Y. (1975). *In:* Nitrogen Fixation by Free-living Micro-organisms (W.D.P. Stewart, ed.), pp. 57-70. Cambridge University Press, Cambridge, England.

Becking, J.H. (1963). *Antonie van Leeuwenhoek,* 29: 326.

Becking, J.H. (1974). *In:* The Biology of Nitrogen Fixation (A. Quispel, ed.), pp. 583-613. North Holland Publishing Company, Amsterdam, The Netherlands.

Becking, J.H. (1976). *In:* Symbiotic Nitrogen Fixation in Plants (P.S. Nutman, ed.), pp. 539-550. Cambridge University Press, Cambridge, England.

Beijerinck, M.W. (1925). *Zentbl.Bakt.Parasitkde.,* II 63: 353-359.

Borkowski, J.D. and Johnson, M.J. (1967). *Biotechnol.Bioeng.,* 9: 635-639.

Carpenter, E.J. and Price, C.C. (1976). *Science,* 191: 1278-1280.

Centifanto, Y.M. and Silver, W.S. (1964). *J.Bacteriology,* 88: 776-781.

Day, J.M. and Döbereiner, J. (1976). *Soil Biol.Biochem.*, 8: 45-50.

Day, J.M., Neves, M.C.P. and Döbereiner, J. (1975). *Soil Biol. Biochem.*, 7: 107-112.

Döbereiner, J. (1974). *In:* The Biology of Nitrogen Fixation (A. Quispel, ed.), pp. 86-120. North Holland Publishing Company, Amsterdam, The Netherlands.

Döbereiner, J. and Day, J.M. (1975). *In:* Nitrogen Fixation by Free-living Micro-organisms (W.D.P. Stewart, ed.), pp. 39-56. Cambridge University Press, Cambridge, England.

Döbereiner, J. and Day, J.M. (1976). *In:* Proceedings of the 1st International Symposium on Nitrogen Fixation (W.E. Newton and C.J. Nyman, eds), pp. 518-538. Washington State University Press, Pullman, Washington.

Döbereiner, J., Day, J.M. and Dart, P.J. (1972a). *Plant Soil*, 37: 191-196.

Döbereiner, J., Day, J.M. and Dart, P.J. (1972b). *J.Gen.Microbiol.*, 71: 103-116.

Ludden, P.W. and Burris, R.H. (1976). *Science*, in press.

Millbank, J.W. (1974). *In:* The Biology of Nitrogen Fixation (A. Quispel, ed.), pp. 238-264. North Holland Publishing Company, Amsterdam, The Netherlands.

Mishustin, E.N. and Shil'nikova, V.K. (1971). Biological Fixation of Atmospheric Nitrogen. The Macmillan Company, London, England.

Newcomb, E.H. and Pugh, T.D. (1975). *Nature*, 253: 533-534.

Okon, Y., Albrecht, S.L. and Burris, R.H. (1976a). *J.Bacteriol.*, in press.

Okon, Y., Albrecht, S.L. and Burris, R.H. (1976b). *J.Bacteriol.*, in press.

Okon, Y., Albrecht, S.L. and Burris, R.H. (1976c). *Applied and Environ.Microbiol.*, submitted.

Okon, Y., Houchins, J.P., Albrecht, S.L.and Burris, R.H. (1976). *J.Gen.Microbiol.*, in press.

Peters, G.A. (1975). *Arch.Microbiol.*, 103: 113-122.

Peters, G.A. and Mayne, B.C. (1974a). *Plant Physiol.*, 53: 813-819.

Peters, G.A. and Mayne, B.C. (1974b). *Plant Physiol.*, 53: 820-824.

Ruinen, J. (1974). *In:* The Biology of Nitrogen Fixation (A. Quispel, ed.), pp. 121-167. North Holland Publishing Company, Amsterdam, The Netherlands.

Ruinen, J. (1975). *In:* Nitrogen Fixation by Free-living Microorganisms (W.D.P. Stewart, ed.), pp. 85-100. Cambridge University Press, Cambridge, England.

Schröder, M. (1932). *Zentbt.Bakt.Parasitkde.* II 85: 177-212.

Silver, W.S., Centifanto, Y.M. and Nicholas, D.J.D. (1963).

Nature, 199: 396-397.
Silvester, W.B. (1976). *In:* Symbiotic Nitrogen Fixation in Plants (P.S. Nutman, ed.), pp. 521-538. Cambridge University Press, Cambridge, England.
Van Hove, C. (1976). *In:* Symbiotic Nitrogen Fixation in Plants (P.S. Nutman, ed.), pp. 551-560. Cambridge University Press, Cambridge, England.
Von Bülow, J.F.W. and Döbereiner, J. (1975). *Proc.Natl.Acad.Sci. U.S.A.*, 72: 2389-2393.

PHYSIOLOGICAL ASPECTS OF N_2 FIXATION IN GRASS-BACTERIA ASSOCIATIONS*

JOHANNA DOBEREINER

Empresa Brasileira de Pesquisa Agropecuaria, Km 47, Via Campo Grande, Rio de Janeiro, Brazil

INTRODUCTION

Although the first results showing nitrogen fixation in tropical grasses were reported several years ago (Rinaudo, 1971; Döbereiner *et al.*, 1972) there is still insufficient information available to give a clear picture of the potential of most of these systems. With the grasses *Paspalum notatum* and *Digitaria* spp., nitrogen fixation and incorporation into plant tissue has now been shown in $^{15}N_2$ experiments (Table I). Extrapolation from acetylene reduction data in grain crops are overestimates because peak activities are limited to short periods

Table I

Incorporation of $^{15}N_2$ in forage grasses[a] (De-Polli *et al.*, 1976)

Plant part	Total N in plants	Atom % ^{15}N excess	$\mu g\ N_2$ fixed/pot	% N_2 fixed
Digitaria decumbens cv transvala				
Root	1.78	0.151	2.69	33.1
Rhizome	2.74	0.146	4.00	49.3
Stem	4.32	0.021	0.91	11.2
Leaf	7.34	0.007	0.51	6.3
Paspalum notatum				
Root	2.89	0.563	16.34	47.5
Rhizome	2.62	0.703	15.28	43.8
Leaf	6.03	0.070	2.97	8.6

[a]Exposure for 72 h to gas mixture enriched with 85.49% $^{15}N_2$.

of the growth cycle. Plant genetics seems to be a promising tool to increase and extend peak activities (Döbereiner, 1976; Bülow *et al.*, 1976) but interpretation of the results will depend on better understanding of the physiological processes involved. A major drawback is the difficulty of properly assessing nitrogenase activities in large intact plants like maize, where a number of results indicate that excised root assays do not give reliable results (Tjepkema and Van Berkum, unpublished; Sloger, 1976).

Supply of Carbon Substrates

Malate the first product of C_4 photosynthesis is also the favoured carbon substrate for *Spirillum lipoferum*, a nitrogen-fixing bacterium most probably involved in association with many grasses. Although it is generally assumed that malate is descarboxylated in the leaf bundlesheath, it has been found in roots of maize, especially when the plants are nitrate-limited (Neyra and Hageman, 1976). High levels of nitrate lowered the size of the malate pool in the roots. Excised sorghum roots treated with malate showed twice as high nitrogenase activities as untreated roots, but glucose had no effect (Table II). Roots

Table II

Effect of bicarbonate on N_2-ase activity on excised sorghum roots and in *S. lipoferum* cultures (Van Berkum *et al.*, 1976; Döbereiner, unpublished)

Carbon substrate	*S. lipoferum*[a] nmol C_2H_4.h per culture Sp7	Sp13	Roots[b] nmol C_2H_4/h per g dw
No addition	0	0	110
$KHCO_3$ 5mM	0	0	223
Glucose 50mM	5	1	111
Glucose+$KHCO_3$	9	8	-
Malate 37mM	95	60	214[c]
Malate+$KHCO_3$	133	148	-

[a]Two ml cultures of strains Sp7 or Sp13 in semisolid N free mineral medium with the given carbon substrates incubated for 24 h at 33°C. One hour C_2H_2 reduction rates determined without disturbing the pellicles (means of 3 replicates).

[b]Roots excised from field grown sorghum at flowering stage exposed for 4 h to the given solutions then preincubated overnight under nitrogen before 2 h C_2H_2 reduction rates were determined under an atmosphere containing 1% O_2 (means of 6 replicates).

[c]5mM malate only.

soaked for 4 h in 5 mM solutions of bicarbonate showed increases in nitrogenase activity similar to those with malate. Dark fixation of CO_2 by the roots or the bacteria to provide malate are a possible explanation. Neyra and Hageman (1977) observed increased malate concentrations in nitrate-deficient maize roots grown in nutrient solution containing bicarbonate. Mulder and Van Veen (1960), on the other hand, observed increased N_2 fixation in nodulated pea plants under such conditions. Nodules of these plants contained higher concentrations of malic and α-ketoglutaric acids, which the authors assumed to be used as carbon skeletons for ammonia incorporation.

The dark CO_2 fixation in legume nodules has now been confirmed by Lawrie and Wheeler (1974), who showed high phosphoenol pyruvate carboxylase and malic dehydrogenase activities in bean (*Vicia faba)* nodules and incorporation of $^{14}CO_2$ into malic acid in detached nodules.

Pure cultures of *S. lipoferum* supplied with malate also showed faster initial growth when bicarbonate was supplied (Table II); in growing cultures the CO_2 from respiration is probably sufficient. These observations explain the experimental difficulties of starting growth of this organism in N free medium. Large inoculants, small amounts of nitrate, ammonia or yeast extract have similar effects (Day and Döbereiner, 1976). Glucose is used very poorly by this organism (Table II) except by a few strains (see discussion on denitrification below) confirming the observations on roots. Numbers of *S. lipoferum* in sorghum roots can be high (1.8×10^5/g fresh roots), ten times higher than in root surface soil (1.7×10^5).

Oxygen requirements and inhibition

For maximal nitrogenase activity, excised sorghum roots show similar oxygen requirements to *Digitaria* roots (Döbereiner *et al.*, 1975). The data in Table III show that sorghum roots preincubated under 0.02 atm O_2 have maximal activity without additional O_2 supply whereas after anaerobic pre-incubation, 0.01 atm O_2 in the gas phase was optimal.

More precise information about the interference of pO_2 in nitrogen fixation by *S. lipoferum* was obtained in experiments carried out with an oxygen-limited chemostat (Döbereiner, Hill and De-Polli, unpublished). A steady state was obtained under air with medium stirred at 420 rpm (OD of 0.3 = 2 mg dry cells/ml; dilution rate = 0.10 h^{-1} at 35°C) with strain Sp 7 (ATCC 29145). A similar steady state was obtained with another strain (Sp80). Increasing the stirrer speed from 420 to 615 rpm did not affect the amount of O_2 in solution, the oxygen electrode remained at zero (less than 0.25 M O_2). However, the rate of nitrogenase activity in the chemostat increased by

Table III

Effect of oxygen partial pressure during preincubation and incubation of excised sorghum roots in N_2-ase activity (nmol C_2H_4/h/g dw roots)
(Podesta, Neyra and Dobereiner, in preparation)

% O_2 added to N_2 for preincubation[a]		
For incubation	0	2
0	152	424
1	358	379
2	240	234
4	119	119

Field grown sorghum roots were removed from the soil, placed immediately into water which was then replaced by N_2 and the given amount of O_2 (in air) injected for an overnight preincubation. Before C_2H_2 injection in the morning, the given amounts of O_2 were added for incubation without further replacement. Two hour C_2H_2 reduction rates were then determined by gas chromatography.

[a]Blanks preincubated in air without O_2 addition for incubation had a mean activity of 37 nmol C_2H_4/h/g dw roots. All data are means of 6 replicate assays.

80%. This increased rate was considered as initial rate for the data in Table IV. Further increase in stirring speed (up to 650 rpm) apparently caused the oxygen solution rate to exceed the saturation of the respiration enzyme systems and dissolved oxygen increased within 7 min to 75 M where it stabilized. When the stirrer speed was returned to 615 rpm, within 6 min most of the O_2 had been consumed and the O_2 electrode returned to 2.5 M O_2; after a further 6 min it returned to zero. These observations indicate a very active respiration system which consumes large amounts of O_2. The effects of increasing amounts of dissolved O_2 on nitrogenase activity are summarized in Table IV: 2.5 M O_2 in solution did not inhibit nitrogenase activity even after 40 min; higher concentrations inhibited nitrogenase activity completely after decreasing time intervals. The higher the O_2 concentration was to which the cultures had been exposed the longer it took for restoration of linear rates. In the first 5 min after increasing the O_2 concentration nitrogenase activity increased even at the highest concentration (75 M) but then it decreased rapidly (data not presented here). The increase in nitrogenase activity resulting from increased O_2 supply might indicate ATP limitation of the system. The final

Table IV

Effect of the oxygen concentration in solution on N_2-ase (C_2H_2) activity in an oxygen limited chemostat culture of *S. lipoferum* *(Döbereiner, Hill and De-Polli, unpublished).

C_2 concentration		Min until N_2-ase activity was nil	Min until linear rates restored	Restored linear rates (% of initial)
Electrode reading	μM**			
4	0	-	-	100
10	2.5***	no inhibition	0	100
20	5.0	35	10	55
80	20	12	15	58
300	75	10	17	45
300	75 under chloramphenicol****	10	not restored	0

*Oxygen limited chemostat culture running under O_2:N_2 mixture of 25:75 with OD = 0.30, D = 0.10 at 35°C in N-free lactate medium. At time zero 15% C_2H_2 was added to the chemostat culture and the stirrer speed increased from 420 to 615 rpm which saturated the respiration enzyme system without any apparent O_2 being in solution. A higher stirrer speed (to 650 rpm) was excessive and the O_2 concentration increased steadily during 7 minutes when it stabilized at 75 μM O_2. The initial C_2H_2 reduction rate was considered the stabilized linear rate which was established 15 minutes after C_2H_2 injection and stirrer increase to 615 rpm, as observed by 5 readings of the C_2H_4 formed, one every two minutes. Readings were continued every 2 min throughout the experiments.

**Concentration of O_2 in solution measured with a Western Biological Sherborne Dorset oxygen electrode calibrated with sterile nutrient solution and an O_2:N_2 gas mixture of 25:75 (reading 1000) and N_2 (reading 4). Lower O_2 concentrations in the gas mixture gave linear readings in between.

***The different readings on the electrode were obtained by changing the stirrer speed manually back and forth to maintain approximately constant readings at the desired O_2 concentrations.

****50 μg of chloramphenicol/ml added to the chemostat 10 min before the O_2 concentration was raised.

rates (30 min after restoration of linear rates) were about half the initial rates. Fifty μg chloramphenicol/ml added 10 min before increasing the O_2 concentrations, in the cultures to 75 μM, completely inhibited restoration of nitrogenase activity after oxygen damage as also observed by Okon *et al.* (1976). The delayed inhibition and slow and incomplete restoration also indicate that at least part of the nitrogenase enzyme is permanently damaged. A bell shaped curve of oxygen inhibition as observed by Drozd and Postgate (1970) with *Azotobacter* does not seem to exist in this organism. Growth with ammonium, as expected, is fastest during vigorous shaking under air as also observed by Okon (1976). Berndt *et al.* (1976) observed similar differences in oxygen requirements for nitrogen-fixing and NH_4^+-grown cells of *Corynebacterium autotrophicum*. The facultative nitrogen-fixing *Klebsiella pneumoniae* was shown to be able to grow in N-free medium supplied with oxygen, in an oxygen limited chemostat but obtaining ATP *via* oxydative phosphorylation (Hill, 1976).

Effects of mineral nitrogen in field experiments

Nitrogenase activity on excised roots of field-grown maize shows a large variation with the growth cycle (Fig. 1) with

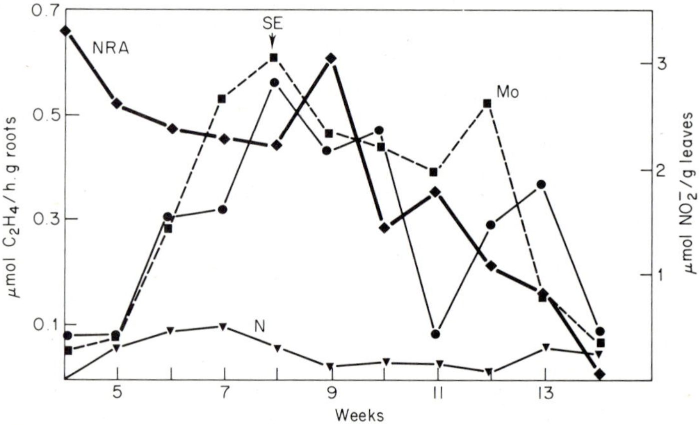

Fig. 1. *Nitrogenase activity in excised roots and nitrate reductase activity in leaves during the growth cycle of maize (Cv Piranao) N_2-ase (C_2H_2) activity determined on excised roots preincubated overnight at pO_2 0.02 atm. Fertilizer application of 40 kg N/ha at planting (■ —— ■); application of 40 kg N/ha at planting and leaf spraying with 1 kg ammonium molybdate/ha twice during the growth cycle (• —— •); application of 100 kg N/ha at planting and 100 kg N/ha at tasseling and Mo spray as above (▼ —— ▼). In vivo nitrate reductase activity determined in the upper half leaves (◆ —— ◆). Silk emergence indicated by SE. All data are means of 12 plants taken from 6 plots. Figure redrawn from Neyra* et al. *(1976).*

maximal activities at silk emergence and half grain filling. Nitrate reductase (NRA) activity in leaves was higher during the beginning of the growth cycle when soil nitrate levels were still higher. Although general trends of NRA and nitrogenase activities seem negatively correlated, reasonable activities of both can occur simultaneously.

Molybdenum increased both activities while high nitrogen fertilizer inhibited nitrogenase and increased NRA. In a similar field experiment with sorghum, steeper peaks but similar effects of Mo and combined nitrogen on nitrogenase and NRA activities were observed (Van Berkum *et al.*, 1976). In maize and sorghum nitrogenase activity on excised roots was completely inhibited by 0.5 mM NH_4^+ but only partially inhibited by 1 mM NO_3^-, (Neyra *et al.*, 1976; Van Berkum *et al.*, 1976). *Spirillum lipoferum* occurrence in roots was initially lower when the plants were grown with high N levels (200 kg N/ha). Once the association was established, the application of the same amount of fertilizer N had no effect on *S. lipoferum* incidence in roots (Podesta *et al.*, 1976). Large variations in nitrogenase activity in excised maize roots could be attributed to differences in *S. lipoferum* incidence between plants (correlation of root activity with enrichment culture activity from roots obtained from the same plants highly significant). Studies of this kind seem fundamental in breeding programmes for simultaneous use of mineral and biologically fixed nitrogen.

Nitrate Reduction and Denitrification by S. lipoferum

Initial observations indicated that *S. lipoferum* cultures grown on nitrate incorporated only a small part of it while the rest seemed to be lost (Day and Döbereiner, unpublished). Estimates of the losses indicated more than 90% when 30 mM nitrate was in the semisolid medium. The importance of this observation, which contrasts to what is known of *Azotobacter paspali* (which has no nitrate reductase, Döbereiner and Day, 1975), lead us to examine in more detail the nitrate metabolism of *S. lipoferum* (Neyra and Van Berkum, 1976; Neyra *et al.*, 1976). Fifty-one strains were grown in semisolid mineral medium containing 10 mM KNO_3. All grew rapidly and none showed any nitrogenase activity, so that it must be assumed that all strains possess an assimilatory nitrate reductase which enables them to use nitrate in preference to N_2 as nitrogen source. No apparent gas production appeared even after shaking. When the cultures were grown with 10 mM NH_4NO_3 and, after two days incubation, handshaken once to mix the *S. lipoferum* pellicle with the medium, some of the strains showed vigorous gas production, indicating that in this medium growth occurs on NH_4^+, and nitrate in the lower layers of the cultures can be used for respiration. Based on this characteristic, the 51 strains could be divided into three groups: 20 strains, including the type strain Sp7 (ATCC 29145) showed gas production and initial nitrite formation, but, after 2 days, no nitrite was left (Group I). Twenty strains produced no gas in these cultures and after 2 days large amounts of nitrite accumulated (Group III). The remaining strains were intermediate; that is, they produced some gas but were unable to reduce all the accumulated nitrite within 2 days (Group II). There were strains from various origins in all three groups but all yeast extract-requiring strains were in Group II and most of the strains isolated from surface sterilized roots, which had been selected for high nitrogenase activity were in Group III (Neyra *et al.*, 1977).

The division of the *S. lipoferum* isolates into at least two groups has been suggested before (Döbereiner *et al.*, 1976; Okon *et al.*, 1976) and has recently been confirmed (Sampaio, Vasconselas and Döbereiner, in preparation). Group I contains the type strain and a number of strains isolated from various tropical regions in Brazil, Peru, Africa and the Philippines. Glucose is not used by these strains; they grow rapidly on semisolid malate medium without a yeast extract supplement, are catalase positive and cells are very motile and small. Group II contains strains from tropical Brazil and Africa but also from temperate regions such as Washington State and South Brazil. The six yeast extract-requiring strains are in this group. These do not grow without yeast supplement even in the presence of nitrate or NH_4^+, they use glucose, are catalase negative and have very large, polymorphic cells. The remaining 5 strains of this group, although intermediate in respect to denitrification, do not require yeast extract and do not use glucose. No other distinctive characteristic of the strains of Group III has been observed except the incapacity to dissimilate nitrite further. Cells in general are larger than the type strain, and have more refractive polyhydroxybutyrate bodies. The denitrification by two of the strains of Group I and lack of this property in two of the strains of Group III has been confirmed (to be published) by the new method of evaluating denitrification by the accumulation of N_2O in the presence of C_2H_2 (Yoshinari and Knowles, 1976). Simultaneous denitrification and nitrogenase activity (C_2H_2) was observed in such semisolid cultures incubated under air and under He. Nitrate-dependent (10 mM KNO_3) increases of nitrogenase activity in anaerobically incubated *S. lipoferum* cultures was also observed by Neyra and Van Berkum (1976) (Fig. 2).

FOOTNOTE

*Contribution of the Programme of International Cooperation in Training and Research on Nitrogen Fixation in the Tropics, sponsored by the Brazilian National Research Council, the Empresa Brasileira de Pesquisa Agropecuaria and the Universidade Federal Rural do Rio de Janeiro.

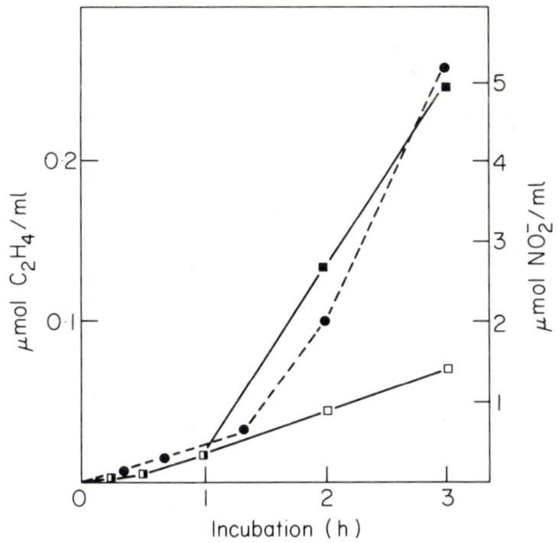

Fig. 2. *Nitrogenase activity dependent on nitrate reduction in near anaerobic* S. lipoferum *cultures. Cultures were grown in semisolid N free malate medium and placed under N_2 and 10% C_2H_2 at time zero. N_2-ase activity with 10 mM KNO_3 added at time zero (■ —— ■); N_2-ase activity without NO_3^- (□ —— □); NO_2^- accumulation in the cultures with KNO_3 (• —— •). Redrawn from Neyra and Van Berkum, 1976.*

REFERENCES

Berndt, H., Ostwal, K.P., Lalucat, J., Schuman, C., Mayer, F. and Schlegel, H.G. (1976). *Arch.Microbiol.*, 108: 17-26.

Bülow, J.F.W. von, Döbereiner, J. and Podesta-Filho, J.A. (1976). XI Reuniao Brasileira de Milho e Sorgo, Piracicaba, Brasil.

Day, J.M. and Döbereiner, J. (1976). *Soil Biol.Biochem.*, 8: 45-50.

De-Polli, H., Matsui, E., Döbereiner, J. and Salatai, E. (1976). *Soil Biol.Biochem.*, in press.

Döbereiner, J. and Day, J.M. (1975). Dinitrogen fixation in the rhizospere of tropical grasses. IBP Conference Nitrogen Fixation in the Biosphere, Edinburgh. (W.D.P. Stewart, ed.), Vol. 1, pp. 39-56. Cambridge University Press.

Döbereiner, J., Day, J.M. and Bülow, J.F.W. von. (1975). Proc. II Internat.Winter Wheat Conf., pp. 221-237, Zagreb, Yugoslavia.

Döbereiner, J., Day, J.M. and Dart, P.J. (1972). *J.Gen.Microbiol.*, 71: 103-117.

Döbereiner, J. (1976). Internat.Symposium Genetic Control of Diversity of Plants, Lahore, Pakistan.

Döbereiner, J., Nery, M. and Marriel, I.E. (1976). *Can.J.Microbiol.*, accepted.

Drozd, J. and Postgate, J.R. (1970). *J.Gen.Microbiol.*, 60: 427-429.
Hill, S. (1976). *J.Gen.Microbiol.*, 93: 335-345.
Lawrie, A.C. and Wheeler, C.T. (1975). *New Phytol.*, 74: 437-445.
Mulder, E.G. and Van Veen, W.L. (1960). *Plant and Soil*, 13: 265-278.
Neyra, C.A. and Hageman, R.H. (1976). *Plant Physiol.*, in press.
Neyra, C.A., Pereira, C.A., Bülow, J.F.W. von and Döbereiner, J. (1976). XI Reun.bras.Milho Sorgo, Piracicaba, Brasil.
Neyra, C.A. and Van Berkum, P. (1976). Proc.VI Reun.Soc.Latino-americana Fisiologia Vegetal, Campinas, Brasil, p. 45.
Okon, Y., Albrecht, S.L. and Burris, R.H. (1976). *J.Bacteriol.*, in press.
Podesta-Filho, J.A., Pereira, P.A.A. and Döbereiner, J. (1976). XI Reun.brasil Milho e Sorgo, Piracicaba, Brasil.
Rinaudo, G., Balandreau, J. and Dommergues, Y. (1971). *Plant and Soil*, Special Vol., pp. 471-479.
Sloger, C. (1976). *Plant Physiol.*, 57: S 533.
Van Berkum, P., Neyra, C.A. and Bülow, J.F.W. von. (1976). XI Reun.brasil.Milho Sorgo, Piracicaba, Brasil.
Yoshinari, T.and Knowles, R. (1976). *Biochem.Biophys.Res.Comm.*, 69: 705-710.

A MODELLING APPROACH OF ACETYLENE REDUCING ACTIVITY OF PLANT-RHIZOSPHERE DIAZOTROPH SYSTEMS

J.P. BALANDREAU[1], C.R. MILLIER[2], PIERETTE WEINHARD[1], P. DUCERF[1] and Y.R. DOMMERGUES[1]

INTRODUCTION

Acetylene reducing activity (ARA) of plant-diazotroph systems whether obligatory or associative is known to be affected by environmental factors, by plant factors and by specific characteristics of associated diazotrophs. Marked and rapid variation in ARA occur due to the number of fluctuating environmental factors acting to limit ARA.

Diurnal ARA variations were reported in the case of symbiotic systems such as legume-*Rhizobium* (Hardy *et al.*, 1968; Mague and Burris, 1972) and non-legume-actinomycetes like organisms (Wheeler, 1971; Bond *et al.*, 1975). In the case of grass-rhizosphere diazotroph systems (Balandreau *et al.*, 1974), variation in the ARA was related to variation in the photosynthetic activity of the plant, rate of translocation of photosynthates to the roots and rate of exudation of compounds from the roots. These processes were all sensitive to light intensity, temperature and other environmental parameters. Variations in ARA from day to day are also known to occur, which result from daily variations in environmental parameters (e.g. Balandreau, 1975). Moreover variation in ARA throughout the plant growth cycle can also be expected (e.g. Hardy *et al.*, 1973; Balandreau and Dommergues, 1973).

Variation in ARA may obscure the interpretation of field experiments where estimations of N_2 fixation are made by extrapolating over long periods of time. Precise estimation of N_2 fixation would necessitate a large number of ARA measurements - at least 8 per day with 5 replicates (Balandreau *et al.*, 1974) for the duration of the experiment.

---

[1] Centre de Pedologie du C.N.R.S., B.P. 5, 54500 VANDOEUVRE-LES-NANCY, France.

[2] Centre National de Recherches Forestieres, Champenoux, 54280 SEICHAMPS, France.

The object of the present investigation is to attempt to correlate observed ARA with soil and climatic factors and with plant growth in order to build a preliminary model (1) allowing the interpolation of a limited number of ARA determinations over periods of time extending from one day to a whole plant growth cycle, (2) predicting variation in ARA from environmental data or simple plant growth parameters.

MATERIAL AND METHODS

ARA was assayed *in situ* using a device designed not to disturb the plant and consisting of two components: an open metal cylinder 700 cm^2 in area and 15 cm in height, which was pushed into the soil to surround an individual plant and a polyethylene bag which was fitted on to the upper rim of the cylinder. Water was placed in the rim of the cylinder to seal the volume containing the plant and enclosed by the bag and the cylinder. Acetylene (10% in volume) was injected into the bag together with propane (tracer gas). After 1 h incubation ethylene production was determined by gas chromatography according to the procedure described elsewhere (Balandreau and Dommergues, 1973). ARA was expressed as nmoles of ethylene evolved/h/700 cm^2 cylinder area). The plant used for these investigations was maize grown in a typical eutrochrept soil located at La Bouzule experimental farm, 20 km north of Nancy, France. A preliminary series of estimations of ARA made in 1971 showed that N_2 fixation by the maize native diazotroph system, was 1-3 kg N_2/ha. growth cycle (Balandreau, 1975).

A second series of ARA measurements, used to build the model of diurnal variation, were taken every 3 h without interruption during 5 days at the beginning of July 1973. Simultaneously the following parameters were monitored:

AT: air temperature measured under shelter.

AH: air humidity measured under shelter.

LE: amount of light energy received by the plant measured with a thermolinear pyranometer, expressed as J cm^{-2}.

UST: temperature of the upper soil horizon (5 cm in depth).

IST: temperature of the intermediate soil horizon (18 cm in depth).

LST: temperature of the lower soil horizon (30 cm in depth).

SW: soil water content expressed on a wet weight basis .

LW: dry leaf weight expressed in g per plant.

A third series of ARA estimations were made to build a model of variation in ARA throughout the whole growth cycle of maize. ARA was estimated every 2 h, one day per week, during the maize growth cycle of 1975.

Non planted controls showed no significant ARA, thus indicating that ARA originated from the native diazotrophs associated to the maize rhizosphere (Fig. 1).

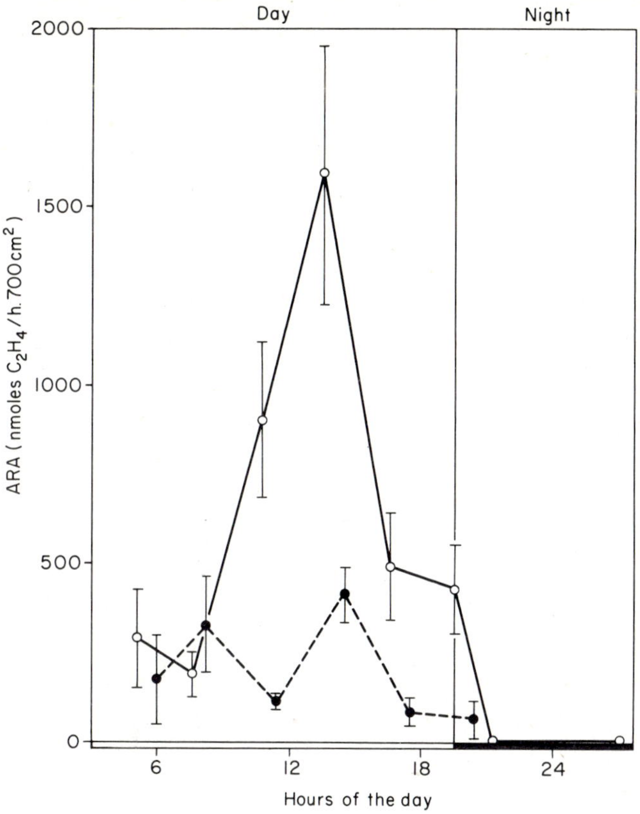

Fig. 1. *Diurnal variations in acetylene-reducing activity (ARA) in the rhizosphere of maize (full line) and in non planted soil (dotted line); measurement performed at flowering stage. Limits show standard error of means.*

MODELLING OF DIURNAL VARIATIONS OF ARA

Fig. 1 is an example of nycthemeral pattern of ARA as observed in the maize field. It shows one mid-day peak at 15 h, very similar to that already reported in the case of rice (Balandreau *et al.*, 1976). A midnight peak also occurred, but only when night temperature was below 18 to 20°C. The occurrence of a midnight peak of ARA was reported by Döbereiner and Day (1974) when temperatures were high enough (20°C) to allow hydrolysis of starch reserves and further translocation of solubilized compounds (West, 1971) towards the sites of utilization by diazotrophs.

Diurnal variations in ARA were significantly and positively correlated (at the .05 probability level) with 4 out of the 7

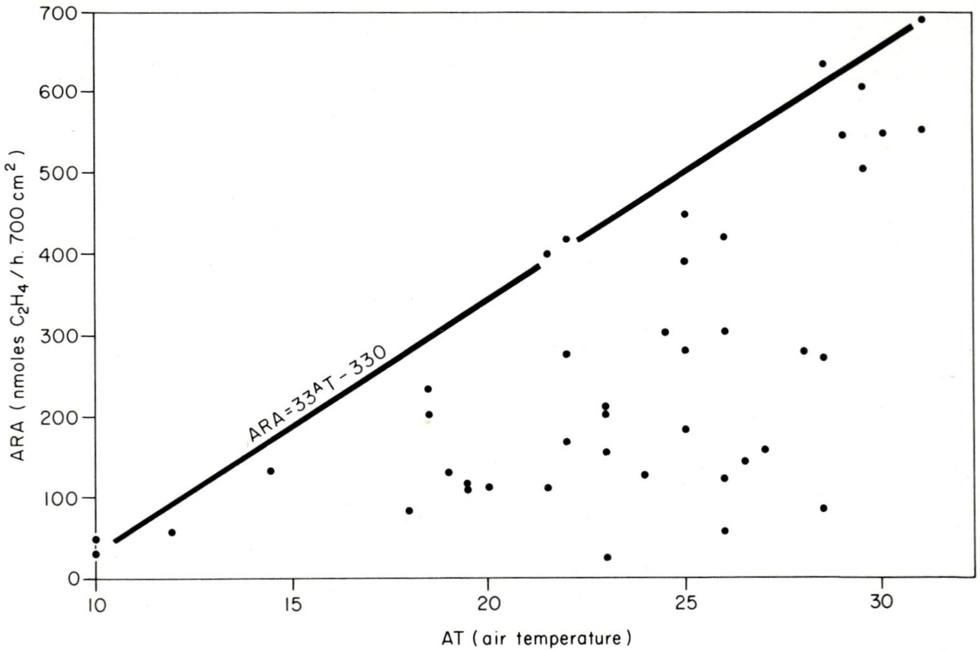

Fig. 2. *Acetylene reducing activity (ARA) plotted against air temperature (AT) measured in July 1973 (investigations on nycthemeral variations).*

environmental parameters monitored, i.e. AT, LE, UST, IST.

When ARA was plotted against one of the environmental parameters, corresponding points appeared to be located between the *x* axis and a straight line, considered the envelope of the points (Fig. 2). This straight line was obtained either graphically or by calculation. The graph was then used to calculate the equation of the straight line:

$$ARA = aAT + b$$

Points on the straight line were related to situations when the parameter studied - e.g. AT in Fig. 2 - was the limiting factor; points below the straight line were related to situations when other parameters were limiting. When IST was lower than 20°C, ARA was nil.

A set of 4 equations was thus obtained:

(I) $ARA = 33\ AT - 330$
(II) $ARA = 40\ UST - 694$
(III) $ARA = 1{,}95\ LE + 140$
(IV) $ARA = 0$, if IST 20.

Resolving equations I, II, III, IV, gave 4 ARA values; the lowest value had to be retained since ARA actually depended on the "most" limiting factor.

This model appeared to describe 70% of ARA diurnal variations, which is not perfect but is an encouraging result.

MODELLING OF VARIATION IN ARA THROUGHOUT THE PLANT GROWTH CYCLE

Calculations by computer are not yet completed. However, preliminary interpretations showed that soil water content (SW) played a major role. (1) When SW was higher than 18%, ARA was maximum if other environmental parameters were not limiting; such values were designed as potential maximum ARA (PMARA). PMARA could be related to LW by a simple equation:

$$PMARA = k.LW$$

k being a constant. (2) When SW was lower than 18%, SW appeared to be a limiting factor. ARA measured at the mid-day peak (MDARA) appeared to be related to SW by the following equation:

$$MDARA = PMARA\ (mSW - p)$$

m and p being constants.

MDARA/PMARA was correlated to SW at the .01 probability level. Actual values of constant k, m and p will be published later.

DISCUSSION

ARA of the maize-rhizosphere diazotroph system was shown to depend primarily upon the stage of growth of the plant. Mid-day ARA (MDARA) was proportional to the leaf weight (LW), in the absence of limitations imposed by other environmental factors.

The influence of environmental parameters on ARA was also significant, but more difficult to elucidate because of interactions. Light energy (LE) was a major factor which was responsible for rapid ARA variations, together with air temperature (AT) and soil temperature (UST, IST). Such variations often resulted in large diurnal variations of ARA. Soil water content varied more slowly so that its effect could be shown clearly only when measurements were made over periods of time greater than 24 h.

Abrantes *et al.* (1975) found that the main factors limiting ARA were air temperature and NH_4^+ content of the soil solution, but that soil water content (SW) was limiting only when the wilting point was reached. Our results confirm the effects of temperature and soil water content reported by Abrantes *et al.*; like these authors, we detected a threshold under which soil water content was limiting.

Surprisingly, in preliminary experiments, no effect of soil combined nitrogen could be detected, possibly because of the spacial heterogeneity of this soil parameter.

The two models of ARA variations related to two time scales (daily and the whole growth cycle) proposed are obviously still in the preliminary stage and require improvements. We are currently trying to develop a unique model integrating the results of the separate preliminary models presented here. Generalisation of such an integrated model to other types of environments and other plant-diazotroph systems should increase the accuracy and precision of evaluations of ARA *in situ*.

Bowen and Rovira (1973) advocated the use of the modelling approach in rhizosphere studies. The preliminary results presented here support the value of such an approach which is a prerequisite for the development of current investigations (e.g. Bülow and Döbereiner, 1975) aimed at the improvement of the efficiency of plant-rhizosphere diazotroph systems.

REFERENCES

Abrantes, G.T.V., Day, J.M., Cruiz, V.F. and Döbereiner, J. (1975). Proc.XVth Congreso Brasileiro de Ciencia do Solo. Campinas, Brazil, 14-21 July 1975.

Balandreau, J.P. (1975). *Rev.Ecol.Biol.Sol.*, 12(1): 273-290.

Balandreau, J.P. and Dommergues, Y.R. (1973). *Bull.Ecol.Res. Comm.(Stockholm)*, 17: 247-254.

Balandreau, J.P., Millier, C.R. and Dommergues, Y.R. (1976). *Appl.Microbiol.*, 27(4): 662-665.
Bond, G., and Mackintosh, A. (1975). *Proc.R.Soc.Lond., B,* 192: 1-12.
Bowen, G.D. and Rovira, A.D. (1973). *Bull.Ecol.Res.Comm.(Stockholm),* 17: 443-450.
Bülow, J.F.W. Von and Döbereiner, J. (1975). *Proc.Nat.Acad.Sci. (U.S.A.),* 72: 2389-2393.
Döbereiner, J. and Day, J.M. (1974). *In:* Proceedings of the Int. Symposium on Nitrogen Fixation. Pullman, in press.
Hardy, R.W.F., Burns, R.C. and Holsten, R.D. (1973). *Soil Biol. Biochem.,* 5: 47-81.
Hardy, R.W.F., Holsten, R.D., Jackson, E.K. and Burns, R.C. (1968). *Plant Physiol.,* 43: 1185-1207.
Mague, T.H. and Burris, R.H. (1972). *New Phytol.,* 71: 275-286.
West, S.H. (1971). Proc.XIth Int.Grassl.Cong., 514-517.
Wheeler, C.T. (1971). *New Phytol.,* 70: 487-495.

SOME REFLECTIONS ON *ALNUS*-TYPE ROOT NODULES

G. BOND

Department of Botany, University of Glasgow, England

THE OCCURRENCE OF *ALNUS*-TYPE NODULES

So far as Western literature is concerned, reports of the presence of root nodules on non-legumes began to appear during the 19th century, and by its end there were records of nodulation in a total of about 12 species belonging to the genera *Alnus*, *Hippophaë*, *Myrica*, *Elaeagnus*, *Ceanothus* and *Casuarina*. Subsequently nodulation was recorded in further species of those genera and also in the additional genera *Coriaria*, *Discaria*, *Purshia*, *Dryas*, *Cercocarpus*, *Shepherdia* and *Arctostaphylos*, and by 1966, when a collaborative survey of nodulation in non-legumes was begun as part of the IBP, there was evidence of a capacity for nodulation in a total of about 120 species belonging to the above 13 genera. As a result of that survey a new nodulating genus, *Colletia*, was discovered, while the grand total of known nodulating species was raised to 160 (Bond, 1976a). In addition information on the regularity of nodulation under field conditions was obtained. Thus the examination of about 1,000 trees of *Alnus glutinosa* growing at various European sites revealed only 14 trees apparently devoid of nodules. The scrutiny of some hundreds of plants of *Myrica gale* in Europe and Canada showed that while the majority were nodulated, an apparent lack of nodules was reported more often than in *Alnus glutinosa*, and was perhaps prone to occur particularly in plants growing in relatively dry situations. A large-scale examination of species of *Casuarina* in Australia, Indonesia, Africa and elsewhere indicated that while some species, e.g. *C. cunninghamiana*, are regularly nodulated, this may not be true of other species, though there is the possibility that nodules are present at deeper, moister levels in the soil which can scarcely be reached by hand excavation.

Already it is possible to say that most if not all countries have some of these plants in their native floras. Nine of the genera are represented in the U.S.A., with a total of upwards of 50 species. However it is extremely likely that further examples of nodulating non-legumes, both at the specific and generic levels, await discovery, particularly in the less-botanised regions, though a complete picture of the extent of non-leguminous nodulation will not be attained until circumstances permit liaison between botanists and microbiologists of all countries.

NODULE AND ENDOPHYTE MORPHOLOGY

The method of origin of the nodule in the mother root (Quispel, 1974) and the anatomy of a lobe of a mature nodule suggest that we are dealing with a structure of root nature, not drastically modified. However, an unfailing feature of these non-legume nodules is the repeated branching of the lobes, resulting in the eventual production of a large nodule cluster. The external inspection of this branching process suggests that it might be due to a dichotomy or trichotomy of the meristem at the tip of the lobe, and indeed McLuckie (1923) figured sections of *Casuarina* nodules which appeared to support that view. Such a process would represent a very drastic modification of the normal method of branching in a root. However, Torrey (1976) has re-examined nodule branching in *Casuarina* and has provided good evidence that despite external appearances the branches in fact arise endogenously. Possibly this is true for other genera also. In passing it may be noted that the meristem cells of non-legume nodules are diploid, as shown by Kodama (1970) for three Japanese species of *Alnus* and by Dr. S. Mian (personal communication) in the author's laboratory for *Alnus glutinosa*, *Myrica gale*, *M. faya*, *M. cerifera*, *Hippophaë rhamnoides* and *Casuarina cunninghamiana*. In all these instances the chromosome number in the nodule tips was the same as in the root tips. This contrasts with the tetraploidy present in the meristems of many legume nodules.

Attempts to elucidate the morphology and taxonomic nature of the endophyte present in cortical cells of *Alnus*-type nodules began in the latter half of the 19th century, when several workers studied sections of nodules under the microscope, and despite the limitations of their optical equipment some of them came fairly close to what we now know to be the truth on this score. Thus Brunchorst (1886-8) recognised that the alder endophyte consisted of fine, septate hyphae bearing terminal vesicles, and though he erred in classing it as a lowly-organised fungus it must be recalled that the existence of finely-hyphal bacteria (*Actinomyces*) had been realised only recently and was possibly unknown to Brunchorst. Shibata (1902) likewise erred in classing the nodules of *Alnus incana* and *Myrica rubra* as examples of mycorrhiza, but he too realised that the endophytes were not typical fungi, and in fact mentioned their resemblances to actinomycetes; incidentally he observed the sub-divisions within the alder vesicles and thus anticipated findings made over 50 years later under the electron microscope. Shibata and Tahara (1917) described the distinctive structural features shown by *Coriaria* nodules, and again assigned the endophyte to the actinomycetes. (In passing it may be noted that of the various *Alnus*-type nodules it is the *Coriaria* nodule that the *Rhizobium*-containing nodules of *Trema* (Trinick and Galbraith, 1976) most closely resemble in gross structure.) In later years Schaede provided papers on endophyte structure in the nodules of *Alnus glutinosa* and *Myrica gale*, the salient points being brought together in his well-known book (Schaede, 1962); he was firmly convinced of the actinomycete-like nature of the endophytes. A discordant note was struck by Hawker and Fraymouth (1951) in reporting their inability to see any definite hyphae in preparations from the nodules of several non-legumes including *Alnus*, and their conclusion that the endophytes are plasmodial organisms.

The advent of the electron microscope greatly facilitated the elucidation of the congested contents of the infected cells of these non-legume nodules (Becking *et al.*, 1964; Gardner, 1976), and finally confirmed the finely-hyphal nature of the endophytes. Although they have sometimes been assigned to the Actinomycetales, the author suggests that the endophytes do not necessarily belong to any of the presently recognised major groups of micro-organisms, since these were formulated without regard to non-legume endophytes.

EVIDENCE FOR NITROGEN FIXATION IN THE NODULES

Experiments similar to those performed a few years earlier with legumes by Hellriegel and Wilfarth, involving a comparison of the ability of nodulated and non-nodulated plants to grow in a nitrogen-free rooting medium, were carried out by Hiltner and his associates over the period 1896 to 1904 with *Alnus glutinosa*, *Elaeagnus angustifolia* and *Shepherdia canadensis*. Though no nitrogen estimations were included, the vigour of growth of the nodulated plants left no doubt that fixation was associated with them, and by inference occurred in the nodules. Over subsequent years similar tests - now including nitrogen analyses - have been made with other nodulating non-legumes, and a recent count (Bond, 1974) showed that a total of 38 species, including 10 of the genera listed earlier, had been so examined, all with positive results, i.e. the nodulated plants, but not those lacking nodules, grew well and accumulated nitrogen. To that total can now be added a species of *Colletia* and two further species of *Myrica* (Bond, 1976a), also a species of *Purshia*

(Bond, 1976b), leaving only *Arctostaphylos* and *Cercocarpus* so far untested by this method, though in respect of the latter Vlamis *et al.* (1964) found that in a nitrogen-deficient soil plant growth was proportional to the extent of nodulation.

The introduction of the ^{15}N technique made it possible to secure direct evidence that the fixation associated with nodulated plants is in fact confined to the nodules. The technique has also been used to test nodules brought in from the field for fixation properties, and in this way evidence of fixation in the nodules of a further 14 species has been secured, as reported by Bond (1974). Hoeppel and Wollum (1971) reported that *Cercocarpus* nodules showed acetylene-reduction. In total there is evidence of fixation in the nodules of 56 species of non-legumes.

Schubert and Evans (1976) have reported that hydrogen evolution during fixation in the nodules of several non-legumes was much lower than in the nodules of most legumes, and concluded that the fixation mechanism operates more efficiently in the non-legumes.

Under greenhouse conditions nodules which fix little or no nitrogen have been obtained in cross-inoculation trials. Thus such nodules are formed when an inoculum containing the endophyte of *Myrica gale* nodules is applied to the roots of seedlings of other species of *Myrica*, and a recent microscopic study (Mian *et al.*, 1976) of these ineffective nodules showed that within them the endophyte fails to form the vesicles which it produces in abundance when associated with its normal host plant.

EFFECT OF COMBINED NITROGEN ON THE SYMBIOSIS

It is well-known that the legume symbiosis functions most effectively in rooting media low in combined nitrogen. That the same is true of non-legumes is indicated in Table I, showing data from a recent paper (Bond and Mackintosh, 1975) which included a literature review. Though the production of whole plant dry matter and the accumulation of nitrogen increased in the presence of nitrate, especially in *Hippophaë*, the production of nodule dry mass and the amount of nitrogen fixed fell in a roughly parallel manner. However it is probably true that at least the higher of the two levels of nitrate-nitrogen supplied in the experiment exceeds that likely to be present in the soil where these species normally grow.

The isotopic analyses showed that in the plus-nitrate plants of both species the label of the nodule-nitrogen was about 1/5th of that in the rest of the plant, which can be explained if the endophyte contains only unlabelled nitrogen obtained by fixation. Such a situation could arise if the membrane which separates the

Table I

Effect of nitrate-nitrogen on growth and fixation; mean results per plant expressed on relative basis[a].

Mg nitrate-N added per litre of culture solution	Increase in nodule dry mass	Increase in dry mass of whole plant	Increase in N content of whole plant	Calculated N fixed during experiment
		Coriaria arborea		
0	100	100	100	100
10	60	99	117	76
25	22	116	135	31
		Hippophaë rhamnoides		
0	100	100	100	100
10	83	142	137	90
25	57	183	176	61

[a]All the plants had been grown in nitrogen-free Peralite for a year, and were then supplied with different levels of sodium nitrate labelled with ^{15}N and grown for a further 10 weeks. The values shown in the second, third and fourth columns from the left are relative increases observed over that period, based on 10 plants per treatment. Table modified from Bond and Mackintosh (1975).

endophyte from the host cell cytoplasm and which represents an invagination of the plasma membrane of the host cell (Lalonde and Devoe, 1976) does not allow the passage of combined nitrogen to the endophyte. Another possibility is that the host plant in some way inhibits the utilisation of fixed nitrogen by the endophyte, as has been suggested by Millbank (1974) in connection with the lichen symbiosis.

POSSIBILITIES FOR EXPLOITATION OF THE FIXATION

Although the fixation associated with these non-legumes is of undoubted ecological importance, as was demonstrated, for example, by Lawrence and his associates in respect of the Glacier Bay region (Crocker and Major, 1955; Lawrence *et al.*, 1967), in the present world economic situation further research on these plants will be better justified if methods for exploitation of the fixation can be devised.

Unlike the legumes, the nodulating non-legumes so far discovered do not of themselves provide food for humans nor, mostly, for animals, although species of *Purshia* and *Cercocarpus* are important browse plants in N. America. There are reports that the planting of shrubby alders to restore worn-out soils has long been practised by farmers in some Asian countries, but such a rotation must be slow in operation and unsuited to densely populated regions. Forestry, which is by nature a slow-moving industry, seems to offer better prospects for exploitation. Tree-forming nodulating species such as alders, casuarinas and myricas could be grown on an increased scale as sources of timber, fuel, wood pulp and other products, without any use of fertiliser nitrogen being necessary. There is also clear evidence from the northern hemisphere that alders planted on infertile soils can in the space of a few years raise the fertility to a level adequate for the establishment of conifers (Tarrant and Trappe, 1971). It is encouraging to know that there is a growing interest in these various possibilities for exploitation. The author acknowledges useful discussion with Professor John C. Gordon on this matter.

REFERENCES

Becking, J.H., De Boer, W.E. and Houwink, A.L. (1964). *Antonie van Leeuwenhoek*, 30: 343-376.

Bond, G. (1974). *In:* The Biology of Nitrogen Fixation (A. Quispel, ed.), pp. 342-378. North-Holland Publishing Company, Amsterdam.

Bond, G. (1976a). *In:* Symbiotic Nitrogen Fixation in Plants (P. S. Nutman, ed.), pp. 443-474. Cambridge University Press, Cambridge.

Bond, G. (1976b). *Proc.R.Soc.Lond.B.*, 193: 127-135.
Bond, G. and Mackintosh, A.H. (1975). *Proc.R.Soc.Lond.B.*, 190, 199-209.
Brunchorst, J. (1886-8). *Unters.Bot.Inst.Tübingen*, 2: 150-177.
Crocker, R.L. and Major, J. (1955). *J.Ecol.*, 43: 427-448.
Gardner, I.C. (1976). *In:* Symbiotic Nitrogen Fixation in Plants (P.S. Nutman, ed.), pp. 485-496. Cambridge University Press, Cambridge.
Hawker, L.E. and Fraymouth, J. (1951). *J.gen.Microbiol.*, 5: 369-386.
Hoeppel, R.E. and Wollum, A.G. (1971). *Can.J.Bot.*, 49, 1315-18.
Kodama, A. (1970). *J.Sci.Hiroshima Univ.*, B2 13: 261-264.
Lalonde, M. and Devoe, I.W. (1976). *Physiol.Pl.Path.*, 8:123-129.
Lawrence, D.B., Schoenike, R.E., Quispel, A. and Bond, G. (1967). *J.Ecol.*, 55: 793-813.
McLuckie, J. (1923). *Proc.Linn.Soc.N.S.W.*, 48: 194-205.
Mian, S., Bond, G. and Rodriguez-Barrueco, C. (1976). *Proc.R.Soc.Lond.B.* , 194: 285-293.
Millbank, J.W. (1974). *In:* The Biology of Nitrogen Fixation (A. Quispel, ed.), pp. 238-264. North-Holland Publishing Company, Amsterdam.
Quispel, A. (1974). *In:* The Biology of Nitrogen Fixation (A. Quispel, ed.), pp. 497-520. North-Holland Publishing Company, Amsterdam.
Schaede, R. (1962). Die Pflanzlichen Symbiosen. 3rd ed. Fischer, Stuttgart.
Schubert, K.R. and Evans, H.J. (1976). *Proc.Natn.Acad.Sci.USA*, 73: 1207-1211.
Shibata, K. (1902). *Jb.wiss.Bot.*, 37: 643-684.
Shibata, K. and Tahara, M. (1917). *Bot.Mag., Tokyo*, 31: 157-182.
Tarrant, R.F. and Trappe, J.M. (1971). *Pl.Soil*, Special Volume, pp. 335-348.
Torrey, J.G. (1976). *Am.J.Bot.*, 63: 335-344.
Trinick, M.J. and Calbraith, J. (1976). *Arch.Microbiol.*, 108: 159-166.
Vlamis, J., Schultz, A.M. and Biswell, H.H. (1964). *J.Range Mgmt.*, 17: 73-74.

CYTOKININS AND NODULATION OF THE NON-LEGUMES *ALNUS GLUTINOSA* AND *MYRICA GALE*

F. BERMUDEZ DE CASTRO, A. CANIZO, A. COSTA, C. MIGUEL AND C. RODRIGUEZ-BARRUECO

Laboratory of Soil Microbiology, Centro de Edafologia y Biologia Aplicada, C.S.I.C. Salamanca, Spain

INTRODUCTION

It is of particular interest from the standpoint of soil fertility that so much of the leaf nitrogen in alder (*Alnus glutinosa* (L.) Gaertn.) should still remain in the leaf at leaf-fall, as found by Rodriguez-Barrueco (1968). He also noticed that, at leaf-fall time, the leaves from *A. glutinosa* still retained their dark green colour. Cytokinins, plant growth hormones, retard chlorophyll degradation when present at certain levels in a plant tissue and therefore might be actually involved in the above phenomena. If present, a chance exists that cytokinins are produced at cytokinin-producing centres of the alder, such as meristematic regions mainly in roots or nodules. That possibility leads us to study the role of cytokinins in the nodulation process of the non-leguminous nitrogen-fixing genera *Alnus* and *Myrica*.

As a first approach, it is of interest to know up to what extent, exogenous cytokinins, mainly kinetin and 2-isopentenyl adenine (2IP), may affect growth and development of plants of *A. glutinosa* and *M. gale* L. when supplied to their rooting medium. In preliminary experiments cytokinins induced pseudonodules (defined below) on the roots of the alder (Rodriguez-Barrueco and Bermudez de Castro, 1973) but plant growth data were not provided. Therefore, the present work extends research on the actual effect of exogenous cytokinins on the above plants. In the present work the word pseudonodule means an abnormal root outgrowth free from a microbial endophyte.

MATERIALS AND METHODS

Seed of *A. glutinosa* and *M. gale* was collected locally and surface sterilized with hypochlorite as described elsewhere (Rodriguez-Barrueco, Ph.D. thesis, Univ. Glasgow, 1967). It was then sown in vermiculite moistened with Crone solution to pH 5.4 comprising the following ingredients in g/l: KNO_3, 1.0; $CaSO_4.2H_2O$, 0.5; $MgSO_4.7H_2O$, 0.5; $Ca_3(PO_4)_2$, 0.25; $Fe_3(PO_4)_2.8H_2O$, 0.25. At the two-leaf stage plants were transplanted singly to each of 35 test tubes, 3 x 19 cm, containing 100 ml of Crone solution at 1/4 strength which had been allowed to sediment before use. Ten days after transplanting, kinetin (Calbiochem) or 2IP (Sigma) were added to quintuplicate groups of plants to 0.12, 0.6 and 1.2 μg/ml. The kinetin and 2IP solutions were in distilled water with a small amount of HCl to increase solubility. The remaining five plants served as untreated controls. Nutrient solutions were renewed every ten days during incubation at room temperature illuminated with principally blue and red light from a Philips HLRG 400 lamp. After 49 days, plants were harvested and data obtained on shoot length, root length, pseudonodule count and plant dry weight.

Once pseudonodules became discernible, sections of roots bearing these structures were evacuated and fixed in FAA (formic acid, acetic acid, ethanol, 3:7:90). The material was dehydrated and then embedded in paraffin (The British Drug Houses Ltd., England). Sections of 6-8 μm thick of the root outgrowths were made with a JUNG microtome. Photographs were taken under a Zeiss phase contrast microscope. The nodules were dissected to see not only the structure of the outgrowths formed, but to make sure that no endophyte was present within the tissue, as aseptic cultural conditions were not used.

RESULTS

Four weeks after transplanting into water culture, small pseudonodules were observed scattered over the entire root systems of the plants treated with 0.6 and 1.2 μg/ml of the two cytokinins, being particularly abundant at the latter level (Fig. 1). At seven weeks from transplanting all the plants treated with cytokinins at any of the three levels tested showed pseudonodular outgrowths. Pseudonodule formation did not occur in the cytokinin-free cultures. The pseudonodules formed had their origin in the regions of active cell division of the parent root cortex (Fig. 2). These cells showed continuous activity giving a mass of undifferentiated cells. Some parts of the large outgrowths showed groups of tracheids giving the aspect of unorganized vascular bundles (Fig. 3). In some instances a spiraloid arrangement of tracheids and parenchymatic cells was observed (Fig. 4). The nodules formed are simple in structure and their inner disarrangement differs from true microorganism-inhabited nodules (Rodriguez-Barrueco and Bermudez de Castro, 1973).

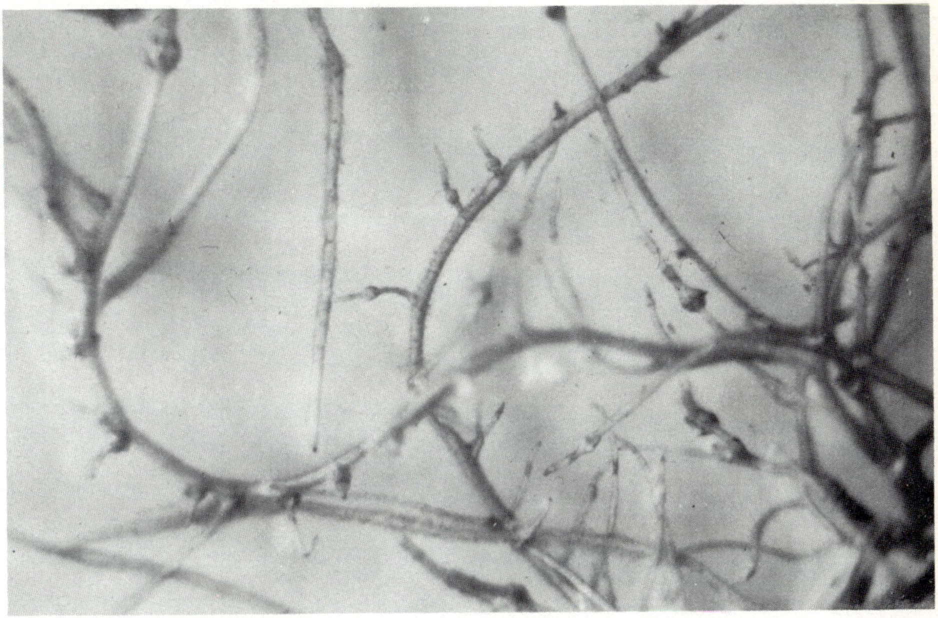

Fig. 1. *View of a pseudonodulated root system of* A. glutinosa *(x1).*

A similar course of events occurred for both *A. glutinosa* and *M. gale*. It should be added that no nodule-roots were formed on the outgrowths formed on *M. gale*, which are normally present in true root-nodules of that plant. At harvest the pseudonodules of both plants were counted and data on length of shoots and roots and plant dry weight taken (Table I and Table II). From those data and their statistical treatment (Fig. 5) it can be seen that the number of pseudonodules attained its maximum at the cytokinin level of 0.6 μg/ml. In all instances the presence of cytokinin in the medium was accompanied by a reduction in height of the plants and length of the roots, as compared with the controls. That reduction gradually increased with the concentration of the cytokinin supplied and became more drastic when kinetin was used. In fact the addition of only 0.12 μg/ml of either cytokinin tested reduced plant growth remarkably. No significant differences were found in plant dry

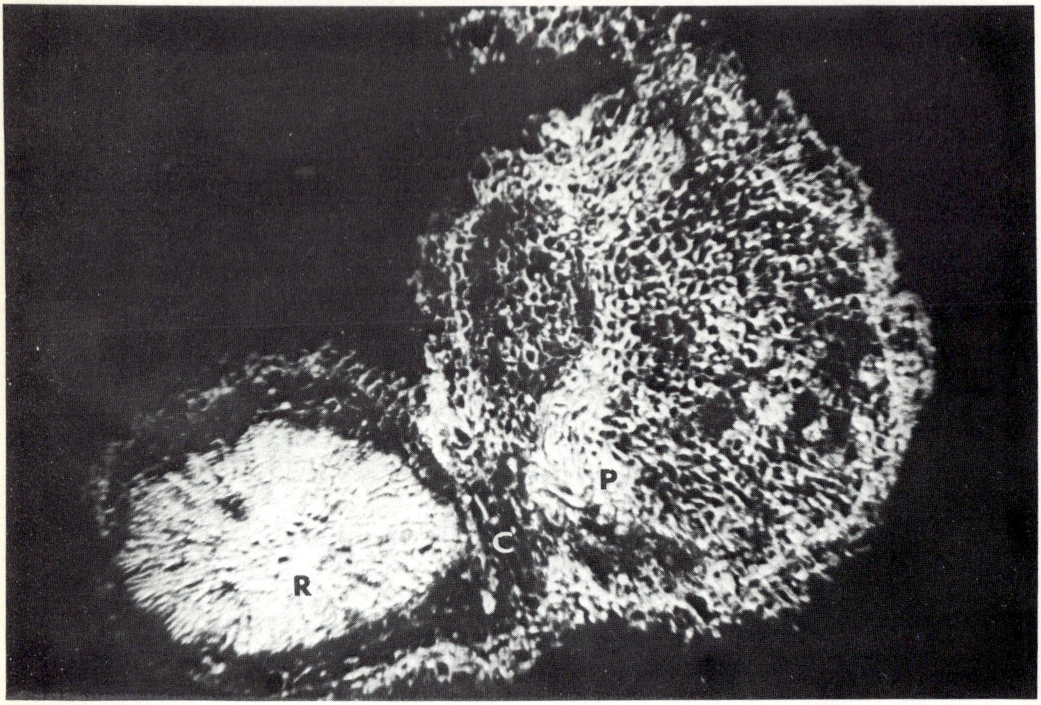

Fig. 2. A. *Cross section of root and attached pseudonodule of* A. glutinosa. *r: parent root; c: cortex; p: pseudonodule. (x40).*

Table I

Mean data of five plants of *A. glutinosa* grown in Crone solution added with kinetin and 2IP respectively. Age of the plants at harvest: seven weeks

Treatment	Dosis μg/ml	No. of pseudo-nodules	Length, cm		Total plant dry weight, mg
			roots	shoots	
Kinetin	0.12	45	9.7	6.0	114.2
	0.60	78	3.6	3.4	38.8
	1.20	39	4.3	3.0	31.7
2IP	0.12	24	8.2	7.2	142.1
	0.60	56	4.5	3.1	48.7
	1.20	53	4.2	3.0	39.7
Control	-	0	15.9	8.8	189.1

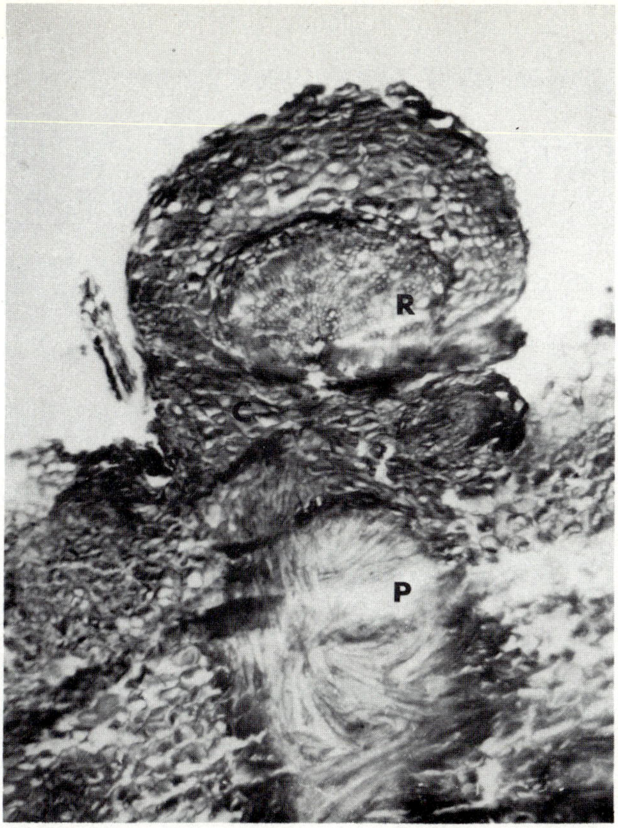

Fig. 2.B. *Cross section of root and pseudonodule of* M. gale. *r: parent root; c: cortex; p: pseudonodule (x25).*

Table II

Mean data of five plants of *M. gale* grown in Crone solution added with kinetin and 2IP respectively. Age of the plants at harvest: seven weeks

Treatment	Dosis µg/ml	No. of pseudo-nodules	Length, cm roots	Length, cm shoots	Total plant dry weight, mg
Kinetin	0.12	13	9.9	5.9	59.7
	0.60	34	6.1	5.1	49.3
	1.20	33	5.8	4.9	36.5
2IP	0.12	10	7.8	11.6	100.6
	0.60	33	5.8	4.9	56.9
	1.20	29	5.2	6.7	50.8
Control	-	0	16.7	17.4	305.3

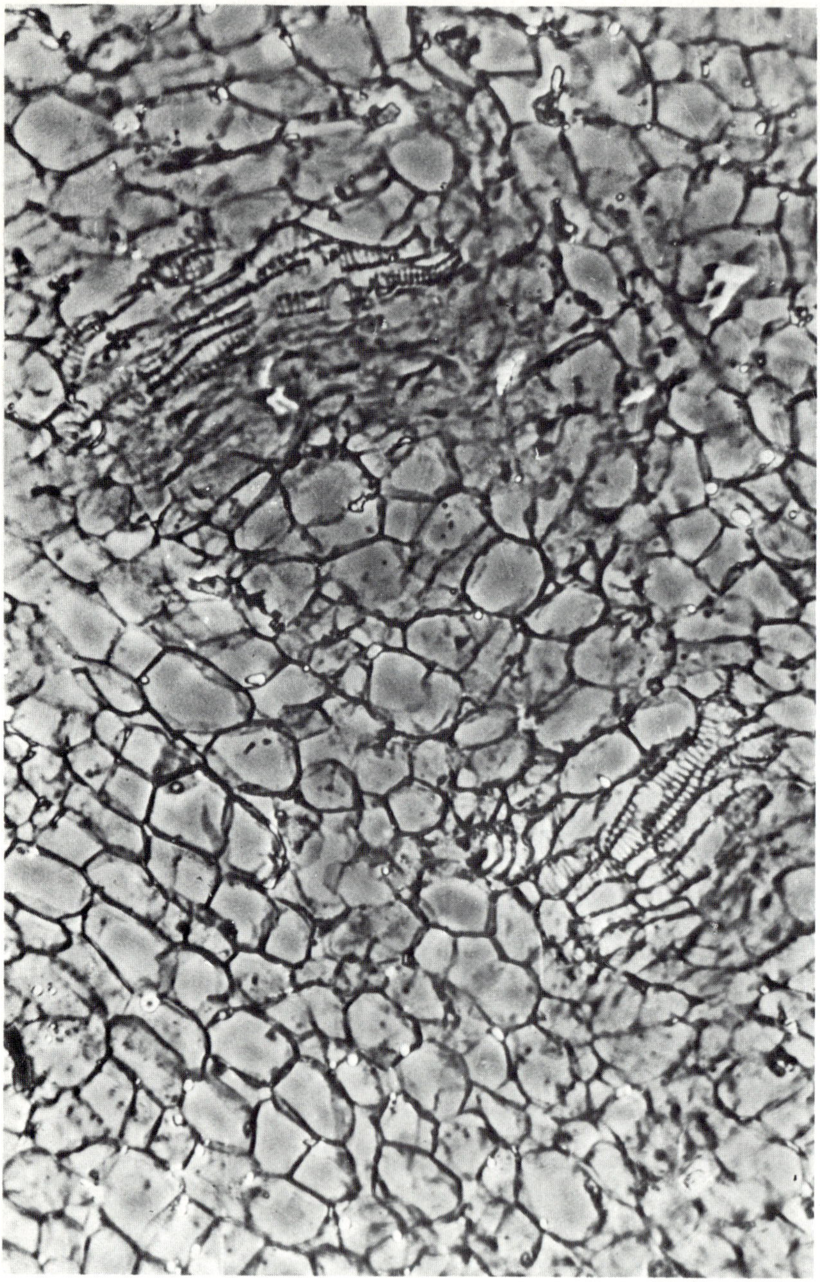

Fig. 3. *Section of a pseudonodule of* A. glutinosa *showing parenchymatic cells and unorganized vascular bundles.*

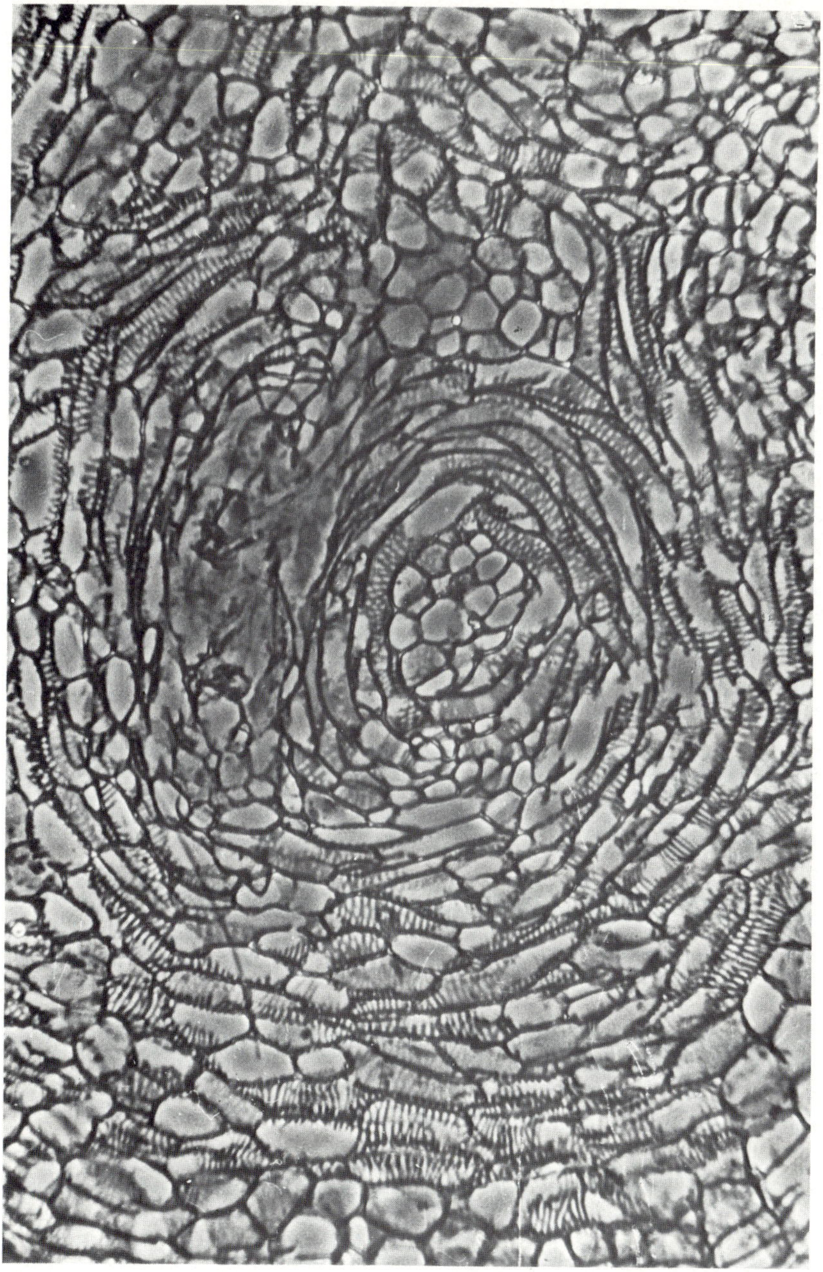

Fig. 4. *Spiraloid arrangement of tracheids in a pseudonodule of* A. glutinosa.

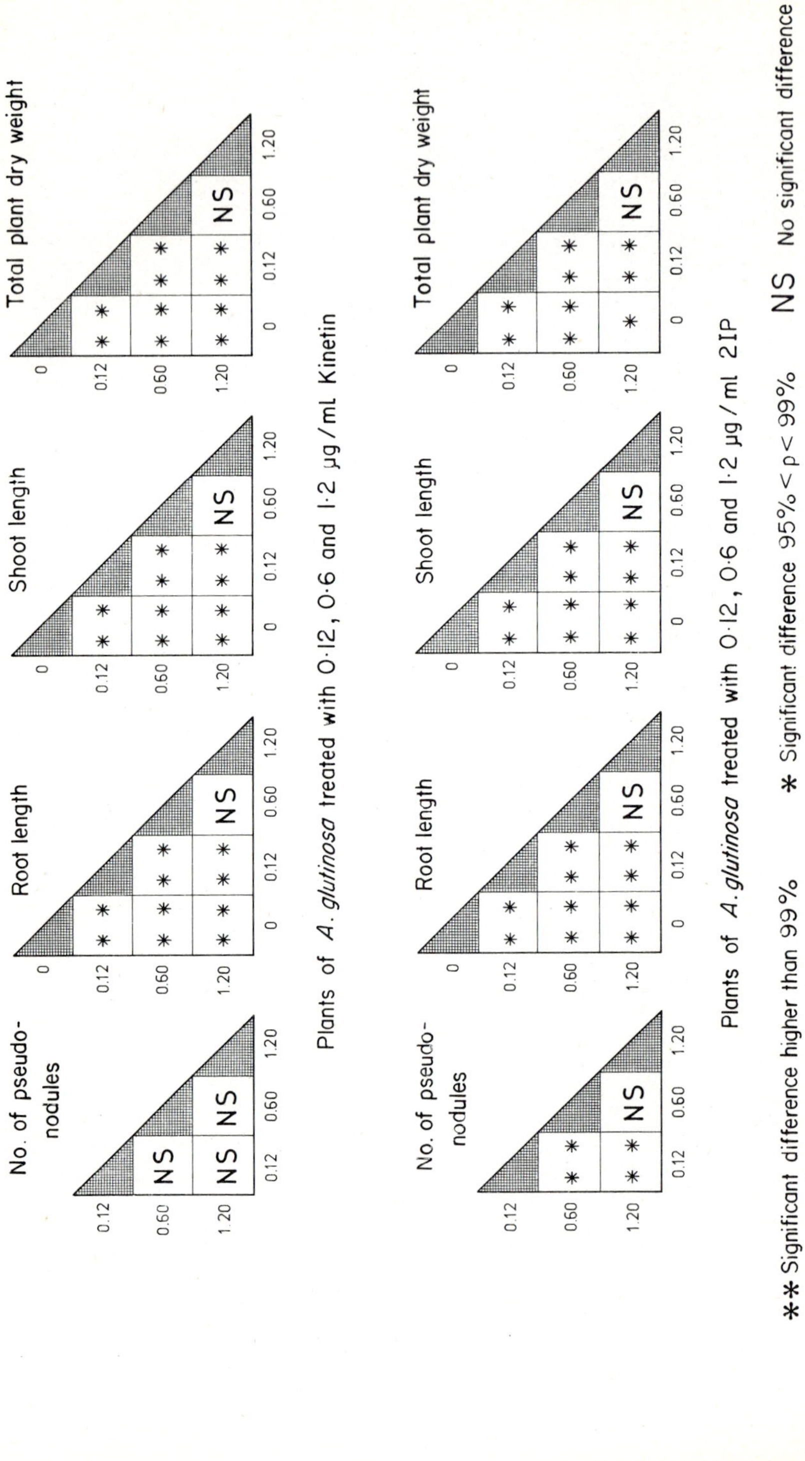
No. of pseudo-nodules
Root length
Shoot length
Total plant dry weight
0
0.12
0.60
1.20
NS
Plants of A. glutinosa treated with 0·12, 0·6 and 1·2 µg/ml Kinetin
Plants of A. glutinosa treated with 0·12, 0·6 and 1·2 µg/ml 2IP
** Significant difference higher than 99%
* Significant difference 95% < p < 99%
NS No significant difference

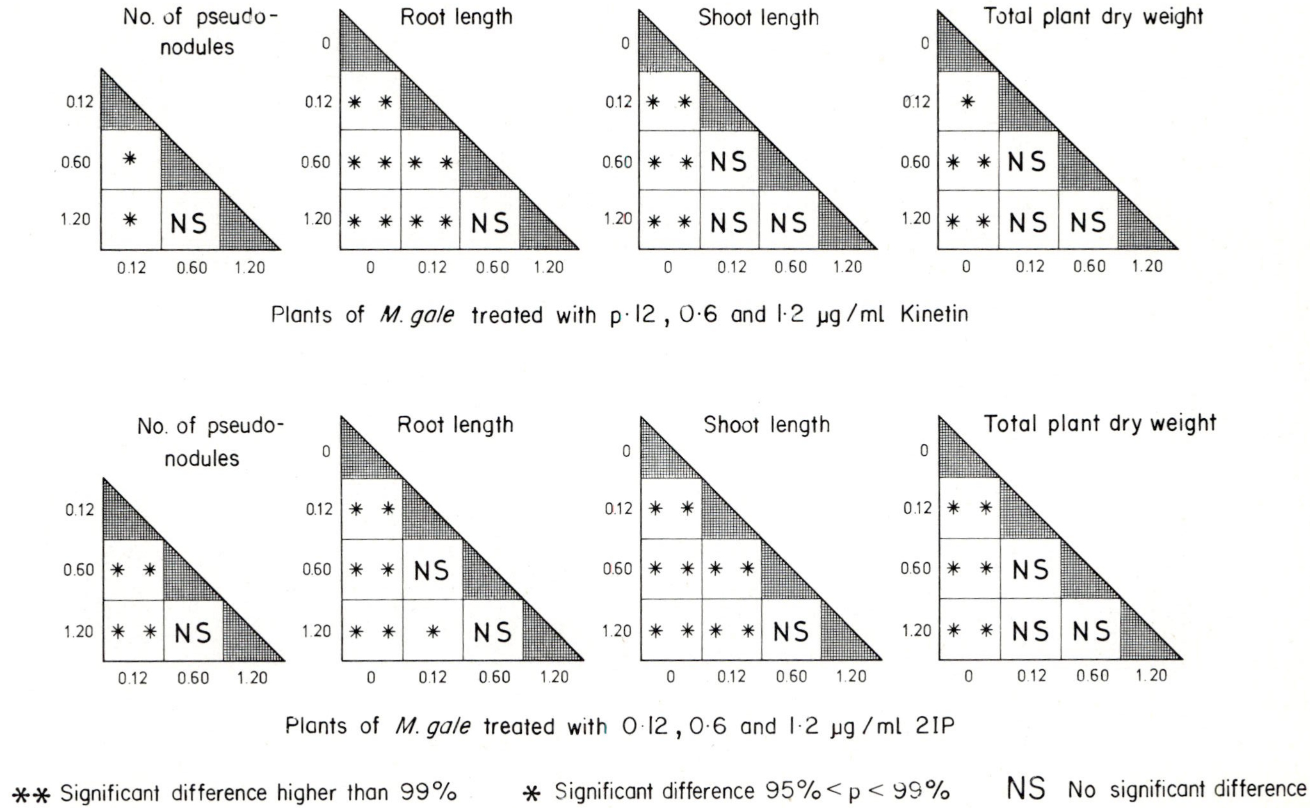

Fig. 5. *Statistical treatment of growth and pseudonodulation data of plants of* A. glutinosa *and* M. gale.

weight and length of shoots and roots between cytokinin levels 0.6 and 1.2. No significant differences occurred between treatments as regards the number of pseudonodules formed on alder when kinetin was used. In the case of *M. gale* some significant difference was found between kinetin level 0.12 and those of 0.6 and 1.2 respectively.

Tables I and II also show that the two cytokinins had opposite effects on root and shoot growth: shoot growth was more drastically affected by kinetin, whereas 2IP showed a higher inhibition of root growth.

Some extra plants bearing pseudonodules were kept growing for a further one month in a cytokinin-free medium. The outgrowths on those plants were reabsorbed by the plants themselves and eventually disappeared. Those plants resumed normal growth, producing normal long roots and shoots, as soon as cytokinin supply was cut.

DISCUSSION

The present results suggest that cytokinins may be involved in nodule formation and development. Tissue differentiation in normal endophyte-inhabited nodules may proceed gradually as more cytokinin is made available by one or both partners of the particular symbiotic system. In our experiments the addition of non-physiological levels of cytokinin to the rooting medium no doubt lead to tumor induction, unorganized tissue controlled by those exogenous levels of cytokinins which were thus supplied in supra-optimal amounts.

The presence of endogenous cytokinin activity in the root nodules of the alder and of *M. gale* has been detected recently by the present authors (Bermudez de Castro and Rodriguez-Barrueco, 1976). That finding and the one reported in the present work point to the need to investigate the translocation of those substances from root nodules to the aerial parts of the plant. The maintenance of chlorophyll in the leaves even after leaf abscission may be related to the presence of those substances in the leaves.

The root growth inhibition shown by the added cytokinins is an effect attributed to that type of substances in respect of many other plants. Miller *et al.* (1955), and Harris and Hart (1964), among others, found that exogenous application of cytokinins was deleterious to the initiation and elongation of the main axis of roots. Previous experience in this laboratory has shown that nodulation is always accompanied by a shortening of the root system of the plants, if compared with the long normal roots of non-nodulated same type of plants. It is also known that addition of 1 μg kinetin/ml to plants causes an inhibition of auxin-induced lateral root initiation in which diploid divisions are inhibited (Torrey, 1961). It is of relevance to add that, in legumes, the initiation of root nodules by *Rhizobium* proceeds mainly on polyploid cortical cells, which represent the most frequent type of cells in the proliferating tissue mass. This contrasts with the diploidy present in the root and nodule meristems of the non-legumes tested in the present work (see G. Bond, this volume). Torrey (1961) suggested that in some way *Rhizobium* overcome the block to mitosis in the mature cortical cells of the legumes, perhaps contributing a kinetin-like factor at the infection stage. However, nodule initiation may not be simply due to a release of hormones by the microorganism. Dekhuijzen and Staples (1968) found that the increased cytokinin content in green islands of rust-infected bean leaves does not originate from the fungus but from the host. Also, the crown gall tumors on bean leaves have been reported to be a response of the host to bacterial infection rather than to the production of a growth factor by the infecting *Agrobacterium* (Lippincott and Lippincott, 1970).

The more drastic effect observed on overall plant growth when kinetin was used can be attributed to its synthetic nature. It has not been found in living tissues as yet.

The remarkable reduction of plant growth shown by the plants at the cytokinin level of 0.12 μg/ml suggests that still lower concentrations would show similar effect. However, the present results do not necessarily imply that endogenous cytokinins may not reach that level at a certain development stage of a true nodulation process. Again, endogenous cytokinins will be gradually synthesized as needed by the symbiotic system, a circumstance which was not accomplished by the procedure followed in our experiment. In any case, the shortening effect on roots was a constant feature observed in all the pseudonodulated plants. Since all the plants at 0.12 μg/ml level formed pseudonodules in a fair amount we cannot say what will be the minimum cytokinin concentration required for pseudonodules to occur. Also, whatever that minimum may be, some degree of root growth inhibition should still be expected.

REFERENCES

Bermudez de Castro, F. and Rodriguez-Barrueco, C. (1976). *In:* "Anuario", Vol. II, pp. 105-114. Centro de Edafologia, CSIC, Salamanca.

Dekhuijzen, H.M. and Staples, R.C. (1968). *Contr.Boyce Thompson Inst.Pl.Res.*, 24: 39-52.

Harris, G.P. and Hart, E.M.H. (1964). *Ann.Bot.N.S.*, 28: 509-526.

Lippincott, B.B. and Lippincott, J.A. (1970). *Plant Physiol.*, 46: 708-714.

Miller, C.O., Skoog, F., von Saltza, M.H. and Strong, F.M. (1955). *J.Amer.Chem.Soc.*, 77: 1392.
Rodriguez-Barrueco, C. (1968). *In:* Proc.VII Symp.Int.Agrochimica, pp.354-360. Salamanca.
Rodriguez-Barrueco, C. and Bermudez de Castro, F. (1973). *Physiol. Plantarum*, 29: 277-280.
Torrey, J.G. (1961). *Exptal.Cell Res.*, 23: 281-299.

ENDOPHYTE AND ASSOCIATION ESTABLISHMENT IN NON-LEGUMINOUS NITROGEN-FIXING PLANTS

J.H. BECKING

Institute for Atomic Sciences in Agriculture, Wageningen, Holland

INTRODUCTION

As in legumes, the establishment of an association in non-legumes is rather complex. It involves a number of successive processes, each being a prerequisite for the other. A proper establishment results in the development of a root nodule and finally the building up of the nitrogen-fixing capacity.

Unlike legumes, however, the symbiont of non-legumes can not be cultivated *in vitro* in pure culture on artificial media, which makes the study of the sequence of the various infection stages even more difficult than in the case of *Rhizobium* in legumes.

The present treatise will discuss the properties of the endophyte, the various infection phases and the growth and functioning of the root nodule. Because the *Alnus* endophyte is the easiest to handle, most of the experimental work refers to *Alnus glutinosa* and its root-nodule symbiont.

ENDOPHYTE

The endophyte of the root-nodules of the five orders of Dicotyledons with nitrogen-fixing capacity (Casuarinales, Fagales, Rhamnales, Coriariales and Rosales) is incontestably actinomycetal. This has been shown both by light and transmission electron microscopy (Becking *et al.*, 1964; Becking, 1968, 1975) and also by scanning electron microscopy (Becking 1975, 1976a,b). On the basis of morphological characteristics and cross-inoculation barriers the endophyte was allocated to belong to a separate family Frankiaceae, which includes a number of species each linked to a specific host. So far the family Frankiaceae consists of one genus *Frankia* and there are 10 species (Becking, 1970, 1974).

Using the chemical criteria of cell-wall composition for delineation and classification of the Actinomycetales (Lechevalier *et al.*, 1966, 1967; Lechevalier and Fekete, 1971), the

Alnus endophyte was shown to contain the cell-wall substances meso-DAP (2,6-diaminopimelic acid), glycine, arabinose, galactose and some lysine (Becking, 1974, 1976a) and it is therefore related to the Actinomycetales species of cell-wall types II (*Micromonospora, Actinoplanes*) and IV (*Nocardia*). The spherical structures of the actinomycete at the tip of the hyphae are homologous to sporangia. Due to the symbiosis the sporangia have lost their function for the propagation and dispersal of the actinomycete, but they are adapted to a special function in the symbiosis of the endophyte (see later). The presence of meso-DAP in the cell wall is characteristic for *Frankia alni* and probably also for some other species, but we could not find it in the cell walls of the endophyte of *Casuarina (Frankia casuarinae)* and *Myrica (F. brunchorstii)*.

The cross-inoculation barriers between the various species and hosts of *Frankia* seem to be rather strict, but there is also some specific adaptation and incompatibility within each host genus. For instance, in the large group of *Alnus, Myrica* and *Coriaria* with species of different geographical origin, there are adaptations to specific host endophytes in these geographical regions. Mutual cross-inoculation results in all degrees of lower compatibility ranging from less effective nitrogen fixation or ineffective root-nodules to complete incompatibility for root-nodule formation. Organisms with such specific adaptations are at the moment not accepted as separate species; first, more should be known about the taxonomy and classification criteria of the endophyte. Recently, it has been shown that an inoculum of *Alnus glutinosa* root nodules (soil grown?) did produce root nodulation in *Myrica gale* plants (water culture?), but not in the reciprocal combination (Rodriguez-Barrueco and Bond, 1976). According to these authors it is, however, not yet known whether the root nodule formation in *Myrica* is due to the *Myrica* endophyte present as contaminant in the *Alnus* root nodules or due to the *Alnus* endophyte itself.

So far all attempts to grow the endophyte on artificial media has failed, but the endophyte of *Alnus glutinosa* root nodules has been grown *in vitro* with *Alnus* root-nodule tissue (Becking, 1965, 1966, 1975, 1976a). In these experiments it has been shown that the transmission of the endophyte to the newly developed callus-tissue cells is generally rather poor and the endophyte is often lost in the subsequent transplants.

GROWTH AND N_2-FIXATION OF THE ENDOPHYTE IN ASSOCIATION WITH ROOT-NODULE CALLUS TISSUE

For *Rhizobium*, evidence has been obtained of N_2 fixation of the symbiont in association with soybean host tissue or cells

(Holsten *et al.*, 1971; Phillips, 1974; Child and LaRue, 1974, 1976) and this phenomenon even occurred with non-host tissue (Child, 1975; Scowcroft and Gibson, 1975).

For the purified endophyte of *Alnus glutinosa* in axenic culture as outlined previously (Becking, 1976a), experiments were set up with sterile alder root-nodule callus tissue grown *in vitro*. Two-month-old explants of *Alnus glutinosa* root-nodule tissue grown in rectangular flasks (so-called Meplats) of 200 ml containing 45 ml medium, were inoculated over the tissue with an axenic culture of the endophyte of *Alnus* obtained from sterile root-nodules of aseptically grown *Alnus glutinosa* seedlings. Unlike earlier experiments (Becking, 1965), where for the tissue culture a rather complex agar nutrient medium containing minerals and sucrose (2%, vol/vol) supplemented with coconut milk (150 ml/l), α-naphtalene-acetic acid (0.1 mg/l) and Ca-panthotenate (2.5 mg/l) was employed, a more simple and completely defined nutrient medium was used. This medium contained per litre: KNO_3 1.0 g, NH_4NO_3 1.0 g, $Ca(NO_3)_2 2H_2O$ 0.5 g, K_2HPO_4 0.3 g, $MgSO_4 7H_2O$ 71 mg, KCl 65 mg, NaFe-EDTA 32 mg, Mn $SO_4 H_2O$ 4.9 mg, $ZnSO_4 7H_2O$ 3.8 mg, H_3BO_3 1.6 mg, KI 0.8 mg, glycine 2.0 mg, thiamine 0.1 mg, pyridoxine 0.1 mg, nicotinic acid 0.5 mg, α-naphtalene-acetic acid (N.A.A.) 2 mg, kinetin 0.5 mg, sucrose 30 g, oxoid agar No. 3 12 g; pH 4.8 to 5.8.

After 4 months of growth, the inoculated root-nodule callus tissue and the non-inoculated controls (both fresh wt 500 - 700 mg) were gassed (after closing the flasks with a serum cap) with acetylene giving the final concentration of 6 to 7% (vol/vol) C_2H_2 in air. Incubation was for 2 h, at 27°C and light intensity 9,000 lux. Several experiments were conducted, but in none of the experiments after incubation a difference in C_2H_4 production between inoculated and non-inoculated flasks could be detected (Fig. 1), indicating that no N_2 fixation occurred. The absence of $N_2(C_2H_2)$ fixation may be due to the high inorganic nitrogen content of the medium used for the cultivation of the callus, but a modified medium with lower nitrogen content of the medium gave a very reduced growth of the callus or no growth at all. As the endophyte is primarily in contact with the callus tissue cells, the lack of N_2 fixation may be also due to other causes belonging to the constitution of the system. Light and transmission electron microscopical examination of the inoculated callus tissue showed the presence of the endophyte at the surface of the callus and also intercellularly but not intracellularly. Only hyphal forms of the endophyte could be found, no vesicular structures were observed.

Several authors have claimed the isolation of the endophyte of *Alnus*. So far these cultures have never been tested for N_2 fixation ability in association with *Alnus* callus tissue. The

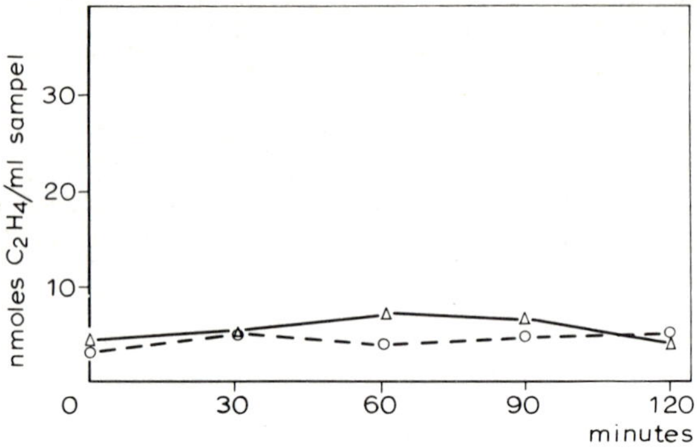

Fig. 1. *Acetylene reduction activity of the association of the alder endophyte (axenic culture) with* Alnus glutinosa *root-nodule callus tissue. (△) infected callus tissue; (○) uninfected callus tissue (control).*

original strain isolated by Von Plotho (1941) is still present in the Culture Collection of Micro-organisms of the Centraalbureau voor Schimmelcultures at Baarn, Holland, under the name *Actinomyces alni* Peklo. Moreover, a large number of strains were obtained from Dr. S. Uemura (see Uemura, 1961) representing his isolates of *Actinomyces alni* of his types I, II and III (strains A-135, A-1032, A-1041, A-151b) and these strains were also inoculated over the callus tissue. Growth of these actinomycete cultures over the callus tissue cultivated in large test tubes (16-17 cm length, 2.6 cm diameter) was in general excellent and growth was not restricted to the tissue, but also in medium (Fig. 2b). However, no nitrogen fixation could be detected with the C_2H_2 reduction assay. Transmission electron microscopical examination of these cultures showed normally only the presence of rods (Fig. 2a) and no vesicular structures were observed. In a few cases, however, the supposed symbiont gave evidence of vesicular structures at the tip of the hyphae (Fig. 2c) or sclerotial-like thickening of the hyphae (Fig. 2d; see also Lechevalier *et al.*, 1973).

Metcalfe and Brown (1957) described two *Nocardia* strains, *Nocardia calcarea* 8863 and *N. cellulans* 8868 (National Collection of Industrial Bacteria (NCIB), Torrey Research Station, Aberdeen, Scotland), able to fix atmospheric nitrogen. Since, the genetic relation (with regard to cell-wall substances) between *Frankia* and *Nocardia*, it was interesting to test these strains, especially as Metcalfe and Brown reported that these strains sometimes lost their ability to fix nitrogen in the laboratory, but that it was regained after a period of storage on sterile soil. But also these strains inoculated on callus tissue gave similar negative results with $N_2(C_2H_2)$ assay.

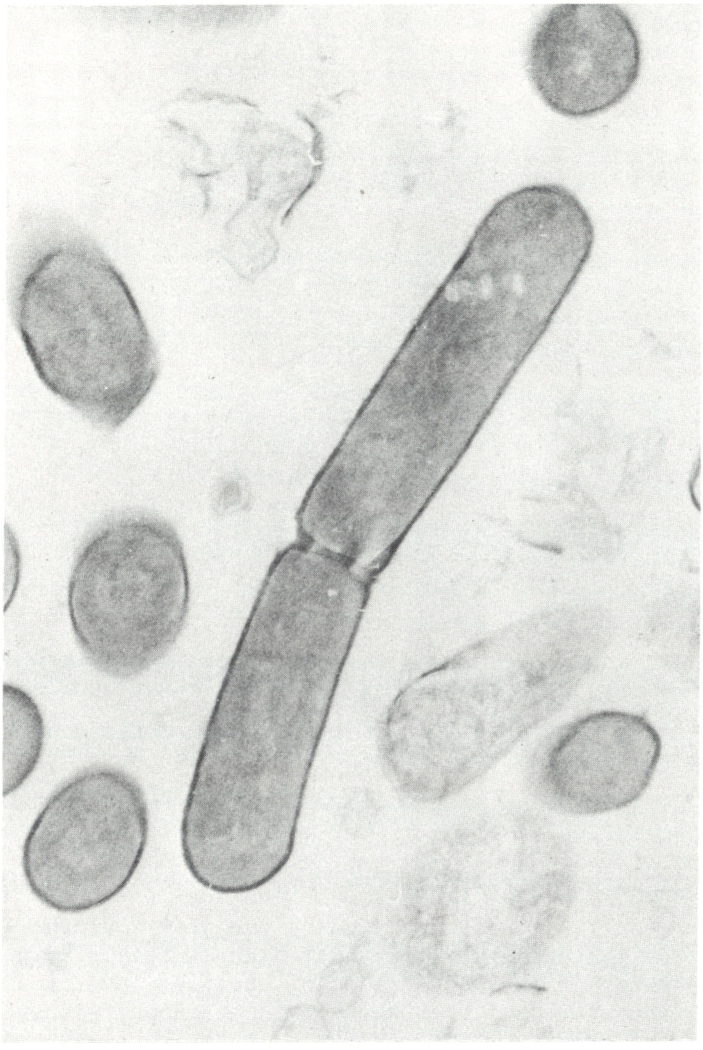

(a)

Fig. 2a. *Transmission electron micrograph of Actinomycetes strain Uemura A-135 developing at the surface of alder callus tissue.* *x 56,000.*

(b)

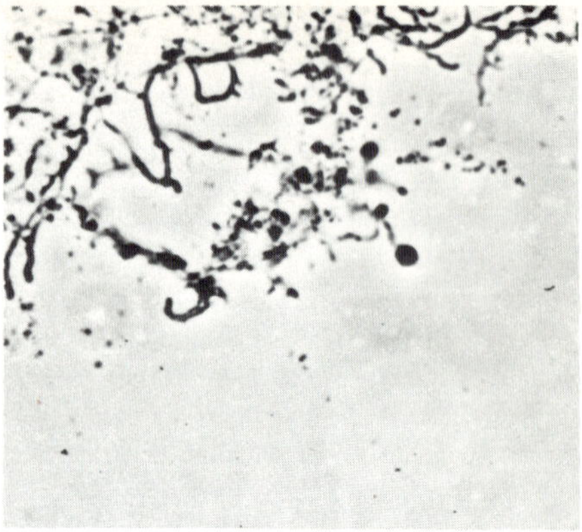

(c)

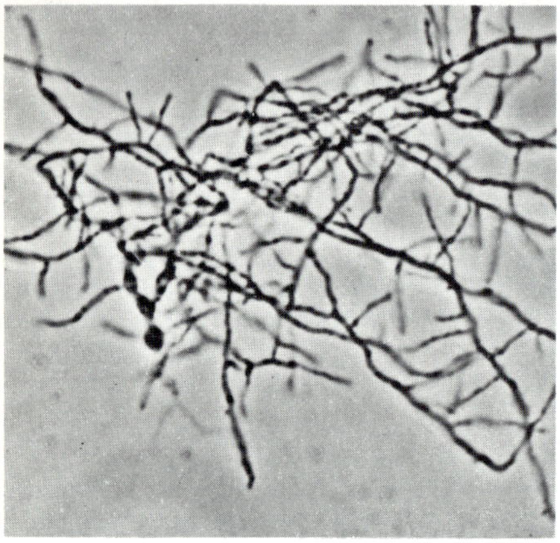

(d)

Fig. 2b. *Actinomycetes strain (Uemura strain A-135) growing in association with* Alnus glutinosa *root-nodule callus tissue x 2.0.*

Fig. 2c. *Actinomycetes strain Uemura A-144a. Phase contrast micrograph x 1,220.*

Fig. 2d. *Actinomycetes strain Uemura A-104l. Phase contrast micrograph x 1,220.*

NODULE INITIATION AND FORMATION

Nodule initiation is preceded by the growth of the endophyte in the rhizosphere of the host plant. Transmission electron microscopy of the root system of sterile *Alnus glutinosa* seedlings inoculated with an axenic *Alnus* endophyte suspension, indicated the growth of the endophyte in the mucilaginous layer around the root hair (Becking, 1975, 1976a).

Prior to entrance in the root hair, and infection of the root, there is extensive curling of root hairs in this region. This is in contrast to non-infected root regions, where the root hairs stay perfectly straight (Fig. 3a,b,c). Not all root hairs becoming curved are actually infected as only a small portion of them contained the endophyte. Hence, the endophyte exercises a chemical influence, probably by production of growth substances like auxins, towards all root hairs around the infected zone. Zones of the root without hairs, which were also commonly observed in *Alnus glutinosa* seedlings, cannot become infected as direct transmission of the endophyte to the root (i.e. not through the way of the root hair) is not possible.

The endophyte enters the root hair solely at the tip, which entrance is preceded by an invagination of the host-cell wall and the outer cytoplasmic membrane (Fig. 3c). The penetration of the endophyte is accompanied by extensive cell-wall formation as response of the host to the endophyte. Due to this action a thick cell wall probably of cellulose or callose nature is produced. This cellulose/callose wall is entirely bounding the infection thread as shown by light and transmission electron microscopy. By the latter technique at the inside the thread wall a thread membrane is clearly visible and a thread matrix, probably of polysaccharide nature, in which the actinomycete cells are completely imbedded.

Along the root hair the infection thread proceeds towards the root accompanied by the root-hair nucleus (Fig. 3c). When the infection thread reaches the main root it induces extensive cell division in the main root in a particular zone of the cortical parenchyma cells. By the re-differentiation of some cortical parenchyma cells becoming meristematic again and by subsequent proliferation of these parenchyma cells, a thickening of the main root is caused (Fig. 3b). In this very early stage of root nodule development in which the cortical parenchyma cell-layer is only a few (3-4) cells thick (see Becking, 1968, Plate 7 and 1975, Fig. 14) some host cells contain already the endophyte in its vesicular form. The thickening of the main root by cortical parenchyma division is not the proper nodule, but only a 'pre-nodule' as from this site a salient root primordium, which was already present, is induced to grow out forming the actual nodule. Host cells containing bacteria-like endophytic structures were so far never observed in the main-root cortex forming the 'pre-nodule'.

In a later stage of root-nodule development, the infection progresses in the lateral root and no further in the main root. That the developing primordium is actually a lateral root is evident from thin sections showing its origin in the protoxylem and protophloem strands within the pericycle of the parent root. By staining the paraffin thin sections with erythrosine and methylene green dyes clearly two endodermal layers staining green were present, one of the main root and one of the lateral root (see Becking, 1968, colour Plate 7). Thus, by

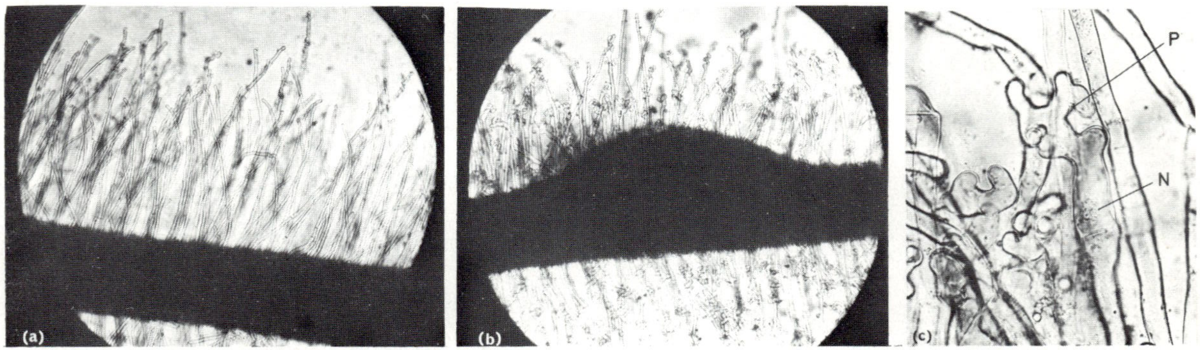

Fig. 3a. *Uninfected root of* Alnus glutinosa *showing straight root hairs. x 90.*
Fig. 3b. *Infected root of* Alnus glutinosa *showing an extensive curling of the root hairs and the development of the 'prenodule'. x 90.*
Fig. 3c. *Tip of a root hair of* Alnus glutinosa *showing the entrance of the endophyte. The tip of the root hair is deformed by the action of the endophyte. At the site of entrance a pocket (P) is formed by invagination of the host-cell wall and from here an infection thread develops inside the host cell which in its growth is accompanied by the host-cell nucleus (N). x 710.*
Note: *the alder roots were growing under monaxenic conditions; inoculation was with an axenic culture of the* Alnus *endophyte. All micrographs of living, unstained preparations; phase contrast.*

contrast with legumes, root nodules of non-legumes are modified lateral roots. A further contrast is the formation first of a 'pre-nodule', before the actual nodule is formed, a succession of events unknown in legumes and which is probably an unique feature of the actinomycete symbiosis of the non-legumes.

ONSET OF N_2 FIXATION

Experiments with *Alnus glutinosa* root nodules using ^{15}N labelled N_2 gas, showed that very young root nodules of alder gave no increase of ^{15}N content in the root in time after exposure to $^{15}N_2$ (Fig. 4).

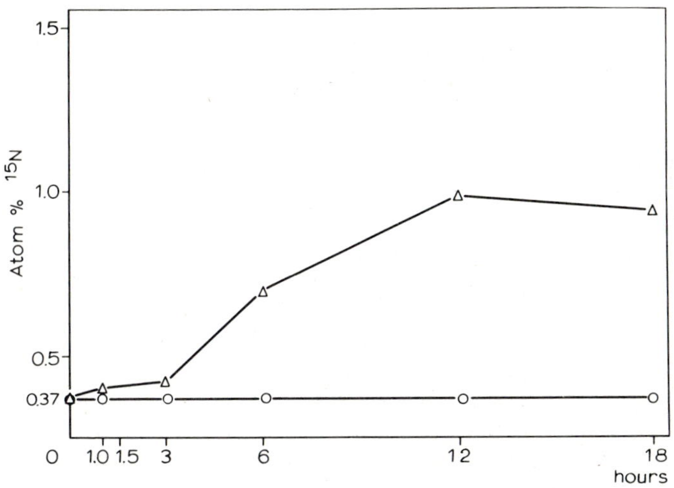

Fig. 4. *Increase in amount of* ^{15}N *in excised root nodules of alder (*Alnus glutinosa*) exposed to an atmosphere containing 19% (vol/vol)* N_2 *(labelled with 35.0 atom %* ^{15}N*), 2 % (vol/vol)* O_2*, 3% (vol/vol)* CO_2*, and 57% (vol/vol) Ar. (△) mature root nodule; (○) very young root nodule.*

This phenomenon was further investigated in more detail in aseptically raised alder seedlings growing in large test tubes (length 17 cm, diameter 0.95 cm). These seedlings were inoculated with an axenic endophyte suspension (see Becking, 1976a), and nodule initiation and development was daily followed by direct observation. Nodule initiation in the test plants was rather irregular and at variable times nodules may develop at the same plant. Nodules may already appear 14 days after inoculation (as visible by the naked eye), but development may extend to 30 days; the average figure for about 30 plants was 20 days. The first external visible signs of root-nodule development are a reddening and later thickening of the main root (Fig. 3a,b). Since root-nodule development is rather unequal, in seedlings of the same age various stages of root nodule development can be observed.

Experiments about the onset of N_2 fixation in such young root nodules were conducted with two-month-old *Alnus glutinosa* seedlings in their 3-4 leaf stage, besides the cotyledons, showing all stages of root-nodule development.

Use was made of the very sensitive C_2H_2 reduction assay for N_2 fixation. From the series a sofar non-nodulated seedling was selected on which the endophyte had only developed in the rhizospere. Moreover, one plant with small nodules (one of 1.3 mm and the others less in diameter) and one with larger nodules (about 2.0 mm in diameter) was selected. The gas volume in the tubes was about 39 ml. The plants were exposed to an atmosphere of 10% (vol/vol) C_2H_2 in air at 23°C. At the end of the experiment the serum caps of the tubes were removed and the atmosphere of the tube flushed with sterile air. Subsequently the tube was plugged again with sterile cotton wool and replaced in the growth chamber. The $N_2(C_2H_2)$ activity was measured with a week interval and the size of the nodules (when present) was registrated by making camera lucida drawings by means of a drawing prism attached to a binocular microscope.

A typical experiment is presented in Fig. 5. As is evident from this graph alder root nodules of a size of 1.3 mm or less gave no N_2 fixation activity and N_2 fixation only started when the root nodules had attained a size of at least 2.0 mm, i.e. when the 'pre-nodule' (the main root thickening) has reached this size. The earlier stages of root-nodule development with no thickening of the main root but only growth of the endophyte in the rhizosphere gave also no evidence of $N_2(C_2H_2)$-fixation activity. In a week time interval, the largest root nodule measuring at the beginning of the experiment 2.0 mm x 1.8 mm, had increased in size to 3.0 x 2.1 mm and at the same time it had more or less doubled its N_2-fixation activity, while the plant with the younger (smaller) root nodules, showed only a slight increase in N_2-fixation activity. In the plant with no root nodules, there was after a week still no visible root nodule and no nitrogen fixation could be detected.

GRADIENT OF N_2 FIXATION WITHIN A SINGLE NODULE LOBE

Light microscopy, but also transmission electron microscopy has revealed various zones of endophyte distribution within singular nodular lobe (Fig. 6a,b). At the extreme tip of the nodule lobe a meristem is active and below this the endophyte is only present in its hyphal form. Subsequently, more below,

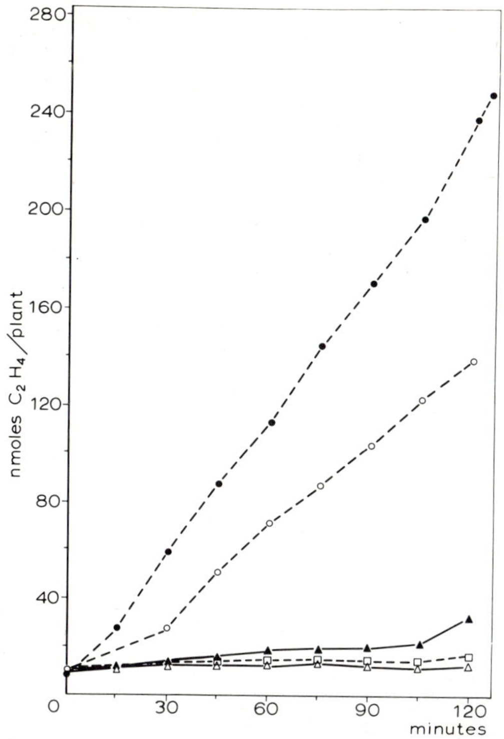

Fig. 5. *Acetylene reduction activities of young root nodules of alder (*Alnus glutinosa*) in various stages of growth. Three seedling plants were tested: Plant* A *nodule (1) 2.0 x 1.8 mm (○); plant* B *nodules (6) 1.3 x 1.0 mm (△); plant* C *nodules absent, but development of the endophyte in the rhizospere (□). The numbers in brackets indicate the number of nodules per plant. With a week time interval the $N_2(C_2H_2)$-fixation activity was remeasured in the three plants (solid symbols). The nodule of plant* A *had now attained a size of 3.0 x 2.1 mm, the nodules of plant* B *had slightly increased, and nodulation in plant* C *was still absent.*

there is a zone with the endophyte mainly in its vesicular form and below this near the base of the nodular lobe there is a reduction of the number of host cells filled with vesicles and an increase of host cells filled with bacteria-like endophyte cells. A schematic presentation of the various zones within a root nodule of *Alnus* is given in Fig. 7.

With the $^{15}N_2$ technique some indication was obtained that in particular, vesicles were associated with nitrogen-fixation

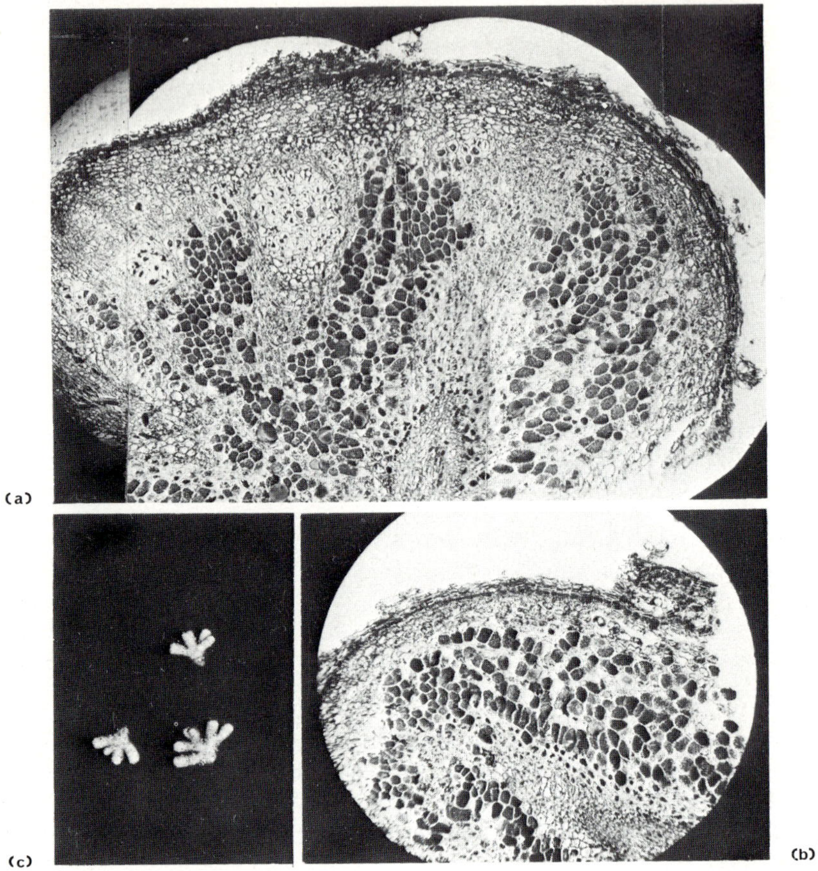

Fig. 6a. *Tip of a broad nodule lobe of a perennial root nodule of* Alnus glutinosa *showing the presence of vascular bundles intersepted with host tissue infected by the endophyte. x 50.5.*

Fig. 6b. *The same as (a) but thinner nodule lobe containing only one vascular bundle (stele). x 50.5.*

Fig. 6c. *General morphology of perennial root nodules of* Alnus glutinosa *(80% of actual size).*

process as nodular slices rich in host cells containing vesicles showed more nitrogen-fixing activity than nodular slices with host cells poor in vesicles. This method is, however, hampered by sample size as the mass spectrometer requires large samples for obtaining enough NH_3-N. For this reason, these zones in a

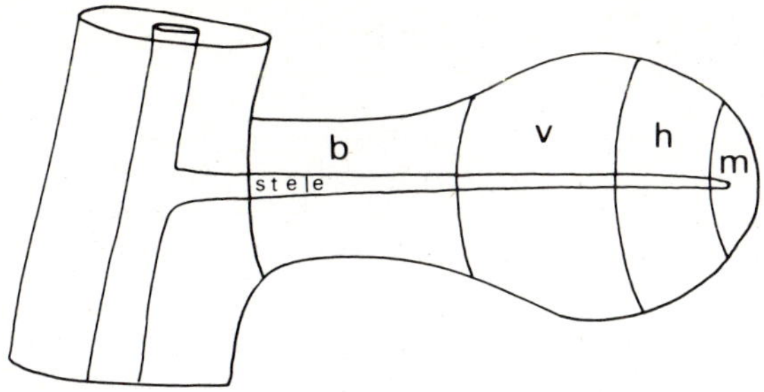

Fig. 7. *Diagram of a root nodule of* Alnus glutinosa *showing the various zones of endophyte infection. Notations: (m) meristem, (h) zone with predominantly hyphae, (v) zone with predominantly vesicle structures, (b) basal part of nodule with senescent vesicles and bacteria-like endophytic cells.*

very large number of different nodules had to be extracted and pooled. However, the far more sensitive acetylene reduction assay offers an opportunity to measure N_2 fixation activities within a single nodule lobe.

Perennial root nodules (see Fig. 6c) of 22-month-old *Alnus glutinosa* plants, kept on water culture, were sampled and one nodule lobe measuring from tip to base 3.5 - 4.0 mm and at the base with a diameter of about 2.1 - 2.4 mm were sliced beginning at the tip in transverse sections. Each section was incubated separately at 23°C in an airtight polypropylene disposal syringe in a gas volume of 1.5 ml containing air with 10% (vol/vol) C_2H_2. At the end of the experiment besides the fresh weight of each nodular section, the size of the sections was also determined with the aid of camera lucida drawings. Moreover, by microscopic examination the presence of the various forms of the endophyte in the slices and their frequency was ascertained.

The results of three typical experiments are given in Table I. In experiment A a single nodule lobe was sliced in three sections and in experiments B and C a nodule was sliced in 5 and 6 parts, respectively. The middle part of the root nodule is the region with most host cells filled with vesicles and as evident from the results presented in Table I the middle part of the root nodule containing the vesicles gave a distinct peak of $N_2(C_2H_2)$-fixing activity. This outcome is again a proof that in particular the vesicles are associated with N_2 fixation process.

Table I

$N_2(C_2H_2)$ fixation in sections of a single nodule lobe of *Alnus glutinosa* subdivided by transverse sectioning

Experiment	Location section	Fresh wt section mg	pmoles C_2H_4/h/ sample	pmoles C_2H_4/h/ mg fr.wt
A. Division in 3 parts (large nodule)[a]	1. tip	6.3	316	50.4
	2. middle	35.4	2466	69.5
	3. base	50.2	913	18.2
B. Division in 5 parts (small nodule)[b]	1. tip	00.3	15.9	63.6
	2.	1.0	13.2	13.2
	3. middle	2.3	362.7	157.7
	4.	3.2	39.3	12.2
	5. base	7.5	11.2	1.5
C. Division in 6 parts (small nodule)[c]	1. tip	0.20	8.0	40.2
	2.	0.32	12.0	37.6
	3.	0.23	49.8	216.6
	4. middle	0.11	64.5	586.3
	5.	0.30	10.6	35.5
	6. base	0.60	6.0	10.0

Notes:

[a] Experiment A is a rough preliminary experiment; Experiment B is much more refined as the sections were much thinner and therefore allow better localization of C_2H_2-reduction activity.

[b] In experiment B the diameter and height of each section was measured and a complete gradient diagram of C_2H_4 production can be drawn. Total nodule length (sum of all heights) was 3.01 mm and the sliced nodule diameter varied from tip to base from 1.4 -2.1 mm. Total fresh wt. nodule was only 14.3 mg.

[c] In Experiment C total length of nodule 1.6 mm, total fresh wt nodule 1.76 mg.

CONCLUSIONS AND SUMMARY

As shown in the preceding sections the endophyte of this common type of non-leguminous plant is actinomycetal. Root-nodule initiation and its further development is a very complex process requiring a number of successive steps such as the establishment of the endophyte in the rhizosphere, the penetration of the endophyte at the tip of root hair and the growth of the infection thread through the root hair towards the main root. When one step of this sequence is not fulfilled then no nodule is formed. For instance, when the growth of the infection thread through the root hair is hampered, this results in abortive infection thread. In fact many infections threads proved to be abortive under normal conditions as only a very few reach the main root.

In the main root the endophyte induces the formation of a 'pre-nodule' by re-differentiation of some cortical parenchyma cells and in the produced host cells, already some vesicular forms of the endophyte are developed. The actual root-nodule is formed later by a lateral root, which is already present as a dormant primordium in the region of infection. The formation of a 'pre-nodule' before the development of the final root nodule is very different from the phases observed in the nodule development of legumes and showed the unique position of these actinomycetes inhabited root nodules of non-legumes.

Small 'pre-nodules' showed no $N_2(C_2H_2)$-fixation activity and this activity is only gradually developed when the nodule has reached a size of about 2.0 mm in length and when it contained a substantial amount of root cells filled with endophytic vesicle structures. The endophyte developing in the rhizosphere prior to injection has no N_2-fixing capacity.

By slicing single nodule lobes and using the very sensitive C_2H_2-reduction assay evidence has been provided that in particular the middle section of the root nodule containing most of the vesicles was very active in $N_2(C_2H_2)$-fixation, indicating incontestably the special role of vesicles of the endophyte in the N_2-fixation process.

ACKNOWLEDGEMENTS

The skilful technical assistance of Mr. T.W. Strubbe is gratefully acknowledged.

REFERENCES

Becking, J.H. (1965). *Nature (London)*, 207: 885-887.
Becking, J.H. (1966). *Annal.Inst.Pasteur Suppl.*, 111: 211-246.
Becking, J.H. (1968). *Stikstof, Dutch Nitrogen Fertilizer Rev.*, 12: 47-74.

Becking, J.H. (1970). *Int.J.Syst.Bacteriol.*, 20: 201-220.
Becking, J.H. (1974). *In:* Bergey's Manual of Determinative Bacteriology (R.E. Buchanan and N.E. Gibbons, eds), 8th ed., pp. 701-706, Williams and Wilkins Co., Baltimore, U.S.A.
Becking, J.H. (1975). *In:* The Development and Function of Roots (J.G. Torrey and D.T. Clarkson, eds), pp. 507-566, Third Cabot Symposium. Academic Press Inc., New York and London.
Becking, J.H. (1976a). *In:* Nitrogen Fixation (W.E. Newton and C.J. Nyman, eds), vol. 2, pp. 581-591, Proc.1st Int. Symposium on Nitrogen Fixation, Pullman, Washington, June 1974, Washington State Univ. Press.
Becking, J.H. (1976b). Proc.Global Impacts of Applied Microbiology, GIAM IV, Sao Paulo, Brazil, July 1973. *In:* Revista de Microbiologia, Brazil (J.S. Furtado, ed.)(in press).
Becking, J.H., Boer, W.E. de and Houwink, A.L. (1964). *Antonie van Leeuwenhoek, J. Microbiol. and Serol.*, 30: 343-376.
Child, J.J. (1975). *Nature (London)*, 253: 350-351.
Child, J.J. and LaRue, T.A. (1974). *Plant Physiol.*, 53: 88-90.
Child, J.J. and LaRue, T.A. (1976). *In:* Nitrogen Fixation (W.E. Newton and C.J. Nyman, eds), vol. 2, pp. 447-455, Proc.1st Int. Symposium on Nitrogen Fixation, Pullman, Washington, June 1974, Washington State Univ. Press.
Holsten, R.D., Burns, R.C., Hardy, R.W.F. and Hebert, R.R. (1971) *Nature (London)*, 232: 173-175.
Lechevalier, H., Lechevalier, M.P. and Becker, B. (1966). *Int. J.Syst.Bacteriol.*, 16: 151-160.
Lechevalier, M.P., Gerber, N.N., Lechevalier, H., Higgins, M. and Hegyi, J. (1967). The Separation of Actinomycetes into Genera by Morphological and Chemical Methods. Workshop Subcomm. on Actinomycetes, Amer.Soc.Microbiol., Institute of Microbiology, Rutgers Univ., New Brunswick, N.J., 16 pp.
Lechevalier, M.P. and Fekete, E. (1971). Chemical Methods as Criteria for Separation of Actinomycetes into Genera. Workshop Subcomm. on Actinomycetes, Amer.Soc. for Microbiol., Institute of Microbiology, Rutgers Univ., New Brunswick, N.J., 23 pp.
Lechevalier, M.P., Lechevalier, H.A. and Heintz, C.E. (1973). *Int.J.Syst.Bacteriol.*, 23: 157-170.
Metcalfe, G. and Brown, M.E. (1957). *J.Gen.Microbiol.*, 17: 567-572.
Phillips, D.A. (1974). *Plant Physiol.*, 53: 67-72.
Rodriguez-Barrueco, C. and Bond, G. (1976). *In:* Nitrogen Fixation and the Biosphere (P.S. Nutman, ed.), pp. 561-565, Intern.Synthesis Meeting, Intern.Biol.Programme Section PP-N, Edinburgh, Sept. 1973. Cambridge Univ.Press, Cambridge, England.
Scowcroft, W.R. and Gibson, A.H. (1975). *Nature (London)*, 253: 351-352.

Uemera, S. (1961). *Sci.Rep.Agric.For.Fisheries Res.Council*, No. 7, pp. 1-90, with 15 plates.
Von Plotho, O. (1941). *Arch.Mikrobiol.*, 12: 1-18.

INFECTION PROCESS OF THE *ALNUS* ROOT NODULE SYMBIOSIS

M. LALONDE*

Botanisch Laboratorium, Leiden, Holland.

INTRODUCTION

Light and electron microscopy studies of nitrogen-fixing root nodules of *Alnus* plants suggested the endophyte to be a pleomorphic actinomycete (Becking *et al.*, 1964; Gardner, 1965; Lalonde and Devoe, 1975; Lalonde and Knowles, 1975a; Van Dijk and Merkus, 1976). Despite numerous claims to the contrary, the isolation, subculture and demonstration of the nodulation of the aseptic host plant with a pure *in vitro* culture of the endophyte has not been achieved according to the strict criterion defined by Koch's postulates. Unexpectedly, and in the absence of pure culture studies, the *Alnus* endophyte and all other non-leguminous root nodule endophytes were speculatively classified as *Frankia* species by Becking (1970b, 1974).

In an early investigation of the *Alnus* symbiosis, Lalonde and Fortin (1972) used the suspension-dilution technique, combined with inoculation tests of aseptically-grown *Alnus* plants, to eliminate endogenous contaminants and to obtain axenic nodules. Recently, Becking (1975, 1976), using the same procedure, corroborated the ability of the suspension-dilution technique to produce *Alnus* root nodules devoid of detectable endogenous contaminants but still containing an infective endophyte. He stated that the serial dilutions of crushed nodule suspensions, lacking detectable contaminants but still able to nodulate aseptic host plants, represent isolates of the *Alnus* organism which should be considered as pure strains which fulfill Koch's postulates. However, such an endophyte suspension, whether pure or not, represents only a special way to purify and conserve (Quispel, 1954; Lalonde and Fortin, 1972; Rogers and Wollum, 1974) the endophyte if growth cannot be detected and if a subsequent subculture of a growing organism cannot be obtained in the absence of a live host plant. The expression

---

* Present and permanent address: Charles F. Kettering Research Laboratory, Yellow Springs, Ohio 45387, U.S.A.

"isolation of the root nodule endophyte" implies the cultivation of a root nodule isolate in absence of live host plant or cells on an artificial growth medium, with the further demonstration of its endophytic nature.

Lalonde *et al.* (1975) reported the isolation of two actinomycetes from both field-collected and axenic *in vitro*-produced root nodules of *Alnus crispa* var. *mollis* host plant. After their transfer to a chemically defined medium, these nodule isolates could not be distinguished from each other on the basis of morphology, cultural reactions and whole cell composition and were considered to be the same species. They were morphologically similar to the root nodule endophyte. Diaminopimelic acid isomers (LL and meso) were present in similar proportions in the nodule endophyte and in the nodule isolates. However, these isolates were not able to nodulate aseptic host plants. Nevertheless, antibodies produced against the whole cells of the two nodule isolates were used in immunolabelling reactions (immunofluorescein and immunoferritin) and demonstrated the homology of these nodule isolates with the *Alnus* nodule endophyte. Within the limit of the immunolabelling technique, these results allow us to by-pass the host nodulation test and to relate the non-infective isolates to the *Alnus* nodule endophyte. Therefore, the main problem of the *Alnus* root nodule symbiosis was suggested to be no longer the isolation of the endophyte but rather removal of its non-infectivity following isolation and cultivation in absence of the host plant (Lalonde *et al.*, 1975).

In order to solve this non-infectivity problem with the endophyte from the *Alnus crispa* var. *mollis* root nodule, we decided to study in detail the normal infection process which takes place primarily at the *Alnus* root hairs and to identify some of the properties needed to produce a successful symbiosis. Furthermore, because it is known that a certain degree of incompatibility exists between North American and European alders with respect to their endophytes (Bond, 1974), we studied the effect of nodulating the European *Alnus glutinosa* host plant with nodules taken from the Canadian *Alnus crispa* var. *mollis* host plant, thus hoping to get a less-compatible system than a normal host-endophyte system. Later on, by comparing the infection process induced in a normal infective system (e.g. *A. glutinosa* endophyte on *A. glutinosa* host plant) with a less-infective system (e.g. *A. crispa* var. *mollis* endophyte on *A. glutinosa* host plant) and with the non-infective isolates (e.g. *A. crispa* var. *mollis* isolates on *A. crispa* var. *mollis* host plant and on *A. glutinosa* host plant), it may be possible to understand the properties which are lacking in the non-infective endophyte isolated from the *A. crispa* var. *mollis* root nodule (Lalonde *et al.*, 1975).

METHODS

The *Alnus glutinosa* seedlings were raised in hydroponic culture as described by Angulo Carmona (1974). However, the seedlings, which were grown in a "nitrogen-plus" nutrient solution (i.e. with NO_3^-) up to the "four leaves stage", were grown in a "nitrogen-minus" nutrient solution for an additional two weeks before their inoculation. The crushed-nodule inoculum was prepared and used following the method of Lalonde and Fortin (1972). The root nodules were fixed and embedded in Epon resin according to Lalonde and Knowles (1975a). Immuno-techniques and samples were those of Lalonde *et al.* (1975).

SURVEY OF PREVIOUS OBSERVATIONS

Observations of the infection process of the *Alnus* host plant was reported previously (Pommer, 1956; Becking, 1968, 1970a; Pizelle, 1972; Angulo Carmona, 1974; Angulo Carmona *et al.*, 1975). From these observations, the general aspects of the process of infection can be summarized as follows. In reaction to the inoculation with a crushed-nodule inoculum or a soil-suspension, the plant root hairs which were initially straight become deformed. The deformation of the root hairs takes place behind the root tips in the elongation-zone of the root where the root hairs are growing. Then, the endophyte penetrates a deformed root hair and propagates in the cortical cells of the root. Stimulated by the microbial intruder, the cortical cells near the infected root hair start to divide causing a slight swelling of the root. This swollen region is called the primary nodule. This primary nodule of *Alnus* is easily detected due to a bright red colouration of a limited zone of the swollen region. This red-coloured zone is called the red infection point. The root hair through which the endophyte entered is associated with this red infection point. During its propagation and differentiation in the root cortex, the endophyte induces the formation of an adventitious lateral root at the pericycle of the root. This induced lateral root develops into a true nodule following the penetration of its cortex by the endophyte present in the primary nodule. Not all primary nodules give rise to a true nodule.

For its internal propagation in the primary nodule and the true nodule, the endophyte, which is seen as hyphae, perforates the host cell wall and enters the adjacent cortical cell (Lalonde and DeVoe, 1975). As the endophyte emerges from the cell wall, the actinomycetal intruder is encapsulated in a pectic matrix probably of host origin (Lalonde and Knowles, 1975b). Surrounding this encapsulation material, an enclosing membrane (which is the invaginated host plasmalemma (Lalonde and DeVoe,

1976) separates the endophyte from the host cell cytoplasm. After its penetration and intracellular surrounding of the host cell nucleus, the filamentous branched and septate endophyte develops the so-called vesicles by the swelling of the hyphae tip (Becking *et al.*, 1964; Gardner, 1965; Lalonde and Knowles, 1975a). Furthermore, for certain types of endophyte, an additional form of hypha, called the spindle, is produced. This spindle is able to differentiate into a sporocyst structure containing spore-like cells (Van Dijk and Merkus, 1976).

OBSERVATIONS OF A NORMAL INFECTIVE HOST-ENDOPHYTE SYSTEM

The normal endophyte of a given host plant is one which occurs in the nodules of a host plant growing in its typical habitat and within its usual geographical distribution. An *Alnus glutinosa* host plant inoculated with a crushed-nodule inoculum prepared from normal *Alnus glutinosa* nodules, is a normal infective host-endophyte system. The next two sections present observations on the process of infection of such an infective host-endophyte system.

Chronology of the Macroscopic Events during the Infection Process

The following chronology of the infection process of a normal host-endophyte system is related to the experimental conditions to which the host plant was submitted (e.g. age of the plant at the time of inoculation, composition, concentration and pH of the growth medium, intensity and duration of the light period, etc.). Another important factor, dealing with the endophyte, is the density of the inoculum (e.g. mg of crushed nodules per plant). In all the experiments, the growth conditions and the density of the inoculum were such that all the possible sites of infection of the root system were nodulated, resulting in the formation of a maximum number of nodules per plant. At time zero during the inoculation of the host plant growing in hydroponic conditions, the root hairs show no deformation. Most of the leaves are yellow, some may be brown and dead due to the prolonged absence of fixed nitrogen in the nutrient solution. At the end of the 1st day, deformation occurs of the root hairs which were growing at the time of inoculation. These growing root hairs include the very young root hairs just appearing on the root and the older ones almost at the end of their growth. This pattern of deformation, observed behind the elongation region of the root, is seen as a "slope" of deformed root hairs. In addition, all the root hairs which are produced after the time of inoculation, are deformed and stay very short. This zone of young, short and deformed root hairs is seen at the base of the slope of deformed root hairs toward the root cap (Fig. 1). At the end of the 6th day, the primary nodule is

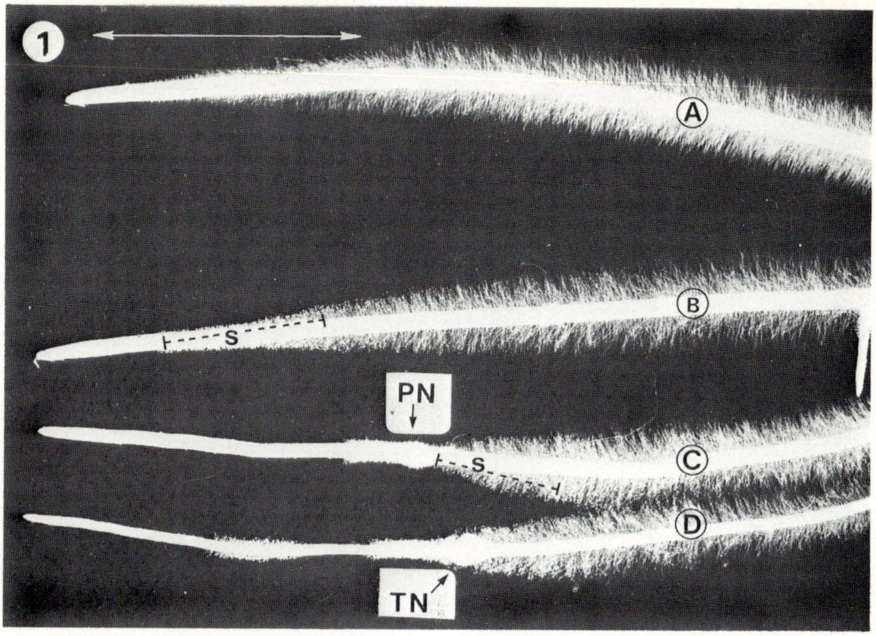

Fig. 1. *Localization of deformed root hairs and nodules on inoculated* Alnus glutinosa *roots. Root A; Control, no deformation of the growing root hairs. Root B; 1st day after inoculation, deformation of all growing root hairs shown as slope (S). Root C; 7th day after inoculation, at the base of the slope (S) of deformed root hairs, a primary nodule (PN) is seen as a swollen zone of the root. Root D; 12th day after inoculation, a true nodule (TN), originating from the pericycle of the root and developing through the root cortex, is visible. Bar is 1 cm.*

usually visible as a red infection point on the top of the swollen zone of the root. Such a primary nodule formation is the first indication that the endophyte was able to penetrate a root hair and to propagate in the root cortex. The primary nodule formation is always limited to the zone of deformed root hairs and is generally seen at the base of the slope in the zone of the youngest root hairs (Fig. 1). The number of primary nodules is generally limited to one per zone of deformed root hairs. At the end of the 12th day, the true nodule, originating from the pericycle of the root and developing through the root cortex, becomes visible. Acetylene assays of a root system bearing 12-day-old nodules shows a detectable nitrogen fixation activity. At the end of the 24th day, small growing green leaves are visible at the top of the plant shoot. The appearance of these green leaves permits a visual assay of the symbiotic capacity of the host-endophyte system.

Chronology of the microscopic events of the infection process

Free stage of the endophyte Following its inoculation of the host root system, the endophyte can be recognised as an encapsulated vesicle attached to its parental hypha. The endophyte then seems to proliferate in the rhizosphere surrounding the elongation region of the root and is presumptuously recognized as bacterial cells (Figs 8, 10, 11, 12 and 13). Such presumptive endophytic bacterial cells seem to feed on products excreted by the growing roots and root hairs.

Exo-encapsulation thread In response to the stimulation of their growth, the bacterial cells excrete compounds which act on the growing roots. These presumptive compounds elicit reactions such as the deformation of the root hairs and the external excretion of blebs containing an electron-dense material (Fig. 6). These excreted blebs originate by the evagination of the so-called external layer which always covers the *Alnus* root epidermis and the root hair cell wall (Figs 7 and 8). These external-layer blebs are liberated in the rhizosphere and they migrate toward the presumptive bacterial endophyte, fuse with it and deposit their electron-dense material on it. This host deposition mechanism rapidly results in the formation of an encapsulation matrix around the presumptive bacterial endophyte which is seen enclosed end-to-end in a tube (Figs 4, 5 and 11). Such a structure is called the exo-encapsulation thread. Thousands of these exo-encapsulation threads can be seen in each zone of deformed root hairs (Figs 2 and 3). Elsewhere on the root, in the absence of a zone of deformed root hairs, the exo-encapsulation thread is never observed. The exo-encapsulation thread seems to be open at one end (Fig. 13) and is attached at the other end to a deformed root hair or to the root epidermis in an elongation-zone of a root (Figs 2, 3, 4 and 11). This encapsulation structure is always surrounded by a layer which is continuous with the external layer covering the root hair cell wall and root epidermis (Figs 9 and 10). This layer may have a key role during the symbiotic host-endophyte interactions of the *Alnus* root nodule.

Penetration in the root hair Light microscopy observations suggest that the penetration of the endophyte into a root hair proceeds by hyphal growth (Fig. 14)(Angulo Carmona, 1974) and electron microscopy observations confirms it (Figs 15 and 16). Following the penetration of a root hair through its cell wall, the bacterial endophyte, which was enclosed in the exo-encapsulation thread, changes morphology to a hypha. As the endophyte

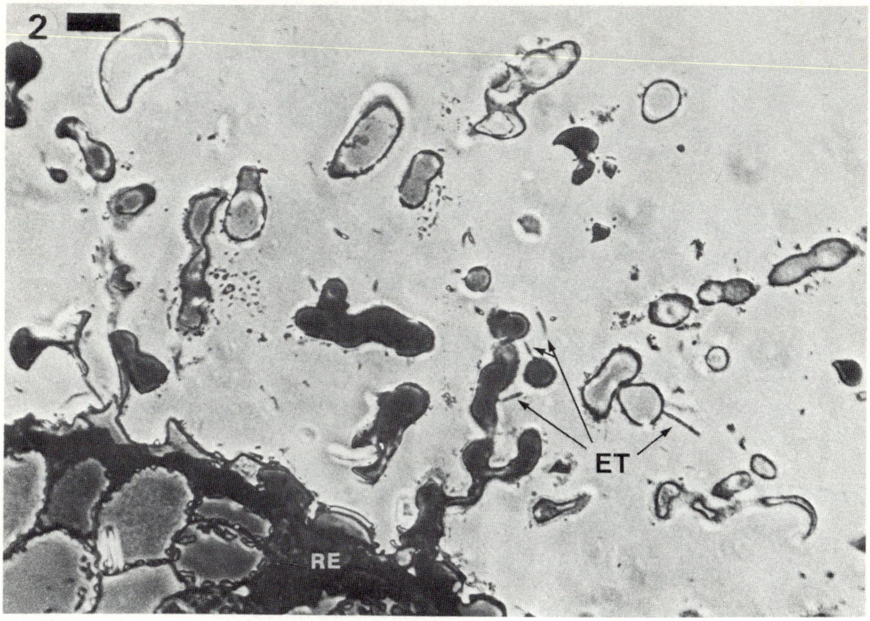

Fig. 2. *Phase contrast microscopy of an Epon section taken in the zone of deformed root hairs of an inoculated* Alnus glutinosa *root. Notice the exo-encapsulation threads (ET) associated with the deformed root hairs. RE: root epidermis. Bar is 30 μm.*

endophyte emerges from the cell wall of the deformed root hair tip, the host cell seems to synthesize, transport and deposit substances for the encapsulation of the hyphal intruder. Surrounding this encapsulation material is a membrane which lies between the hyphal endophyte and the host cell cytoplasm. This enclosing membrane is probably the invaginated plasmalemma of the root hair cell. Tannin deposits are observed in the cytoplasm of the infected root hair (Figs 15 and 16).

During the penetration of the root hair cell wall, the host cell nucleus migrates toward the microbial intruder. This encapsulated microbial intruder, seen as a single hypha or as strands of hyphae, migrates toward the base of the root hair cell accompanied by the host nucleus (Figs 15 and 16). On reaching the cell wall at the base of the root hair cell, the hyphal endophyte may start to produce vesicles. Then the hyphal endophyte perforates the host cell wall and spreads into the neighbouring dividing cortical cells of the root and forms vesicles at the tips of the parental hyphae near the cell wall of

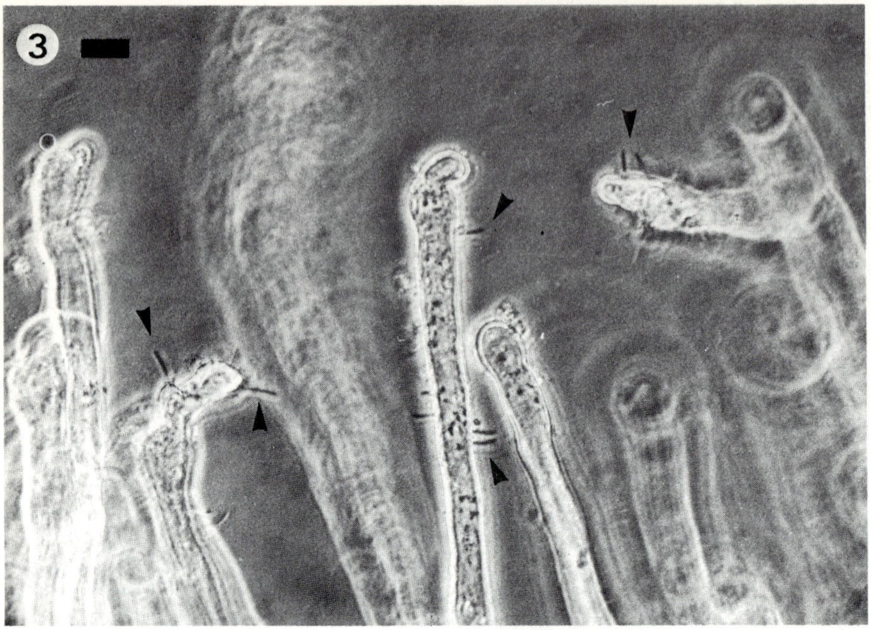

Fig. 3. *Phase contrast microscopy of exo-encapsulation threads (arrows) associated with growing and deformed root hairs of* A. glutinosa. *Bar is 15 μm.*

the infected cells. The primary nodule is born. These encapsulated hypha or hyphae are similar to the other encapsulated hyphae observed elsewhere in the infected cells of the nodule, consequently the utilization of the term "infection thread" for the designation of the encapsulated hyphae of an infected root hair cell should be avoided.

The penetration of a root hair seems to originate from an exo-encapsulation thread, but only one of the thousands of threads observed gives rise to the penetration of a deformed root hair. This observation suggests that the penetration of the root hair cell wall, probably an enzymatic process, is a very critical step in the process of infection. It seems also that once the penetration of a root hair has succeeded, a second penetration of a root hair of the same zone of deformed root hairs is no longer possible due to an induced host defense mechanism (e.g. tannin deposition, etc.). However, from time to time, two primary nodules can be observed in the same zone of deformed root hairs of a root. In these cases, the development of the two primary nodules is synchronous, indicating that

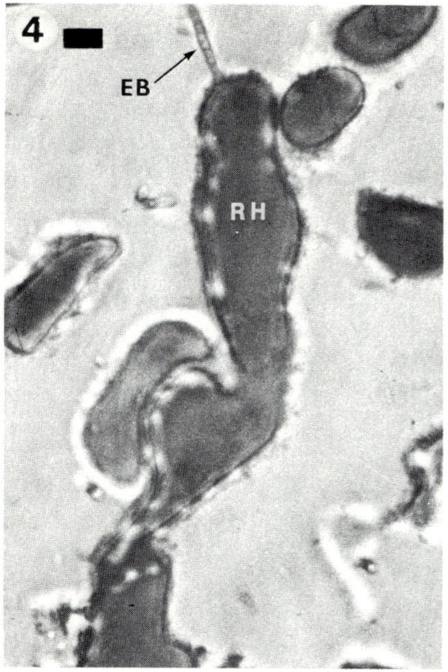

Fig. 4. *This Epon section of an exo-encapsulation thread associated with a deformed root hair shows presumptive endophytic bacterial cells (EB) inside the thread. RH: root hair. Phase contrast microscopy. Bar is 5 μm.*

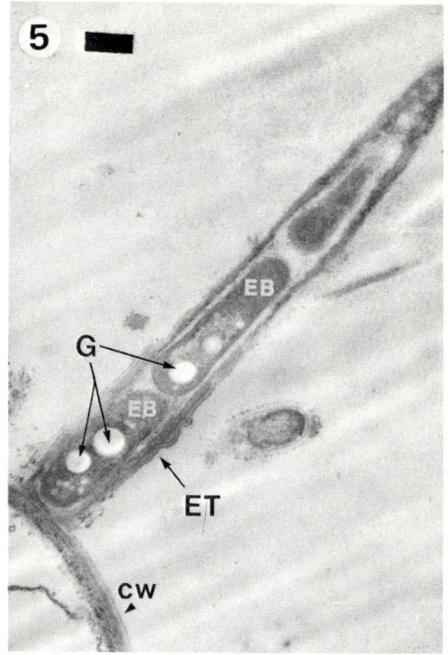

Fig. 5. *Electron microscopy of an exo-encapsulation thread (ET) associated with a deformed* Alnus *root hair. Presumptive endophytic bacterial cells (EB) are resolved inside the thread. Notice the translucent granular material (G) in the cytoplasm of these encapsulated bacterial cells. CW: root hair cell wall. Bar is 1 μm.*

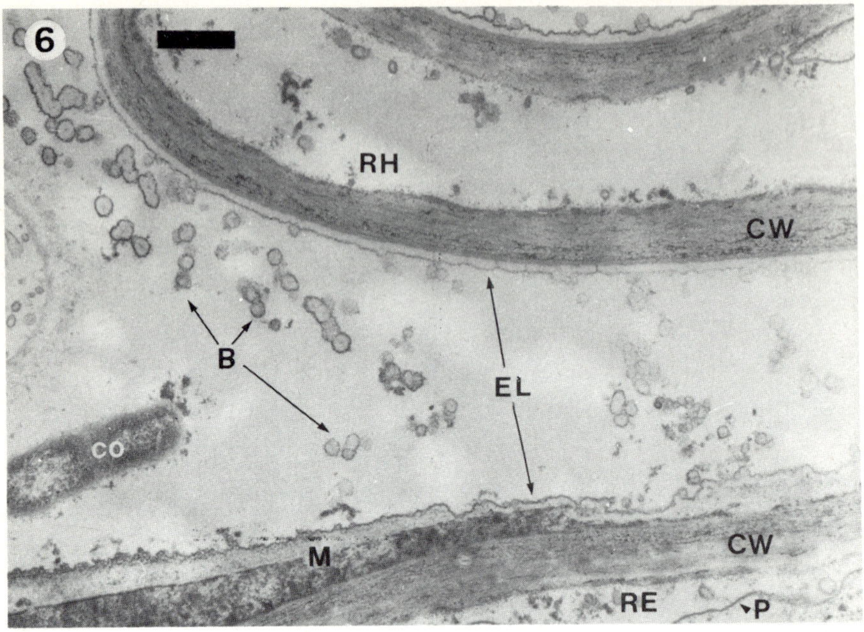

Fig. 6. *Electron microscopy of a deformed root hair (RH) observed near a root epidermis cell (RE). Notice the external layer (EL) covering the cell wall (CW) of the root hair and of the root epidermal cell. Notice also the numerous electron-dense blebs (B) originating from the external layer (EL). M: root mucigel; P: host plasmalemma; CO: bacterial contaminant. Bar is 1 μm.*

the process of penetration took place simultaneously thus avoiding the host defense mechanism induced by the penetrating endophyte.

OBSERVATION OF AN ABNORMAL HOST-ENDOPHYTE SYSTEM

The abnormal host-endophyte system is represented by the *Alnus crispa* var. *mollis* root nodule endophyte, taken from its Canadian *Alnus* host plant and inoculated on the European *Alnus glutinosa* host plant. It is unlikely that the European *A. glutinosa* endophyte can be present, as a contaminant or as a dominant endophyte (Rodriguez-Barrueco and Bond, 1975), in the Canadian *A. crispa* var. *mollis* root nodules used in the preparation of the crushed-nodule inoculum. However, unwanted contamination of the nodules could occur during the growth of the host plant. Therefore, the specific immuno-marker of the non-

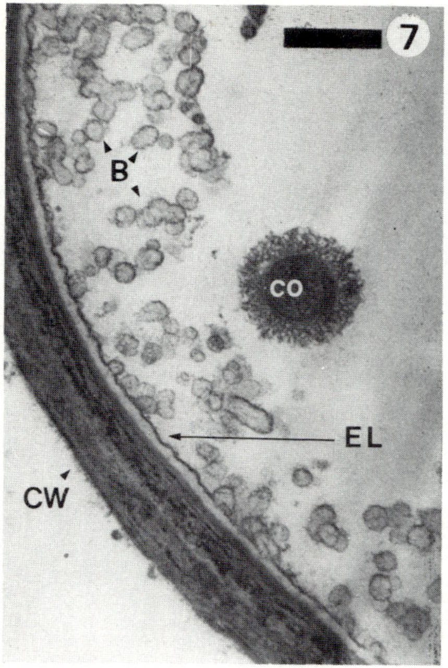

Fig. 7. *Production of membrane-bound blebs (B) by the evagination of the external layer (EL) covering the cell wall (CW) of a deformed root hair. Notice that the electron-dense blebs are not migrating toward the bacterial contaminant (CO). Bar is 1 μm.*

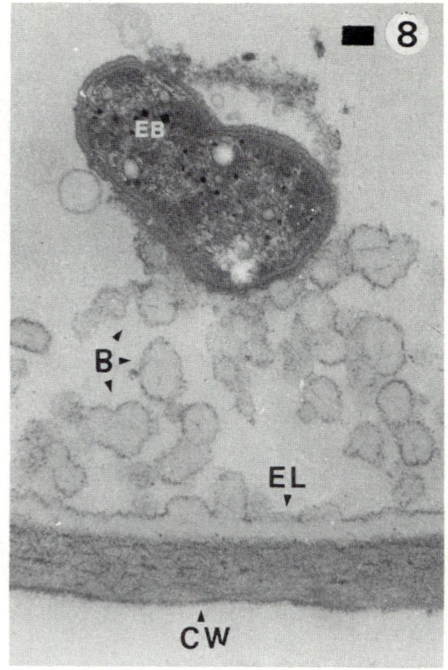

Fig. 8. *Membrane-bound blebs (B), originating from the external layer (EL) covering a root hair cell wall (CW), are migrating toward a presumptive endophytic bacterial cell (EB) in order to encapsulate it. Bar is 0.2 μm.*

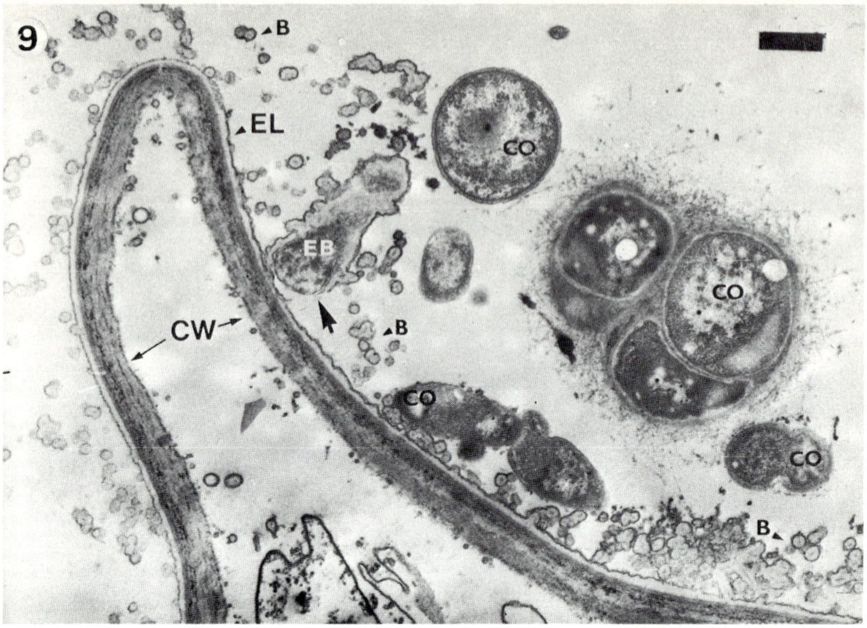

Fig. 9. *Exo-encapsulation (arrow) of a presumptive endophytic bacterial cell (EB) within the mixed microbial population associated with a deformed root hair. Notice that the other microbial contaminants (CO) are not recognized by the electron-dense blebs (B). CW: root hair cell wall; EL: external layer. Bar is 1 μm.*

infective *A. crispa* var. *mollis* endophyte was used to demonstrate that the endophyte observed in the nodules induced on the *A. glutinosa* host plant was the normal *A. crispa* var. *mollis* endophyte (Lalonde and Quispel, unpublished). These immunofluorescence reactions are summarized in Table I and clearly demonstrate the limited homologous antigenic properties of both root nodule endophytes. Adsorption of the common antigenic activity thus permits the specific recognition of the *A. crispa* var. *mollis* endophyte versus the *A. glutinosa* endophyte.

The *Alnus glutinosa* host plant inoculated with *Alnus crispa* var. *mollis* crushed-nodule inoculum of comparable density to the one used with the normal host-endophyte system, shows a typical and extensive deformation of the growing root hairs associated with the elongation-zones of the root system. The exo-encapsulated threads were numerous and only associated with the zones of deformed root hairs. However, the number of primary

Table I

Immunofluorescence reactions of *Alnus* nodule endophytes stained with FA conjugates of non-infective *Alnus crispa* var. *mollis* isolates[a]

Antigen Pectinase treated[b]	FA conjugate against:	FA reaction[c]
A. crispa endophyte from *A. crispa* nodule	Isolates[d]	4+
	Isolates/Ads. *A. crispa*[e] endophyte	-
	Isolates/Ads. *A. glutinosa* endophyte	3+
	Normal serum	-
A. glutinosa endophyte from *A. glutinosa* nodule	Isolates	1+
	Isolates/Ads. *A. glutinosa* endophyte	-
	Isolates/Ads. *A. crispa* endophyte	-
	Normal serum	-
Presumptive[f] *A. crispa* endophyte from *A. glutinosa* nodule	Isolates	4+
	Isolates/Ads. *A. crispa* endophyte	-
	Isolates/Ads. *A. glutinosa* endophyte	3+
	Normal serum	-

[a]See Lalonde *et al.* (1975) for details of the sampling and staining techniques.

[b]All endophyte suspensions were pectinase-treated (see Lalonde and Knowles, 1975b) in order to remove the pectic capsule which shows autofluorescence.

[c]Staining-reaction symbols. -, no fluorescence; 1+ to 4+, increasingly good specific FITC apple-green fluorescence.

[d]A 1:1 mixture of field nodule isolate FA and of axenic nodule isolate FA, diluted to 1:6.

[e]A 1:1 mixture of both FA conjugates, adsorbed with an endophyte suspension of *A. crispa* var. *mollis* root nodule.

[f]An endophyte suspension prepared from the nodules induced on *A. glutinosa* host plant by the inoculation with crushed-nodules of *A. crispa* var. *mollis* host plant.

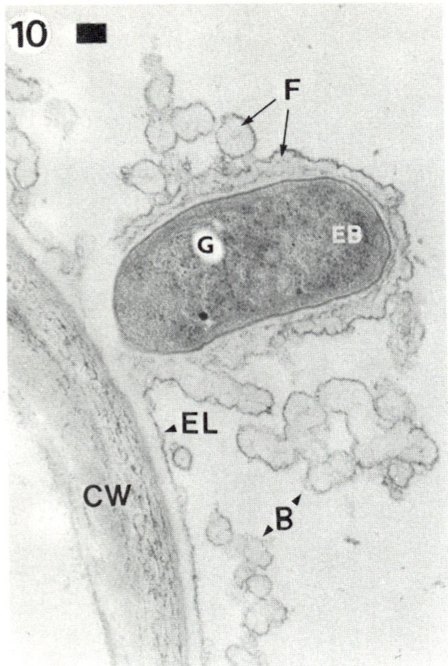

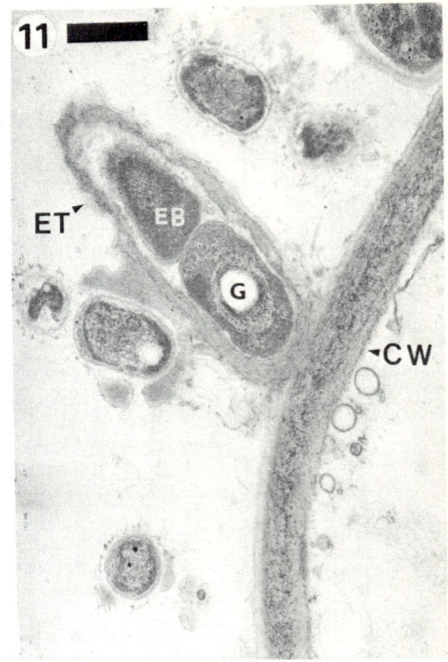

Fig. 10. *Formation of an exo-encapsulation thread around a presumptive endophytic bacterial cell (EB). Notice the continuity of the external layer (EL) covering the root hair cell wall (CW), with the layer bordering the developing encapsulation thread structure. The membrane-bound blebs (B) seem to fuse with the thread-bordering layer and to discharge a fibrillar material (F) onto the presumptive endophytic bacterial cell (EB). Notice also the translucent granule (G) in the cytoplasm of this bacterial cell. Bar is 0.2 μm.*

Fig. 11. *Longitudinal ultra-thin section of an exo-encapsulation thread (ET), containing presumptive endophytic bacterial cells (EB), at the point of contact with the cell wall (CW) of a deformed root hair. G: translucent granule. Bar is 1 μm.*

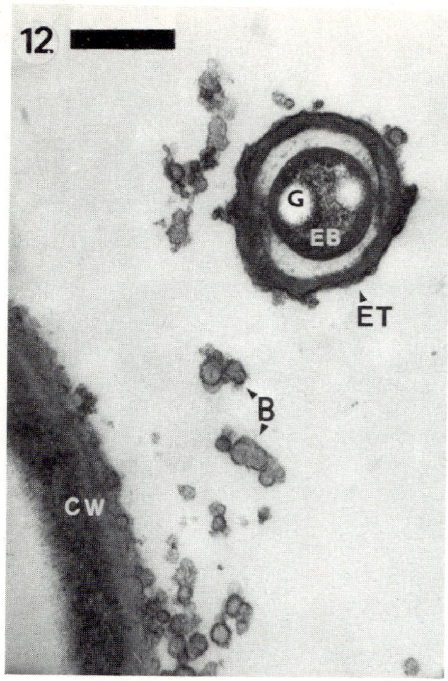

Fig. 12. *Transversal sectioning of an exo-encapsulation thread (ET) nearby a root hair cell wall (CW). Electron-dense blebs (B) are migrating toward the presumptive endophytic bacterial cell (EB) enclosed in the thread structure. G: translucent granule. Bar is 1 μm.*

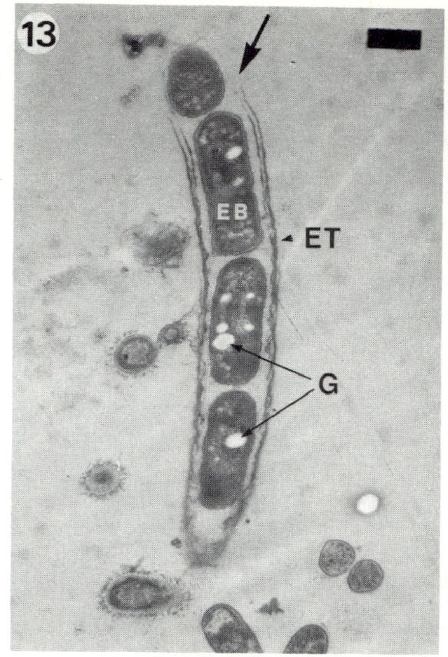

Fig. 13. *Electron microscopy of an exo-encapsulation thread (ET) enclosing presumptive endophytic bacterial cells (EB). Notice the thread structure is open at one end (arrow). G: translucent granules. Bar is 1 μm.*

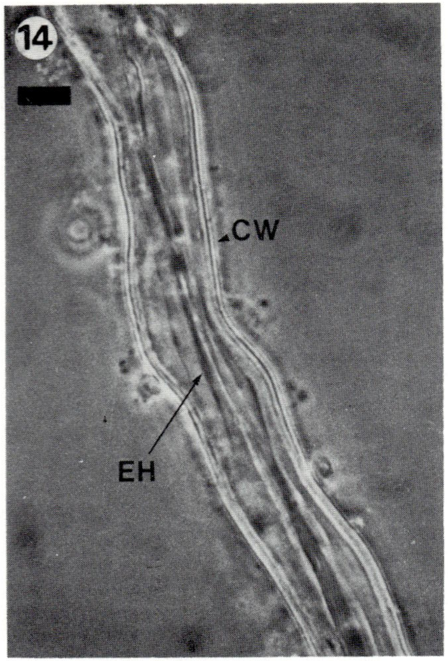

Fig. 14. *Phase contrast microscopy of* Alnus *root hair infected by a penetrating hyphal endophyte (EH). CW: root hair cell wall. Bar is 10 μm.*

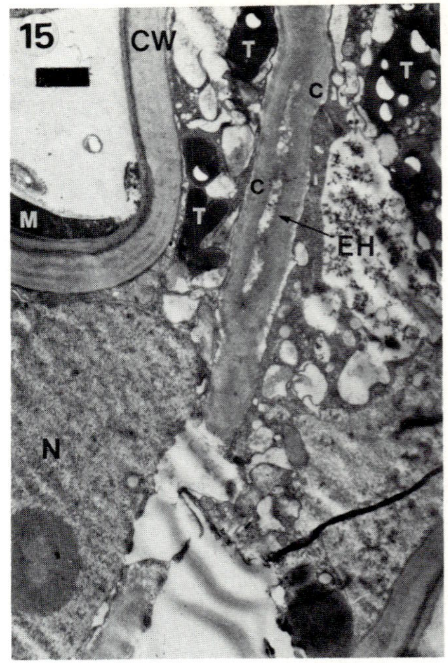

Fig. 15. *A partially-seen* Alnus *root hair infected by a hyphal endophyte (EH). Notice the host cell nucleus (N) near the encapsulated (C) endophyte hypha (EH). Notice also the tannin deposits (T) in the root hair cytoplasm. M: root mucigel; CW: root hair cell wall. Bar is 1 μm.*

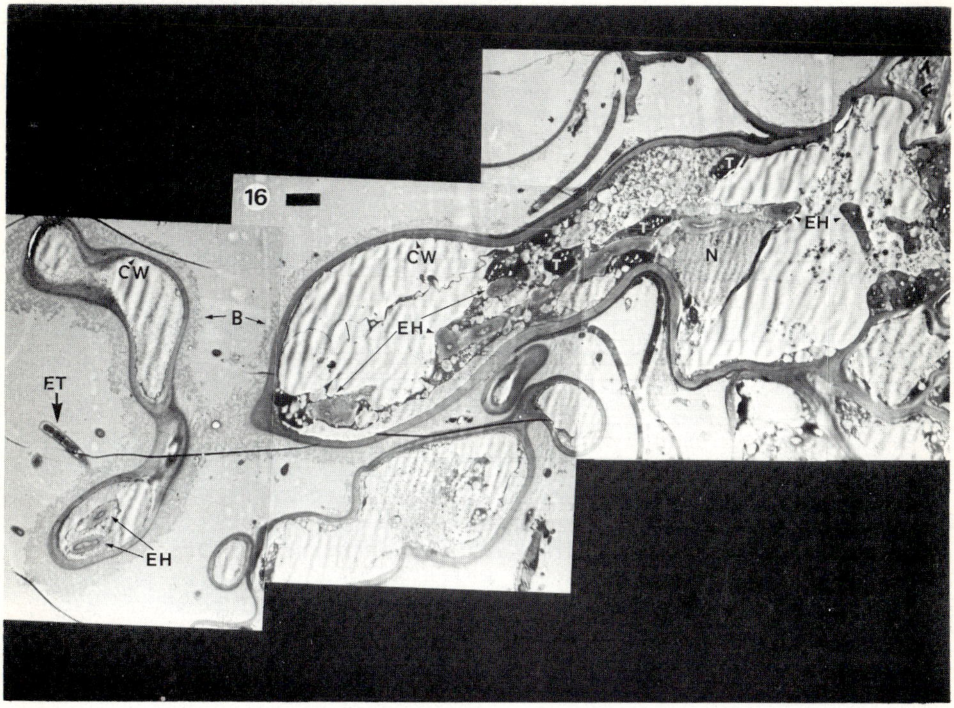

Fig. 16. *Electron microscopy mosaic of the infection of an* Alnus glutinosa *root hair inoculated with an* Alnus crispa *var.* mollis *endophyte. An exo-encapsulation thread (ET) enclosing presumptive endophytic bacterial cells is associated with the tip of the deformed and infected root hair. Numerous electron-dense blebs (B) are seen nearby the root hair cell wall (CW). The penetrating endophyte, seen as encapsulated hyphae (EH), is migrating in the root hair cytoplasm containing abundant tannin deposits (T). At the base of the root hair cell, the host nucleus (N) is associated with the spreading hyphal endophyte. Bar is 5 μm.*

nodules per root system was drastically reduced and the appearance of such primary nodules was delayed by 1 to 2 weeks as compared with the normal host-endophyte system. This delay carried through (but without additional slowness) to the other steps of the infection process.

The ultrastructural development of the *A. crispa* var. *mollis* endophyte nodulating the *A. glutinosa* host plant (Lalonde and Quispel, unpublished) was similar to the normally induced one (Lalonde and Knowles, 1975a). The symbiotic activity of the abnormal system (expressed by the acetylene reduction activity per fresh weight of nodules) was equivalent to the normal host-endophyte system when the delay of 1 to 2 weeks in nodule development was taken into account for estimating developmental age.

DISCUSSION AND CONCLUSION

The production of exo-encapsulation threads appears to be a critical step in the process of infection of the *Alnus* host plant. This exo-encapsulation phenomenon is associated with the presence of an endophyte inoculum and suggests that the endophyte cells are needed for the induction of this mechanism. This induction may be the result of the growth of the presumptive bacterial endophyte in the neighbourhood of the growing root hair in the zone of elongation of a root. The recent papers of Becking (1975, 1976) indicate that prior to the infection, the *Alnus glutinosa* endophyte can grow in the mucilaginous layer covering the root and root hairs. The ultrastructural observations of Greaves and Darbyshire (1972) on the mucilaginous layer of plant roots, suggest that a distinct external boundary (external layer) is characteristic of the mucigel of non-axenic roots.

The present ultrastructural observations of the process of infection suggest that the exo-encapsulation mechanism is reactive against a certain type of cell, presumably the endophytic bacterial cell. The other microbial cells of the mixed microbial population growing in the *Alnus* rhizosphere do not seem to be recognized by the electron-dense blebs. These electron-dense blebs are formed by the evagination of the external layer covering the root hairs and root epidermis and are thus bordered by a part of this layer. These blebs seem able to migrate toward the presumptive endophytic bacterial cells, to recognize them and to encapsulate them. For such a recognition mechanism, we may speculate that the external surface of the growing endophytic bacterial cells display characteristic components which can be recognized by the external layer. This external layer is reactive towards the ruthenium red-osmium tetroxide stain, indicating an acidic polysaccharide nature (Luft, 1971). It is quite probable that this external layer can show antigenic properties. With the importance given to the theory of "cross-reactive antigens in the microbial invasion of a host" (Dazzo and Hubbell, 1975), it seems worthwhile to check whether a lectin-like substance, extracted from the *Alnus* seeds, can interact with this external layer covering the host root hairs and with the presumptive bacterial endophyte observed in the zone of deformed root hairs.

The present paper is not the first description of the presence of an exo-encapsulation thread structure associated with the infection process of a non-leguminous host plant. The recent paper of Fletcher and Gardner (1974) indicates that "rod-shaped bacteria-like structures attached end-on to the root hairs of *Myrica gale*" were previously observed by Fletcher in 1953. Unfortunately, no importance was given to this crucial observation. The *Myrica gale* symbiosis should therefore be assessed for the presence of an exo-encapsulation thread structure associated with the infection process.

The penetration of the root hair cell wall seems to be another critical step in the process of infection of the host plant. The penetration phenomenon seems to be associated with the presence of an exo-encapsulation thread attached to a deformed root hair. If the growing endophytic bacterial cells, enclosed in the exo-encapsulation thread, are able to synthesize cell wall-degrading enzymes, they may be efficient in perforating the near-by host cell wall and in penetrating into the root hair cytoplasm. One mechanism for host-endophyte recognition could act at this stage, *viz.* induction and/or inhibition of cell wall-degrading enzymes.

During the penetration of the deformed host root hair, the bacterial endophyte is changed into a hyphal endophyte. This change probably depends on the environmental conditions (e.g. oxygen tension, etc.) prevailing in the root hair cytoplasm. Furthermore, in order to repel the penetrating microbial intruder, the host cell encapsulates it. This encapsulation material is probably of pectic nature as found for the capsule surrounding the endophytic cells in the true nodule (Lalonde and Knowles, 1975b). However, the spreading endophyte may react by the production of pectolytic enzymes, causing the degradation of the host deposited pectic capsule. The existence of this host deposition-endophyte degradation phenomenon is supported by the recent finding that the *Alnus glutinosa* root nodules are the site of a high pectolytic activity (Schouten and Lalonde, unpublished). These pectolytic enzymes were extractable from the root nodule and were shown to be able to degrade the pectic capsule surrounding the *Alnus glutinosa* endophyte. Pectolytic activity was not detectable in the non-nodulated *Alnus* roots or root apex.

The infection process of the *A. glutinosa* host by the *A. crispa* var. *mollis* endophyte was characterized by a delay of 1 to 2 weeks in the appearance of the primary nodules, these being also less numerous than normally. However, exo-encapsulation threads were normally induced and produced. The delay in the nodulation is probably the expression of a certain inability on the part of the endophyte to penetrate the root hair cell wall of the abnormal host plant.

Following the above description and discussion of the infection process of an infective and of a less-infective host-endophyte system, it will be interesting to determine if the non-infective endophyte, isolated from the *Alnus crispa* var. *mollis* root nodule (Lalonde *et al.*, 1975), is able to induce the production of exo-encapsulation threads and to penetrate a deformed root hair on its appropriate host plant.

ACKNOWLEDGEMENTS

The author acknowledges Dr. A. Quispel, State University of Leiden, Holland, for providing laboratory facilities. I thank Miss P.V. v. Spronsen for her help at the electron microscope. I am indebted to Dr. R. Knowles, MacDonald College of McGill University (Canada) for providing darkroom facilities and for revising the manuscript. This work was supported by a NATO Postdoctoral Fellowship.

REFERENCES

Angulo Carmona, A.F. (1974). *Act.Bot.Neerl.*, 23: 257-303.

Angulo Carmona, A.F., Van Dijk, C. and Quispel, A. (1975). *In:* Symbiotic Nitrogen Fixation in Plants (P.S. Nutman, ed.), pp. 475-483. International Biological Programme, Vol. 7, Cambridge University Press, Cambridge.

Becking, J.H. (1968). *Stikstof.*, 12: 47-74.

Becking, J.H. (1970a). *Plant Soil.*, 32: 611-654.

Becking, J.H. (1970b). *Int.J.Syst.Bacteriol.*, 20: 201-220.

Becking, J.H. (1974). *In:* Bergey's Manual of Determinative Bacteriology (R.E. Buchanan and N.E. Gibbons, eds), 8th ed., pp. 701-706. The Williams and Wilkins Company, Baltimore.

Becking, J.H. (1975). *In:* The Development and Function of Roots (J.G. Torrey and D.T. Clarkson, eds), 3rd Cabot Symposium, pp. 506-565. Academic Press, New York.

Becking, J.H. (1976). *In:* Proc. 1st Int.Sympos. on Nitrogen Fixation (W.E. Newton and C.J. Nyman, eds), Vol. 2, pp. 581-591. Washington State University Press, Pullman, Washington.

Becking, J.H., DeBoer, W.E. and Houwink, A.L. (1964). *J.Microbiol.Serol.*, 30: 343-376.

Bond, G. (1974). *In:* The Biology of Nitrogen Fixation (A. Quispel, ed.), pp. 342-378. North-Holland Publishing Co., Amsterdam.

Dazzo, F.B. and Hubbell, D.H. (1975). *Appl.Microbiol.*, 30: 1017-1033.

Fletcher, W.W. and Gardner, I.C. (1974). *Ann.Micro.*, 24: 159-172.

Gardner, I.C. (1965). *Arch.Mikrobiol.*, 51: 365-383.

Gardner, I.C. (1975). *In:* Symbiotic Nitrogen Fixation in Plants (P.S. Nutman, ed.), pp. 485-495, Vol. 7. International Biological Programme, Cambridge University Press, Cambridge.

Greaves, M.P. and Darbyshire, J.F. (1972). *Soil Biol.Biochem.*, 4: 443-449.

Lalonde, M. and DeVoe, I.W. (1975). *Arch.Mikrobiol.*, 105: 87-94.

Lalonde, M. and DeVoe, I.W. (1976). *Physiol.Plant Path.*, 8: 123-129.

Lalonde, M. and Fortin, J.A. (1972). *Can.J.Bot.*, 50: 2597-2600.

Lalonde, M. and Fortin, J.A. (1973). *Can.J.Bot.*, 19: 1115-1118.

Lalonde, M. and Knowles, R. (1975a). *Can.J.Microbiol.*, 21: 1058-1080.

Lalonde, M. and Knowles, R. (1975b). *Can.J.Bot.*, 53: 1951-1971.

Lalonde, M., Knowles, R. and Fortin, J.A. (1975). *Can.J.Microbiol.*, 21: 1901-1920.

Luft, J.H. (1971). *Anat.Rec.*, 171: 347-368.

Pizelle, G. (1972). *Bull.Soc.Bot.Fr.*, 119: 571-580.

Pommer, E.H. (1956). *Flora*, 143: 603-634.

Quispel, A. (1954). *Acta Bot.Neerl.*, 4: 495-511.

Quispel, A. (1974). *In:* The Biology of Nitrogen Fixation (A. Quispel, ed.), pp. 499-520. North-Holland Publishing Co., Amsterdam.

Rodriguez-Barrueco, C. and Bond, G. (1975). *In:* Symbiotic Nitrogen Fixation in Plants (P.S. Nutman, ed.), Vol. 7, pp. 561-565. International Biological Programme, Cambridge University Press, Cambridge.

Rogers, R.D. and Wollum II, A.G. (1974). *Soil Sci.Soc.Am., Proc.*, 38: 756-759.

Van Dijk, C. and Merkus, E. (1976). *New Phytologist*, 76: in press.

NITROGENASE ACTIVITY OF NODULE HOMOGENATES OF *ALNUS GLUTINOSA:* A COMPARISON WITH THE *RHIZOBIUM*-PEA SYSTEM

A.D.L. AKKERMANS, J. VAN STRATEN and W. ROELOFSEN

Laboratory of Microbiology, Wageningen, Holland.

INTRODUCTION

The agricultural importance of legumes has led to intensified research concerning the *Rhizobium*-legume symbiosis. As a result, basic research on nitrogen fixation by actinomycetes in root nodules of non-leguminous shrubs and trees has been somewhat neglected, which might lead to a one-sided view on symbiotic nitrogen fixation.

In the present paper, attention will be focused on the nitrogenase system in root nodules of *Alnus glutinosa*. These nodules are perennial, branched clusters of lobes. The micro-symbiont, "*Frankia alni*" (Becking, 1970), is an actinomycete-like organism that forms hyphal clusters within the host cells. The hyphal tips at the periphery of the clusters develop into spherical vesicles. Certain strains of this organism are further characterized by the ability to form spore-like bodies called "granules" (Schaede, 1933; van Dijk and Merkus, 1976).

Since the vesicle clusters are the most pronounced structures of the micro-symbiont in the nodules, it was suggested that nitrogen fixation might be related to these structures (Schaede, 1933; Becking *et al.*, 1964). Evidence for this suggestion came from the observation that vesicle clusters, like heterocysts in blue-green algae (Stewart *et al.*, 1969), strongly reduce tetrazolium salts (Akkermans, 1971; see also Quispel, 1974a,b). Moreover, the vesicle clusters with tetrazolium-reducing activity are localized within the upper 3 mm of the lobes, and this region contains almost all of the nitrogenase activity.

In addition to the micro-chemical test of localizing reducing sites in alder nodules, a method was developed to obtain purified suspensions of intact vesicle clusters from nodule homogenates. Using this method, vesicle-cluster suspensions as well as nodule homogenates showed low, but significant nitrogenase activity (Akkermans, in press). The rapid loss of activity during homogenization was more pronounced with nodules of alder than with nodules of leguminous plants, and might be related to the presence of large quantities of phenolic compounds in the former. A further explanation of the low values found by alder nodule homogenates might be that permeability to ATP or ADP is restricted. For that reason a treatment was included with EDTA and toluene which, in bacteroid suspensions from peas, were found to enhance nitrogenase activity to a large extent (van Straten and Roelofsen, 1976). No further work on nitrogenase activity of alder nodule homogenates is available in the literature. This means that all conditions for the preparation of the homogenates and for the incubation mixture of the acetylene-reduction assay had to be determined before evidence could be given about the localization of the nitrogenase within the nodule tissue. In the present study, factors affecting nitrogenase activity of alder nodule homogenates and vesicle cluster suspensions were investigated, using the nitrogen-fixing system of pea bacteroid suspensions as a reference.

MATERIALS AND METHODS

Plant material

Nodule material of variable age and size was collected from field-grown black alder plants (*Alnus glutinosa)* and from alder seedlings cultivated in the greenhouse on a Crone's nutrient solution without combined nitrogen.

Peas (*Pisum sativum* cv. *Rondo)* were inoculated with *Rhizobium leguminosarum,* strain PRE, and cultivated as described by van Straten and Roelofsen (1976).

Preparation of alder nodule homogenates

Following the determination of nitrogenase activity in nodulated root pieces or intact seedlings, the nodules were excised and weighed. A 0.5 g portion was wetted and left between wet filter paper. After 20 min, the nodules and 2 drops of antifoam were added to a homogenizing flask of a Virtis homogenizer ("45" Hi-speed). A condom was wrapped around the flask and closed around the chuck cap of the homogenizer to prevent air from entering. After flushing with argon for 5 min, 30 ml of buffer was added. Unless stated otherwise, the buffer contained 0.05 M Tris-HCl (pH 8.0), 0.3 M sucrose, 5 mM 1,4-dithioerythritol (DTE), 25 mM sodium dithionite and 4% soluble polyvinyl-pyrrolidone (PVP). The gas phase continued to be flushed and the sample was homogenized for 1 min at 45000 rpm at 20± 2°C. The homogenate was removed from the flask and filtered through a 100 μm nylon filter ("Monyl", Züricher

Beuteltuchfabrik A.G., Switzerland) in a Sartorius filter holder (SM 165 10) which was kept anaerobic. The homogenate obtained in this way was free from non-homogenized nodule fragments and contained intact vesicle clusters, hyphal fragments, isolated vesicles and plant-cell debris. The filtrate was used either directly or after centrifugation for 10 min at 27000 x g at 20°C and resuspension of the pellet in 20 ml of fresh buffer. Vesicle clusters were isolated by subsequent anaerobic filtration of the 100 μm-filtrate through a 20 μm filter.

Isolated vesicles and vesicle clusters were counted microscopically by using a haemocytometer.

Preparation of pea bacteroid suspensions

Bacteroid suspensions were prepared and treated with EDTA and toluene as described by van Straten and Roelofsen (1976).

Preparation of cell-free extracts of Azotobacter vinelandii

Cell-free extracts were prepared by disintegrating the cells with a French Press, following centrifugation of the crude extracts (Brotonegoro, 1974).

Acetylene reduction assay

Intact nodules Nodulated root pieces from field-grown plants were incubated in 120 ml vessels and young seedlings (whole plants) in 4800 ml vessels under 9% acetylene in air. Ethylene production was measured with a gas chromatograph after 6 and 12 min. Only nodules producing at a rate of more than 1000 nmol of ethylene/g fresh nodules/h were used for making nodule homogenates.

Nodule homogenates Suspensions were incubated in 16 x 125 mm Hungate tubes (Bellco, Glass, Inc.) at 24°C. The composition of the incubation mixture (2 ml) varied as described in Results and Discussion. The gas phase was 10% acetylene in argon. Ethylene production was first measured following 20 min incubation. Results are expressed as nmoles of ethylene/g fresh weight nodule/h.

Micro-autoradiographic localization of $^{13}N_2$ fixation

$^{13}N_2$ was prepared by L. Lindner and J. Helmer at the Instituut voor Kernphysisch Onderzoek at Amsterdam. Autoradiographic techniques were applied according to Wolk *et al.* (1970).

RESULTS AND DISCUSSION

Acetylene reduction by alder nodule homogenates

It is well-known that alder nodules, like other parts of the plant, contain large amounts of phenolic compounds that rapidly oxidize to form red-brown complexes after disruption of the tissue. Although the nature and the natural function of these compounds are still unknown, phenolic compounds in general are strong inhibitors of enzymes in cell-free extracts. By using oxygen uptake of nodule homogenates as a qualitative measure of the concentration of auto-oxidable compounds, we found that more than 95% of these compounds were removed by addition of 4% PVP.

Preliminary experiments had shown that pressing alder nodules under anaerobic conditions, as is generally done with pea nodules, resulted in a complete loss of nitrogenase activity. In contrast, homogenization of the nodules gave positive results. The negative results may be explained either by insufficient removal of inhibiting compounds from the nodule squash or by interference by air still present in the tissue during pressing. However, evacuating the nodules before homogenization did not enhance the nitrogenase activity of the homogenates. In contrast, storage of the excised nodules for 20 min between wet filter paper before homogenization, resulted in a 30% increase of the activity of the homogenates compared to the control.

Effect of sucrose Sucrose is commonly added to cell suspensions to protect them against osmotic shocks. Sucrose has also served as a protective compound during determination of nitrogenase activity in soybean breis (Bergersen, 1966a,b), soybean bacteroid suspensions (Bergersen and Turner, 1967, 1968; Bergersen *et al.*, 1973) and lupin bacteroid suspensions (Kennedy, 1970).

Nodules were therefore homogenized with or without 0.3 M sucrose. Following filtration, each filtrate was divided into two equal parts and centrifuged. The pellets were resuspended in buffer with or without 0.3 M sucrose. The results in Table I show clearly that sucrose is needed during homogenization whereas addition of sucrose during incubation did not increase the activity. Similar results were obtained with mannitol. In contrast, sucrose had an inhibiting effect on acetylene reduction by EDTA-toluene-treated pea bacteroid suspensions of strain PRE. In another experiment with nodule brei of *R. leguminosarum*, strain PF_2, which is characterized by the formation of spherical bacteroids on pea (van der Berg, to be published), sucrose inhibited acetylene reduction by 65%.

Incubation conditions The low values obtained so far with alder were probably not due to an inhibitory effect of the Tris buffer because HEPES gave no higher results.

As both the choice and the concentration of DTE had been somewhat arbitrary, the use of DTE was compared to that of mercaptoethanol. The nodules were homogenized in the presence of

Table I

Effect of 0.3 M sucrose during preparation and incubation on acetylene reduction by alder nodule homogenates and pea bacteroid suspensions

Nodule material	Incubation time minutes	nmol of C_2H_4/g of fresh wt nodule/h			
		++[a]	+-[a]	--[a]	-+[a]
Alder homogenates[b]	0 - 20	375	438	75	48
	20 - 40	327	408	60	30
	40 - 60	282	342	45	27
Pea bacteroids[c]	0 - 10	18900	-	24500	17000

[a] ++ sucrose present during preparation and incubation; +- sucrose present during preparation only; -- no sucrose added; -+ sucrose added to the incubation mixture only.

[b] Incubation mixture (2 ml): 25 mg of fresh wt nodule, 100 μmol of Tris-HCl (pH 8.0), 50 μmol of $Na_2S_2O_4$, 5.0 μmol of $MgCl_2$, 11.2 μmol of ATP and 40 mg of PVP.

[c] EDTA-toluene-treated bacteroid suspensions. Incubation mixture (1 ml): 20 mg of fresh wt nodule, 50 μmol of Tris-HCl (pH 7.2), 20 μmol of $Na_2S_2O_4$, 7.5 mol of $MgCl_2$, 2.8 μmol ATP, 11.7 μmol creatine phosphate, 0.03 mg of creatine phosphokinase and 10 mg of PVP.

5 mM DTE. Following filtration, the filtrate was divided into two equal parts and centrifuged. The pellets were resuspended in either DTE or mercaptoethanol giving final concentrations of 2.5, 5.0 and 12.5 mM. As shown in Fig. 1 increasing concentrations of DTE had a destabilizing effect on the nitrogenase activity. As compared to DTE, all concentrations of mercaptoethanol could be considered as inhibitory. In another experiment, DTE below 2.5 mM gave the same nitrogenase activity as the control. We therefore decided to incubate without DTE.

Nitrogenase activity had generally been measured in the presence of 11.2 μmol of ATP, 18.4 μmol of creatine phosphate and 0.05 mg of creatine phosphokinase per assay (2 ml). An experiment was set up in which phosphokinase and creatine phosphate were either omitted or present in the above-mentioned concentrations. Table II, ethylene production was higher in their absence. A concentration series of creatine phosphate from 0

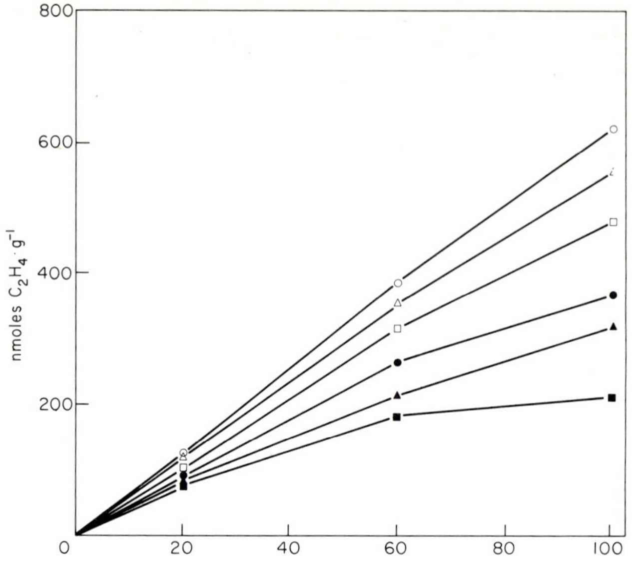

Fig. 1. *Effect of different concentrations of dithioerythritol (DTE) and mercaptoethanol during incubation on acetylene reduction by alder nodule homogenates. The incubation mixture (2 ml) contained 31 mg of nodule tissue, 100 μmol of Tris-HCl (pH 8.0), 50 μmol of $Na_2S_2O_4$, 5.0 μmol of $MgCl_2$, 5.6 μmol of ATP, 18.4 μmol of creatine phosphate, 0.03 mg of creatine phosphokinase and 40 mg of polyvinylpyrrolidone DTE concentration: 2.5 mM (○ —— ○), 5.0 mM (△ —— △) or 12.5 mM (□ —— □). Mercaptoethanol concentration: 2.5 mM (● —— ●), 5.0 (▲ —— ▲) or 12.5 mM (■ —— ■).*

to 9.2 μmol with 0.05 mg phosphokinase, per assay gave highest ethylene production at zero and the lowest concentration. Omission of phosphokinase had no effect, and we concluded that creatine phosphate is an inhibitor of the nitrogenase activity. This effect was not found with related compounds, such as creatine and creatinine. An inhibitory effect of creatine phosphate on nitrogenase activity was also observed in cell-free extracts of *Azotobacter vinelandii*, when creatine phosphokinase was omitted from the incubation mixture. In addition, the added ATP-generating system kept on functioning normally in cell-free extracts of azotobacter when they were mixed with inactive alder nodule homogenates. Therefore, alder nodule homogenates do not appear to contain compounds that block the creatine phosphokinase reaction. The results obtained with the ATP-generating system and alder nodule homogenates are in contrast to results

Table II

Effect of creatine phosphate and creatine phosphokinase on acetylene reduction by alder nodule homogenates

Creatine phosphate mol per assay	Creatine phosphokinase mg per assay	nmol of C_2H_4/g of fresh wt nodule/h a)
18.4	0.05	540
18.4	-	612
-	0.05	1302
-	-	1353

a) based on 20 to 40 min incubation period. The incubation mixture (2 ml) contained 35 mg of fresh wt nodule, 100 μmol of Tris-HCl (pH 8.0), 50 μmol of $Na_2S_2O_4$, 5.0 μmol of $MgCl_2$, 11.2 μmol of ATP and 40 mg of polyvinylpyrrolidone.

obtained with pea bacteroid suspensions. The latter showed a clear need for creatine phosphate. The use of other ATP-generating systems to get higher nitrogenase activity of alder nodule homogenates is currently being studied. Different concentrations of ATP and $MgCl_2$ were studied in the absence of an ATP-generating system. Fig. 2 shows that 5.6 μmol of ATP per assay (2 ml) was optimal at a concentration of 2.5 mM $MgCl_2$. At higher $MgCl_2$ concentrations both the optimum ATP concentration and the nitrogenase activity were different, as will be described elsewhere.

The experiments described so far were all carried out with 25 mM dithionite. This concentration was far from optimal (Table III), and 100 mM is recommended for further experiments.

Disintegration and fractionation of alder nodule nomogenates
The experiments described above were intended for finding optimum incubation conditions of the homogenates. Therefore, the procedure of the preparation of the homogenates was kept constant.

As longer homogenization time will result in a further disintegration of the micro-symbiont of the nodules, the effect of prolonged homogenization was then studied. After different homogenization times, aliquots were taken from the homogenizing flask and incubated for the acetylene-reduction assay. The total volume in the flask was kept constant by addition of equal volumes of fresh buffer. Optimum activity was found after 4 to 7 min homogenization (Table IV). Longer times had an adverse effect. During the first minute most of the vesicle clusters were released from the host cells.

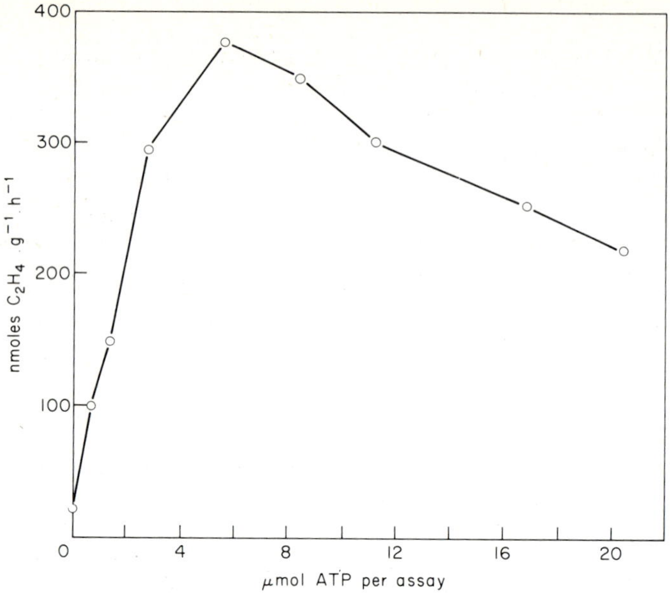

Fig. 2. *Effect of ATP concentration on acetylene reduction by alder nodule homogenates. The incubation mixture (2 ml) contained 10 mg of nodule tissue, 100 μmol of Tris-HCl (pH 8.0), 50 μmol of $Na_2S_2O_4$, 2.4 μmol of $MgCl_2$ and 20 mg of polyvinylpyrrolidone. C_2H_4 production was measured after 20 min.*

Table III

Effect of altering concentrations of dithionite during incubation on acetylene reduction by alder nodule homogenates

$Na_2S_2O_4$ mM	nmol of C_2H_4/g of fresh wt nodule/h[a)]
22.5	339
50	500
75	532
100	562
150	513
200	417

[a)] based on the 0 to 20 min incubation. The incubation mixture (2 ml) contained 19 mg of fresh wt nodule, 100 μmol of Tris-HCl (pH 8.0), 5.0 μmol of $MgCl_2$, 5.6 mol of ATP and 40 mg of polyvinylpyrrolidone.

Table IV

Effect of homogenization time on acetylene reduction by alder nodule homogenates

Homogenization time in min	nmol of C_2H_4/g of fresh wt nodule/h	
	Experiment 1	Experiment 2
0.29	96	177
0.50	390	300
1	693	570
2	792	711
4	819	867
7	888	759
10	855	678

The activity of the nodules used in experiments 1 and 2 was 12482 and 6760 nmol of C_2H_4/g fresh wt nodule/h, respectively. The dithionite concentration during homogenization and during the acetylene reduction assay was 100 mM. With the exception of the dithionite concentration, the incubation mixture was the same as given in Table I. The activity of the homogenates was expressed per g of original nodule tissue.

In order to get information about the nitrogenase activity of different structures of the micro-symbiont, nodule homogenates (homogenization time 1 min) were further fractionated. As shown in Table V, most of the activity was in the fraction below 20 m. No activity was detected in the supernatant after centrifugation for 10 min at 27000 x g. The fraction between 20 and 100 μm, which contained intact vesicle clusters, had only very low activity. The results are in agreement with earlier observations (Akkermans, to be published), but in the present study higher nitrogenase activities of homogenates were obtained. Further separation by low-speed centrifugation showed that nitrogenase activity was associated with relatively small particles.

The results obtained so far can be explained in different ways:

a) nitrogenase is not restricted to the vesicle clusters;

b) it is restricted to the clusters but inaccessible to the assay systems *in vitro*; the activity can be measured once the clusters are fragmented;

c) the fraction between 20 and 100 μm contains mostly inactive clusters, while the young, active clusters are destroyed during homogenization. However, sonication of the homogenates resulted in a complete loss of activity.

Table V

Acetylene reduction by different fractions of alder nodule homogenates

Fraction of homogenate	nmol of C_2H_4/g of fresh wt nodule/h
Filtrate 100 μm filter	426
Residue on 20 μm filter	15
Filtrate 20 μm filter	435
Pellet 20 μm filtrate (10 min 27000 x g)	300
Supernatant 20 μm filtrate (10 min 27000 x g)	0
Pellet 20 μm filtrate (0.5 min 715 x g)	54
Supernatant 20 μm filtrate (0.5 min 715 x g)	237

Incubation conditions as given in Table I.
Ethylene production was measured after 20 min.

Effect of EDTA-toluene, cellulase, pectinase and lysozyme As mentioned before, treatment of pea bacteroids with EDTA and toluene resulted in a rapid increase in nitrogenase activity, which was explained as a change in permeability of the cell membranes resulting in enhanced transport of either ATP or ADP. However, a similar treatment of alder nodule homogenates resulted in a decrease in activity (Table VI). Since toluene had a negative effect, other compounds specifically attacking cell walls were tested (Table VII). Lysozyme had a strong inhibitory effect and almost stopped nitrogenase activity after 50 min. Pectinase had a negative effect during the first 20 min incubation only, while cellulase showed a negative but reversible effect on nitrogenase activity. Additional work with cell-wall degrading enzymes is in progress.

Micro-autoradiographic localization of $^{13}N_2$ fixation

Localization of the site of nitrogen fixation was further investigated by using the radio isotope ^{13}N. Recent work by Wolk *et al.* (1974) and Thomas *et al.* (1976) had shown the value of this isotope in N_2 fixation.

Excised alder nodules were incubated in 1 ml vials and incubated in a gas phase containing 20% O_2, 20% N_2 and 60% He. The radio-activity of the N_2 at the beginning of the incubation was 20 mCi. After 10 min, the gas phase was removed by evacuation and the radio-activity of the tissue measured. The rate

Table VI

Effect of EDTA and/or toluene treatment on acetylene reduction by alder nodule homogenates

Homogenate treated with	nmol of C_2H_4/g of fresh wt nodule/h
Buffer	431
Buffer + EDTA[a)]	184
Buffer + toluene[b)]	156
Buffer + EDTA[a)] + toluene[b)]	186

[a)] Final concentration 1 mM.

[b)] A total of 0.2 ml/ml of homogenate.

Pretreatment with EDTA and toluene: 2 min. When toluene was present, the aqueous phase was removed for acetylene reduction test.

The incubation mixture (2 ml) contained 20 mg of fresh wt nodule, 100 μmol of Tris-HCl (pH 8.0), 100 μmol of $Na_2S_2O_4$, 5.0 μmol of $MgCl_2$, 5.6 μmol of ATP and 40 mg of polyvinylpyrrolidone.

Ethylene production was measured after 20 min.

Table VII

Effect of cellulase, pectinase and lysozyme on acetylene reduction by alder nodule homogenates

Treatment[a)]	nmol of C_2H_4/g of fresh wt nodule/h		
	0 - 20	20 - 50	50 - 100 min[b)]
-	198	180	170
Cellulase (10 mg/ml)	159	192	200
Pectinase (10 mg/ml)	122	109	106
Cellulase and pectinase (each 10 mg/ml)	111	109	108
Lysozyme (0.25 mg/ml)	129	46	4

[a)] Enzymes were added to the incubation mixture at the start of the acetylene reduction test.

[b)] Time intervals in minutes during which ethylene production was calculated.

The incubation mixture (2 ml) contained 17 mg of fresh wt nodule, 100 μmol of Tris-HCl (pH 8.0), 100 μmol of $Na_2S_2O_4$, 5.0 μmol of $MgCl_2$, 5.6 μmol of ATP and 40 mg of polyvinylpyrrolidone.

of fixation was 0.4 μmol of N_2 fixed/g fresh wt nodule/h. Simultaneous assay of the acetylene reduction by another aliquot of the nodule material gave a rate of 1.8 μmol C_2H_4/g fresh wt nodule/h over 10 min, which almost agrees with the data obtained with $^{13}N_2$. After different times of incubation up to 10 min in $^{13}N_2$, the nodules were evacuated and frozen in liquid nitrogen. Hand-made sections of the nodules were freeze-dried and exposed to Ilford G-5 emulsion, as described by Wolk *et al.* (1974). After development of the emulsion layer, the number and the location of the positron tracks were determined microscopically.

Assuming that all labeled products would be present in the vesicle clusters after 10 min a theoretical maximum of about 10 tracks per vesicle should be obtained. However, the results so far obtained showed that the actual number of detectable tracks on the radiogram was 100 times lower.

An attempt was made to repeat the experiments described by Wolk *et al.* (1974) with *Anabaena cylindrica*. Again, very few tracks could be distinguished. Since tracks are extremely difficult to locate, especially when nodule sections are exposed to an emulsion layer, track autoradiography in this context is not necessarily reliable.

The direct measurement of the built in ^{13}N might be a better approach to localizing the site of N_2 fixation in intact nodules. After homogenization, the suspension was fractionated by filtration and the fixed ^{13}N was determined in the various fractions. After 10 min, 94% of the label was present in the fraction below 0.8 μm. This result can be interpreted in two ways:

a) the label had already diffused from the larger fraction, e.g. vesicle clusters, into the soluble fraction because of normal transport or because of physical disruption;

b) nitrogen fixation does indeed take place in the fraction below 0.8 μm.

The use of $^{13}N_2$ may prove to be of great value in the localization problem and further experiments are in progress.

CONCLUSION

It appears that the physiology of the alder symbiont differs from that of the leguminous nodule symbiont. Especially interesting is the inhibitory effect of creatine phosphates, added as part of the artificial ATP-generating system, on nitrogenase activity of alder nodule homogenates. No explanation can yet be given.

As far as known, this is the first reported successful attempt to apply the acetylene reduction assay to homogenates of non-leguminous actinomycete-containing root nodules under anaerobic conditions. So far, we have been unable to locate nitrogenase within the micro-symbiont. Further development of fractionation procedures together with finding optimum conditions for acetylene reduction by nodule homogenates should be useful in solving this problem.

ACKNOWLEDGEMENT

Thanks are extended to L. Lindner and J. Helmer of the Instituut voor Kernphysisch Onderzoek, Amsterdam, for preparation and use of $^{13}N_2$.

REFERENCES

Akkermans, A.D.L. (1971). Ph.D. thesis, Univ. of Leiden, The Netherlands, 83 pp.
Becking, J.H. (1970). *Intern.J.Syst.Bacteriol.*, 20: 201-220.
Becking, J.H., Boer, W.E. de and Houwink, A.L. (1964). *Antonie van Leeuwenhoek*, 30: 343-376.
Bergersen, F.J. (1966a). *Biochem.Biophys.Acta*, 115: 247-249.
Bergersen, F.J. (1966b). *Biochem.Biophys.Acta*, 130: 304-312.
Bergersen, F.J. and Turner, G.L. (1967). *Biochem.Biophys.Acta*, 141: 507-515.
Bergersen, F.J. and Turner, G.L. (1968). *J.Gen.Microbiol.*, 53: 204-220.
Bergersen, F.J., Turner, G.L. and Appleby, C.A. (1973). *Biochem. Biophys.Acta*, 292: 271-282.
Brotonegoro, S. (1974). Ph.D. thesis, Communications Agricultural University Wageningen, The Netherlands, 76 pp.
Dijk, C. van and Merkus, E. (1976). *New Phytol.*, 77: 73-91.
Kennedy, I.R. (1970). *Biochem.Biophys.Acta*, 222: 135-144.
Quispel, A. (1974a). *In:* The Biology of Nitrogen Fixation (A. Quispel, ed.), pp. 499-520. North Holland Publ.Co., Amsterdam, Oxford.
Quispel, A. (1974b). *In:* The Biology of Nitrogen Fixation (A. Quispel, ed), pp. 719-745. North Holland Publ.Co., Amsterdam, Oxford.
Schaede, R. (1933). *Planta*, 19: 389-416.
Stewart, W.D.P., Haystead, A. and Pearson, H.W. (1969). *Nature*, 224: 226-228.
Straten, J. van and Roelofsen, W. (1976). *Appl.Environm.Microbiol.*, 31: 859-863.
Thomas, J., Wolk, C.P., Shaffer, P.W., Austin, S.M. and Galonski, A. (1975). *Biochem.Biophys.Res.Commun.*, 67: 501-507.
Wolk, C.P., Austin, S.M., Bortins, J. and Galonski, A. (1974). *J.Cell.Biol.*, 61: 440-453.

SUBJECT INDEX

A

R